Physics at the Large Hadron Collider

Edited by

Amitava Datta
Indian Institute of Science Education and Research, Kolkata
Haringhata, P.O. Mohanpur, Nadia, West Bengal 741246, India

Biswarup Mukhopadhyaya
RECAPP, Harish-Chandra Research Institute
Chhatnag Road, Jhunsi, Allahabad 211019, India

Amitava Raychaudhuri
Harish-Chandra Research Institute
Chhatnag Road, Jhunsi, Allahabad 211019, India

Indian National Science Academy
A Platinum Jubilee Special Issue

Editors:

Amitava Datta
Indian Institute of Science Education
and Research
Kolkata

Biswarup Mukhopadhyaya
Harish-Chandra Research
Institute
Chhatnag Road, Jhusi
Allahabad - 211 019, India

Amitava Raychaudhuri,
Director
Harish-Chandra Research
Institute
Chhatnag Road, Jhusi
Allahabad - 211 019, India

Datta, Mukhopadhyaya & Raychaudhuri: Physics at the Large Hadron Collidor

Softcover re-print of the Hardcover 1st edition 2009

First Published 2009

ISBN 978-81-322-1728-2

This edition is manufactured in India for sale only in India, Pakistan, Bangladesh, Nepal and Sri Lanka and any other country as authorized by the publisher.

This edition is published by Springer (India) Private Limited, (a part of Springer Science+Business Media), Registered Office: 3rd Floor, Gandharva Mahavidyalaya, 212, Deen Dayal Upadhyaya Marg, New Delhi - 110 002, India.

Printed in India by Rajkamal Electric Press, Plot No. 2, Phase-IV, HSIDC, Kundli, Haryana.

FOREWORD

The Indian National Science Academy was established in January 1935 with the objective of promoting science in India and harnessing scientific knowledge for the cause of humanity and national welfare. In 1968, it was designated as the adhering organisation in India to the International Council for Science (ICSU) on behalf of the Government of India. Over the years, the Academy has published a number of journals, volumes, biographical memoirs, etc.

The year 2009–10 will be specially celebrated to mark the Platinum Jubilee of the Academy. Many programmes are planned in different centres in India on this occasion. In addition, the Academy has decided to publish a number of special volumes on different subjects ranging from the earth sciences to the life sciences. This volume on the 'Physics at the Large Hadron Collider' is one in this collection.

The Large Hadron Collider (LHC) is a huge international experimental facility to probe the fundamental structure of matter. This particle accelerator in CERN (Geneva), now in its final stage of test-runs, will yield highly energetic protons which will collide with protons in a counter-rotating beam to produce new particles never seen before. The much-awaited results are expected to shed light on unsolved issues in particle physics as well as in cosmology. There have been valuable Indian contributions to this effort through, among other things, the fabrication of magnets for the accelerator itself as well as participation in the development of the detectors and software for the ALICE and CMS experiments at the LHC. Together with this, theoretical research by Indian particle physicists have inspired new ways for probing at the LHC those models now considered viable for our understanding of the building blocks of Nature.

It is my expectation that the articles by experts collected in this volume will give a flavour of this frontier subject to scientists across different disciplines. Students and researchers active in this area will also find this collection of use for their work. It is also hoped that the contributions will encourage further such collaboration of the Indian scientific community in major world facilities.

M. Vijayan
President
Indian National Science Academy

PREFACE

This collection – 'Physics at the Large Hadron Collider' – is one of several such volumes to be published by the Indian National Science Academy to mark its Platinum Jubilee. The anniversary falls in an epoch when particle physics is awaiting a major step forward. The Large Hadron Collider (LHC) at CERN, Geneva will soon be in operation. It will collide a beam of high energy protons with another similar beam circulating in the same 27 km tunnel but in the opposite direction resulting in the production of many elementary particles, some never created in a laboratory before. The LHC will have an unprecedentedly high centre of mass energy of 14 Tera electorn Volts (TeV) and a luminosity of $10^{34}\mathrm{cm}^{-2}\mathrm{s}^{-1}$, which is expected to be upgraded to the order of $10^{35}\mathrm{cm}^{-2}\mathrm{s}^{-1}$ after the first few years. These collisions will be studied by sophisticated state-of-the-art detectors which are also huge in size. It is widely expected that the LHC will discover the Higgs boson, which gives masses to all fundamental particles via spontaneous symmetry breaking (SSB). It is worth noting that the first application of SSB in particle physics was due to Y. Nambu, which has most recently been recognised through the award of the Nobel Prize in physics for 2008. It is also anticipated that results from the experiments will address many other open issues of this subject such as CP-violation, new theories, e.g., supersymmetry, extra-dimensional models, the formation of a quark-gluon plasma, etc. For the latter, highly energetic heavy ions will be made to collide using the LHC and a dedicated experiment, ALICE, will look for the tell-tale signals. It is therefore quite timely to bring out a compilation of articles on various aspects of the Large Hadron Collider to mark the Platinum Jubilee of the Academy.

CERN, the European Centre for Particle Physics, has been at the forefront of research and the major international hub for many years. It is now embarking on a new journey through the launch of the LHC, which took more than a decade to be built. India enjoys an 'Observer status' at CERN – one of eight Observer states, Member states of CERN being all European – and there is substantial Indian participation in the LHC, both in the fabrication of components for the machine itself and in two of the four major experiments to be carried out there.

As far as fabrication is concerned, a number of components and subsystems, including precision magnet positioning stand jacks, superconducting corrector magnets and quench heater protection supplies, are part of the Indian contribution. A number of organisations including RRCAT, BARC, IGCAR, VECC, BHEL, ECIL, ATL, and IGTR have taken part in the endeavour. The overall Indian contribution to the accelerator aspect of LHC is valued at around 43 million Swiss Francs. These contributions are especially noteworthy because they have to satisfy critical time deadlines and exacting levels of precision.

The Indian involvement in the two LHC experiments, namely, CMS and ALICE, is decidedly significant. This includes detector building, installation, analysis software, physics simulation, etc.

The Indian component of CMS is contributing to the tracker, electromagnetic as well as hadron calorimeters, and designs related to magnets. On the physics side, Indian scientists are taking an active role in the search for Higgs, supersymmetry and various new particles. The participating institutions include TIFR, BARC, University of Delhi, Panjab University, and Visva-Bharati.

The ALICE experiment, where the search for quark-gluon plasma forms the main thrust of the heavy ion collisions, Indian institutions such as the VECC, SINP, IOP, IIT Bombay, and the Universities of Jammu, Rajasthan, Aligarh, and Panjab are involved. They have been instrumental in the designing of photon multiplicity detector devices as well as muon chambers.

In the LHC grid computing front, regional Tier II centres have been set up in TIFR and VECC, while 17 Tier III centres are distributed across the country. During the operation of the experiments and data collection as well as analysis, these centres will have active roles to play.

Parallely, theoretical physicists from India have established an acclaimed role for themselves in the prediction, interpretation and application of data from the LHC in relation to diverse particle physics models. Numerous contributions have been made from HRI, TIFR, IISc, Jadavpur University, SINP, VECC, IACS, IISER Kolkata, PRL, IMSc, IOP, IIT Kanpur, IIT Gauhati, NEHU, and the Universities of Calcutta, Delhi, and Mumbai. The contributions made by scientists from here relate to the Higgs boson, quantum chromodynamics, implications of supersymmetric theories, TeV-scale extra dimensions, and a number of other issues connected with the ongoings at the LHC.

This volume begins with an article by Choudhury and Mukhopadhyaya summarising the motivations and physics behind high energy colliders in general, and the LHC in particular. This is followed by a contribution from Gurtu which brings out the status of the Standard Model of elementary particle physics as probed by the earlier accelerators, namely, LEP, SLC, and the Tevatron. Expectations from the LHC are also touched upon.

The accelerator physics issues that arise when a state-of-the-art machine such as the LHC is to be designed and how they are addressed are dealt with by Brüning, Chohan, and Chattopadhyay. A very formidable challenge is also posed by the software and computing requirements of such a major facility, leading, for example, to distributed computing environments like the GRID. This is covered in the following article by Banerjee.

The Standard Model, though in remarkable agreement with experiments, is still missing one critical element – the Higgs boson responsible for spontaneous symmetry breaking and mass generation. This is naturally going to be at the top of the LHC agenda. The status of this subject is presented in the article by Djouadi and Godbole. In the following contribution, Mellado focusses on the CMS and ATLAS experiments describing their structure in some detail and stressing how they will be searching for the Higgs boson. In the next paper, Mathews and Ravindran discuss the important corrections to the lowest order predictions of the Higgs boson production and decay channels that arise through strong interaction effects. Without the incorporation of these contributions any analysis of the LHC results would not be reliable.

The success of the Standard Model notwithstanding, our understanding of quantum chromodynamics, the theory of strong interactions is incomplete. Was there formation of a quark-gluon plasma at the very early stages of the evolution of the Universe? With the accessibility of collisions of highly energetic heavy ions LHC should be able to throw light on this question. Bhalerao and Gavai give an introductory review of this subject. The ALICE experiment, especially the Indian contribution, is covered in the article by Nayak and Sinha.

Spontaneous symmetry breaking effectively adds a longitudinal degree of freedom to gauge bosons, the hallmark of a massive spin-1 boson. This leaves its imprint in gauge boson scattering processes as discussed in the next article by Rindani.

Among the many extensions of the Standard Model the most studied is undoubtedly supersymmetry which postulates a symmetry between bosons and fermions. LHC is expected to observe the much awaited signals of supersymmetry. The many alternative search modes for supersymmetry are discussed in the article by Datta, Guchait, and S. Roy. Supersymmetry addresses the vexing problem of naturalness inherent in the Standard Model, its main motivation in particle physics. It also offers candidates for dark matter which fill the gap between the amount detected gravitationally and the observed matter in the Universe. LHC will provide a window for studying the dark matter predictions of supersymmetry and other models as covered in the article by Baer and Tata. At the LHC, τ leptons will be produced in the decays of Higgs bosons as well as other standard or non-standard particles. The polarisation of the τ provides a handle to distinguish between different production mechanisms and is a valuable tool for the LHC experiments as discussed by Guchait and D.P. Roy.

The possibility of extra compact dimensions has been an area of much recent activity. The recent interest has been sparked off by the realisation that the compactification scale need not be at the Planck scale, as earlier thought, but could be much lower, resulting in testable predictions at LHC energies. An introduction to this subject has been provided by Sreerup Raychaudhuri. This is followed by a more detailed discussion by Dobrescu.

Regretably this volume does not include any discussion of a few important topics, e.g., the precision study of the top quark at the LHC.

We wish to thank the authors for responding positively to our invitation to contribute to this volume. We acknowledge help from Nishita Desai and Soumitra Dasgupta for the cover design. Ram Lal Awasthi, Nabanita Bhattacharyya, Sandhya Choubey, Ananda Dasgupta, Nishikanta Khandai, and Abhoy Pratap Singh have also extended their help in the preparation of this volume.

We thank the Indian National Science Academy for bringing out this volume. In particular, our thanks are also due to Professor Alok K. Gupta for his constant encouragement of this endeavour.

The work of AD has been supported by a grant from the Department of Science and Technology, India while BM and AR have been helped through the 'RECAPP' and 'Neutrino Physics' projects at HRI funded by the Department of Atomic Energy, India.

January 2009

Amitava Datta
IISER, Kolkata

Biswarup Mukhopadhyaya
RECAPP, HRI, Allahabad

Amitava Raychaudhuri
HRI, Allahabad

ABBREVIATIONS

ALICE	A Large Ion Collider Experiment, CERN
ATLAS	A Toroidal LHC Apparatus, CERN
BARC	Bhabha Atomic Research Centre, Mumbai
BHEL	Bharat Heavy Electricals Limited
CMS	Compact Muon Solenoid, CERN
ECIL	Electronics Corporation of India Limited
HRI	Harish-Chandra Research Institute, Allahabad
IACS	Indian Association for the Cultivation of Science, Kolkata
IGCAR	Indira Gandhi Centre for Atomic Research, Kalpakkam
IGTR	Indo German Tool Room
IISc	Indian Institute of Science, Bangalore
IISER	Indian Institute of Science Education and Research
IIT	Indian Institute of Technology
IMSc	Institute of Mathematical Sciences, Chennai
IOP	Institute of Physics, Bhubaneswar
LEP	Large Electron and Positron Collider, CERN
LHC	Large Hadron Collider, CERN
NEHU	North-Eastern Hill University, Shillong
PRL	Physical Research Laboratory, Ahmedabad
RECAPP	Regional Centre for Accelerator-based Particle Physics, HRI, Allahaabd
RRCAT	Raja Ramanna Centre for Advanced Technology, Indore
SINP	Saha Institute of Nuclear Physics, Kolkata
SLC	The Stanford Linear Collider, SLAC
TIFR	Tata Institute of Fundamental Research, Mumbai
VECC	Variable Energy Cyclotron Centre, Kolkata

CONTENTS

(Selected figures printed in colour appear at the end of the Volume)

The Large Hadron Collider:Overall Motivation

Debajyoti Choudhury[a] *and Biswarup Mukhopadhyaya*[b]

[a]*Department of Physics and Astrophysics, University of Delhi, Delhi-110007, India*

[b]*Regional Centre for Accelerator-based Particle Physics, Harish-Chandra Research Institute, Chhatnag Road, Jhunsi, Allahabad-211019, India*

The motivations for building particle accelerators in order to probe the elemental structure of all matter is explained. After establishing the need to develop colliding beam type accelerators, the nature and utilities of the Large Hadron Collider is discussed. The related physics issues are outlined in a non-technical manner.

Ever since the early phases of civilisation, questions such as 'what is everything in the universe made of?' or 'what is the nature of the fundamental constituents of all matter?' have haunted mankind. In course of time, ontological debates on the subject have been overtaken by experimental investigations, together with efforts to connect the outcome of experiments with a theoretical framework, so that the observations are explicable as well as predictable. Of course, the foray into the micro-world owes much to the development of technology and that is why the most remarkable progress has taken place since early twentieth century.

The epoch-making experiment in this context is the one carried out under the direction of Ernest Rutherford in 1909, when he bombarded alpha particles on gold foils and concluded from the results that the atom has two sectors within, namely a tiny, positively charged nucleus where most of its mass lies, and a planetary system of negatively charged electrons which accounts for most of the 'volume' of the atom. The question of how such a system could be stable led to quantum mechanics, a theoretical system that revolutionised our understanding of the realm of the very small. Among many things, quantum mechanics taught us that every object has wave as well as particle nature. Thus, when we use a particle to probe the fundamental structure of matter, how small a distance can be studied depends on the smallness of λ, the wavelength of the probe (the alpha particle in Rutherford's experiment), which is related to its momentum (p) by the celebrated de Broglie equation

$$\lambda = \frac{h}{p} \tag{1}$$

where h is Planck's constant, the fundamental quantity that is part and parcel of the entire formulation. Thus the smaller the distance scale one is trying to investigate, the more energetic must the probe be. It follows immediately that, while Rutherford's original strategy of probing small structures by colliding matter (read small particles) against matter still remains the accepted paradigm, probing even smaller distances requires incident particles of much higher energy than what Rutherford could achieve in his time. While the distance scale to be probed by Rutherford was of the order of 10^{-8} metres, the quest for fundamental structures has now taken us to the domain of 10^{-19} metres. To learn further, we therefore want the energies of incident particles to be higher by eleven or twelve orders of magnitude, something that can be delivered only by extremely sophisticated technology. Thus the role of particle accelerators has posed a crucial challenge to basic as well as applied science. The Large Hadron Collider (LHC) is the latest and the most gigantic step in this direction.

First of all it is important to realise that accelerator experiments can be of two varieties, namely of the fixed target and the colliding beam types. A fixed target experiment is one where accelerated particles (which typically are charged objects—such as electrons and protons—as electric charge facilitates acceleration) are used to hit some dense material whose protons and neutrons act as the 'targets'. This ensures a sizable luminosity (defined in terms of the number of particles encountering each other per unit area per unit time).

Now the relevant quantity in determining the energy reach of an accelerator is the centre-of-mass energy (E_{cm}), the total energy of the two particles in a reference frame where they appear to approach each other with equal and opposite momenta[1]. For the spe-

[1]Basically, the energy reach is dictated by how massive are the particles that are produced. This is but a consequence of the famous mass-energy equivalence in Einstein's special theory of relativity.

cial case of particles of the same type colliding with each other with the two beams carrying the same energy, this frame is identical to the laboratory frame and $E_{cm} = 2\,E$ where E is the energy of each beam. On the other hand in a fixed target experiment, even for an ultra-relativistic beam, we only have $E_{cm} \approx \sqrt{2\,M\,E}$ where M is the mass of the target particle and E the energy of the incident particle as measured in the laboratory frame.

It is thus obvious that, in order to enhance E_{cm} by a factor of ten, one has to increase the beam energy a hundred times in a fixed target experiment, while a ten-fold enhancement suffices in the case of colliding beams. In spite of the loss of assured luminosity and the technological challenge of making up for the loss by packing the beams densely enough, the colliding beam or collider, is therefore a better choice when very high energies are to be attained. Consequently the quest for the elementary constituents of nature has become heavily dependent on high energy colliders, as the answers to most questions that are being asked seem contained in the energy range between 100 GeV and 1 TeV.

Elementary particles, as they are known till date are classified by the different types of interactions they participate in. We are aware of four fundamental forces of nature so far, namely gravitational, electromagnetic, strong and weak. Of these gravity has an insignificant role in interactions between subatomic particles, unless one goes down to a length scale of 10^{-32} metres (a far cry from current technological abilities). Electromagnetic interactions affect all charged particles in the microscopic as well as macroscopic world, while the strong and weak forces operate within very short ranges only, the former being responsible for holding protons and neutrons together in nuclei, and the latter being instrumental for processes such as nuclear β-decay. The "elementary" particles which participate in the strong forces are called hadrons, all of whom are found to consist of spin-1/2, pointlike particles called quarks (along with another set of spin-1 particles—aptly called gluons—serving to hold them together). Quarks have electric charges that are fractions of that of a proton, and are so far known to be of six varieties, namely up, down, charm, strange, top and bottom. There are similarly six spin-1/2 particles of the non-strongly interacting type, which are called leptons. These include three particles with identical charge, namely the electron, the muon and the tau, and three neutral partners, each of which is termed a neutrino (and identified by their charged partners namely ν_e, ν_μ and ν_τ).

The strong, electromagnetic and weak forces are mediated by spin-1 particles: the photon for electromagnetism, the gluon for the strong force and the $W^{\pm}$ and the Z^0 for the weak force. Together with an additional spinless particle called the Higgs boson which is as yet unseen but expected to be an essential component of nature, these are believed to constitute the universe as we know it today, at the fundamental level (Fig. 1).

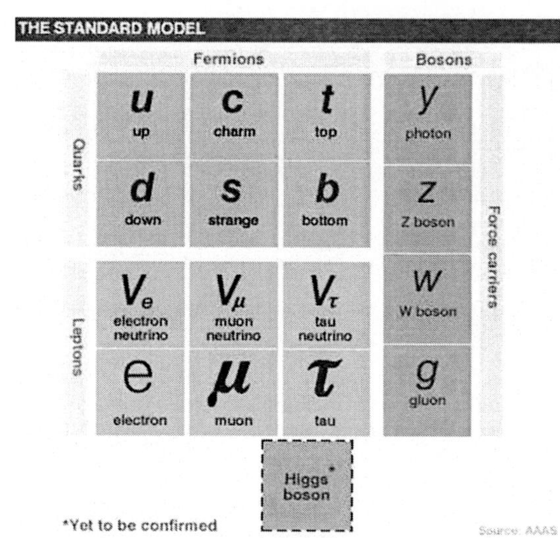

Figure 1. The complete set of elementary particles, as they are known to exist till date

In order to describe interactions among these particles, which should be at the root of all 'that has been and may be again', one requires a framework that is consonant with both the special theory of relativity as well as quantum theory. At subatomic scales the particles whirl about with speeds close to that of light; at the same time, the ultra-short distance scales necessitate a quantum mechanical framework, including the provision of creation and annihilation of particles in elementary processes. The above features are all realised in relativistic quantum field theories (QFTs) and thus QFTs have turned out to be the theoretical language of high energy particle physics.

A highly successful QFT that describes the strong, electromagnetic and weak interactions has been in existence for about four decades now and is called the Standard Model (SM) of particle interactions. It contains a complete description of strong interactions in terms of quarks and gluons, which works quite well for energy scales above a GeV or so[2]. SM dealing with strong interactions is called quantum chromodynamics (QCD),

[2]At energies below this scale, we are limited by our computa-

a fancy name inspired by a unique property of quarks and gluons, called the 'colour' degree of freedom.

The advent of QCD has also familiarised us with a special feature of theories of its kind. It transpires that the strength of interactions between quarks and the gluons, the mediators of the strong force, diminishes as one considers phenomena at higher and higher energies. Thus, at energies well above a TeV or so, the quarks can be looked upon as free from the shackles of the strong force binding them to the neighbouring quarks in nucleons, and the participation of one of them in the strong force in collider phenomena can be discussed without worrying about its neighbours. This feature is called 'asymptotic freedom', and the Nobel Prize for Physics in 2004 went to H. Politzer, D. Gross and F. Wilczek for its discovery in the early 1970s.

The description of weak and electromagnetic forces is intertwined in a framework called the 'electroweak' theory of S. Glashow, A. Salam and S. Weinberg (GSW) who shared the Nobel Prize for 1979. In the low-energy limit, it yields an enormously successful theory of electromagnetic interactions, namely quantum electrodynamics (QED)[3], which had emerged years earlier through the works of Dirac, Feynman, Schwinger, Tomonaga and others. The remaining part of the GSW model describes weak interaction in a comprehensive manner. One crucial part of this description is an explanation of the fact that the weak force is a short-range force. This is ensured through an ingenious theoretical scheme which requires the existence of a spinless neutral particle, the Higgs boson, mention of which has been made earlier.

The remarkable success of the SM is based on the following achievements:

- Strong interaction processes at high energies are extremely well-explained.

- The weak force is predicted to be mediated by two massive spin-1 particles, namely the W and the Z (both with masses close to 100 GeV/c^2). Both of these have been experimentally observed, and their properties match the predictions of the theory to the limits of current sensitivities.

- The theory of electromagnetism, whose success provided the paradigm of QFTs, is reproduced in an uncontaminated form in the low-energy limit.

- The predictions of the SM have been verified, not only to the lowest order *(of perturbative calculations)*, but to higher orders as well, thereby upholding the quantum field theoretical *nature* of the SM to a high degree of precision.

The role of high-energy colliders in establishing the *bona fide* of the SM can hardly be overemphasized. Obviously, stable and charged particles are best suited for collision, and one is thus inclined to use electrons, protons and the corresponding 'antiparticles', namely positrons and antiprotons. It is the proton-antiproton collision experiment at CERN, Geneva, that led to the discovery of the W and the Z particles, the carriers of the weak force. Similarly the top quark (t), the most massive elementary particle known so far (the mass being about 172 GeV/c^2) has been discovered in proton-antiproton collision at the Fermilab Tevatron experiment. The features of the colliders that made these discoveries possible are (a) the high centre-of-mass energy attained and (b) especially in the case of the t, the fact that the protons and its constituents participate in strong interaction, a force to which the top quark is also susceptible, so that the production rate is enhanced by the strength of interaction.

Side by side electron-positron colliding machines have been instrumental in establishing aspects of the SM that would have otherwise evaded our attention. The point to note in this context is that, while a hadron (proton-antiproton or proton-proton) collider produces 'new' particles abundantly, it also produces a lot of other things at the same time. A huge background, arising from such extraneous processes, thus tends to submerge any signal one is looking for. While the high rate of production and the facility to generate high energies still qualifies hadron colliders as extremely effective machines, one may have to supplement their findings with studies in environments that are relatively 'cleaner' in terms of both backgrounds and theoretical uncertainties[4]. Such an environment is provided

tional abilities (especially in perturbative analyses, where one can make order-by-order predictions, with progressively higher orders altering the results less and less), although in recent years, several non-perturbative techniques such as sum rules and lattice gauge theories have allowed us to register significant progress.

[3]The successes of QED include the agreement of the theoretical calculation of the Lamb shift, or the modification of energy levels of electrons in an atom due to subleading electromagnetic effects, with the experimental results to a high degree of precision. A similar agreement has been found for the magnetic dipole moment of the electron. Also remarkable is the accuracy of the prediction for the variation of the electromagnetic coupling with the scale of probing (analogous to the familiar charge screening, though not limited to a medium but applicable to measurements in vacuo as well).

[4]Note that our understanding of strong interactions at the 1 GeV energy scale, and hence that of the structure of the proton is still imprecise. This in turn leads to theoretical uncertainties in the rates of strong processes requiring the participation of protons

by electron-positron colliders. It is the Large Electron Positron (LEP) collider at CERN that allowed one to study in great detail the properties of non-strongly interacting particles such as the W and the Z, confirming that their properties agreed closely with what is expected in the GSW framework. The so-called quantum effects of the SM, meaning the effects in next-to-leading orders, were also verified quite precisely at LEP which ran in two phases (from 1989 till 1993 and from 1995 till 2001). This went a long way in convincing physicists that after all, three out of the four fundamental forces can be described at the sub-microscopic level by consistent QFTs of pointlike particles.

One central pillar of the GSW model, however is yet to be found. That is the Higgs boson, the electrically neutral, weakly interacting particle that is responsible for the mechanism that renders the W- and Z-bosons massive, and consequently the weak force short-ranged. In addition we have an explanation of why the quarks and leptons have non-vanishing masses of various magnitudes, by assuming that the Higgs boson is there and all these particles interact with it.

Since the entire edifice of the SM is likely to crumble without the Higgs particle, it is of great urgency to verify its existence in collider experiments. The LEP and the Fermilab Tevtaron have looked for it, but without any success so far, and it is generally believed that a machine that can achieve still higher energy and at the same time a very dense flux of colliding particles is required for the purpose. Remarkably, though the Higgs boson itself is not expected to participate in strong interactions, the quarks and gluons involved in hadron collisions can conspire to give rise to processes that lead to its production.

In spite of its elusive character, certain expectations about the properties of the Higgs particle have developed; the most important of these concerns its mass, a quantity that essentially tells us how much energy is required to produce it. The viability of the GSW model as a QFT, where predictions can be made in a systematic, order-by-order fashion, suggests that the Higgs mass is likely to lie within a TeV/c^2. If it is more massive, then its interaction will cease to be weak and the general prejudice of the physics community is contrary to such a possibility. In a still more startling revelation a heavier Higgs leads to unitarity violation, meaning a loss of conservation of total probability in WW scattering, thereby coming into conflict with the basic tenets of quantum mechanics. The LEP experiment (and partially the Fermilab Tevatron) on the other hand, have yielded data that tell us how much the Higgs boson can 'indirectly' affect various weak processes without being actually produced. Such indirect evidence, in spite of large uncertainties, points to the likelihood of a Higgs boson being within a mass range of about 200 GeV/c^2. This brightens the prospects of discovering the Higgs boson at a hadron collider that can attain sufficient energy. While the Fermilab Tevatron has been colliding protons against antiprotons at 1.96 TeV, it should be noted that the effective reach of any hadron collider is significantly lower than the nominal value as the constituent quarks and gluons (which essentially participate in the high-energy collision) carry only a fraction each of the energy of the proton. A machine with higher 'effective' reach has therefore been sought, and such endeavours have culminated in the design of the LHC.

The LHC is a proton-proton collider where each proton is expected to carry an energy of 7 TeV in the steady condition. Thus the target centre-of-mass energy is 14 TeV. It is a storage ring type accelerator where the rings carrying the colliding beams have a radius of 4.3 km. Although it is of advantage in some ways to have a proton-antiproton collider, the task of both producing and storing antiprotons is far too challenging at such high energy, where an intensely collimated beam is demanded. The entire assembly is situated underground, using the tunnels and other assemblies of the erstwhile LEP, thereby saving enormously on both costs and construction effort. The depth of the storage ring depth varies between 50 and 150 m, and it goes around the neighbourhood of the city of Geneva, and passing partly through Switzerland and partly, France.

The colliding beams are made to cross at four places where the all-important collisions occur. A detector is built around each such collision point. These are the four main LHC detectors, carrying the acronyms CMS (Compact Muon Solenoid), ATLAS (A Toroidal LHC Apparatus), LHCb and ALICE (A Large Ion Collider Experiment). Unlike the first two which are aimed at understanding the SM and the physics beyond as a whole, the last two are of more specialised nature. Much of the rest of the discussion in this article is centred around the projected findings of the CMS and ATLAS detectors. As for the last two, the purpose of LHCb is to explore the properties of hadrons comprising the bottom quark at high energy, an endeavour that may uncover interesting features of some symmetries of nature as well as their violation. ALICE on the other hand, aims to observe a new likely phase of matter, namely a plasma consisting of quarks and gluons.

In addition to these four, two rather small experiments, namely the LHCf (LHC forward) and the

and antiprotons

TOTEM (TOTal Elastic and diffractive cross section Measurement) serve a unique purpose. Situated close to ALICE and CMS respectively, these will study very forward physics that is not accessible to the general purpose detectors. Concentrating on particles that are deflected only very slightly from the beamlines (in fact these consist the overwhelming majority), these "Roman pots" can measure the trajectories with an accuracy of better than 0.1 mm. Apart from serving to calibrate the luminosity to great accuracy, these experiments are going to study the structure and interactions of the proton in an energy regime hitherto unexplored, thereby shedding light on many well-motivated theoretical constructs.

What do these big detectors (CMS, ATLAS, etc.) detect? While many details about them are discussed in other articles in this volume, let us note here that heavy particles produced in the collider are destined to decay into lighter ones, the decay process being usually rather fast (with lifetime $\lesssim 10^{-12}$ s). Therefore the detectors are mostly able to record only stable particles which are produced in decays, exceptions being muons and pions which, in spite of their penchant for decays, move fast enough in the detector environment to have their lifetime time-dilated, so that their presence can be recorded directly.

Thus the objects of detection include all charged (quasi)stable particles which can leave ionising tracks, (quasi)stable strongly interacting particles that can be absorbed in a dense material, electrons and photons which are light and are thus promptly absorbed via electromagnetic interactions, and muons which live long enough and are just heavy enough to survive absorption before their tracks are seen in the outer part of the detector. In addition the innermost chambers of the detector contain contraptions to track down particles that are unstable but survive just that extra bit longer to leave 'kinks' in their tracks, thus unleashing ringtones of their decay. Hadrons containing the bottom quark are the primary members of this club.

It is essentially with the above capabilities that the LHC detectors set out to observe the Higgs boson. Exactly what kind of final states will unequivocally point towards the Higgs depends, of course, on its mass, something that is important in deciding how it will decay after production. In the absence of a clear guideline on the Higgs mass (despite some indirect indications), it is therefore the task of theorists and experimentalists together to enlist the different final states corresponding to various possible masses and the resulting decay channels, and estimate the probabilities of observing such final states. In each case one needs to eliminate backgrounds or fake events that resemble the expected signals and devise means of isolating them from the real signals. Various final states have thus been identified as signatures of the Higgs boson, such as

- Four leptons (specifically, electrons or muons), with the energies and momenta of two pairs related in a particular manner (two 'invariant mass peaks' in technical language).

- Two photons with energies and momenta correlated in a similar fashion.

- A definite sign of a W-particle with certain kinematic characteristics.

- Pairs of hadrons containing the bottom quark, seen simultaneously with some other specified particles (a W boson as identified from its decay products, for example).

- Two highly energetic jets in opposite directions, moving almost parallel to the beam pipe, together with pairs of bs/taus/photons in the region between them.

If the Higgs boson is discovered in any one or more of these channels, then one will have 'completed' the GSW part of the SM, knowing why the weak force is short-ranged, how the quarks and leptons have the various masses that they appear with and also how the underlying QFT turns out to be a consistent one. In addition some other goals of the LHC in perfecting our understanding of the SM are:

- Producing a large number of top quarks. The top being the heaviest of quarks, with a mass exceeding those of the carriers of the weak force, a detailed study of its behaviour will provide a much better understanding of the entire framework of the electroweak theory.

- Knowing more about strong interactions. A plethora of hadrons are going to be produced at the LHC, thanks to the strong force with which the participating protons strike each other. A painstaking analysis of the recorded data is likely to tell us quite a lot about how quarks and gluons are contained within hadrons, and also how strong interaction modifies weak and electromagnetic processes at high energy.

- Looking for hitherto unexplored facets of the W and Z particles, such as 'anomalous couplings'.

- Finding out more abut the properties of hadrons containing the bottom quark and looking for quark-gluon plasma, as has been mentioned before.

However the purpose of such an enormous experimental set-up is not confined to the verification and perfection of some laws of physics whose foundations have been already laid. One major target of the experiment is to look for physics beyond the GSW model of electroweak interactions. Let us try to understand in the next few paragraphs how this aspiration for 'new laws of physics' has taken root in the physicist's mind.

First of all, in spite of its enormous success, the GSW structure rests on too many free parameters. Various masses, interaction strengths, mixing angles etc. built into the theory appear with their 'God-given' magnitudes, without any ostensible correlation. There are about twenty such free parameters without any explanation of their values, all extracted from experimental data. It is often felt that the capacity of the SM to explain a large volume of data is essentially due to such a large number of free parameters. These parameters may have emerged from some yet unknown underlying principle, and thus the GSW scenario perhaps hides more than it tells us in terms of the masses of various particles, the strengths of their various interactions and the degrees of mutual overlap of the different quantum mechanical states.

In particular it is not understood why quarks and leptons occur in three 'families' – that is to say, for example, why the electron, the muon and the tau have identical properties in every way excepting that their masses are progressively higher. A similar observation applies to the quarks also – it seems that nature has neatly divided its spin-1/2 building blocks into three different stories (where the spin denotes the intrinsic angular momentum in units of $h/2\pi$, h being the Planck's constant) all of identical layout, with just a step-by-step increase in height (read mass). No explanation of this is yet available within the SM.

Also, the fact that the weak force is left-right asymmetric is an enigmatic property whose origin is not understood. We know that the elementary particles have 'spins' associated with them, and that relative to its direction of motion, a particle can be 'spinning' in either the left- or the right-handed sense. Surprisingly, the weak force makes a distinction between the two kinds of 'handedness', so that left- and right-handed electrons (or other spin-1/2 particles) suffer the weak force to different degrees. This parity (P)-violating nature of the weak force (postulated first by T. D. Lee and C. N. Yang and experimentally verified by C.S. Wu), is unique amongst all the fundamental forces of nature. Moreover parity violation is, in some sense, 'maximal'. Weak interactions come in two related flavours, namely 'neutral current' (NC) processes wherein the identities of the particles being scattered remains unchanged and 'charged-current' (CC) wherein the identities do change. The latter were the first to be identified and nuclear β-decays constitute the most famous example. The aforementioned maximal violation of parity (discovered first by E. C. G. Sudershan and R. Marshak and confirmed later by R. Feynman and M. Gell-Mann) is exemplified by the fact that only left-handed electrons are emitted in β-decays as well as any other CC process.

A further symmetry is violated by the weak force and once again maximally by the CC processes. This relates to 'charge conjugation' (C). Whereas the electromagnetic forces between particles remain unchanged if the charges of all the participating particles were reversed (i.e. the particles turned into their antiparticles), the same is not true of the weak force, especially if the handedness remains unchanged. For example, while left-handed electrons can be emitted in a nuclear beta decay, left-handed positrons may not be.

On the other hand if all left-handed particles were mapped into right-handed ones, and particles were at the same time turned into antiparticles, then the overall character of the fundamental interactions seemed to remain invariant. In other words, the weak interactions, though violating each of C and P maximally, somehow seemed to respect the combination CP. However such a fond hope was also belied in the first half of the 1960s in certain experimental situations pertaining to what are known as K-mesons. Curiously, compared to the maximal nature of parity or charge conjugation violation, CP is violated only by a very small amount—at the one per mille level. This again was a puzzle leading to various speculations about the origin of CP violation, including the postulation of yet another new force. More conservatively though, the GSW model was shown (by M. Kobayashi and T. Maskawa) to admit CP violation, but only if three families of quarks were postulated (at the time only two families were known, whether of quarks or of leptons). Fortunately enough, members of the third family were discovered soon after, culminating finally in the relatively recent discoveries of the top quark and the tau-neutrino. While this, at first glance, would seem to have vindicated the proposal of Kobayashi and Maskawa, it needs to be realized that their model was essentially an *ad hoc* one, explaining one piece of data at the cost of several new parameters. In recent years though, several predictions of this model (pertaining to B-meson decays) have been brilliantly corroborated by the BaBar and Belle collaborations at the two B-factories. The success of the Kobayashi-Maskawa scheme model is reflected in the Nobel Prize for 2008, given to the two persons who proposed it.

Yet, problems persist. The KM model despite its success, remains a model and does not offer any deep understanding of the puzzle of fermion masses and mixing or CP violation. On the other hand, CP violation is a key ingredient to our very existence, for it is necessary for the observed dominance of matter over antimatter in the universe. The GSW scenario, however cannot account for the requisite degree of CP-violation for explaining the matter-antimatter asymmetry of the universe. Is there, then a source of CP violation in physics going beyond the GSW model? Questions such as this are the rationale behind the LHCb experiment that will probe B-mesons of various hues, some for virtually the first time ever and others to an unprecedented degree.

Various observations of the above kind have led physicists to believe that the SM is but an 'effective theory', behind which lie some new laws of physics. The exact nature of such underlying physics is, however yet unknown. Therefore we have been desperately awaiting hints to come from high-energy particle accelerators. The expectations are mounting as the LHC is being turned on, for thus begins the real exploration of the TeV scale.

But why do we expect the new physics clues to lurk around the TeV scale? Strangely enough, the most crucial hint in this connection is provided by the yet elusive Higgs boson, the central pillar of the GSW theory itself.

As has been repeatedly said, the GSW framework is a QFT. Such a theory has certain inner compulsions, based on the very logic which gives it a solid foundation. One of these compulsions is the shift in the mass of a particle as a result of interactions with others. It may be recalled that an electron moving in a solid develops an 'effective mass' different from its original mass, which is a result of its interaction with the solid core. In a similar vein, an elementary particle, even in vacuum, is susceptible to a shift in its mass, through interaction with particles produced out of vacuum itself—something which is always possible in a relativistic framework so long as the particle carries enough energy required for the process.

Now interactions of the above kind do not alter the masses of the electron and other spin-1/2 particles by any appreciable amount. This is due to some special symmetry identified in QFTs of spin-1/2 particles, called chiral symmetry. This symmetry ensures that the potential mass shifts due to various sources tend to cancel each other. In a similar manner, the carriers of the strong, weak and electromagnetic forces are immune to large mass corrections due to the very symmetries that are at the foundation of the GSW framework.

The Higgs boson, however is not blessed with any of these prospects, and the underlying dynamics of the very structure it supports threatens to revise its mass upwards by a very large amount – all the way to the maximum energy upto which the GSW theory is valid. If this limit is the energy at which gravity overtakes the other interactions in strength, then the Higgs mass becomes heavier by some sixteen orders of magnitude than what it is presumed to be. This is indeed a paradoxical feature of the GSW model, which can be resolved only by assuming that the large corrections to the Higgs mass cancel each other accidentally, through the 'fine-tuning' of otherwise unrelated parameters.

A solution to this puzzle is often sought through the bold hypothesis that the upper limit of validity of the GSW theory is itself in the TeV range of energy. If that is indeed the situation, then the corrections to the Higgs boson mass will be restricted within this limit, thus making the theory sensible. This will require (a) some new laws of elementary particle physics around the TeV scale and (b) some feature built into these laws, which can contain the Higgs mass corrections within acceptable values without the need to fine-tune parameters.

Some possibilities in this direction are discussed in this volume. The point to note, however is that all of these entail some new observations in accelerators that can attain the TeV scale. One may observe new particles, new laws of particle interactions or the manifestation of hitherto unnoticed symmetries in the world of elementary particles. For the verification of either of the above, a collider experiment at the TeV scale is the ideal testing ground.

As we have mentioned earlier, a proton-proton collision actually means a variety of processes involving whatever is contained within each proton. But these constituents, namely quarks and gluons, can each carry only a fraction of the proton energy. Therefore, if the actual collisions at the fundamental level have to take place with centre-of-mass energies on the order of a TeV, the protons have be more energetic. On careful analysis of the roles of the quarks and gluons in producing the Higgs boson or traces of new physics, and the probability with which these quarks and gluons carry particular fractions of the proton energy, it has thus been found optimal to tune the centre-of-mass energy of the colliding protons at the LHC at 14 TeV.

Another enigmatic feature of the universe we live in

may be connected with the LHC experiment. Observational cosmology has established that only about 4% of whatever contributes to the energy density of the universe is 'visible' or detectable matter (remembering again that mater is a form of energy). Of the remaining amount, 24% consists of 'dark matter', meaning massive objects whose presence is not registered in any detector commonly used in terrestrial or astronomical observations. The rest of the energy is called 'dark energy' whose only noticeable role is to produce a tiny but finite acceleration of the universe as it expands.

While the origin of 'dark energy' is extremely ill-understood till this date, sources of dark matter may lie in the new physics laws that are supposed to transcend the SM. Particle detectors depend on either strong or electromagnetic interaction. If there are particles which take part in weak interaction alone, then it is impossible to detect them in the usual accelerator experiments or with telescopes that normally bring us information about the outer space. Only very specially designed experiments which employ the weak process can reveal their trace, and that too with considerable limitations.

Of course, particles which interact just weakly are contained in the GSW framework itself. These are the three species of neutrinos, spin-half particles that have tiny but non-zero masses. While they would have been the obvious candidates for the dark matter content of the universe, such choice is disallowed by some recent observations (mainly from structure formation in the universe). The current conclusion is that a weakly interacting particle needs to be much more massive than neutrinos in order to be a viable candidate for providing the dark matter content of the universe. And most tantalisingly, the mass range of 100 GeV/c^2 – 1 TeV/c^2 seems to be appropriate for such candidates in most theories.

It is thus natural to connect the dark matter candidate to possible new physics around the TeV scale. If this supposition is indeed true, then one can shed light not only on why the Higgs boson and all mysteries about it hover around that magic scale, but also on how a massive invisible particle that accounts for a large part of our universe is accommodated within the framework of elementary particle physics. And since the LHC is determined to probe the specific energy scale in question, it is natural to develop expectations about the signature of dark matter in the laboratory.

But how will the LHC reveal the existence of something that cannot be seen with the usual detecting devices? In order to answer this question, we should recall that any physical process has certain conserved quantities, including energy and momentum. Thus the sum of the momenta of all particles in the initial state must equal that of all particles in the final state *in every direction*. In a proton-proton collision process, neither of the colliding particles has any momentum in the plane transverse to the line of their approach. It can therefore be said with certainty that the transverse momenta of all the particles produced in the collision will also add vectorially to zero.

When some of these final state particles are invisible to the detectors, then there remains an unbalanced or 'missing' momentum in the transverse direction. If one is able to subtract fake contributions to this quantity via. for example, momentum mismeasurement, one can draw realistic conclusions about particles produced in the collision, which are indeed invisible, and thus are potential candidates for the dark matter content of the universe. Since the order of magnitude of mass of the dark matter candidate particle(s) is within the reach of the LHC, the experiment thus shimmers with the expectation of revealing the invisible. With this in view, those theories of new physics around the TeV scale, which predict massive invisible particles are discussed with special attention in the context of the LHC.

While various theoretical as well as experimental details related to the LHC will be amply covered in the other articles in this volume, it may be in order to end this discussion with some remarks on where we may expect to stand after the machine has completed its run. It is almost universally agreed that, after running for more than a decade, the LHC will unravel the Higgs boson if it exists within the commonly stipulated mass limit of about one TeV/c^2. If that happens, the ultimate victory of the GSW theory will be achieved. On the other hand, the failure to discover the Higgs after the entire run of the LHC may not cause the GSW picture to be discarded altogether, but one will then have to explain the mechanism that makes the weak force short-ranged. A way out is, for example, to postulate not the Higgs boson with whose notion we are commonly accustomed, but some equivalent of it, which interacts strongly. In addition the modified properties of the Higgs (or an assembly of objects that appears Higgs-like) will be reflected in some new traits in high-energy scattering processes involving the spin-1 carriers of the weak force. Measuring the observable quantities related to such scattering is thus another goal of the LHC.

And, then there is the top quark, the heaviest particle in the SM pantheon, but for possibly the Higgs itself. Its very heaviness suggests that it may play a crucial role in anything to do with masses and/or the Higgs and indeed many speculative alternates to the Higgs mechanism do assign a special role to the top. Thus, a refined measurement of the properties of the top quark, such as its mass and interaction strength, is expected to yield clues on whether something new is in store in this sector. This has not been possible until now since so far only the Tevatron has sufficient energy to produce the top, and the production rate is not copious. With the LHC being a 'top factory' (producing nearly 20 million top quarks per years), the exhaustive analysis of top quark properties is expected to yield rich dividends.

If neither the Higgs nor any new feature in weak boson or top scattering is seen, then of course one will be forced to the conclusion that something is seriously wrong with whatever we have thought about the weak and electromagnetic forces for the last four decades. While such a prospect for the LHC is the experimentalist's nightmare, it will force theorists to start completely afresh, creating an almost unprecedented challenge for them.

However it will be far from the end of business for particle physicists even if the Higgs is found in due course. First of all, one needs to establish that the properties of the Higgs boson are in conformity with the predictions of the GSW theory. This is not an easy task in the LHC environment where any process involving the strong interaction has too large an amount of theoretical uncertainty to make any measurement precise. Under such circumstances, one will feel the necessity of a high energy electron-positron collider to test the properties of the Higgs boson (or any new physics) in conditions less vitiated by uncertainties and backgrounds.

Of course, the discovery of some new physics at the LHC, over and above the Higgs, will bring in a most exciting era, where the process of redefining the building blocks of the universe will be pursued with renewed vigour. On the contrary, if the LHC discovers the Higgs but fails to find anything else, then the issue on the stability of the Higgs mass will still remain unresolved.

On the whole, given the richness of possibilities that it is equipped to explore and the enormous technological leap that it is, the LHC deserves recognition as a landmark in the history of fundamental physics. It is only in the next few years that we will know which direction it gives to our knowledge of the basic building blocks of nature, and the nature of forces that makes these blocks accessible to human experience.

Acknowledgement: DC thanks Pratishruti Saha for a careful reading of the manuscript and the Department of Science and Technology, India for support under project number SR/S2/RFHEP-05/2006. The work of BM has been supported by funding made available from the Department of Atomic Energy, Government of India, for the Regional Centre for Accelerator-based Particle Physics, Harish-Chandra Research Institute.

The Standard Model and Beyond–LEP/SLC/Tevatron and the LHC

A. Gurtu

TIFR, Mumbai-400005, India

The current status of the standard model of particle physics is described, in particular the recent progress made in its consolidation at LEP, SLC and the Tevatron. With the advent of the LHC, at CERN, science stands poised for a generational leap in the understanding of the universe at both the sub-nuclear and the cosmological level. Thus the LHC will act both as an ultra powerful microscope and telescope simultaneously, recreating conditions as they existed soon after the ig ang. Speculations about the supersymmetric nature of dark matter would be confronted with direct experiment. Perhaps the most exciting and uni ue feature of the LHC is the opportunity it will provide to directly study interactions at the tera energy scale. Something has to give at this scale, some new phenomena have to be lurking in the shadows. The presence or absence of the Higgs and S S will both be momentous discoveries. Of course one hopes for the totally unexpected which would be the real icing on the cake.

1. Introduction

The Standard Model (SM) of particle physics is one of the most developed and tested theories in the history of science, and attempts to describe the universe at the most fundamental level, both at the consituent level and in terms of the forces between them. While both theoretical developments as well as experimental observations accumulated over a period of time, the advent of the SM as we know it today may be dated from the late 1960s.

The coherent unification of electromagnetism with the theory of weak interactions, now known as electroweak (EW) theory by Salam [1] and Weinberg [2], may be considered to be the turning point. Soon thereafter its renormalisability was proven by 't Hooft and it became an established theory. The 1970s also saw the advent of quantum chromodynamics (QCD) as the theory of strong interactions, with an octet of gluons mediating the strong force between quarks and binding the nucleons.

The work of Kobayashi and Maskawa arguing for minimum of three doublets of constituent particles (quarks and leptons) in order to account for CP violation set the stage for later discoveries which completely vindicated this line of work.

The parallel string of experimental discoveries more than kept pace with the theoretical developments. The prediction of the EW theory that there must be a neutral current, Z^0, in addition to the well known charged current ($W^{\pm}$) was confirmed in a bubble chamber experiment at CERN, setting the seal on the veracity of the EW theory. This was followed by the discovery of the fourth quark, called charm in 1974, thus completing two doublets of quarks. The third charged lepton, the τ was discovered soon after thus necessitating the existence of 3 lepton doublets. This was followed up with the discovery of the fifth quark, the bottom in 1977 at Fermilab and the gluon in $e^+e^- \to 3$ jets interactions at PETRA. The discovery of directly produced $W^{\pm}$ and Z^0 in the early 1980s at the CERN $p\bar{p}$ collider crowned the success of the EW theory. The final icing on the cake was the discovery of the top quark at Fermilab in 1994/95, which completed the experimental observation of all the 3 doublets of quarks and leptons as well as the force particles, the gluon and the W and Z, and the photon of course had been known for a long time.

The Z^0 factories SLC at Stanford and LEP at CERN have played a crucial role in experimental consolidation of the SM. In particular LEP, with its large Z^0 statistics set new standards for the precision determination of its properties. After its energy was raised above the W-pair production threshold it allowed a precision measurement of the W properties too, and finally raising its energy to the highest possible, around 209 GeV, allowed a direct search for the Higgs boson upto hitherto unexplored mass limts. While the SLC did not compare well with LEP in terms of statistics it had the great advantage of having polarised beams which enhances the sensitivity of the data to EW mixing. Thus the determination of the EW angle from the SLC is the single most precise value of this quantity.

In this article the experimental results will be reviewed in the chronological order, from LEP100 and SLC, then the discovery of the top quark, followed by results from LEP200. After that a brief description of what one expects from the Large Hadron Collider (LHC) will be given, with emphasis on a few most salient topics.

2. SLC and LEP100

While the Z^0 and $W^\pm$ were discovered at the CERN $p\bar{p}$ collider it was clear that the number of events that could be studied would always remain small and in order to make detailed studies of the production and decay of these particles one would need Z^0 and $W^\pm$ "factories". LEP was designed as a 27 km circular e^+e^- collider with initial energy around the Z^0 mass, with a later energy upgrade to become a $W^\pm$ factory with CM energy above the W^+W^- threshold. The SLC was the world's first e^+e^- Single-pass Linear Collider with CM energy focussed on the Z^0, but with an added advantage of polarised beams, which strongly helps in the study of certain EW quantities.

A detailed account of the work carried out at LEP100 and SLC is given in a review by the LEP/SLC groups [3] and in the W and Z sections of the review of particle properties [4].

The location of the LEP storage ring and the placement of the four experiments is shown in Fig. 1 and the layout of the SLC is shown in Fig. 2.

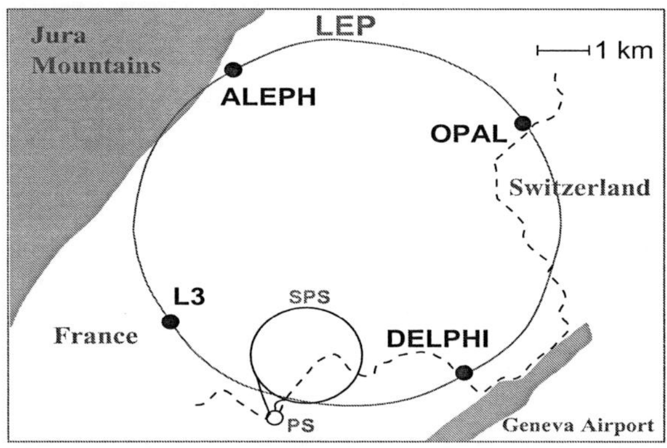

Figure 1. The LEP storage ring and location of the four experiments, ALEPH, DELPHI, L3 and OPAL

The main physics process studied at LEP and SLC is the production of a fermion anti-fermion pair, $e^+e^- \to f\bar{f}$ where f maybe one of the leptons or a quark, as shown in Fig. 3. All quark anti-quark pairs, except $t\bar{t}$ can be produced at these energies, the top quark being heavier than kinematically allowed. The study of heavy (bottom) quark production leads to indirect information about vertex corrections involving heavier particles as depicted in Fig. 4.

The main topics studied at LEP100 and SLC are the production and decay properties of the Z^0. As one scans

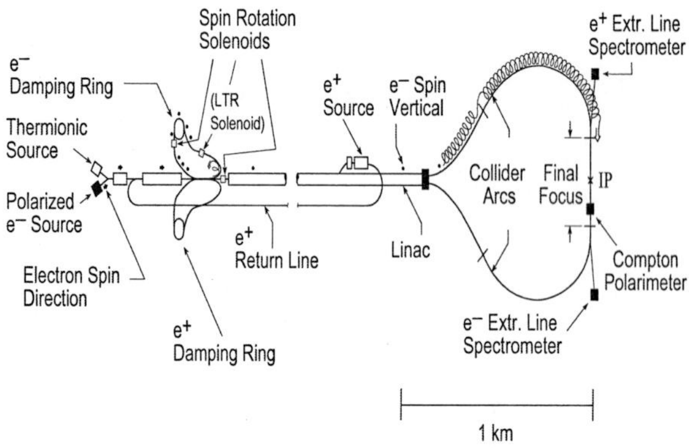

Figure 2. The layout of the SLC at Stanford. There was only a single experiment at the intersection point of the beams

the CM energy range from a few GeV below the Z^0 mass to above it, the cross section traces out the "lineshape" of the Z. The variation of the hadronic cross section in e^+e^- interactions from low to the highest energies is shown in Fig. 5. Differentiating between "forward" and "backward" cross sections (where "forward" means the fermion follows the direction of the electron) one can determine the asymmetry in the production process. A measurement of the lineshape allows one to determine the Z^0 mass, total width and its partial decay widths into various channels. The asymmetry measurements allow one to separately determine the vector and axial-vector couplings of the Z^0 to leptons and quarks and test the predictions of the EW theory in detail. At LEP one has also made a measurement of the τ-polarisation and used it to obtain additional information on τ-asymmetry parameter.

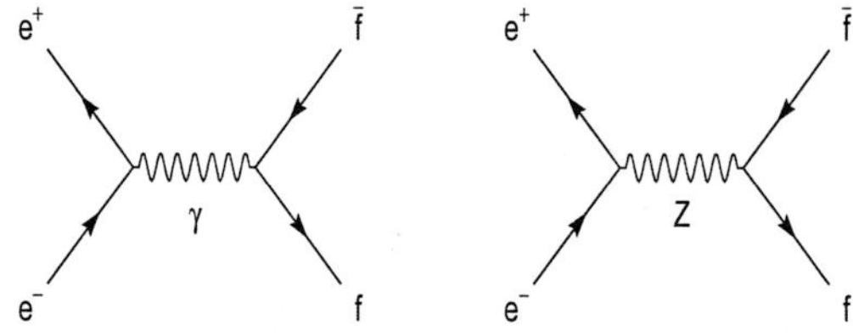

Figure 3. The lowest order s-channel Feynman diagrams

2.1. Z^0 Lineshape and Asymmetry Studies

While the SLC ran at a single energy very near the Z^0

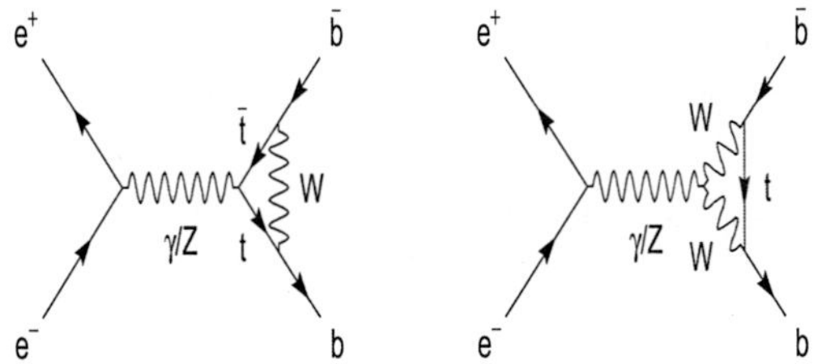

Figure 4. Vertex corrections in the process $e^+e^- \to b\bar{b}$

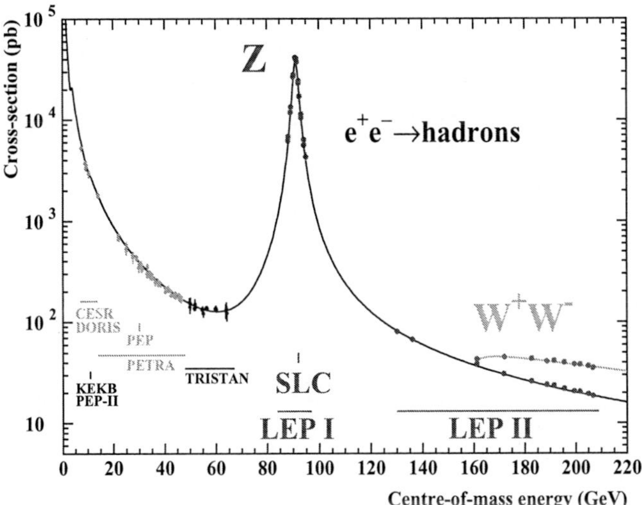

Figure 5. Hadronic cross section vs energy in e^+e^- interactions

pole, LEP ran at various energies at one GeV intervals within ±3 GeV around the pole and scanned the lineshape of the Z^0.

The shape of the cross-section variation around the Z^0 peak can be described by a Breit-Wigner *ansatz* with an energy-dependent total width [5–7]. The three main properties of this distribution, viz. the position of the peak, the width of the distribution and the height of the peak, determine respectively the values of M_Z, Γ_Z and $\Gamma(e^+e^-) \times \Gamma(f\bar{f})$, where $\Gamma(e^+e^-)$ and $\Gamma(f\bar{f})$ are the electron and fermion partial widths of the Z. The quantitative determination of these parameters is done by writing analytic expressions for these cross sections in terms of the parameters and fitting the calculated cross sections to the measured ones by varying these parameters, taking properly into account all the errors. Single-photon exchange (σ^0_γ) and γ-Z interference ($\sigma^0_{\gamma Z}$) are included, and the large (∼25 %) initial-state radiation (ISR) effects are taken into account by convoluting the analytic expressions over a 'Radiator Function' [5–9,26] $H(s, s')$. Thus for the process $e^+e^- \to f\bar{f}$:

$$\sigma_f(s) = \int H(s,s')\, \sigma^0_f(s')\, ds' \tag{1}$$

$$\sigma^0_f(s) = \sigma^0_Z + \sigma^0_\gamma + \sigma^0_{\gamma Z} \tag{2}$$

$$\sigma^0_Z = \frac{12\pi}{M_Z^2}\,\frac{\Gamma(e^+e^-)\Gamma(f\bar{f})}{\Gamma_Z^2}\,\frac{s\,\Gamma_Z^2}{(s-M_Z^2)^2 \;+\; s^2\Gamma_Z^2/M_Z^2} \tag{3}$$

$$\sigma^0_\gamma = \frac{4\pi\alpha^2(s)}{3s}\, Q_f^2 N_c^f \tag{4}$$

$$\sigma^0_{\gamma Z} = -\frac{2\sqrt{2}\alpha(s)}{3}\,(Q_f G_F N_c^f \mathcal{G}_V^e \mathcal{G}_V^f) \times \frac{(s-M_Z^2)M_Z^2}{(s-M_Z^2)^2 + s^2\Gamma_Z^2/M_Z^2}\,, \tag{5}$$

where Q_f is the charge of the fermion, $N_c^f = 3$ for quarks and 1 for leptons and $\mathcal{G}_V^f$ is the vector coupling of the Z to the fermion-antifermion pair $f\bar{f}$.

Since $\sigma^0_{\gamma Z}$ is expected to be much less than σ^0_Z, the LEP collaborations have generally calculated the interference term in the framework of the SM. This fixing of $\sigma^0_{\gamma Z}$ leads to a tighter constraint on M_Z and consequently a smaller error on its fitted value. It is possible to relax this constraint and carry out the fit within the S-matrix framework.

The QED radiative corrections have been explicitly taken into account by convoluting over the ISR and allowing the electromagnetic coupling constant to run [10]: $\alpha(s) = \alpha/(1-\Delta\alpha)$. Weak radiative corrections that depend upon the assumptions of the EW theory and on the values of $M_{\rm top}$ and $M_{\rm Higgs}$ are accounted for by absorbing them into the couplings, which are then called the *effective* couplings $\mathcal{G}_V$ and $\mathcal{G}_A$.

$\mathcal{G}_V$ and $\mathcal{G}_A$ are complex numbers with small imaginary parts. As experimental data does not allow simultaneous extraction of both real and imaginary parts of the effective couplings, the convention $g_A^f = {\rm Re}(\mathcal{G}_A^f)$ and $g_V^f = {\rm Re}(\mathcal{G}_V^f)$ is used and the imaginary parts are added in the fitting code [8].

Defining

$$A_f = 2\frac{g_V^f \cdot g_A^f}{(g_V^f)^2 + (g_A^f)^2} \tag{6}$$

the lowest-order expressions for the various lepton-related asymmetries on the Z pole are [11–13] $A_{FB}^{(0,\ell)} = (3/4)A_eA_f$, $P(\tau) = -A_\tau$, $P(\tau)^{fb} = -(3/4)A_e$, $A_{LR} = A_e$. The full analysis takes into account the energy dependence of the asymmetries. Experimentally A_{LR} is defined as $(\sigma_L - \sigma_R)/(\sigma_L + \sigma_R)$ where $\sigma_{L(R)}$ are the $e^+e^- \to Z$ production cross sections with left- (right)-handed electrons.

The definition of the partial decay width of the Z to $f\overline{f}$ includes the effects of QED and QCD final state corrections as well as the contribution due to the imaginary parts of the couplings:

$$\Gamma(f\overline{f}) = \frac{G_F M_Z^3}{6\sqrt{2}\pi} N_c^f(\left|\mathcal{G}_A^f\right|^2 R_A^f + \left|\mathcal{G}_V^f\right|^2 R_V^f) + \Delta_{ew/\mathrm{QCD}} \quad (7)$$

where R_V^f and R_A^f are radiator factors to account for final state QED and QCD corrections as well as effects due to non-zero fermion masses and $\Delta_{ew/\mathrm{QCD}}$ represents the non-factorisable EW/QCD corrections.

The e^+e^- final state
Unlike other final states, this state has contributions from $t-$channel and $s-t$ interference and these amplitudes are not amenable to fast calculation. Thus the non-s channel part of this cross section is calculated using the SM programmes ALIBABA [14] and TOPAZ0 [15], for a given M_{top} and M_{Higgs}, and added to the $s-$channel cross section as for other channels. The theoretical uncertainties and the uncertainties due to errors on top and Higgs masses are folded into the error calculation.

Errors dur to uncetainty in beam energy [16–21]
Systematic errors on beam energy may be due to the absolute energy scale, energy-point to energy-point errors due to non-linear response of magnets or higher order effects relating the dipole field and beam energy, and energy reproducibility errors due to uncertainties in temperatures, tidal effects, RF status, etc. A detailed model was developed which took into account these factors, including leakage currents produced by trains running nearby, the earth-tide effects due to the sun and the moon. A covariance matrix for energy errors for LEP running between 1993 and 1995 was provided by the LEP Energy Working Group [16].

The choice of fit parameters
The parameter set $M_Z, \Gamma_Z, \sigma^{\circ}_{\mathrm{hadron}}, R(\mathrm{lepton}), A_{FB}^{(0,\ell)}$, where $R(\mathrm{lepton}) = \Gamma(\mathrm{hadrons})/\Gamma(\mathrm{lepton})$, $\sigma^0_{\mathrm{hadron}} = 12\pi\Gamma(e^+e^-)\Gamma(\mathrm{hadrons})/M_Z^2\Gamma_Z^2$ was chosen by the LEP collaborations for fitting the data. The main advantage is that these parameters form the least correlated set of parameters thus making it easier to combine the data from the four experiments.

Thus one starts with the general fit in which lepton universality is not assumed and there are three $R(\mathrm{lepton})$ and $A_{FB}^{(0,\ell)}$ parameters, making a total of nine. Having ascertained the validity of lepton universality a five parameter fit is carried out.

Combining results of LEP and SLC collaborations
With the huge amount of statistics collected by each of the LEP experiments the main errors are due to systematics, many of which are common to the different experiments. The experimental systematic errors common among LEP experiments are due to the LEP energy calibration uncertainties. Other systematic errors that are common to LEP and SLC experiments are due to theoretical uncertainties, in the luminosity determination using small angle Bhabha scattering, estimating non-s channel contribution to large angle Bhabha scattering, calculation of QED radiative effects and parametrisation of the cross section in terms of the parameter set used.

All the theory related systematic errors utilise SM programmes which need some basic inputs and all LEP collaborations used identical values for these: $M_Z = 91.187$ GeV, the Fermi constant $G_F = (1.16637 \pm 0.00001) \times 10^{-5}$ [22], $\alpha^{(5)}(M_Z) = 1/128.877 \pm 0.090$ [23], $\alpha_s(M_Z) = 0.119$ [24], $M_{top} = 174.3 \pm 5.1$ GeV [24] and $M_{Higgs} = 150$ GeV.

Methodology and results of LEP combination
Each LEP experiment provided the results from a nine-parameter fit result using the variables: $M_Z, \Gamma_Z, \sigma^{\circ}_{\mathrm{hadron}}, R(\mathrm{e}), R(\mu), R(\tau), A_{FB}^{(0,e)}, A_{FB}^{(0,\mu)}, A_{FB}^{(0,\tau)}$, together with the full 9×9 covariance. A grand covariance matrix, V, was constructed using the four covariance matrices as its diagonal components and filling the remaining off-diagonal elements with common systematic errors. A combined 9-parameter set was the obtained by minimising $\chi^2 = \Delta^T V^{-1} \Delta$, where Δ is the vector of residuals of the combined parameter set to the results of individual experiments.

After verifying that the fit parameters for the three leptons are compatible, each LEP experiment assumed lepton universality and obtained a 5-parameter fit and the corresponding error matrix. These were then combined as described above to obtain the LEP combined 5-parameter values.

Results of the fits as described above are given in Table 1.

Some important physics related quantities can be derived using the above fitted values. These are the Z

Parameter	Fit not assuming lepton universality	Fit assuming lepton universality	Derived not assuming lepton universality	Derived assuming lepton universality
M_Z(GeV)	91.1876 ± 0.0021	91.1875 ± 0.0021		
Γ_Z(GeV)	2.4952 ± 0.0023	2.4952 ± 0.0023		
σ^0_{had}(nb)	41.541 ± 0.037	41.540 ± 0.037		
R_e	20.804 ± 0.050			
R_μ	20.785 ± 0.033			
R_τ	20.764 ± 0.045			
R_ℓ		20.767 ± 0.025		
Γ_e(MeV)			83.92 ± 0.12	
Γ_μ(MeV)			83.99 ± 0.18	
Γ_τ(MeV)			84.08 ± 0.22	
Γ_ℓ(MeV)				83.985 ± 0.086
Γ_{had}(MeV)				1744.4 ± 2.0
Γ_{inv}(MeV)				499.0 ± 1.5
N_ν(light)				2.9840 ± 0.0082
$A_{FB}^{(0,e)}$	0.0145 ± 0.0025			
$A_{FB}^{(0,\mu)}$	0.0169 ± 0.0013			
$A_{FB}^{(0,\tau)}$	0.0188 ± 0.0017			
$A_{FB}^{(0,\ell)}$		0.0171 ± 0.0010		

Table 1. Model independent combined LEP fit results.

decay width into invisible particles and, from this the number of light neutrino species into which the Z can decay. As there is one neutrino species per generation, this also indicates strongly that the total number of matter species may be limited to three, unless additional neutrinos have a mass far heavier than the known neutrinos.

Study of τ-polarisation in $Z \to \tau^+\tau^-$ at LEP and the availability of beam polarisation at the SLC enables one to obtain very accurate information on the asymmetry parameters of Z decays into various modes and helps one to fix the value of the effective mixing angle $\sin^2\theta^{\rm lept}_{\rm eff}$. Figure 6 summarises these measurements and also depicts the variation of $\sin^2\theta^{\rm lept}_{\rm eff}$ with the Higgs mass. One can notice the inconsistency between the measurement from SLC and from the b-asymmetry measurement at LEP.

2.2. Z^0 Decays to Heavy Flavours (b- and c-quarks)

The LEP experiments have measured the ratios of partial widths $R_b = \Gamma(Z \to b\bar{b})/\Gamma(Z \to \text{hadrons})$ and $R_c = \Gamma(Z \to c\bar{c})/\Gamma(Z \to \text{hadrons})$ and the forward-backward (charge) asymmetries $A^{b\bar{b}}_{FB}$ and $A^{c\bar{c}}_{FB}$. The SLD experiment at SLC has measured the ratios R_c and R_b and, utilising the polarisation of the electron beam was able to obtain the final state coupling parameters A_b and A_c from a measurement of the left-right forward-backward asymmetry of $b-$ and $c-$quarks. The high precision measurement of R_c at SLD was made possible owing to the small beam size and very stable beam spot at SLC, coupled with a highly precise CCD pixel detector. Several of the analyses have also determined other quantities, in particular the semileptonic branching ratios, $B(b \to \ell^-)$, $B(b \to c \to \ell^+)$ and $B(c \to \ell^+)$, the average time-integrated $B^0\overline{B}^0$ mixing parameter $\overline{\chi}$ and the probabilities for a c–quark to fragment into a D^+, a D_s, a D^{*+} or a charmed baryon. All these quantities are correlated with the EW parameters, and since the mixture of b hadrons is different from the one at the $\Upsilon(4S)$, their values might differ from those measured at the $\Upsilon(4S)$.

The LEP Electroweak Heavy Flavour Working Group has developed [25] a procedure for combining the measurements taking into account known sources of correlation. The combining procedure determines fourteen parameters: the six parameters of interest in the EW sector R_b, R_c, $A^{b\bar{b}}_{FB}$, $A^{c\bar{c}}_{FB}$, A_b and A_c and, in addition $B(b \to \ell^-)$, $B(b \to c \to \ell^+)$, $B(c \to \ell^+)$, $\overline{\chi}$, $f(D^+)$, $f(D_s)$, $f(c_{\rm baryon})$ and $P(c \to D^{*+}) \times B(D^{*+} \to \pi^+ D^0)$ to take into account their correlations with the EW parameters. Before the fit both the peak and off-peak asymmetries are translated to the common energy

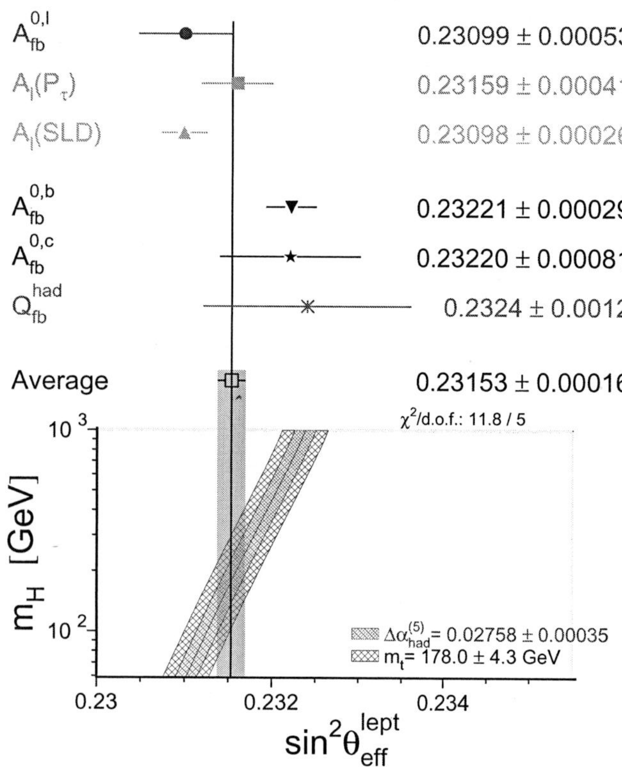

Figure 6. $\sin^2\theta_{\rm eff}^{\rm lept}$ derived from various asymmetry measurements at the Z-pole

$\sqrt{s} = 91.26$ GeV using the predicted energy dependence from ZFITTER [26].

The results are given in Table 2.

2.3. Search for Higgs at LEP

A systematic search for the SM and non-SM Higgs bosons has been made at LEP. The main production mechanism for the SM Higgs is the Higgs-strahlung process:

$$e^+e^- \to Z^\star \to HZ$$

and all the possible detectable decay modes of H and Z have been used in the search. While some initial hints of a Higgs signal with mass around 115 GeV was seen, in the final combined paper by the LEP collaborations, a 95% lower limit on the mass is given [27]: 114.4 GeV. The reconstructed Higgs mass with loose and tight cuts is shown in Fig. 7 and the likelihood function for observing a Higgs is shown in Fig. 8.

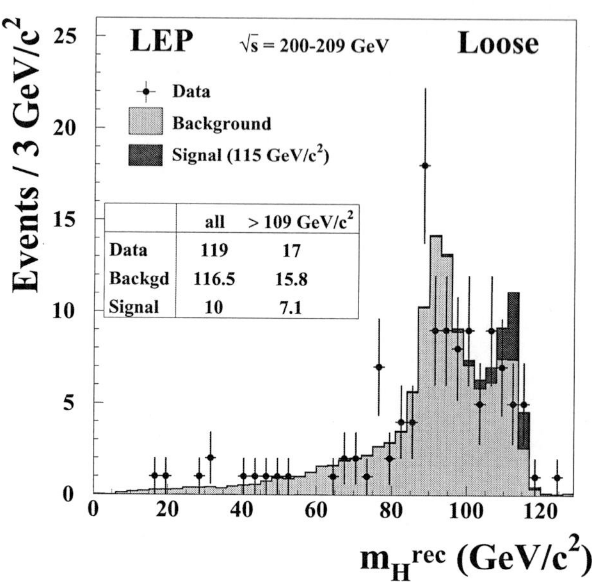

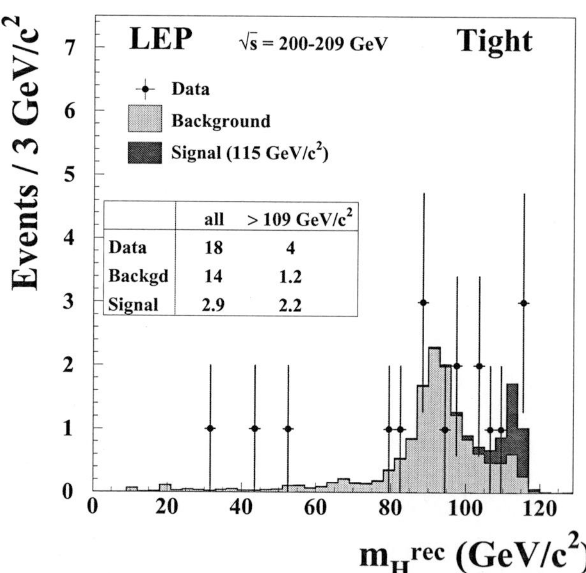

Figure 7. Reconstruted Higgs mass using loose and tight cuts (all LEP)

3. Discovery of Top Quark at Fermilab

The CDF [28] and D0 [29] collaborations discovered the top quark in the mid-1990s. The initial studies were made in RUN I, with a centre-of-mass energy of 1.8 TeV and subsequent RUN II studies are at 1.96 TeV. The production cross section at 1.96 TeV is around 7 pb, with almost 85% contribution due to the quark-

$R_b^0 = 0.21629 \pm 0.00066$	$R_c^0 = 0.1721 \pm 0.0030$
$A_{FB}^{0,b} = 0.0992 \pm 0.0016$	$A_{FB}^{0,c} = 0.0707 \pm 0.0035$
$A_b = 0.923 \pm 0.020$	$A_c = 0.670 \pm 0.027$
$B(b \to \ell^-) = 0.1071 \pm 0.0022$	$\mathrm{B}(b \to c \to \ell^+) = 0.0801 \pm 0.0018$
$B(c \to \ell^+) = 0.0969 \pm 0.0031$	$\overline{\chi} = 0.1250 \pm 0.0039$
$f(D^+) = 0.235 \pm 0.016$	$f(D_s) = 0.126 \pm 0.026$
$f(c_{\text{baryon}}) = 0.093 \pm 0.022$	$P(c \to D^{*+}) \times \mathrm{B}(D^{*+} \to \pi^+ D^0) = 0.1622 \pm 0.0048$

Table 2. Results from global fit to measurements on Z decays to b- and c-quarks.

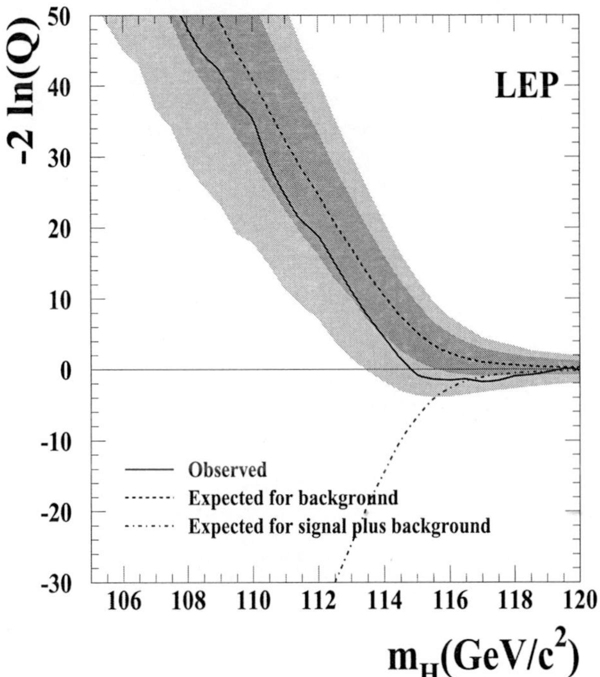

Figure 8. Log likelihood function as a function of Higgs mass

antiquark annihilation. The dominant decay mode of the top quark is $t \to bW$. Thus the decay signatures of $t\bar{t}$ production are the presence of b-quarks and high p_T leptons if the Ws decay leptonically. The all hadronic final state is more problematic to identify above background and analyse. The latest results on the $t\bar{t}$ production cross section in RUN II [30] are $\sigma(p\bar{p} \to t\bar{t}) = 7.0 \pm 0.3(stat) \pm 0.4(syst) \pm 0.4(lumi)$ pb for CDF and $\sigma(p\bar{p} \to t\bar{t}) = 7.8 \pm 0.5(stat) \pm 0.6(syst) \pm 0.5(lumi)$ pb for D0; and the Tevatron combined top mass value is $172.4 \pm 0.7 \pm 1.0$ GeV.

4. Standard Model Fit to all Electroweak Data

With only the Higgs boson as the missing piece of the SM predictions, it is obvious to attempt to predict its mass by fitting all data within the SM framework, having it as one of the free parameters. The data used in the fit is shown in Fig. 9 and the χ^2 variation as a function of the Higgs mass is shown in Fig. 10.

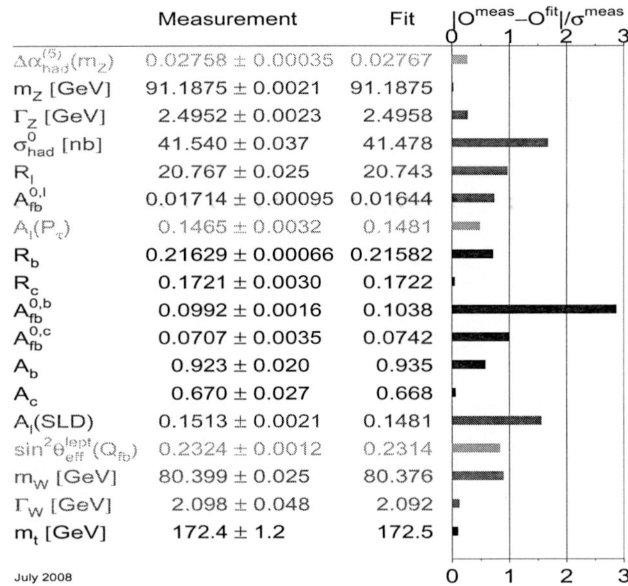

Figure 9. Electroweak data used in the SM fit

5. Enter the LHC Programme

The world stands today on the verge of a new era of scientific discovery. The brand new atom smasher under construction at the European Laboratory of Particle Physics (CERN) located a few kilometres out-

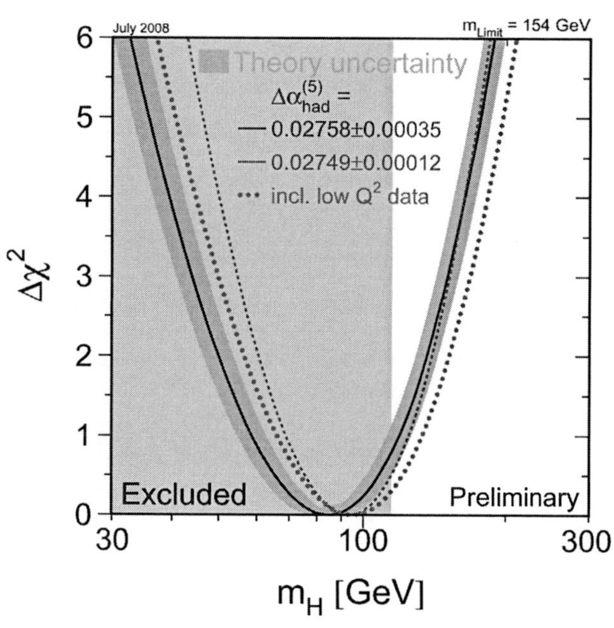

Figure 10. Electroweak data used in the SM fit

side Geneva, Switzerland, will begin operation in 2009. Conceptualised around a quarter century back, approved for construction in the mid-1990s and now almost a decade in the making, this technological marvel of a machine which accelerates counter rotating beams of protons in two steel pipes 27 km in circumference is ready to operate. It will become the highest energy particle accelerator in the world with 7 times as much energy compared to the existing accelerator, Tevatron, at Fermilab, USA.

Its scientific goals are truly stupendous, ranging from understanding the microcosm of the sub-nuclear world to attempting to answer the question what was the universe like at the very beginning of time, a few moments after the Big Bang. Is the universe really filled with an all pervading Higgs field, as postulated by theorists, or is there some other explanation for the origin of mass itself? Does dark matter, which constitutes 25% of the universe, really consist of the so-called supersymmetric (SUSY) particles which form an integral part of the modern so-called theory of everything that scientists are working overtime on? These questions relating to the nature of the universe are some of the most fundamental questions that have been asked by humankind ever since men started to wonder about the world around them. And the new atom smasher called the LHC has the capability of answering them. Science will not be the same after a few years, whether the answers to the above questions are in the affirmative or not, in which case there will be a paradigm shift in theory itself.

There are four experimental collaborations in the LHC programme: two large experiments, ATLAS and CMS, which are geared more towards proton-proton collisions; ALICE which is meant to study heavy ion collisions; and LHCb which will mainly study b-quark production and decay. India is participating in CMS and ALICE experiments and in this article Indian participation in CMS will be discussed briefly.

5.1. The CMS Detector and Indian Contributions

A view of the CMS detector is shown in Fig. 11.

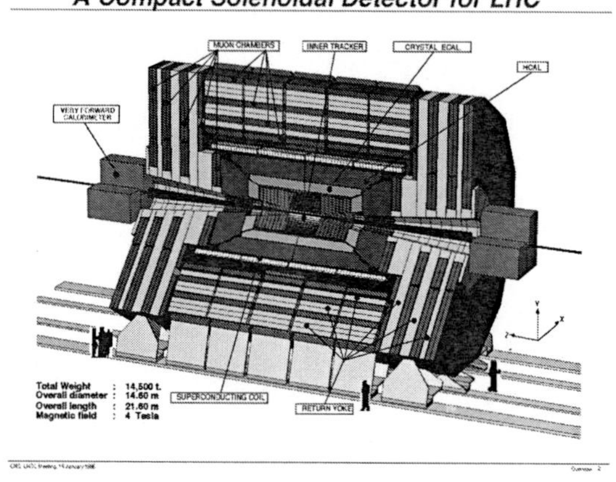

Figure 11. CMS detector

It is a typical high energy physics detector, cylinderical in geometry, surrounding the beam pipe at the intersection region. Closest to the beam pipe there is a silicon pixel vertex detector followed by an all-silicon microstrip tracker. Next is the electromagnetic calorimeter consisting of lead-tungstate crystals. This calorimeter will provide an excellent electromagnetic energy resolution that could be critical in the detection and mass measurement of a low mass Higgs as indicated by fit to the EW data within the SM framework. Next is the hadron calorimeter whose active elements are plastic scintillator tiles and the absorber is brass. All these detectors are surrounded by a 4 tesla superconducting solenoidal magnet that will enable precise momenta of charged particles as well as sweep away the very low momentum uninteresting debris and thus reduce background. Just outside the magnet is the outer hadron calorimeter (HO), consisting of layers of plastic scintillator, to measure the remnant hadronic energy after the particles have passed through the magnet coil that acts as another layer of absorber. Indian groups were responsible for the R&D and fabrication of this detector. The outermost layers are four muon detection layers with drift tubes as the detector elements. A similar layering of detectors is present in the forward-backward directions, with an additional element: the silicon pre-shower detector whose active elements are silicon microstrip detectors. This detector is necessary to distinguish between a γ and a π^0, in order to effectively identify the Higgs $\rightarrow \gamma\gamma$ decay mode, which will be the only way to detect a light Higgs. Indian groups have participated in providing 25% of the detectors for this. The fabrication took place in BEL, Bangalore. The placement of the HO and silicon pre-shower detector within the CMS detector is shown in Figs. 12 and 13.

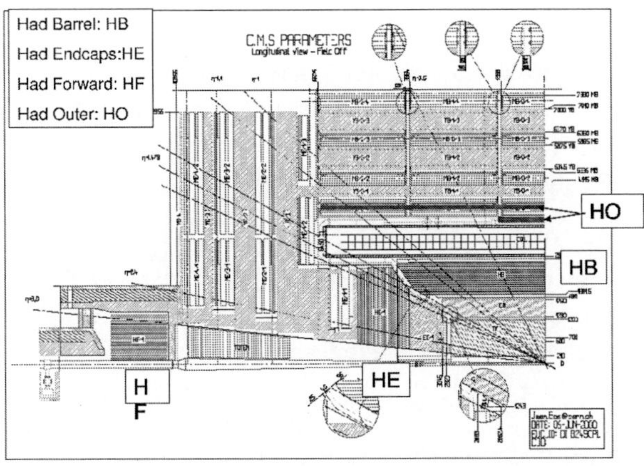

Figure 12. Quarter longitudinal view of the CMS detector showing HO placement

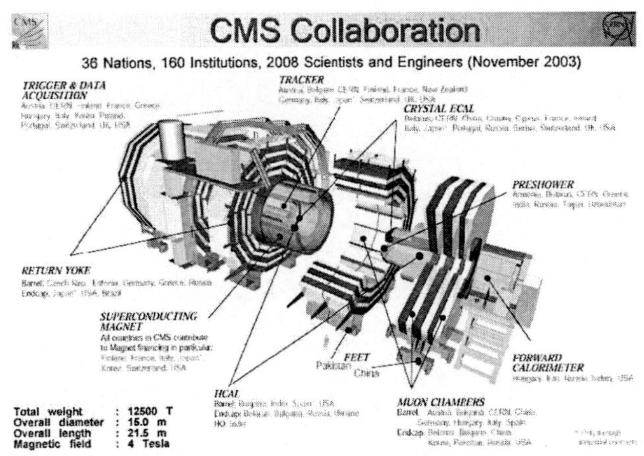

Figure 13. Opened out view of the CMS detector, showing placement of the silicon pre-shower detector

The HO detector has been fully installed and commissioned within the overall CMS detector and the silicon PSD will be installed during the 2008–09 shut down, both as per CMS schedule.

Peta-bytes of data per year will start flowing once the LHC starts operation. In order to reconstruct raw data in the form of pp interactions, called events, to generate simulated events, and to allow teams spread worldwide to analyse this data to extract physics, the only possible scenario was to go in for GRID technology. This is the child of the worldwide-web which was invented at CERN for making available high energy data to various geographically separated laboratories. The GRID does one better: with the help of suitable middleware it enables relatively cheap computer farms located worldwide to act as a coherent computing engine. Such a tier-2 centre is now becoming operational at TIFR, being connected to CERN by a 1 Gbps link.

Indian scientists in CMS have been very actively preparing for physics analyses using simulated data, working out and fine tuning algorithms which would sift the wheat from the chaff in real data. A number of studies have been carried out for detection of different types of Higgs, of SUSY, studying the W and Z, etc. Thus Indian groups are well prepared for extracting exciting science when real data becomes available.

5.2. Physics at the LHC

While there have been accelerators before, it is for the first time in history that the TeV energy scale will become available for systematic scientfic exploration. One will get definitive answers to the most pressing questions of the day: experimental evidence on the EW symmetry breaking via search for the Higgs and on the existence of supersymmetric particles, which form an essential building block of the most ambitious theories of everything or supergravity theories. This would also throw light on the nature of dark matter; if SUSY is discovered in its popular (R-parity conserving) form, then the lightest SUSY particle could account for the dark matter.

Apart from the above two critical issues, detailed studies of the top quark would become possible, and also of the b-quark, decays of which may still bring surprise discoveries. The programme of colliding heavy ions (Pb on Pb) would extend that energy reach far beyond what is available today at RHIC and again could be decisive in confirming the existence of quark-gluon plasma and studying its properties.

Finally the most exciting new physics would probably come in the form of totally unexpected discoveries upon breaching the tera-energy scale.

As there are only theories and models about physics beyond the SM at these energy scales, below we will only mention two topics: how the SM Higgs would be discovered and how the first evidence of SUSY could be detected.

5.3. The Standard Model Higgs at the LHC

As mentioned earlier, indirect experimental evidence points strongly to a Higgs of relatively low mass, < 154 GeV at 95% confidence level. And LEP has excluded a mass < 114.4 GeV. Moreover, within the minimal SUSY extension of the SM (MSSM) the lightest neutral Higgs must be low mass, less than around 140–150 GeV. At such low mass the only decay mode that will be detectable above QCD background is the $H \to \gamma\gamma$ mode. Thus a lot of effort has gone into improving the electromagnetic energy resolution of the detectors; in particular CMS went in for lead tungstate crystals for this purpose. For an assumed Higgs mass of 130 GeV, the effective mass distribution of two photons in the CMS detector is shown in Fig. 14 for an integrated luminosity of 100 fb^{-1}.

On the other hand, if the Higgs is heavier than 180 GeV it can decay into a ZZ pair, whose leptonic decay modes provide a gold plated signature for detecting Higgs. Upto about 400 GeV Higgs mass an integrated luminosity of 20 fb^{-1} is sufficient to discover and measure the Higgs mass. Beyond that around 100 fb^{-1} would be required. This is shown in Fig. 15.

While the discovery of a neutral Higgs would establish its existence, it would take some while to make detailed studies to determine if its a SM or beyond SM Higgs (in MSSM there are three neutral Higgs). Discovery of a charged Higgs (as in MSSM) would clearly indicate the presence of physics beyond the SM.

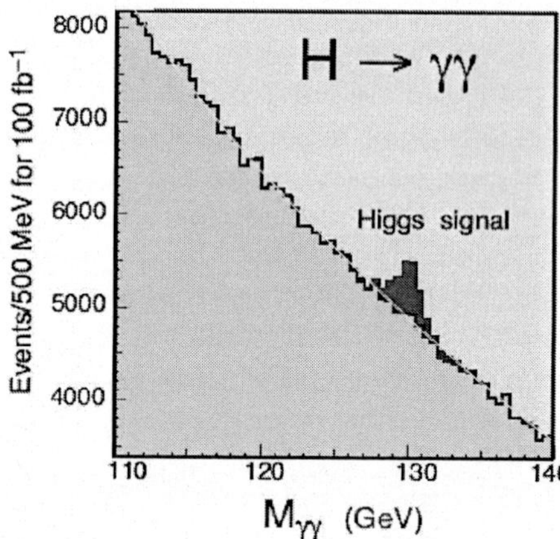

Figure 14. Higgs $\to \gamma\gamma$ reconstruction in CMS. Higgs mass = 130 GeV

5.4. SUSY at the LHC

One of the continuing themes in science has been the quest for unification. Electricity and magnetism unified into the electromagnetic theory of Maxwell. Then the unification of this with weak interaction theory to give us the EW theory. A further unification of EW with the theory of strong interactions, QCD is accommodated within the supersymmetric framework. The two become unified at around 10^{15} GeV mass scale. Within this framework each known particle has a supersymmetric partner with the spin differing by half unit. Thus the normal fermions have scalar superpartners and the normal bosons have spin half superpartners. The spectrum is shown in Fig. 16.

Under the popular assumption of R-parity conservation s-particles would be produced in pairs and at the end of the decay chain the lightest s-particle would escape undetected. As the masses of these particles would be at least a few hundred GeV the typical signature of such an event would be large amount of missing energy. A typical decay chain is as shown in Fig. 17.

Leptons being easy to identify and measure, decay chains involving leptons is a clean way of identifying s-particle producing events. A characteristic signature is a sharp drop in the effective mass of the dilepton pair called the dilepton edge. The mass of the s-particle can be inferred from this. An example is shown in Fig. 18.

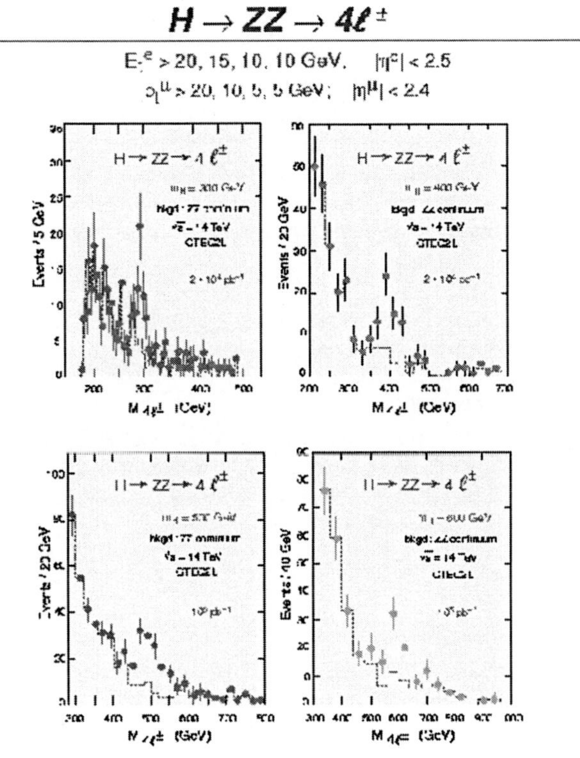

Figure 15. Higgs $\rightarrow ZZ \rightarrow \ell eptons$ reconstruction in CMS

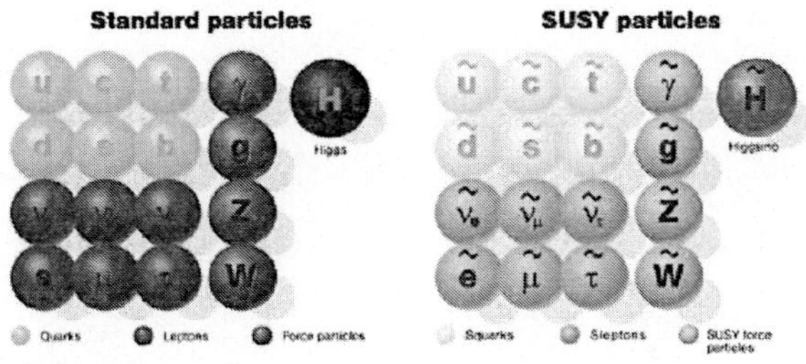

Figure 16. SUSY spectrum

Successive application of this method can allow one to reconstruct the masses of the particles.

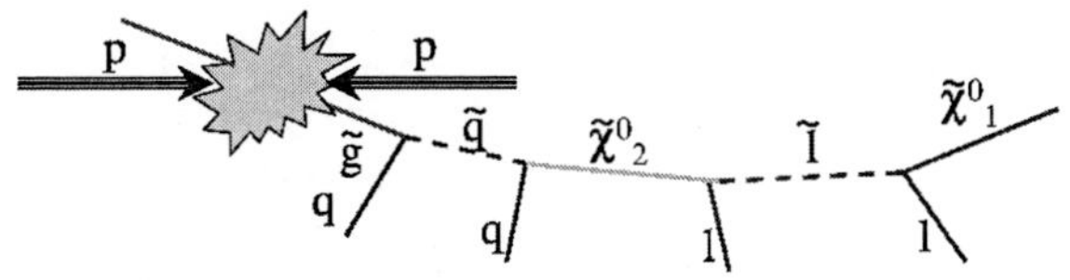

Figure 17. SUSY decay chain

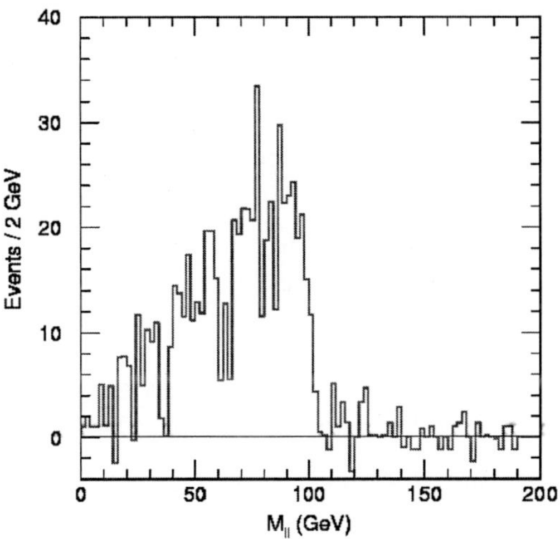

Figure 18. SUSY discovery by leading di-lepton leading edge

6. Summary

The scientific world is at a major junction in history. With the turning on of the LHC in 2009 many outstanding questions about the universe would become addressable, among them the origin of mass and the nature of dark matter. One looks forward to many years of exciting discoveries.

REFERENCES

1. A. Salam, *Elementary Particle Theory*, ed. N. Svartholm Alm uist and Wiksells, Stockholm, 1968 , **367**
2. S. Weinberg, Phys. Rev. Lett.**19** 1967 1264
3. The ALEPH, DELPHI, L3 and OPAL Collaborations,

the LEP Electroweak Working Group, the SLD Electroweak and Heavy Flavor Working Group, Phys. Rep. **427** “2006“ 257

4. C. Amsler *et al*, Phys. Lett. **B667** “2008“ 1
5. R.N. Cahn, Phys. Rev.**D36** “1987“ 2666
6. F.A. “erends *et al*, “*Z* Physics at LEP 1”, CERN Report 89-08 “1989“, “ol. 1, eds. G. Altarelli, R. “leiss, and C. “er“egnassi, p. 89
7. A. “orrelli *et al*, Nucl. Phys. **B333** “1990“ 357
8. D. “ardin and G. Passarino, ““pgrading of Precision Calculations for Electroweak Observables,” `hep-ph/9803425`,
   D. “ardin, G. Passarino, and M. Grünewald, “Precision Calculation Project Report,” `hep-ph/9902452`
9. D. “ardin *et al*, “. Phys. **C44** “1989“ 493“Comp. Phys. Comm. **59**, “1990“ 303“
   D. “ardin *et al*, Nucl. Phys. **B351** “1991“ 1“ Phys. Lett **B255** “1991“ 290 and CERN-TH/6443/92 “1992““ Comp. Phys. Comm. **133**, “2001“ 229
10. G. “urgers *et al*, “*Z* Physics at LEP 1”, CERN Report 89-08 “1989“, “ol. 1, eds. G. Altarelli, R. “leiss, and C. “er“egnassi, p. 55
11. M. Consoli *et al*, “*Z* Physics at LEP 1”, CERN Report 89-08 “1989“, “ol. 1, eds. G. Altarelli, R. “leiss, and C. “er“egnassi, p. 7
12. M. “ohm *et al*, *ibid*, p. 203
13. S. Jadach *et al*, *ibid*, p. 235
14. W. “eenakker, F.A. “erends, and S.C. van der Marck, Nucl. Phys. **B349** “1991“ 323
15. G. Montagna *et al*, Nucl. Phys. **B401** “1993“ 3“Comp. Phys. Comm. **76** “1993“ 328“ Comp. Phys. Comm. **93** “1996“ 120
16. R. Assmann *et al* “Working Group on LEP Energy“, Euro. Phys. Jour. **C6** “1999“ 187
17. R. Assmann *et al* “Working Group on LEP Energy“, “. Phys. **C66** “1995“ 567
18. L. Arnaudon *et al* “Working Group on LEP Energy and LEP Collaborations“, Phys. Lett. **B307** “1993“ 187
19. L. Arnaudon *et al* “Working Group on LEP Energy“, CERN-PPE/92-125 “1992“
20. L. Arnaudon *et al*, Phys. Lett. **B284** “1992“ 431
21. R. “ailey *et al*, ‘LEP Energy Calibration’ CERN-SL-90-95-AP, Proceedings of the “2nd European Particle Accelerator Conference,” Nice, France, 12–16 June 1990, pp. 1765-1767
22. T. van Ritbergen and R. Stuart, Phys. Lett. **B437** “1998“ 201“Phys. Rev. Lett.**82** “1999“ 82
23. S. Eidelman and F. Jegerlehner, “. Phys. **C67** “1995“ 585“M. Steinhauser, Phys. Lett. **B429** “1998“ 158
24. Particle Data Group “D.E. Groom *et al*“, Euro. Phys. Jour. **C15** “2000“ 1
25. The LEP Experiments“ ALEPH, DELPHI, L3, and OPAL Nucl. Instr. & Meth.**A378** “1996“ 101
26. D. “ardin *et al*, “. Phys. **C44** “1989“ 493“Comp. Phys. Comm. **59** “1990“ 303
27. ALEPH Collab., R. “arate *et al*, Phys. Lett. **B526** “2002“ 191,
    DELPHI Collab., J. Abdallah *et al*, Euro. Phys. Jour. **C31** “2003“ 421
    L3 Collab., M. Acciarri *et al*, Phys. Lett. **B517** “2001“ 319
    OPAL Collab., G. Abbiendi *et al*, Euro. Phys. Jour. **C26** “2003“ 479
    ALEPH, DELPHI, L3 and OPAL Collaborations & the LEP Working Group for Higgs “oson Searches, Phys. Lett. **B565** “2003“ 61
28. F. Abe *et al*, Phys. Rev. D**50** “1994“ 2966“Phys. Rev. Lett.**73** “1994“ 225
29. S. Abachi *et al*, Phys. Rev. Lett.**74** “1995“ 2632
30. Erich W. “arnes, “Top “uark Physics”, Plenary talk at the 34th Intern. Conf. on High Energy Physics, Philadelphia, PA, 29 July to 5 August, 2008

Accelerator Physics and Technological Challenges of the LHC

Oliver Brüning, Vinod Chohan[a] *and Swapan Chattopadhyay*[b]

[a] *CERN, Geneva, Switzerland*

[b] *Cockcroft Institute, Daresbury, Cheshire, United Kingdom*

The LHC at CERN has completed its construction in summer 2008. It is just entering into its commissioning phase in preparation for collider operation for science in 2009. The first beams were already observed in an inaugural commissioning run on September 10, 2008. An inaugural ceremony for the collider organised on October 21, 2008 celebrated the achievement of bringing the LHC to reality by an international team of scientists with support from governments of nations around the globe contributing to the programme. As we anticipate the non-trivial task of a careful, detailed and prolonged commissioning of the collider, it is time to take stock of the achievements to date and the future potential of the LHC, highlighting contributions of our colleagues from India in particular and the sociology of global collaboration.

1. Introduction

The key objective for the Large Hadron Collider (LHC) is to explore the validity of the standard model at unprecedented collision energies, with sufficiently high collision rates and statistics that allow the discovery of new particles, such as the Higgs boson and supersymmetric particles. The LHC performance is measured by its centre of mass (CM) collision energy and the number of events it can deliver to its experiments. A successful operation of the LHC collider requires particle collisions with CM energies above 1 TeV (almost an order of magnitude higher than the current collider energy frontier in the Tevatron collider) and an event rate of more than 1 hadronic event per collision and roughly 30 million collisions per second spaced by intervals of 25 ns.

Since E.O. Lawrence's invention of the cyclotron, the particle accelerator technology has advanced by leaps and bounds, in enabling us to envision and construct the LHC, seven orders of magnitude higher in energy and five orders of magnitude larger in size than the original cyclotron (Fig. 1).

A collider can, in principle, be designed for a range of different particle species. Existing collider machines deploy beams of electrons, positrons, protons, anti-protons or ions. For example, the Tevatron collider, which currently defines the energy frontier for particle colliders, operates with proton and anti-proton beams and the last collider project at CERN, the Large Electron Positron (LEP) collider generated beam collisions between electron and positron beams. Each particle species has its own advantage and disadvantage and the choice of particles must be carefully tailored to the key objective of the collider project and plays a central role in the collider design. For example, Lepton collider machines, such as LEP, generate collisions between elementary particles with precisely defined CM collision energies. They are therefore well suited for high precision experiments. The beams of hadron colliders such as the Tevatron at Fermilab and the LHC project at CERN, on the other hand, consist not of elementary particles but are rather composites of smaller constituents. In its key operation mode the LHC deploys two beams of protons, which are not fundamental particles, but consist of quarks and gluons. The collisions in the LHC therefore occur between pairs of quarks and gluons each carrying only a fraction of the total proton energy and the CM energy of these collisions can vary significantly between different collisions. Hadron beam collisions therefore are not well suited for high precision experiments but offer a tremendous discovery potential which is well suited for the key objective of the LHC: the discovery of new particles whose properties (and mass) are not yet known. Another advantage for using proton beams is that protons are relatively heavy particles that loose only a small fraction of their energy during acceleration in form of synchrotron light. This feature allows the utilisation of superconducting magnet technology and thus, the construction of a reasonable size efficient circular machine where the particle beams have a chance to collide with each other at each turn.

The main drawbacks for using proton beams in a circular collider are the need for higher beam energies (only a fraction of the beam energy contributes to the CM collision energy). Using two counter rotating proton beams in a collider requires two separate vacuum chambers with magnetic fields of opposite polarity for

Figure 1. *Top:* First Cyclotyron: 1930 E.O. Lawrence, 11 cm diameter, 1.1 MeV protons. *Bottom:* The LHC, 2008 9 km diameter, 7 TeV protons, after 80 years, 10^7 times more energy and 10^5 times larger

the two counter rotating beams (a common magnetic field would deflect the two counter rotating beams in opposite directions). The only option for avoiding the construction of 2 separate vacuum systems would be the use of protons and anti-protons, a solution that has been adopted for the Tevatron collider at Fermilab. However, the currently achievable production rates for anti-protons are too low for the design performance of the LHC.

2. The LHC performance goals and constraints

The key design parameters for the LHC are the generation of CM collision energies above 1 TeV and an event rate of more than 1 hadronic event per beam crossing. Recognising that each proton consists of three quarks plus gluons the proton beam energies should be significantly higher than the target CM collision energy. The minimum required beam energies for the LHC are thus 5 TeV. However, the number of collisions with CM energies above 1 TeV increases with higher beam energies. The design beam energy for the LHC was therefore set slightly higher at 7 TeV.

The number of events that can be delivered to the experiments is given by the product of the event cross-section and the machine luminosity L which is entirely determined by the proton beam parameters

$$L = \frac{f_{rev} \cdot n_b \cdot N^2}{\sigma_x \cdot \sigma_y} \cdot F(\phi, \sigma_{x,y}, \sigma_s) \; , \tag{1}$$

where σ_x and σ_y are the transverse RMS beam sizes at the Interaction Points (IPs), f_{rev} the revolution frequency, n_b the number of particle bunches, N the number of particles within each bunch and F a geometric reduction factor for collisions at a crossing angle that depends on the crossing angle ϕ, the transverse RMS beam size and the RMS bunch length σ_s. In order to provide more than one hadronic event per beam crossing the design luminosity has been set to L = 10^{34} $\text{cm}^{-2}\text{s}^{-1}$ leading to a design bunch intensity of 1.15×10^{11} protons per particle package (ppb), 2800 particle packages (called bunches), a transverse RMS beam size of 16 μm, an RMS bunch length of 7.5 cm and a total crossing angle of 320 μrad at the IPs. The LHC features six experiments: two high luminosity experiments, ATLAS [1] and CMS [2], requesting CM collision energies above 1 TeV, two supplementary low scattering angle experiments near ATLAS and CMS, LHCf [3] and TOTEM [4] respectively, one B-meson experiment, LHCb [5] and one dedicated ion physics experiment, ALICE [6,7]. Figure 2 shows a schematic layout of the LHC collider.

3. The LHC within the existing CERN infrastructure

In order to make best use of the existing infrastructure at CERN the LHC machine is being built in the existing 27 km long LEP [8] tunnel. Approximately 22 km of the LEP tunnel consist of curved sections that allow the installation of bending dipole magnets. The remaining 5 km of the LEP tunnel consist of 8 straight sections that provide space for the installation of the experiments, injection and extraction elements for the proton beams, acceleration devices and dedicated 'cleaning' insertions that protect the superconducting magnets from stray particles.

Not all of the space in the arcs of the LEP tunnel can be used for the installation of dipole magnets. In addition to bending fields a storage ring requires a focusing mechanism that keeps the particles centred on the design orbit. Most modern storage rings use the concept of strong focusing [9,10] where dedicated quadrupole magnets provide magnetic field components that are proportional to the particles deviation of the design orbit. The field pattern is designed such that the resulting

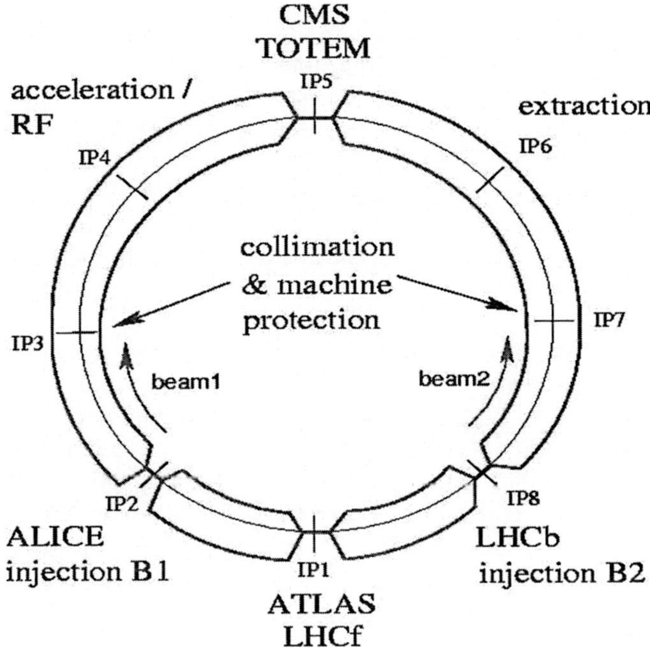

Figure 2. The schematic layout of the LHC collider

Lorentz force deflects stray particles back towards the design orbit and prevents their trajectories from diverging from the design orbit. Rather, it forces the particles to oscillate around the design orbit as they circulate in the storage ring. The number of transverse oscillations per revolution is referred to as the machine tune or 'Q' and presents a key parameter in the design and operation of a storage ring. The stronger the transverse focusing the smaller are the oscillation amplitudes (and thus the transverse RMS beam sizes) and the larger are the machine tunes.

The accelerator magnet design becomes easier and less expensive for small apertures of the magnets. In order to facilitate the magnet design one is therefore inclined to increase the number of focusing elements in the machine in order to minimise the transverse beam sizes. The price to pay with this approach is that not all the space in the tunnel can be used for the installation of dipole magnets and design of a storage ring requires a careful trade off between maximising the space for dipole installation (maximum beam energy reach) and providing sufficient space for the transverse focusing (smaller transverse beam sizes and more efficient magnet designs).

The LHC adopted a design where approximately 80% of the length of the arcs is actually filled with dipole magnets and where the maximum transverse RMS beam size in the arcs can be kept below 1.3 mm. Keeping 7 TeV proton beams on a closed orbit inside the LHC machine implies in this case the use of magnetic bending fields of 8.4 T which requires the use of superconducting magnets at the limit of the existing magnet technology (previous superconducting storage rings use maximum bending fields of ca 5 T). Confining two counter rotating proton beams into the existing LEP tunnel requires separate magnet apertures with opposite dipole field orientations for the two beams. In order to fit these two magnet systems into the existing LEP tunnel (internal tunnel diameter of only 3.76 m) and to minimise the cost and infrastructure requirements for the two storage rings the LHC adopted a novel 2-in-1 magnet design where the two magnetic coils share a common infrastructure and cryostat [11].

4. Challenges for the LHC magnet design

Figure 3 shows the schematic cross section of the novel 2-in-1 magnet design for the main LHC magnets.

LHC DIPOLE : STANDARD CROSS-SECTION

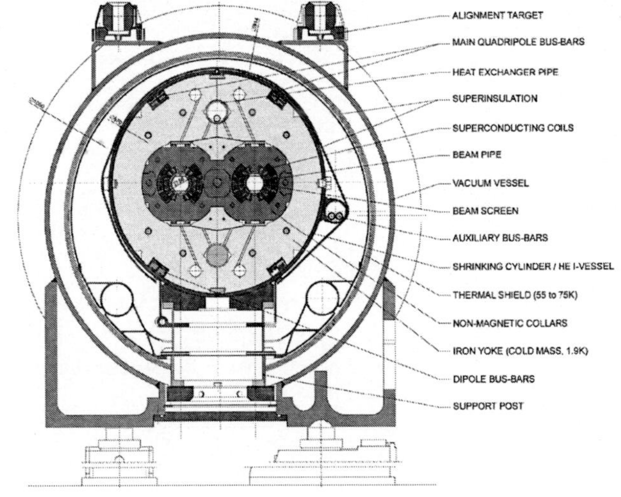

Figure 3. The schematic cross section of the 2-in-1 magnet design for the main LHC magnets

While the 2-in-1 magnet design provides a compact structure (cryostat diameter of 0.914 m) that allows the installation of 2 separate beam apertures into the existing LEP tunnel, it also couples the construction constraints of the 2 magnetic units imposing new challenges and tighter tolerances for the magnet production. The

LHC is the first particle collider that uses this magnet design and the magnet construction could therefore not build on existing experience from previous projects.

In order to minimise the number of magnet interconnections and thus, the lost space for dipole field installations, the LHC adopted a design option of 15 m long dipole magnets. The main LHC dipole magnets are more than a factor 2 longer than dipole magnets in previous accelerator projects (approximately 6 m for the Tevatron [12] and HERAp rings [13]) and weigh approximately 35 tons. The large dimensions of the LHC magnets impose tighter geometric constraints for the magnet construction and new limitations and challenges for the magnet transportation and installation as compared to previous magnet productions. Figure 4 shows the schematic layout of the periodic magnet structure inside the LHC arcs. Each half-cell consists of 3 bending dipole magnets (Main Bends) and one quadrupole magnet (Main Quadrupole). Each arc consists of 46 such half-cells.

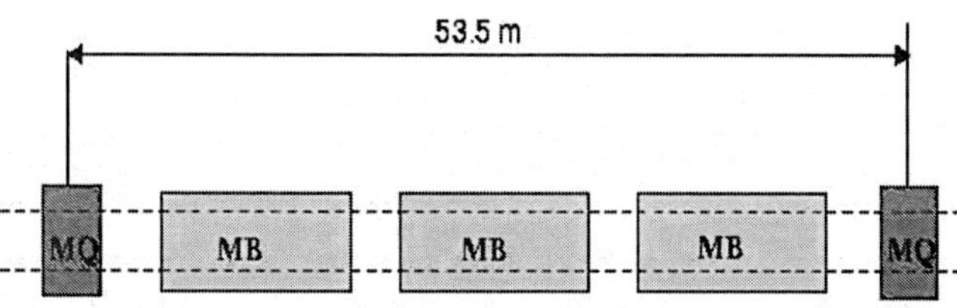

Figure 4. Periodic structure of the magnet installation in the LHC arcs. Each dipole magnet (MB) has a length of 15 m. The quadrupole magnets have a length of 3.4 m yielding a total length of 53.5 m for the basic periodic structure

Figure 5 shows an LHC dipole on the CERN site ready for installation on the back of a truck and Fig. 6 shows the tight manoeuvring in the LHC tunnel during installation.

The superconducting material used for the LHC magnets is NbTi. Like all superconducting materials NbTi is only in a superconducting state provided the key operational parameters, temperature, current density and ambient magnetic field are below the critical values required for sustaining a superconducting state. The critical values define a critical surface in the three-dimensional parameter space of temperature, current density and ambient magnetic field. Figure 7 shows the critical surface for NbTi. An operating magnet field of 8.4 T requires very low operating temperatures and relatively small current densities in the superconducting cables. The operating temperature for the LHC was chosen at 1.9 K allowing a current density between 1.5 kA and 2 kA inside the superconducting cables. The magnets are cooled using liquid He and the choice of an operating temperature below 2 K offers the additional benefit of a high thermal conductivity of He that facilitates the cooling of the magnet coils. However, operating the magnets at a temperature of 1.9 K and an ambient magnetic field of 8.4 T implies only very small margins during the operation of the magnets and even small particle losses inside the magnets, or any other sources for fluctuations in the magnet temperature, can lead to the loss of the superconductive state of NbTi.

Figure 5. An LHC dipole on the CERN site ready for installation on the back of a truck

If such a transition occurs during the magnet operation NbTi becomes normal conducting and the Ohmic losses lead to a further increase in the operating temperate and an unstable set of operating parameters. This process is called a magnet quench. All magnets in the LHC are designed to withstand a magnet quench and quenching the magnets prior to their installation presents a central acceptance test for all magnets. However, in order to minimise the likelihood for this process during operation, the LHC has two dedicated 'cleaning' sections where dedicated absorbers remove stray particles from the beams before they can reach the superconducting magnets in the tunnel.

Figure 6. Example for the tight manoeuvring in the LHC tunnel during installation. The picture shows the installation of the low-B triplet magnets near the experimental detectors

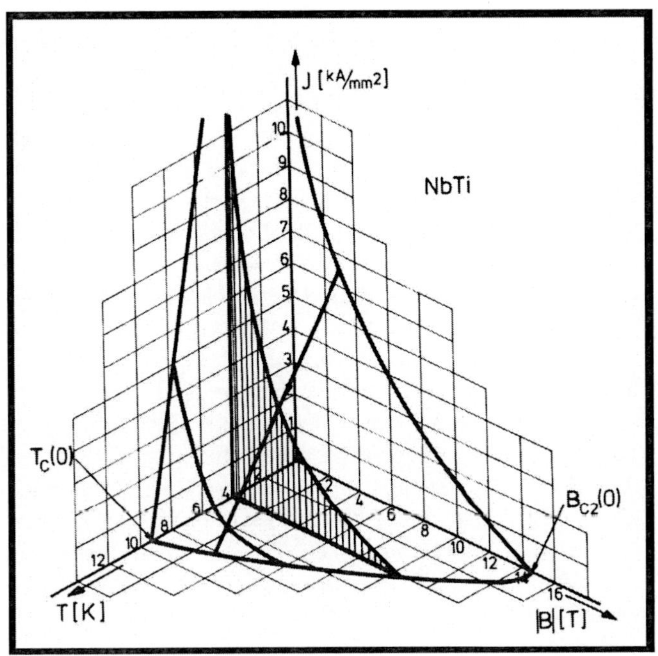

Figure 7. Critical surface for NbTi. The shaded area indicates the preferred operating temperature for most existing superconducting accelerators. The LHC will operate the magnets at a temperature of 1.9 K

5. Challenges to get the LHC ready: Cryomagnet tests at CERN

The LHC magnets constitute roughly 50% of the LHC machine costs; together with cryogenics, this figure comes to $\sim$ 66% of the total material costs of around 3300 MCHF. From the first pre-series production cryomagnet arrival in $\sim$ 2001 to the recent installation of the last dipole in the tunnel in April 2007, the testing of all the cryomagnets of the LHC was a $\sim$ 5 year long major task prior to connection, cool-down and hardware commissioning of the LHC systems in the tunnel. The LHC essentially consists of two interleaved synchrotron rings of 26.7 km circumference. The main elements of these rings are the two-in-one superconducting dipoles and quadrupoles operating in superfluid helium at a temperature of 1.9 K. Cryomagnet assemblies include 1232 dipoles (with correctors), 360 Short Straight Sections (SSS) integrated with quadrupoles and higher order poles which are needed for the different accelerator lattice functions and 114 matching and Dispersion Suppressor region magnets integrated in Special SSS (IR-SSS).

Testing, training and qualification of these magnets under cryogenic conditions was a prerequisite to their installation in the tunnel; these tests were not feasible at the manufacturers' premises.

The testing and qualification activities of a magnet was intended to verify its cryogenic, mechanical and electrical insulation integrity, qualify the performance of magnet protection systems, train the magnet up to the nominal field or higher so as to minimise training of magnets in the tunnel, characterise the intended magnetic field, accept the magnet based on its quench and training performance and generally, ensure that the magnet met its design criteria. These may be categorised broadly within the five phases namely, to connect, cool down, cold test, warm-up and to disconnect respectively.

The SM18 magnet test facility was assembled at CERN to accomplish the goal of testing the 1706 cold masses produced in Europe since 2001 for the LHC [11].

The test facility is equipped with 12 test benches and the necessary cryogenic infrastructure to perform the power tests and magnetic measurements for qualifying these magnets. Testing of the first series production magnets commenced in $\sim$2001. Since early 2003, the test facility was operated round the clock to meet the target to complete the testing of all the magnets required for the LHC by December 2006. The construction of all the 12 test benches was only completed around June 2004 and full usage started soon after. The cryomagnets were all successfully tested by February 2007, within budget and nearly in time.

For these tests, considerable challenges had to be faced and overcome since 2002; in particular, the majority of staff for tests and measurement purposes was provided by India on a rotating, one-year-stay basis, as part of the CERN-India Collaboration for the LHC. This was complemented by some CERN accelerator operation staff. While only 95 dipoles were tested till 2003, the efforts and innovative ideas coming from the Operation Team contributed significantly to the completion of tests of all 1706 cryomagnets. These included the improvements and management of the tests work flow as well as the test rates. Amongst these, certain pivotal ideas to stream-line the tests methodology as proposed and implemented successfully by the Indian Associates deserve a special mention. The following gives a broad insight into this as well an overall view of the tests operation, together with an indication of some of the operation-related results from the tests programme.

The workforce in the SM18 test facility consisted of three teams, with the tests and measurement Operation Team as the pivotal entity supported by the Cryogenics Team and the Magnet Connect/Disconnect Team (called ICS). The Operation Team consists mainly of associates from the Department of Atomic Energy (DAE), India, along with a number of regular CERN employees. The other two teams consist of contract employees from industrial consortia. A CERN team called Equipment Support looked after the improvements, exploitation and the troubleshooting of tests hardware and software on an on-call basis. A sub-team of ICS handles the movement of magnets within the test facility by means of a remotely controlled vehicle named ROCLA. All these teams worked in mutual collaboration to complete the magnet tests by February 2007.

Figure 8. President of India with the Indian Magnet Tests Team in SM18 at CERN May 2005

6. Tests concerns and hurdles

Like any facility of unique, one-off requirements, SM18 had also its own characteristic issues, ranging from personnel logistics to infrastructure limitations. Following is a brief account of some of the major issues and challenges that had to be addressed in the routine operation of the facility.

Personnel logistics issues: In early 2002, for financial, technical and organisational reasons, the outsourcing of the tests operation was no more an option. Moreover, due to various factors, only 7 non-experienced CERN staff members from accelerator operation could be assigned to run the SM18 test facility. However, for an anticipated round the clock operation of the facility with 12 test benches, a minimum of 4 persons per shift was necessary, thereby demanding minimum staff strength of 24. It was at this time that DAE, India, offered technical human resources for SM18 operation. India already had a collaboration agreement with CERN since the nineties for the LHC, including a 10 man-year arrangement for tests and measurements during the magnet prototyping phase. Subsequently, over 90 qualified personnel from 4 different Indian establishments participated in the LHC magnet tests on a one-year rotational basis. The technical acumen and success of the early group of Indian Associates lent credence and confidence that the tests activity could be successfully carried out in this manner. The strict, one year rotation was a condition desired by India, leading to the necessity of a large number of persons participating in the programme. Figure 8 shows the President of India visiting the Indian Associates of the LHC cryomagnet tests team in SM18.

The Indian technical engineers, being not directly related to operation or CERN type of activities, had to get familiar with the magnet tests work before being productive. This essentially necessitated a continuous mentoring programme, hence, limiting the number of 'trained staff' at any time. Preparing the work shift schedule with the limited experienced personnel and keep within the CERN rules and regulations was a major hurdle. Arranging proper facilities for the Indian associates to make them 'feel at home' in Europe was also an equally challenging task.

Novelty aspects: Considering that the LHC cryomagnets were unique, they were tested with a research and development mindset by magnet and equipment specialists during the initial phases of SM18 operation. Partially automated test systems (for one magnet at a time) which existed then were considered adequate for usage by the experts. When the Indian and accelerator operation staff took over the running of the facility, the system appeared to be more or less a 'black box' where, not many details of the test systems and test sequences were provided. Testing of the SSS magnets was a challenging task until the end of 2004 while all necessary information got collected and collated; similarly, testing of the special SSS magnets was a grey area even till beginning of 2006; the special SSS magnets have a large variety of structures, types, temperature regimes and their complexity made the collection of all the relevant information required for tests an extremely complex task. Even the role and responsibilities of the Operation Team had to be properly defined during the early phases.

Magnet qualifying criteria: During the early phases, each dipole was trained to reach its ultimate field (about 8% above the one required for the LHC), which was a major time consuming activity. Extensive magnetic, special measurements and thermal cycles were carried out in majority of the magnets. Qualification of 'poorly performing' magnets was another laborious task, whereby the magnet was removed from the test bench, fitted with anticryostats and quench location instruments and brought for re-testing at a later date.

Co-ordination of teams: Language was the biggest obstacle in proper co-ordination of activities of different teams involved in magnet testing. Indian associates, all non-French speaking, found it difficult to verbally communicate with other teams which were exclusively French speaking. For this major issue, an innovative solution had to be found and implemented

Nature of industrial contracts of other teams: The nature of consortium contracts was also a hurdle. It had been observed that many times, the work slowed down during the weekends because the contractual working hours of the ICS/ROCLA team got exhausted for the week; magnets were not moved, connected or disconnected. Likewise, lack of suitable technical support in case of malfunctioning of certain systems during outside normal hours was also a factor which affected the overall performance.

Infrastructure limitations: The test facility in SM18 is organised in 6 clusters of two test benches each, total 12 benches. However, for space and costs reasons, each cluster has a common power converter, one set of data acquisition system and one set of quench heater power supplies, shared by both benches. This meant that at any given time, these resources could only be utilised by one of the two benches in a cluster.

The cryogenic infrastructure had limited resources which could not feed to the simultaneous demand from all the 12 benches. This put forth a limit on the number of magnets at superconducting temperatures concurrently, the number of training quenches allowed within a specified time period as well as a precise number of magnets using the cryogenic cool-down or warm-up resources [14]. Water resources (to cool down the power converters and other auxiliary systems) are also limited. These constraints necessitate the operation team to optimise all the work by following a complex set of rules and by exercising judicious judgement.

Sometimes due to some imposed factors, the shared resources were blocked. For example, when some special tests were conducted on a magnet, exceptional priority was assigned to this bench, which affected operation on many other benches due to the interlinking of various resources.

The synchronous, cog-wheeling approach foreseen initially [15] was never applied in routine operation because of varying performance of magnets; rather, the 'asynchronous' approach managed by the operation team yielded the desired magnet test rates, aided by the fine trimming of the magnetic and quench performance programmes [16].

7. Early tests performance

Magnet tests work began in $\sim$ 2001 with two benches and a limited cryogenic infrastructure. The work environment that existed till late 2002 was not favourable for a time limited and challenging activity like this. Tests were conducted mainly with laboratory type of systems and mobile racks which were not suitable for round the clock operation. The first sets of dipoles consisting of 30 samples from each of the three suppliers (called the pre-series magnets) were required to be tested elaborately with full magnetic measurements and many other extensive tests. In the early phases of testing till end-2003, due to the lack of readiness of all test benches and cryogenic feed boxes, adequate information, supporting tools and operational experience, only 95 dipoles (including pre-series ones) could be tested [17].

Figure 9 shows the time required for testing the dipoles during the early stages of 2001–2. With such a low testing rate it would have been impossible to meet the target. Hence, it was imperative to formulate proper throughput strategies and to develop supporting tools for enhancing the throughput; this necessitated

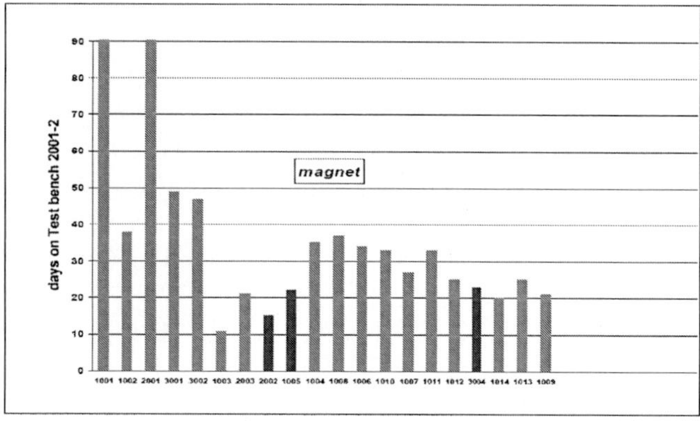

Figure 9. Bench occupancy during 2001–02

an ambitious figure of 16–18 magnet tests per week, higher than anticipated, in order to complete the tests of all magnets by December 2006. This also entailed an extensive study [18] resulting in the application of a selective and reduced magnetic measurement effort.

8. Tests operational strategies and tools

In the attempt to overcome the inherent hurdles and to attain maximum throughput, some effective management principles had to be addressed, necessary supporting tools developed and significant level of operator empowerment had to be efficiently implemented, based on several innovative ideas and techniques. Feedback based on operational experience was given due importance in framing the strategies. Furthermore, the web-based network backbone of CERN and computer facilities have been widely used for developing the supporting tools.

Most important innovations and strategies which essentially helped in achieving a high throughput included the introduction of a template based tests approach, web-based tools for tests management, magnet training rules and criteria for 24-hour operator decision taking and empowerment, general and cryogenic priorities handling by the shift crew, thermal cycle criteria and so forth.

For the final, smooth operation of the facility with 12 benches, it was necessary to ensure a minimum staff strength of 24 at any time, comprising at least 15 experienced staff; however, the staff strength had to be appropriately adjusted according to the expected departures, arrivals and experience as and when required. This aspect was even further exemplified by the projected work load while the full 12 benches were still under construction on a cluster by cluster basis in 2003–4. To ensure all this, the number of Indian associates inducted into the project at any given time had to be carefully defined and planned, considering the strict one year rotation as well as input of the additional CERN staff during 2005 due to the year-long accelerator shutdown. Figure 10 gives a histogram of the total staff strength during the peak period 2005–6, and depicts the intricacies of manpower management. The staff strength was projected to drop steadily after December 2006, the scheduled deadline for the completion of all tests. Mentoring of newly inducted associates was designed to be an 'on the job' and a continuous process, increasing the number of personnel per shift during the process to ensure that the throughput was not affected.

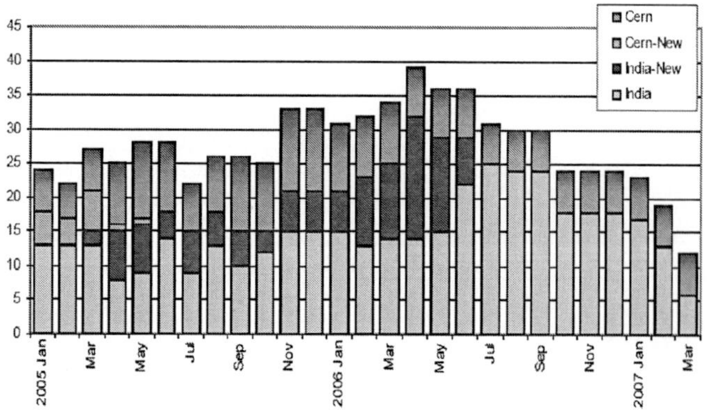

Figure 10. Variation of operation staff strength 2005–06

On the initiative of the Operation Team, a number of new features to aid the magnet tests had been brought in since mid-2003; the whole process of operation for magnet tests underwent a renaissance from crude manual data logging to a more efficient, sophisticated and highly automated tests management system.

A *To-Do-List* was created, which described the minimum set of tests to be performed on a magnet [19]. The tests were sequentially numbered and prefixed with the nature of tests i.e. *Preparatory test* (PREP) or *Power test* (PT). The *To-Do-List* approach weaned away the R&D culture in magnet tests and evaluation to a very stable and clear cut approach that could be handled by the Operation Team. Operation methods necessary for conducting each test in the *To-Do-List* were systematically prepared and reviewed to avoid the human errors in magnet testing to the maximum possible extent.

The Magnet Tests Report templates were designed in a manner for ease of use, with operational notes/checklists appended wherever necessary. The flow of tests in the template obeyed the *To-Do-List* to ensure that the tests were carried out systematically, efficiently and in a failsafe, sequential manner.

A new operation website was developed where all important documentation like operation methods, manuals, presentations, various template files, troubleshooting procedures, shift-plan and so forth could be obtained with minimum effort. This site immensely helped in easing the training of fresh staff as well as in managing the daily operation activities. The Indian Associates were the exclusive contributors to these very significant and essential documentation production and continual mentoring activities.

A web-based system using HTML and ASP codes called the SM18 Test Management System (SMTMS) was developed with all the data relevant to magnet tests stored in this system [20]. Based on the *To-Do-List*, the web-based retrieval from SMTMS permitted the automatic generation of the test sequences and reports such as the CDPT (CryoDipole Power Tests, which contains the training history), MAPS (Magnet Appraisal and Performance Sheet, which is a single page tabulation of the goodness of the magnet) and so forth. This enabled a fast, reliable and error-free generation of crucial data pertaining to the magnet tests. With SMTMS, it was also possible to keep track of times taken for the various phases in magnet tests; all persons directly concerned could keep track of the tests progress from varied geographical locations in CERN and outside [21].

An electronic log-book was implemented using the CERN network backbone in providing web-based applications. Apart from ensuring easy access and usage by all SM18 operation or support personnel, this helped in categorising and recording the different faults that occurred during the course of magnet testing.

To ensure smooth interaction between the various teams during the different stages of preparation before testing as well as at the end of the tests, a web based tool in the form of an Electronic-Workflow manifest called the e-traveller was created [20,22]. The interface of this tool with mobile phones alerted and informed relevant teams (via short message service in appropriate languages) about the need for their services on a particular magnet. This helped the Indian associates to overcome the difficulties in verbal communication with the other teams but maintained the work rhythm as well as keeping an automatic record of the tests phases.

9. Magnet training criteria for tests

In order to attain a high throughput, it was necessary to reduce the number of training quenches per magnet, both from the point of view of limited cryogenic resources as well as the time involved. During 2003, the Operation team had observed that the majority of the magnets cross their nominal field (8.33 T or 11850 A) in the second ramp (Fig. 11), whereas not much additional information on the 'goodness' of the magnet was available from the third and higher quenches [23]. Based on this, a new training rule named the 'Two-Quench Rule' was accepted by the magnet experts [16], under which it was recommended to do only two training quenches in each magnet provided it crossed the nominal field with a small margin. Later on, this rule was complemented by the so called 'Three-Quench Rule', under which a magnet was accepted if it crossed a field of 8.6 T (12250 A) in the third quench even if it had not passed the preceding rule. This strategy drastically reduced the overall cold tests time, thereby resulted in a high throughput. Likewise, the introduction of a Rapid On Bench Thermal Cycle (ROBTC) for magnets with poor performance in the first run was another major step towards reducing the overall magnet test time. These new rules, along with a 24-hour decision taking by the operator on the goodness of the magnet by analysing the results and using the MAPS, helped in achieving a higher throughput. ROBTC and MAPS are discussed in detail elsewhere [24,25]. The criteria for arriving at precise MAPS formulation were based on clear-cut rules and magnet specifications as well reviews, e.g. [26].

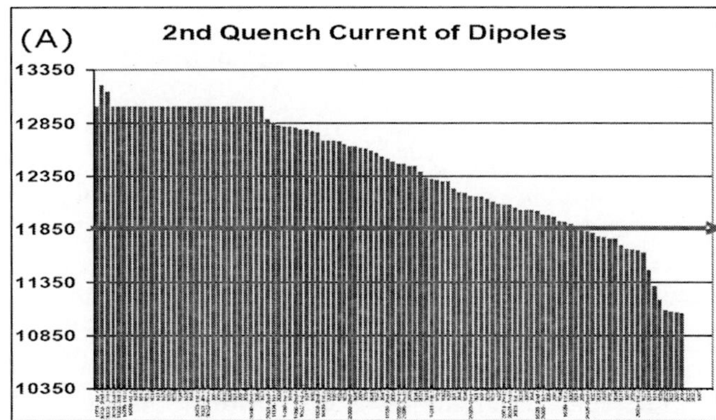

Figure 11. 2nd Quench current of dipoles till December 2003

Magnet	300–80 K (hours)	80–4.2 K (hours)	4.2–1.9 K (hours)	1.9–300 K (hours)
Dipole	16	10	4	15
Quadrupole	8	7	3	12
Special SSS	8	7	3	12

Figure 12. Average cooling and warm-up times (2005)

10. Overall and cryogenic priority handling in tests

Overall priority allocation becomes critical for maximising the throughput from a constrained system with limited resources. In this context, operation team empowerment for deciding and setting the overall and cryo priorities has played a crucial role in maximising the throughput through effective and clash-free resource management.

The limited cryogenics infrastructure [17] in SM18 could support only 6 magnets at a time out of the total 12 that could be in the cooling-down, warming-up or cold test phase. To effectively utilise even this 50% capacity, the operation team has to make careful priority decisions keeping in mind the average time requirement for cooling down/warming up of the particular type of magnet (Fig. 12) along with the constraints in the number of magnets that can co-exist simultaneously within each cryo regime, such as

- 3–5 magnets at 1.9 K
- Up to 2 magnets in 300 K to 80 K phase
- Up to 2 magnets in warm up phase
- 2 magnets in 80 K to 4 K phase
- Maximum 3 magnets simultaneously in cool down and warm up phases put together
- Minimum of 20 minutes delay between two quenches.

The operation team initiated a priority change based on the following broad guidelines [27]:

- A magnet under warm-up phase shall be assigned highest priority (1 or 2), allowing it to go out as fast as possible
- Due consideration shall be given to a cooling down magnet assessing the overall situation for the next 12 hours
- Magnets already at 1.9 K shall be given next higher priority (2–5) with maximum of 3 magnets getting the major share of cryo cool-down/warm-up resources (85 g/s for each magnet out of the total 300 g/s gaseous helium) and a fourth one with the remaining resources
- Priority numbers 6–8 could be assigned amongst the magnets cooling from 80 K down to 4.4 K
- The remaining priorities were allotted to the other magnets considering their exact status and the time that would elapse before they required the resources.

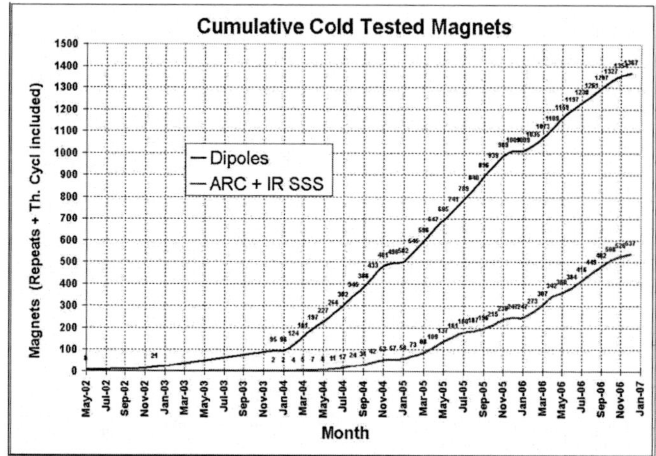

Figure 13. Cumulative cold tested magnets

11. Tests results and first tunnel commissioning

Figure 13 depicts the cumulative number of magnet tests, including repeats, since 2002. While the throughput was low till end-2003, it picked up sharply after the introduction of throughput strategies and tools. The plateau regions at the end of each year are due to the annual cryogenic infrastructure shutdown of typically seven weeks. Figure 14 gives further details of the magnets tested each year. It segregates the number of dipoles, arc SSS and IR-SSS tested each year, along with the cumulative number of magnets tested in that year. Starting with the meagre 21 magnets tested in 2001–2 and 76 in 2003, 456 magnets were tested in 2004. This count went up to an all time high of 703 magnet tests during 2005. During 2006, 648 magnet tests have been carried out; while this may appear low compared to 2005, it was a remarkable achievement taking into account the fact that the majority of the Special SSS magnets were also tested during 2006. Testing of the Special SSS magnets was a major time consuming activity in logistics and magnet training; each of the 114 magnets needed a special, dedicated to-do-list. Often, each special magnet was trained until it reached the ultimate field and elaborate magnetic measurements were also required [28]. Average repeat rates for the dipole, arc SSS and Special SSS magnets have been around 9%, 12.5% and 12.8% respectively, not counting the repaired and renamed magnets. In addition, ~3% of the dipoles and ~6% of the SSS had to be repaired or rejected after the cold tests due to unacceptable quench performance. The latter type of issues, observed early in the project, confirmed the need to systematically test all the LHC magnets under cryogenic conditions.

Period	Dipoles tested (Reqd. Number = 1232)			ARC-SSS + 500 tested (Reqd. Number = 392)			IR-SSS (Reqd.Number = 82)			Total magnets tested
	Fresh	Repeat	Total	Fresh	Repeat	Total	Fresh	Repeat	Total	
Year 2002	21	0	21	0	0	0	0	0	0	21
Year 2003	74	0	74	2	2	2	0	0	0	76
Year 2004	356	45	401	49	3	52	3	0	3	456
Year 2005	468	45	513	148	17	165	20	5	25	703
Year 2006	326	32	358	187	37	224	59	7	66	648
Total	1245	122	1367	386	57	443	82	12	94	1904

Figure 14. Magnets tested in each year

Magnetic measurements were performed on ~18% of dipoles, ~13% of arc SSSs and ~31% of Special SSSs. Often, exceptional tests were performed by the magnet experts on the Special SSSs, needing a considerable amount of time and data analyses.

Overall, about 38% of the total number of tested dipoles reached nominal field without a training quench. About 9% of the dipoles were tested for a second time after a thermal cycle, mostly to further investigate weak quench performance.

In the LHC tunnel, during the hardware commissioning of one of the first sectors in February 2008, the first natural quenches occurred at around 9.8 kA (at an equivalent energy of ~5 TeV). The other LHC sectors have yet to hardware commissioned at the time of writing this paper.

12. Tests concluding remarks

To complete in ~5 years the tests of all the LHC cryomagnets well in time before the LHC installation and hardware commissioning in the tunnel, several innovative ideas, strategies, tools and techniques were introduced and implemented by the magnet tests operation team. The results and statistics of magnet tests underline the significance of them in the successful completion of the tests. While many challenges were met and overcome in operation, delays in magnet delivery issues particularly since mid-2006 remained beyond the control of the operation team. Nevertheless, all magnet tests for the LHC were completed by February 2007. The LHC magnet tests operation has also been a singular and very successful example of a large scale collaborative effort in terms of human resources; over 90 persons from India have spent one year each at CERN since 2001 and hence, it remains a unique example in international collaboration of that scale in the particle accelerator domain. Figure 15 shows the last group of Indian Associates who participated in this massive tests effort.

Figure 15. Indian Tests Team in October 2006, nearly at the end of the 5-year collaboration for the LHC magnet tests

13. Challenges in the LHC cryogenics

Ten years of multidisciplinary R&D for the LHC have resulted in significant advances in cryogenic engineering of large helium – particularly helium II – systems. The installed cryogenic system of the LHC is the largest in the world in terms of refrigeration capacity, with an equivalent to 144 kW at 4.5 K [29,30]; working under normal operation needs about 400,000 litres of superfluid helium for the 25 km of superconducting magnets below 2 K, implying a cryogen inventory of ~100 tons of helium.

The LHC magnets are cooled with pressurised superfluid helium, which has some interesting properties that make it a unique engineering material. Best known is its very low bulk viscosity which allows it to permeate the smallest cracks. This is used to advantage in the magnet design by making the coil insulation porous and enabling the fluid to be in contact with the strands of the superconductor. It also has a very large specific heat, 100,000 times that of the superconductor per unit mass and 2000 times per unit volume. Hence, superfluid helium provides very high thermal conductivity. Just to illustrate the size and complexity of the systems, some examples are appropriate; during the cool-down of the first octant of the LHC in 2007, ~1200 tons of liquid nitrogen were required (equivalent of 64 trucks of 20 tons each) for the pre-cooling of this octant from room temperature to 80 K. From 80 K to 4.5 K, cool-down was carried out with the refrigerator plant, needing about three weeks with about 4700 tons of material to be cooled. Lastly, from 4.2 K to 1.9 K, cold compressors at 15 mbar were employed, needing four days for achieving this cool-down.

14. Total stored energy and machine protection issues

Generating the required dipole field of 8.33 T for the nominal LHC operation with 7 TeV proton beams requires a magnet current of 11.85 kA [11]. With 1232 magnets and an electrical inductance of L = 98.7 mH per magnet, this implies a total stored electromagnetic energy of 8.5 GJ for the dipole circuits alone (E = 0.5 * L * I2). 1 MJ is sufficient energy for melting 2 kg of Cu. The total stored electromagnetic energy inside the LHC dipole magnet chain exceeds the stored energy of previous superconducting storage rings by more than an order of magnitude (HERA: E = 0.7 GJ [13]) and presents a significant damage potential to the LHC equipment. In case of a magnet quench this electromagnetic energy needs to be extracted and dissipated in a controlled way before any of the magnet equipment is damaged. These protections is achieved by separating the main LHC magnet circuits into 8 independent powering sectors (stored electromagnetic energy comparable to that in previous superconducting storage rings) and by dissipating the energy during a quench into dedicated dump resistors and bypass quench diodes. One challenge for the LHC operation will be to synchronise the powering of the independent magnet sectors to the required accuracy. Existing storage rings avoid this synchronisation problem by powering all main magnets in series in a central circuit. The LHC will enter into new territory in this respect due to 8 independent powering sectors.

The stored beam energy provides another source for potential equipment damage during the LHC operation.

The LHC beam parameters translate to a total circulating beam current of approximately 0.5 A that corresponds to total stored beam energy of 370 MJ at 7 TeV. In case of problems during the machine operation the beams have to be quickly removed from the machine before the stored beam energy can damage any of the LHC hardware in the tunnel. An elaborate machine protection system, that constantly monitors all critical beam parameters and the beam losses along the storage ring, plays therefore a central role in the LHC machine design.

15. Other accelerator physics issues

The beam lifetime in the LHC is expected to be limited by beam-beam interaction, rest gas collisions, achievable vacuum levels (cryo pump), dynamical non-linear resonances and the resulting dynamic aperture of stability, the limitations of realistic number of corrector circuits and tolerances, dynamic effects and persistent currents. The challenge of adjusting the circuit settings and the need for non-destructive measurements and observables are higher than ever.

The resonances arising in single particle dynamics of a circulating particle in the collider, the various collective effects and instabilities (e.g. intra-beam coulomb scattering and beam-wall electromagnetic interaction), fluctuations in the power converter and ambient noise and vibrations will cause emittance growth of the proton beams leading to loss of luminosity.

The LHC needs a very effective collimation system as it needs to absorb stray particles. The cleaning inefficiency, specification of the required opening tolerances, hardware tolerances (e.g. surface flatness and temperature margins during operation) and operational tolerances all call for a very well designed LHC collimation system.

In order to protect the magnets, the LHC requires a dedicated and special Magnet Quench Protection System including voltage tabs, dedicated heater systems and beam loss monitors. The total number of input signals and reliability requirements compare close to safety related to flying a plane

The total event pile up rate and topology of hadronic showers in the experimental detectors are important concerns, needing detector commissioning with low luminosities.

The radiation inside the detector and central tracker lifetime are important issues requiring collider operation with lead for cool down before shut down.

16. Outlook and upgrade options

The LHC team is already considering various upgrade scenarios with collaboration from Europe (CARE) and US (LARP). The upgrades will comprise of: (a) IR and detector upgrades and (b) injector complex upgrade options of the original proton synchrotron PS2 and the linac systems. The radiation damage limit at an integrated luminosity of 700 fb^{-1} will require replacing the interaction region magnets any way by 2012, allowing us to design new tighter collision focus and incorporate novel techniques of crossing angle geometry and transversely deflecting "crab" cavities to reach the ultimate LHC luminosity going beyond 10^{34} to possibly 10^{35} $\mathrm{cm}^{-2}\mathrm{s}^{-1}$and beyond. The crab cavities allow us to compensate for the strong beam-beam interactions at the 4 primary IPs and 30 long range collisions per IP as shown in Fig. 16.

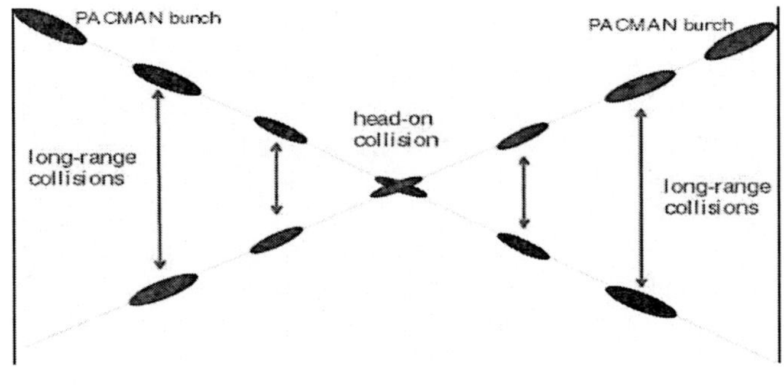

Figure 16. Beam crossing at a crossing angle

The ultimate luminosity of the LHC will possibly be determined by a process known as the Electron-Cloud Effect, depicted in Fig. 17, where photo-emission and secondary electron emission from the surrounding vacuum chamber in presence of beam leads to an amplification process ultimately shutting down proton beam sustenance in the collider. Such a phenomena might be expected at a luminosity of between 10^{35} to about 10^{36} $\mathrm{cm}^{-2}s^{-1}$.

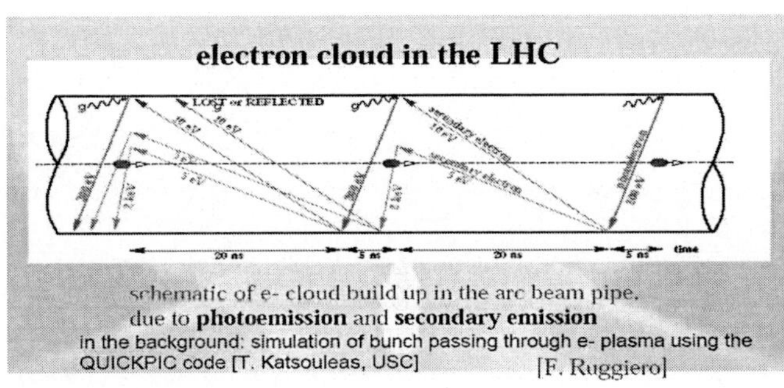

Figure 17. The electron cloud effect

17. Summary and outlook

We have given an impression of the currently configured LHC and its potential future upgrades. Most appropriately for this publication, we have highlighted at some length, at the risk of losing some technical readers, the sociological dimension of the international collaboration that defines the LHC. In particular, in the critical labour-intensive area of bench-testing and qualifying the pioneering superconducting magnet test programme, the contributions of Indian colleagues are inestimably immense. May the success of the LHC be a beacon for further successful international multinational scientific collaborations and herald the arrival of a large productive nation such as India on the international arena of large facilities.

REFERENCES

1. Armstrong, W. W. *et al.* The ATLAS Collaboration ATLAS Technical proposal for a general-purpose pp experiment at the Large hadron collider at CERN 1994 http //cdswebdev.cern.ch/record/290968
2. Della Negra, M, Petrilli, A, Herv, A. & Fo, L. CMS Physics Technical design report olume I Detector Performance and Software 2006 http //cdswebdev.cern.ch/record/922757
3. Mukari, ., Itow, . & Sako, T. LHCf Technical design report 2006 http //cdswebdev.cern.ch/record/926196

4. fferardi, ff. TOTEM Technical design report ff2004fff Total cross-section, elastic scattering and diffractive dissociation at the LHCff httpff/cdswebdev.cern.ch/record/704349
5. Amato, S. *et al.* ffThe LHCb Collaborationff CERN/LHCC 98-4, 1998
LHCb Technical proposal, A Large Hadron Collider beauty experiment for precision measurements of CP violation and rare decays
6. Carminati, F. *et al.* ffThe ALICE collaborationff
J. Phys. G **30(2004)**, 1517–1763
7. Alessandro, ff. *et al.* ffThe ALICE collaborationff
J. Phys. G **32(2006)**, 1295–2040
8. LEP Design Reportffhttpff/cdsweb.cern.ch/record/102083 ff1984ff
9. Christofilos, N. C. ffS patent 2,736,799 ff1950fff reprintedff Livingston, M. S. *The Development of high energy accelerators* ffDover, New ffork, 1966ff
10. Courant, E. D. & Snyder, H. S. *Ann. Phys.* **3(1958)**, 1–48
11. ffrüning, O. S *et al.*ffLHC Design report, CERN-2004-003, June 2004
LHC Design report ffol. 1ff The LHC Main ringff httpff/cdsweb.cern.ch/record/782076 ff2004ff
12. A Report on the Fermi national accelerator laboratory superconducting accelerator, May 1979, ffatavia Illinois ff.S.A.
13. HERA Proposal, DESff HEA 81/10, July 1981
14. J. Axensalva *et al.*, Proc. ICEC20, ffeijing, May 2004
Cryogenic infrastructure for testing LHC series superconducting magnets
15. J. Axensalva *et al.*, Proc. ICEC20, ffeijing, May 2004
Control system and operation of the cryogenic test facilities for LHC series superconducting magnets
16. Ph. Lebrun, private communication, 2004,
CERN EDMS 811861, CERN, Geneva, January, 2004ff httpff/edms.cern.ch/file/811861/1
17. ff. Chohan *et.al,* Proc. EPAC 2004, Lucerne, Switfferland, July 2004
Testing of the LHC magnets in cryogenic conditionsff current experience and near future outlook
18. L.ffottura, S.Fartoukh *et al,* Proc. EPAC 2004, Lucerne, Switfferland, July 2004
A strategy for sampling the field ffuality of the LHC dipoles
19. ff. Chohan, Proc. WAO 2005, Fermi National Accelerator Laboratory, ffSA, May 2005
Operation before LHC operationff The SM18 magnet tests facility
20. G.H. Hemelsoet *et.al,* Proc. ICALEPCS 2005, Geneva, Oct.2005.
Cryogenic magnet tests for the LHCffprocess operation using web-based tools and facilities
21. ff. Chohan, Proc. PAC 2005, Tennessee, May 2005.
Testing of the LHC magnets in cryogenic conditionsff operation challenges, status and outlook
22. A. Tovar-Gonffaleff *et.al,* Proc.CEC-ICMC, ffeystone, Colorado, Aug. 2005
Automatic management systems for the operation of the cryogenic test facilities for LHC series superconducting magnets
23. ff. Chohan, CERN EDMS 806640, CERN, Geneva, Dec.2003
Recent experience from operation & near future outlook, follow-up review of reception tests of cryomagnets
24. S.R.Shimjith *et.al.* CERN EDMS 811706
Rapid On-ffench thermal cycle for testing LHC cryomagnetsff A Review, httpsff/edms.cern.ch/file/811706/1/ROffTC.doc
25. E. ffandaswamy *et.al.* CERN EDMS 811589
Stripping or standby - An operational procedure ffMAPSff for goodness evaluation of LHC magnets, httpsff/edms.cern.ch/file/811589/1/MAPSfinal.pdf
26. L.Rossi, Streamlined series programme ff2003ff Review of reception tests of LHC cryomagnets, CERN, Geneva, 23 July 2003
27. Sampathkumar *et.al.* CERN EDMS 811583
Cryo priority handlingff A necessity for optimising resources in SM18 magnet test facility, httpsff/edms.cern.ch/file/811583/1/Cryo_priority.pdf
28. S.Sanfilippo *et.al.* Proc. LHC Project workshop, Chamonix ffff, January 2006
Transfer function of the ffuadrupoles and β-ffeating
29. S. Claudet, Proc. EPAC 2006, Edinburgh, ffff, Jun 2006, pp.3626
Design, construction, installation and first commissioning results of the LHC cryogenic system
30. P.Lebrun, Cryogenics for the Large Hadron Collider 16th International conference on magnet technology, Ponte ffedra ffeach, FL, ffSA, 26 Sep - 2 Oct 1999, httpff/cdsweb.cern.ch/record/411139

Challenges to Software/Computing for Experimentation at the LHC

Sunanda Banerjee

Fermilab, P.O. Box 500, Batavia, Illinois 60510, USA

The demands of future high energy physics experiments towards software and computing have led the experiments to plan the related activities as a full-fledged project and to investigate new methodologies and languages to meet the challenges. The paths taken by the four LHC experiments ALICE, ATLAS, CMS and LHCb are coherently put together in an LHC-wide framework based on Grid technology. The current status and understandings have been broadly outlined.

1. Introduction

The Large Hadron Collider (LHC) [1] is constructed and commissioned at CERN, Geneva and will soon start providing the world with physics information at the scale of a few TeV. The exciting physics which is waiting the human world to explore has posed enormous challenges to the experimenters not only in the detector technology but also in the field of software and computing. The four experiments ALICE [2], ATLAS [3], CMS [4] and LHCb [5] are built to look for complementary physics processes.

ALICE is a dedicated experiment to study heavy ion collision looking for the physics of quark-gluon plasma. LHCb is also a specialised experiment to study the physics of b-hadrons and is aimed to look for the mystery of matter-antimatter asymmetry and more precisely at the effect of CP violation. ATLAS and CMS are dedicated to look into 4×10^{14} beam crossings per year in the pursuit of the Higgs boson and new physics processes in proton-proton collision at a centre of mass energy of 14 TeV. Though the aim of the four experiments are rather orthogonal, all these experiments have something in common. The cross sections for the signal processes in all the experiments are very small and one has to filter out such final states from a vast amount of background events. Software and computing for these high energy experiments have to face challenges on several fronts. The traditional approaches may not be safe enough to cope up with these challenges.

Several R&D activities were carried out by these experiments over the past couple of decades to address the demands on software and computing. The experiments have developed their own computing model and the common aspects are steered by a combined effort led by a worldwide collaboration. This has brought together all the expertise in software and computing from all over the world and has resulted to a novel situation not only in the field of high energy physics but also in the general field of computing.

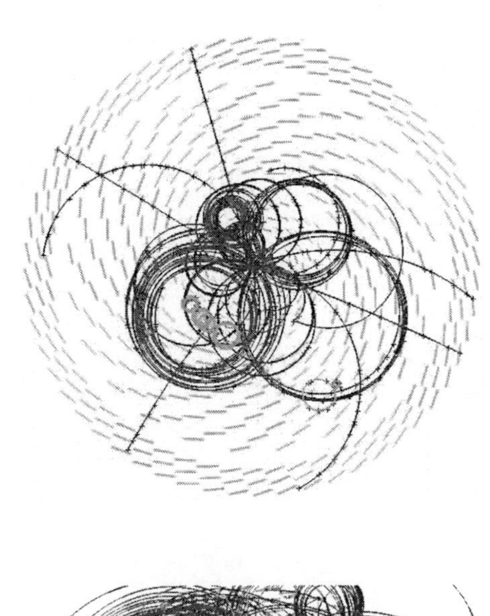

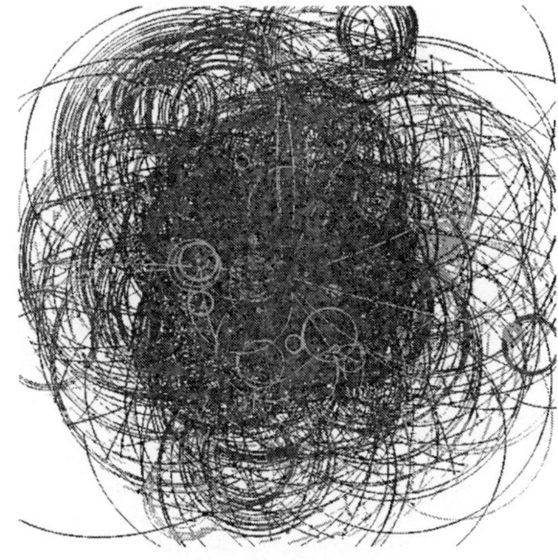

Figure 1. A typical event simulated for the CMS detector: (a) a sliced view of the event; (b) projection of the complete event in a plane perpendicular to the beam direction

Event output rate	10^9 events/year
Data written to tape	1 petabyte/year
Life time of an experiment	2–3 decades
Processing capacity	$> 10^{13}$ instructions/second
Typical network bandwidth	Hundreds of Mbits/second
Number of users	$\sim$ 2000 physicists
Number of software developers	$\sim$ 100 individuals

Table 1. Typical requirements of a LHC experiment towards software and computing.

2. Challenges

Challenges posed to software and computing project of the LHC experiments can be classified into the three following categories:

2.1. Complexity

Complexity arises due to the high energy and high luminosity of the LHC machine as well as the intricacies of the detectors. The high energy of the machine will result in a large number of particles produced per interaction. In the high luminosity mode of the LHC, there will be typically 20 overlapping uninteresting collisions obscuring a signal event. Bunch crossing time of the LHC is 25 nsec and this will cause overlapping from subsequent bunch crossing in the signals of the slower detectors. So one needs to extract information from typical events as in Fig. 1.

To cope up with this complexity the detectors need to have adequate granularity. This will result in larger number of detector channels. The LHC detectors are expected to have typically 2 orders of magnitude more channels than a detector of earlier generation. This increase in the number of channels is to keep the occupancy of a given channel at an acceptable level. The triggers for these experiments have to choose correctly at a level of 1 in 400,000 or better. The high level triggers are software based and this demands highly accurate and robust software.

2.2. Scale

One requires to stretch technology to the limit for data volume, processing time, network bandwidth, etc. Some typical requirements are summarised in Table 1.

One has to remember that there are additional requirements on CPU power and data processing capabilities for simulation studies. So each experiment can have easily 100 petabytes of data during the life time of the LHC.

2.3. Geographical spread

Each of these experiments comprises a large collaboration of physicists and engineers ($\sim$ 2500 for the CMS experiments) from many institutes ($\sim$ 175 for CMS) distributed in many countries ($\sim$ 45 for CMS). Figure 2 shows participating laboratories to the CMS. There will be major challenges associated with (a) communication and collaboration at a distance; (b) distributed computing resources; (c) remote software developments and physics analyses.

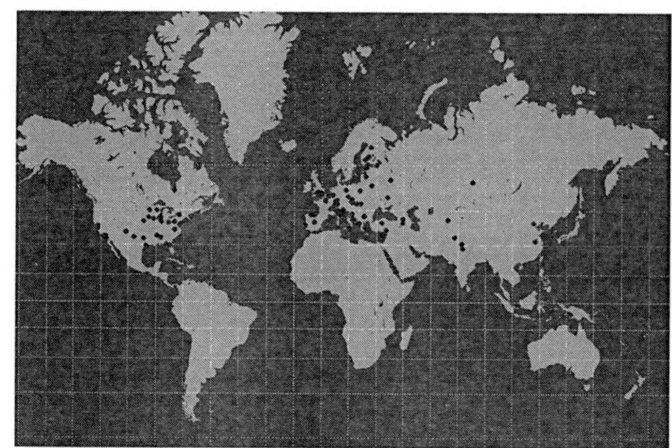

Figure 2. Distribution of collaborating laboratories in the CMS experiment

3. Experiment's Requirements

The online systems of the LHC detectors will identify and record the most interesting events out of the interactions which will be happening at the rate of 40 MHz. The maximum output rate is limited to few 100 Hz and is driven by the final tape writing capability. Figure 3 summarises the Level 1 trigger rate as a function of the event size for the four LHC experiment. The trigger rate is the highest for the LHCb experiment while the event size is the largest for ALICE. ATLAS and CMS experiments have large event size as well as large trigger rates. ALICE foresees two modes of operation with different sets of requirements: heavy ion (HI) mode with Pb-Pb collision and proton-proton mode (pp).

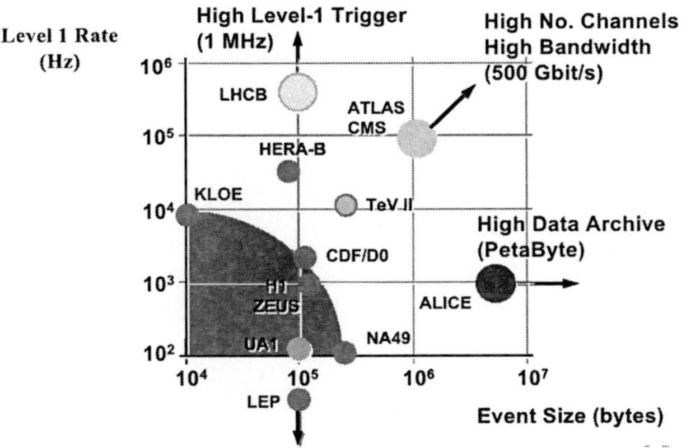

Figure 3. Level 1 trigger rate as a function of the event size for different experiments

The raw data collected by the experiments will be processed by the reconstruction programmes which will do pattern recognition and geometrical fitting to correlate the hits in the detectors to trajectories of the produced particles. The results of the processed data are again recorded in the data storage system. The processed data may consist of several types of data with different storage requirements in terms of capacity, bandwidth, latency and lifetime. Typical examples for the types of processed data are "reconstructed data" (RECO), "event summary data" (ESD), "analysis object data" (AOD), "event tag data" (TAG).

The final interpretation of the data as well as the understanding of the detectors require in addition an equivalent amount of simulated event samples. Each experiment has its own plan for Monte Carlo production. ALICE and CMS expects to produce the same number of Monte Carlo events as the real p-p or heavy ion data while ATLAS and LHCb plan for about 20%. The expected event rates and sizes are summarised in Table 2.

In addition to event data, each experiment requires non-event data for reconstruction and simulation. There are two types of non-event data – static and dynamic. The static data include information of all sub-detector construction, the results of their integration in the form of geometry and electronics information and also configuration of the front-end electronics. The dynamic data comprise calibration and alignment results and are associated with an "Interval of Validity" (IOV).

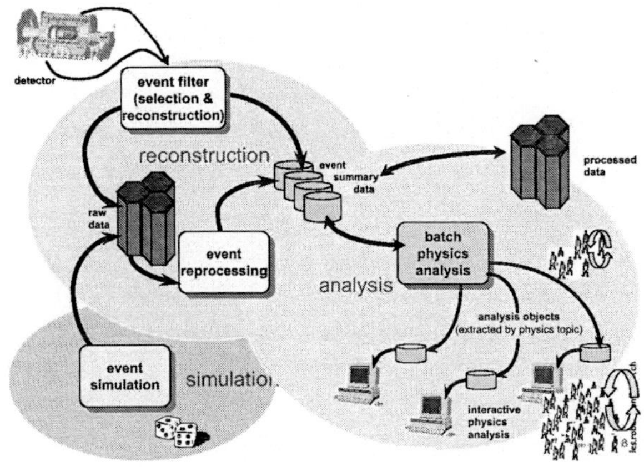

Figure 4. The main data flow in a high energy physics experiment

4. Choice of Software

Figure 4 shows the data flow in a typical high energy physics experiment. Data read out from the detector as well the ones generated by the simulation programmes are usually reconstructed by a common reconstruction programme which interprets the event in terms of basic physics objects like electrons, photons, muons and jets. These reconstructed events are then analysed by hundreds of physicists by a number of analysis programmes performed either interactively or with a set of batch processes. The role of software in such a high energy physics experiment is

- To filter data coming from the detector
- To monitor the performance of the detector
- To calibrate and to align the detector components
- To simulate detector response from known physics processes
- To carry out user analyses leading to physics results.

In traditional approaches, these tasks are viewed as separate programmes for filtering, reconstruction, generation of events, simulations and analyses. For a large

	Rate [Hz]	RAW [MB]	ESD, RECO [MB]	AOD [kB]	MC [MB]	MC % of real
ALICE HI	100	12.5	2.5	250	300	100
ALICE pp	100	1.0	0.04	4	0.4	100
ATLAS	200	1.6	0.5	100	2	20
CMS	150	1.5	0.25	50	2	100
LHCb	2000	0.025	0.025		0.5	20

Table 2. Event rates and sizes of the 4 LHC experiments.

and complex environment, the LHC experiments create a software architecture that allows the same programme modules to be used in several different computing environments in a flexible and user friendly manner. In addition, the users should have the liberty to substitute their own versions of certain programme modules while using the standard versions of the rest of the software.

Apart from the correctness and reliability of the code as demanded by the high level triggers, the main concern on the software side is the time scale of the experiments. The maintainability of the code over a period of 2–3 decades and managing a software developed by hundreds of physicists and engineers is the biggest factor in deciding the software technology to be followed.

In the field of software, the experiments were using procedural methods based on code written in FORTRAN to take care of detector optimisation studies. All the detector Technical Design Reports were written and submitted to the LHC committee from these studies. CMS has used the programme package CMSIM for simulation as well as reconstruction, whereas ATLAS have used several packages DICE, ATRECON, etc. for simulation and reconstruction work. Standard data flow diagrams have been used for software design work towards domain decompositions.

For softwares surviving the LHC era, one requires an engineered software framework

- To handle the complexity of the detector and the data
- For reliability and maintainability of the software over a long project life-cycle
- To serve data efficiently to a worldwide distributed collaboration
- For an efficient and cost-effective data analysis.

Several studies in the past have led the experiments to choose Object Oriented Software with modern programming language, methods and tools. Many R&D projects have been undertaken to study the model, framework, products and tools.

In a procedural approach, one needs central control, global synchronisation, explicit knowledge of dependencies and with FORTRAN codes some preemptive allocation of resources. With object oriented programming, there is no need of central steering, nor of global action, nor external synchronisation. The framework need not be changed if an application class is added or modified. However moving to the new paradigm requires a change in thinking which leads to retraining of the physicists.

The basic building blocks used in high energy physics software are summarised in Fig. 5. One identifies in each of these blocks those specific to an experiment, those common to all high energy physics experiments and the ones which are not specific to high energy physics. To minimize risk and to guarantee higher quality, the experiments have decided to choose industry standard, widely used and well supported technologies wherever applicable and to have common projects for solving common HEP specific problems.

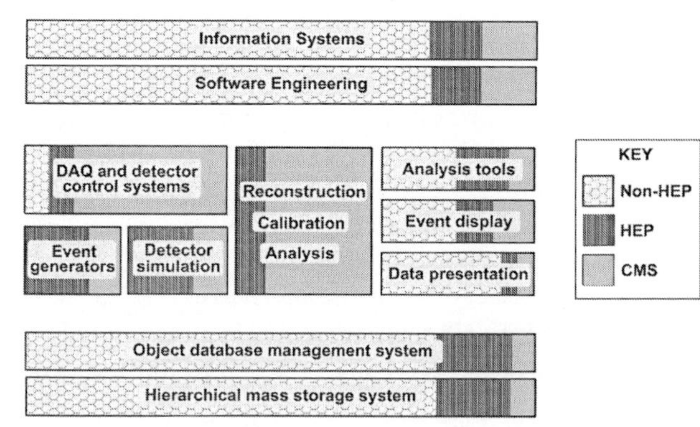

Figure 5. Building blocks for high energy physics software

There have been several HEP wide common software packages: CLHEP (Class Library for High Energy

Physics), LHC++, ROOT, SRT (Software Release Tool for configuration management) etc. In addition, there were several common R&D projects:

- Video conferencing support
- Detector simulation (GEANT4)
- Persistent storage (RD45)
- Globally Interconnected Object Database (GIOD)
- High Energy Physics VISualisation (HEPVIS)
- Model Of Networked Analysis at Regional Centres (MONARC)
- Software development environment (SDE).

The LHC experiments made some specific choices for their initial software developments. They started with some short-term decisions to make progress in the understanding of the processes and the products. They unanimously chose cvs/SRT (for code management), C++ language, OO paradigm, UML (Unified Method) design method. For the operating systems, several alternative platforms have been tried out, namely Solaris operating system, NT and Linux. ATLAS chose ASP for software process and StP as a case tool while CMS decided to use Evolutionary Booch for software process and Rational Rose as a case tool. For input/output, some of the experiments tried OODBMS (Object Oriented Data Base Management System), while others focused on relational data base system and HEP specific IO packages developed inside the ROOT framework. Regarding data analysis packages, several alternatives have been tried out, namely LHC++, ROOT, etc. Different scripting languages like Perl, Python, etc. were also tried.

The common project on video conferencing support started in July 1997 and the first phase was completed during early 1999. It set up two demonstrations using Codec and Packet and provided support to HEP community. During its second phase (lasting about a year) the project equipped several rooms at CERN and started a service. Video conferencing is already used a lot in these collaborations and has been found to be extremely valuable in international collaborations of this size.

The detector simulation tool kit, GEANT4, is a successor of its FORTRAN predecessor GEANT3. It has been completely redesigned with object oriented technology. It has several new features: compliance with CADD (Computer Aided Design and Drawing), multiple physics processes, "fast simulation", different visualisation schemes, etc. The first version was released in January 1999 with $\approx$ 300 thousand lines of C++ code and 600 classes. This package has been adopted as the basic simulation toolkit in the 3 out of 4 major LHC experiments.

The common project on persistent storage (RD45) provided a system to store "objects" using standard technologies of OODBMS and MSS (mass storage) and language and DB features for transparent event level access. It provided some production service to different experiments like BaBar, NA45 and test beams of ATLAS and CMS. It made significant R&D activities on a marriage between OODBMS and MSS (Objectivity DB with HPSS), development of DB administrative tools, HEP specific OODBMS class libraries, etc. Eventually the industrial solution of OODB system could not be used in the LHC experiments, but the results of this R&D programme has been extremely useful in defining the top level interfaces for all IO system covering event objects and calibration objects.

The project on GIOD used existing leading edge hardware and software systems to study integrity, efficiency, scalability of interconnected object database. It used Caltech HP Exemplar (for a parallel architecture), HPSS (for mass storage), Objectivity DB (and also Versant for OODBMS), C++ and Java (as computing language) and ATM local area network and high speed wide area network. It observed almost linear throughput up to 150 clients for event level parallelism and in the process developed a Java3D event viewer.

The project on HEPVIS identified common elements of interactive detector and event visualisation and analysis systems. It provided OO class library based on mainstream technologies for generic graphics classes for detectors and events and components for the construction of portable graphical user interfaces. The project eventually provided the basic architecture of event display programmes for a number of experiments.

The project on networked analysis at regional centres (MONARC) had the mandate of understanding ensemble of centres as "distributed data analysis system", of identifying and developing candidate sites for regional computing centres. The results of this project led to the final choice of computing models of the experiments.

For the detector specific programmes, ATLAS started the project ARVE (Atlas Reconstruction and Visualisation Environment) way back in 1995. The working version used events input from old simulation programmes based on GEANT3. It underwent several improvements on the various components and effort was made to integrate it with GEANT4 outputs. The project called CHAOS (CompreHensive Atlas Object oriented Simulation), started in a complete Object

Oriented way through the process of writing up the requirement document followed by the phases of analysis and design. The alpha version of the product was first made available by the end of 1999 and then it evolved to a more complete simulation programme for ATLAS. ATLAS collaboration tried both ROOT and LHC++ for analysis tools, and eventually came up with ROOT as the final choice of analysis tools.

CMS likewise launched the ORCA project (Object oriented Reconstruction for Cms Analysis) in 1998. The first version of the code was released in December 1998 and was used to understand High Level Triggers for the TriDAS (Trigger and Data Acquisition System) Technical Design Report in 2001. The CRISTAL (Cooperating Repository and an Information System for Tracking Assembly Life cycle) project looked into recording the specifications and quality assurance test results of various components during the production of a detector. It was used for the electromagnetic calorimeter. Various test beam studies used OO tool kit for simulation and analyses. The project on CMS detector simulation started with the OSCAR project using GEANT4. The proof of concept of the project was achieved by middle of 1998 through a description of the barrel detectors with a pre-release version of GEANT4.

After a period of intensive R&D programme, the experiments came up with their final solutions for the initial analysis of the LHC data. Though they differ in some of the choices which are sometimes dictated by the goal of the experiments, they use several common features and benefit from the common projects and interactions among the four experiments. There is a consensus to use C++ as the main programming language and Python or CINT as the scripting language, Object Oriented analysis and design. There is a development of a core software library which provides several mathematical functions, fitting, histogramming and a variety of useful functionalities. The data management relies on ROOT input/output package for bulk data and Relational Data Base Management System (RDBMS) for conditional data. Event simulation services are taken care of with a generator library and dedicated tool kits Geant4, Fluka for detector simulation. There are additional tools for error reporting system, quality assurance and documentation. The LHC has established a well tested, well documented software system through all these years consolidated effort.

5. Choice for Computing

Because of the complexities and challenges, all major LHC collaborations made software and computing into separate projects. The scope of the projects was to address

- Computing aspects of the design, construction, evaluation and calibration of detectors including event reconstruction and physics analyses
- Storage, distribution, access and processing of petabytes of event data
- The computing, networking and "remote collaborative" infrastructure for the items mentioned above.

The experiments wrote up computing technical proposals [6–8] summarising the thinking of the goals and ways of achieving them. The proposals were submitted to the LHC committee and were reviewed externally. The proposals were updated in every 2–3 years and finally the technical design reports [9–12] were written up. The main considerations in the considerations were:

- The LHC will produce 15 petabytes of data every year
- Access to the experimental data will be provided to 5000 scientists in some 500 institutions worldwide
- The LHC experiments are estimated to run for 15–20 years
- Analysis of data requires 100,000 CPUs at 2004 measures of processing powers
- Scalability is the critical issue and the main concern is data access and mass storage.

A traditional approach is to centralise all of these computing facilities in one location near the experiments. But to scale up such a facility with increasing need is not an easy task and serving a community of physicists world wide would have been extremely challenging. In view of these, a distributed model of data storage and analysis is envisaged. This choice has been driven from a study of the trend in the prices of the computing components over the years, which is given by Moore's law. The costs of CPU, disk storage, memory, mass storage system and network are falling exponentially, each following a somewhat different slope. Figure 6 shows the trend of the cost of CPU (desktop and cheaper versions adequate for data handling), data storage (for sequential as well as random access), wide area networking over years available at the time of writing the technical proposals. Computer speed doubles every 18 months and networking speed doubles in every 9 months. So one would expect the same money will buy 60 times more computing power and 400 times higher bandwidth in networking over a period of 10 years.

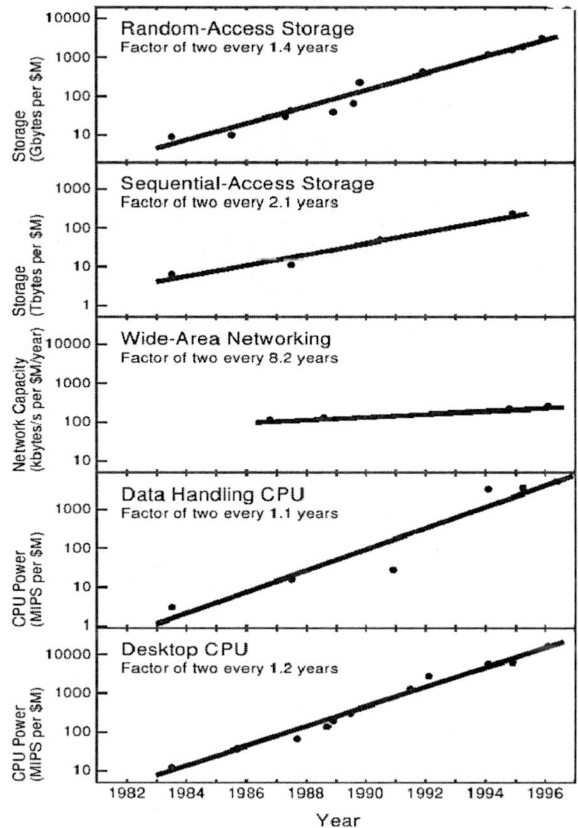

Figure 6. Trend in the cost of data storage, networking, CPU as a function of time

The Internet provides universal connectivity. This changes in modes of working and problem solving in a teamwork. Fast and dependable worldwide networks have opened up the possibility of tying together geographically distributed resources. The success of the worldwide web, discovered at CERN during the LEP experiments, for sharing information has spawned a push for a system to share resources. The network has become the virtual bus of a virtual computer. "Grid computing" is the key word for computing in the LHC era.

The simulation work of the MONARC project has led to the idea of a hierarchical distributed tier structure for the LHC computing (Fig. 7). All data collected by the experiments will first arrive at a Tier-0 centre at the experimental site where the first offline treatment will be applied to the raw data. The data link speed is driven by the size of events and rate of data taking (a few hundred MB/sec). The entire processed data (also part of the raw data) will be distributed to each of the major (inter-)national computing centre serving as Tier-1s. The networking speed between the Tier-0 and Tier-1 centres is expected to be in excess of 2.5 Gbits/sec. The clusters of Tier-1s are in turn connected to national regional computing centres, Tier-2s again with a high bandwidth network. Physicists working in the universities are served by local clusters (Tier-3) which are in turn benefit from the computing and storage capabilities of Tier-2s. The "Grid" is necessary to tie the distributed resources together.

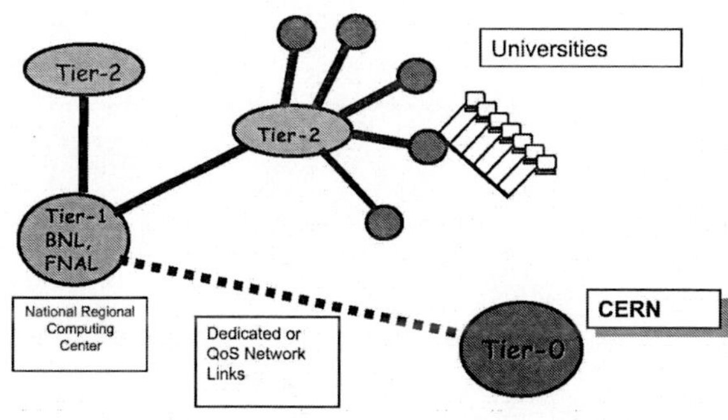

Figure 7. The planned architecture of computing during the LHC era

The design of the computing model assumes the ratio of resources available at CERN and outside is roughly 1:2 and the distribution between all Tier-1 and all Tier-2 centres is roughly even. The different tiers are chosen for the service hierarchy. Tier-0 is supposed to provide data acquisition, initial processing, long-term data curation and distribution of data to Tier-1 centres. Tier-1 centres, which should be "online" to the data acquisition process are supposed to manage grid enabled data service through huge mass storage system and to serve data-heavy analysis for large regions. Tier-2 centres provide computing power for simulation and for both batch and interactive analysis of different physics groups.

The choice of distributed computing system comes naturally from the trend in development in this field.

Computing hardware is becoming more and more commoditised. So computational and storage resources are appearing everywhere. There are giant clusters and mass storage systems in big computing centres, large clusters in research and engineering laboratories. Even the personal notebooks are becoming enormously powerful devices with reasonably large storage system. Network bandwidth is growing to match this – even at home and on the move. So it is just good sense to share all of this knowledge processing capacity.

Distributed computing will ease maintenance and upgradation of the facility. The responsibilities are simply shared among the providers. There will not be a single point of failure. Load balancing and providing access of data to all users can be more easily ensured. However, this comes with some concerns which the management has to deal with. It needs adequate bandwidth between contributing resources. A coherent software version needs to be maintained in different centres starting from operating system, application and service software. This has to be done over a heterogeneous hardware environment. The greatest concern is regarding protection of data and computing resources.

6. Grid Computing

Large scale computing has evolved over the pas decades from the use of super-computers to use of large clusters to Grid computing (Fig. 8). The name "Grid" has been chosen in analogy with electric power grid. The vision is to plug-in computer for processing power just like plugging a toaster for electricity. The concept has been around for decades with the banes of distributed computing, meta computing and so forth. The key difference between Grid and earlier ideas is the realisation of the vision on a global scale.

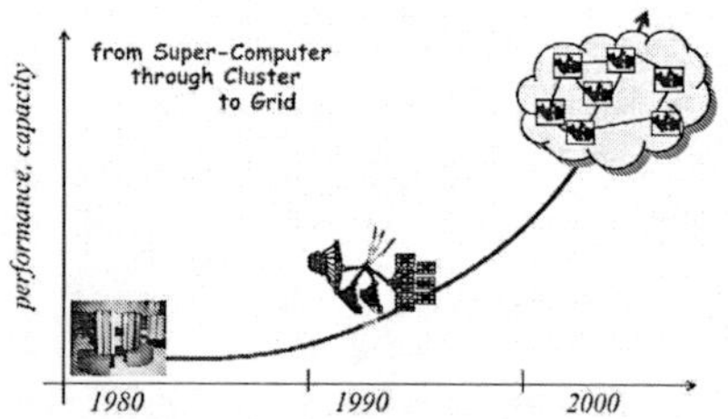

Figure 8. The evolution of performance capacity of a computing system during the past decades

The name Grid is used to describe many functions:

- To scavenge unused cycles (not so easy to scavenge unused storage)
- To share data (needs standardised ways of querying and mining related data bases)
- To carry out remote instrument control
- To link researchers to test equipments, sensors and computers.

Grid provides an environment where virtual organisations are built. This gives easy and flexible collaboration-sharing knowledge and computing resources. It creates relationships among otherwise independent communities. From a users view point, it is a virtual computer which minimizes time to complete the user application while transparently managing access to inputs and resources. From a programmer's view point, grid is a toolkit of applications and Application Programming Interfaces (APIs) which provide transparent access to distributed resources. From an administrator's view point, grid is an environment to monitor, manage and secure accesses to geographically distributed computers, storage and networks.

Grid requires flexible, secure, coordinated resource sharing. It enables "virtual organisations" to share geographically distributed resources. The potential problems may arise in resource sharing, coordinated problem solving and in multi-institutional virtual organisations. Automatisation is the key word for grid computing. This needs emphasis on the middleware, namely in workload management, job submission protocol and monitoring software.

Figure 9 gives an example of work-flow in a grid environment. The user sends a job in the job description language specifying the application programme and list of data files to be used. The job generator then uses the Grid File Catalog to generate the jobs which then goes to the Grid resource broker. The resource broker in turn uses the information service and sends these jobs to the appropriate grid computing centre on the basis of availability of the appropriate data and the workload of the system. The computing element in the centre then sends the jobs to the working nodes which perform the task with the necessary input/output service from the storage elements. The results are then sent back to the grid information service as well as to the user.

7. The LHC Computing Grid

Inputs from the four LHC experiments have been reviewed and combined to form a central infrastructure

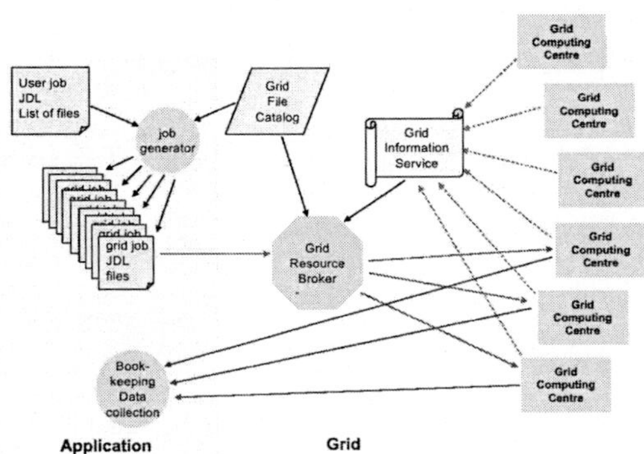

Figure 9. An example of a Grid application

for the LHC computing. The combined project, the LHC Computing Grid (LCG) [13] has been designed to provide the basic services to the LHC experiments. The Grid services include

- Information system
- Security framework which includes the security model, infrastructure for providing Grid certificates and also the policies and processes
- Virtual organisation management services which include user registration, user authentication and authorisation
- Grid operation toolkit which includes monitoring tools as well as tools for accounting and user support
- Resource broker and Grid catalog
- Installations of applications software.

In addition there are services for the sites related to compute resources, storage management and data transfers.

Information about available and pledged computing resources has been collected by the Worldwide LCG (WLCG) collaboration. All the experiments together will utilise 142 M SPECint2000[1] equivalent CPU power, 57 petabytes of disk storage space and 53 petabytes of mass storage space. The CPU power and disk storage space will be distributed among CERN, Tier-1 and Tier-2 centres in the ratio of 17.6:39.4:43.0 and 12.3:54.4:33.3, respectively. Mass storage facility will be available at CERN and Tier-1 centre in the ratio of 34:66.

WLCG will rely on two major science Grid infrastructures: OSG (Open Science Grid) serving USA and EGEE (Enabling Grid for E-Science) serving the rest of the world. Tier-1 centres have been set up at 11 laboratories all over the world and they are often geared to serve more than one experiment with priority. The list of these 11 centres is summarised in Table 3. There are more than 100 Tier-2 centres (Fig. 10) in operation already with two in India. One of them located at Mumbai is for the CMS experiment while the other in Kolkata is dedicated to the ALICE experiment. The number of Tier-2 centres is growing with time.

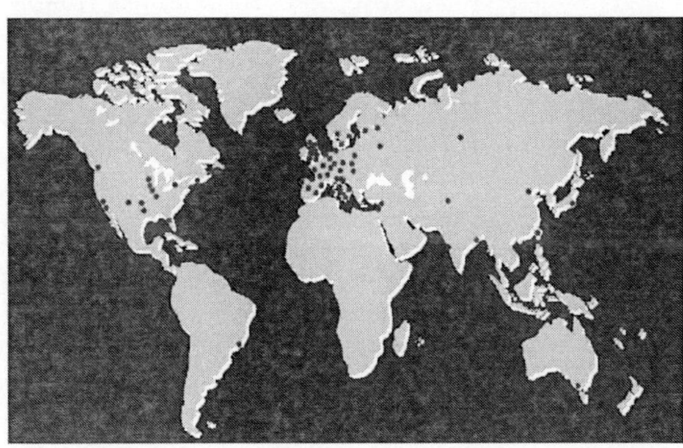

Figure 10. Distributions of Tier-2 centres all over the world

WLCG has identified several sources for grid middleware: grid management as well as file access and storage management. The Grid community has done well in identifying the standard organisations. But it needs to define a standard yet. It may be too early to form standards. It needs greater experience in working with large grids to come up with a broad consensus.

Security may be the biggest risk faced by Grids. Even a denial of service exploit could close down the whole Grid. One can test the readiness to react quickly. But a security model can be tested well by the so called "bad guys". There has been a great deal of effort to increase the level of security in the Grid services of today. One

[1]SPECint2000 is an integer benchmark suite maintained by the Standard Performance Evaluation Corporation (SPEC). The measure has been found to scale well with typical HEP applications. A powerful Pentium 4 processor delivers 1700 SPECint2000.

Tier-1 Centre	Experiment served with priority			
	ALICE	ATLAS	CMS	LHCb
TRIUMF, Canada		X		
Karlsruhe, Germany	X	X	X	X
IN2P3, Lyon, France	X	X	X	X
CNAF, Bologna, Italy	X	X	X	X
NIKHEF, Netherlands	X	X		X
Nordic Data Grid Facility	X	X	X	
ASGC, Taipei		X	X	
Rutherford Laboratory, UK	X	X	X	X
BNL, USA		X		
FNAL, USA			X	
PIC, Barcelona, Spain		X	X	X

Table 3. Distribution of Tier-1 centres and their priority list.

needs much more experience in the core of Grid infrastructure and at the level of interactive applications.

Computing Grid provides an exciting phase in High Energy Physics Computing. To an optimist this is the best thing since WWW. It will solve all the computational and data problems in HEP computing. For a Grid pessimist, "the Grid is merely an excuse by computer scientists to milk the political system for more research grants so they can write yet more lines of useless codes". Alternately it is "a distraction from getting real science done". A realist should look at it in a way – "The Grid can solve our problems because we design it to. We must work closely with the developers as it evolves, providing our requirements and testing their deliverables in our environment."

8. Summary

The challenges to software and computing for the LHC experiments have been identified and used to define the computing models of the LHC experiments. The ground work has been done: software process is defined and tools like ROOT and GEANT4 are now deployed by the LHC experiments in their basic software components for analysing the LHC data.

All the experiments are confident to move towards the new object oriented paradigm and the new technologies. The transition towards the OO framework has been proven to be successful. Good progress has been made for common as well as experiment specific projects. The milestones defined in the computing technical proposals have been met so far through a number of data challenges and the experiments are confident to have an adequate computing framework at the start of data taking in the LHC.

REFERENCES

1. The LHC - The Large Hadron Collider, *http://lhc.web.cern.ch/lhc/*
2. ALICE Technical Proposal, CERN/LHCC 95-71, LHCC/P3, December 15, 1995
3. ATLAS Technical Proposal, CERN/LHCC 94-43, LHCC/P2, December 15, 1994
4. CMS Technical Proposal, CERN/LHCC 94-38, LHCC/P1, December 15, 1994
5. LHCb Technical Proposal, CERN/LHCC 98-4, LHCC/P4, February 20, 1998
6. ALICE Computing Technical Proposal, CERN/LHCC 96-43, December 19, 1996
7. ATLAS Computing Technical Proposal, CERN/LHCC 96-43, December 19, 1996
8. CMS Computing Technical Proposal, CERN/LHCC 96-45, December 19, 1996
9. ALICE Technical Design Report of the Computing, CERN/LHCC 2005-18, June 15, 2005
10. ATLAS Computing Technical Design Report, CERN/LHCC 2005-22, July 4, 2005
11. CMS, The Computing Project, Technical Design Report, CERN/LHCC 2005-23, June 20, 2005
12. LHCb Computing, Technical Design Report, CERN/LHCC 2005-19, June 20, 2005
13. LHC Computing Grid, Technical Design Report, CERN/LHCC 2005-24, June 20, 2005

Electroweak Symmetry Breaking at the LHC

Abdelhak Djouadi[a] *and Rohini Godbole*[b]

[a]*Laboratoire de Physique Théorique, Université Paris–Sud, F–91405 Orsay Cedex, France, Physikalisches Institut, University of Bonn, Nussallee 12, 53115 Bonn, Germany*

[b]*Center for High Energy Physics, Indian Institute of Science, Bangalore-560012, India*

One of the major goals of the Large Hadron Collider is to probe the electroweak symmetry breaking mechanism and the generation of the masses of the elementary particles. We review the physics of the Higgs sector in the Standard Model and some of its extensions such as supersymmetric theories and models of extra dimensions. The prospects for discovering the Higgs particles at the LHC and the study of their fundamental properties are summarised.

1. Introduction

Establishing the precise mechanism of the spontaneous breaking of the electroweak gauge symmetry is indeed a central focus of the activity in the area of high energy physics and, certainly, one of the primary goals of the Large Hadron Collider, the LHC, which will soon start operation. In the Standard Model (SM), electroweak symmetry breaking (EWSB) is achieved via the Higgs mechanism [1,2], wherein the neutral component of an isodoublet scalar field acquires a non-zero vacuum expectation value. This gives rise to non-zero masses for the fermions and the electroweak gauge bosons, which are otherwise not allowed by the $SU(2)\times U(1)$ symmetry. In the sector of the theory with broken symmetry, one of the four degrees of freedom of the original isodoublet field, corresponds to a physical particle: the Higgs boson with $J^{PC}=0^{++}$ quantum numbers under parity and charge conjugation [3,4].

Clearly, the discovery of this last missing piece of the SM is a matter of profound importance. In fact, in spite of the phenomenal success of the SM in explaining the precision data [5], the SM can not be considered to be established completely until the Higgs particle is observed experimentally and, further, its fundamental properties such as its mass, spin and other quantum numbers, as well as its couplings to various matter and gauge particles and its self-couplings are established. These studies are important not only to establish the SM as the correct theory of fundamental particles and interactions among them, but also to achieve further clarity into the dynamics of the EWSB mechanism.

Indeed, in spite of the success of the idea of spontaneous symmetry breaking (in fact, partly honoured by the Nobel prize of 2008) in different areas of physics, very little is known about its realisation in particle physics via the Higgs mechanism. The many important questions which one would like answered are: does the dynamics involve new strong interactions and is the Higgs a composite field? if elementary Higgs particles indeed exist in nature, how many fields are there and in which gauge representations do they appear? does the EWSB sector involve sizable CP violation? etc.

Theoretical realisations span a wide range of scenarios extending from weak to strong breaking mechanisms, including the so-called Higgsless theories in extra dimensional models. As far as the representations of the gauge group are concerned, there is again a whole range starting from models involving light fundamental Higgs fields, arising from an SU(2) doublet, such as in the SM and its supersymmetric extensions which include two–Higgs doublets in the minimal version, to those containing additional singlet fields or higher representations in extended versions in unified theories and/or alternative theories such as little Higgs models.

Furthermore, the link between particle physics and cosmology means that the EWSB mechanism can have implications for the generation of the baryon–antibaryon asymmetry in the early universe and could play an important role in the annihilation of the new particles that are responsible for the cosmological dark matter and thus impact their density in the universe today. In fact, possible CP violation in the Higgs sector can have a direct bearing on the two cosmology issues mentioned above. An understanding of the EWSB mechanism at a more fundamental level might also hold clues about why the three generations of quarks and leptons have masses which differ from each other; the so-called flavour issue.

A complete discussion of Higgs physics thus touches upon almost all the issues under active investigation in theoretical and experimental particle physics.

2. Electroweak Symmetry Breaking Mechanism

2.1. The Higgs Boson in the SM

In the SM there exists only one isodoublet complex scalar field and, thus there are initially four real scalar fields [1–4]. After spontaneous EWSB, we are left with one physical degree of freedom, the Higgs scalar and the other three would–be Nambu-Goldstone bosons are absorbed to build up the longitudinal components of the $W^{\pm}, Z$ gauge bosons and generate their masses. Yukawa interactions of the fermions with the same scalar field give rise to the fermion masses. The Higgs scalar has $\mathrm{J}^{\mathrm{PC}} = 0^{++}$ assignments of spin, parity and charge conjugation quantum numbers. The Higgs couplings to the fermions and gauge bosons are related to the masses of these particles and are thus decided by the symmetry breaking mechanism. In contrast, the mass of the Higgs boson itself is completely undetermined in the model. There are, however, both experimental and theoretical constraints on this fundamental parameter, which we will summarise below.

One available direct information on the Higgs mass is the lower limit $M_H \gtrsim 114.4$ GeV at 95% confidence level (c.l.) established at LEP2 [6]. The collaborations have also reported a small, $\lesssim 2\sigma$, excess of events beyond the expected SM backgrounds, consistent with a SM-like Higgs boson with a mass $M_H \sim 115$ GeV [6]. In addition to this, the Tevatron physics potential for the discovery of Higgs particles looks promising, with the coming larger data sets. In particular, evidence for the SM Higgs boson could be obtained if the mass is near the observed experimental lower limit from LEP of about 115 GeV or if it is near 160 GeV. In fact, with the run-II data collected by both the experiments, corresponding to 2.5 fb^{-1}, the observed upper limits are a factor 3.7 (1.1) higher than the expected SM Higgs cross section at $M_H = 115\,(160)$ GeV at 95% c.l. [7].

Furthermore the high accuracy of the electroweak data measured at LEP, SLC and the Tevatron [8] provides an indirect sensitivity to M_H: the Higgs boson contributes logarithmically, $\propto \log(M_H/M_W)$, to the radiative corrections to the W/Z boson propagators. A recent analysis, which uses the updated determination of the top quark mass (172.4 GeV), yields the value $M_H = 84^{+34}_{-26}$ GeV, corresponding to a 95% c.l. upper limit of $M_H \lesssim 154$ GeV [5]. A very recent analysis, using a new fitting programme gives the more precise value $M_H = 116.4^{+18.3}_{-1.3}$ GeV [9].

Figure 1 shows the global fit to the electroweak data and the ensuing limit on the Higgs mass. The limit increases to 185 GeV when the LEP2 direct search limit of 114 GeV is included. If the Higgs boson turns

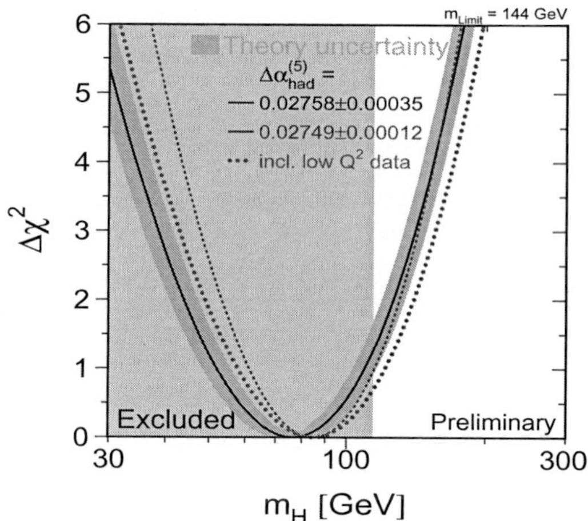

Figure 1. Experimental limits on the mass of the Higgs boson in the SM from a global fit to the electroweak precision data; the excluded region from direct Higgs searches is also shown [5]

out to be significantly heavier than these upper limits, $M_H \gtrsim 200$ GeV, there should be an additional new ingredient that is relevant at the EWSB scale which the next round of experiments should reveal.

From the theoretical side, interesting constraints can be derived from assumptions on the energy range within which the SM is valid before perturbation theory breaks down and new phenomena would emerge. For instance, if the Higgs mass were larger than ~ 1 TeV, the W and Z bosons would have to interact very strongly with each other so that their scattering at high energies respects unitarity. Imposing the unitarity requirement in the high-energy scattering of gauge bosons leads to the bound $M_H \lesssim 700$ GeV [10]. If the Higgs boson were too heavy, unitarity would be violated in these processes at energies above $\sqrt{s} \gtrsim 1.2$ TeV and new phenomena should appear to restore it. It is interesting to note, as an aside, that just the requirement of perturbative unitarity in WW scattering leads to a model with exactly the same particle content and couplings as the SM [11].

Another important theoretical constraint comes from the fact that the quartic Higgs self-coupling, which at the scale M_H is fixed by M_H itself, grows logarithmically with the energy scale. If M_H is small, the energy cut-off Λ at which the coupling grows beyond any bound and new phenomena should occur, is large; if M_H is large, the cut-off Λ is small. The condition $M_H \lesssim \Lambda$ sets an upper limit on the Higgs mass in the SM, the so-called triviality bound. A naive one-loop analysis assuming the validity of perturbation theory [12] as well as lattice simulations [13] lead to the

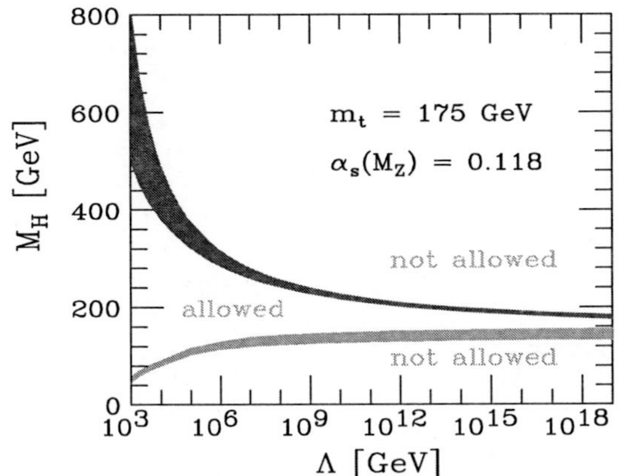

Figure 2. Theoretical upper and lower bounds on the Higgs mass in the SM from the assumption that the SM is valid up to the cut-off scale Λ [14]

estimate $M_H \lesssim 630$ GeV for this limit. Furthermore, loops involving top quarks tend to drive the coupling to negative values for which the vacuum is no longer stable.

Requiring the SM to be extended to, for instance, the GUT scale $\Lambda_{\rm GUT} \sim 10^{16}$ GeV and including the effect of top quark loops on the running coupling, the Higgs boson mass should lie in the range 130 GeV $\lesssim M_H \lesssim 180$ GeV [14] (Fig. 2). In fact, in any model beyond the SM in which the theory is required to be weakly interacting up to the GUT or Planck scales, the Higgs boson should be lighter than $M_H \lesssim 200$ GeV. Such a Higgs particle can be thus produced at the LHC.

Once its mass is fixed the profile of the Higgs particle is uniquely determined and its production rates and decay widths are fixed. As its couplings to different particles are proportional to their masses, the Higgs boson will have the tendency to decay into the heaviest particles allowed by phase space. The Higgs decay modes and their branching ratios (BR) are briefly summarised below; see Ref. [15] for details.

In the "low-mass" range, $M_H \lesssim 130$ GeV, the Higgs boson decays into a large variety of channels. The main mode is by far the decay into $b\bar{b}$ with BR$\sim 90\%$ followed by the decays into $c\bar{c}$ and $\tau^+\tau^-$ with BRs$\sim 5\%$. Also of significance is the top-loop mediated decay into gluons, which occurs at the level of $\sim 5\%$. The top and W-loop mediated $\gamma\gamma$ and $Z\gamma$ decay modes, which lead to clear signals are very rare with BRs of $\mathcal{O}(10^{-3})$.

In the "high-mass" range, $M_H \gtrsim 130$ GeV, the Higgs bosons decay into WW and ZZ pairs, one of the gauge bosons being possibly virtual below the thresholds. Above the ZZ threshold, the BRs are 2/3 for WW and 1/3 for ZZ decays, and the opening of the $t\bar{t}$ channel for higher M_H does not alter this pattern significantly.

In the low-mass range, the Higgs is very narrow, with $\Gamma_H < 10$ MeV, but this width increases, reaching 1 GeV at the ZZ threshold. For very large masses, the Higgs becomes obese, since $\Gamma_H \sim M_H$ and can hardly be considered as a resonance.

The BR and total decay widths are summarised in Fig. 3, which is obtained from a recently updated version of the code HDECAY [16] and where the new value $m_t = 172$ GeV is used as an input.

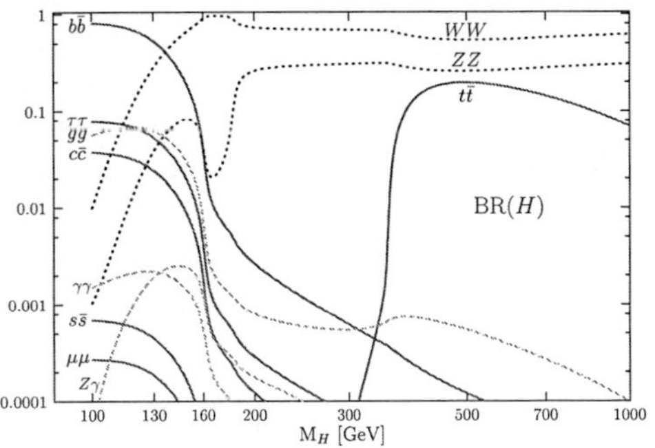

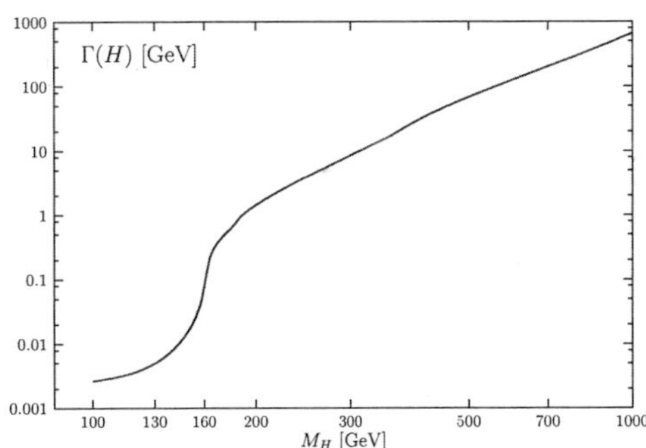

Figure 3. The decay branching ratios (top) and the total decay width (bottom) of the SM Higgs boson as a function of its mass [16]

The SM in spite of its spectacular success, is ridden with two well known problems, which are the major stumbling blocks while trying to extend the validity of the SM to the GUT scale $\Lambda_{\rm GUT}$. The first one is the so-called naturalness problem: the radiative corrections to M_H being quadratically divergent push the Higgs boson mass to be the order of these large scales. The second problem is that the running of the three gauge couplings of the SM is such that they do not meet at a single point and thus do not unify at the GUT scale.

Low energy supersymmetry solves these two problems at once [17]: supersymmetric particle loops cancel exactly the quadratic divergences and help stablise the Higgs boson mass at the weak scale, and they contribute to the running of the gauge couplings to allow their unification at $\Lambda_{\rm GUT}$. In addition, it allows for a good candidate for the dark matter in the universe.

2.2. The Higgs Particles in the MSSM

The minimal supersymmetric extension of the SM (MSSM), requires the existence of two isodoublet Higgs fields to cancel anomalies and to give mass separately to up and down–type fermions. Two CP-even neutral Higgs bosons h, H, a pseudoscalar A boson and a pair of charged scalar particles, $H^{\pm}$ are introduced by this extension of the Higgs sector [3,17,18]. In fact, in this case, the scalar potential does not involve an arbitrary self coupling λ as is the case with the SM, but involves only the gauge couplings and as a result the mass of the lightest Higgs boson h is bounded from above. Besides the four masses, the properties of the Higgs sector in the MSSM are determined by two more parameters: a mixing angle α in the neutral CP-even sector and the ratio of the two vacuum expectation values $\tan\beta$. The value of the latter lies in the range $1 \lesssim \tan\beta \lesssim m_t/m_b$.

Supersymmetry leads to several relations among these parameters and only two of them, taken in general to be M_A and $\tan\beta$ are in fact independent. These relations impose a strong hierarchical structure on the mass spectrum, $M_h < M_Z, M_A < M_H$ and $M_W < M_{H^{\pm}}$, which however is broken by radiative corrections as the top quark mass is large; see Ref. [19] for a review. The leading part of this correction grows as the fourth power of m_t and logarithmically with the SUSY scale or common squark mass M_S; the mixing (or trilinear coupling) in the stop sector A_t plays an important role. For instance, the upper bound on the mass of the lightest Higgs boson h is shifted from the tree level value M_Z to $M_h \sim$ 130–140 GeV in the maximal mixing scenario where $X_t = A_t - \mu/\tan\beta \sim 2M_S$ with $M_S = \mathcal{O}(1 \text{ TeV})$ [19]; see left panel of Fig. 4. The masses of the heavy neutral and charged Higgs particles are expected to range from M_Z to the SUSY breaking scale M_S.

The pseudoscalar Higgs boson A has no tree level couplings to gauge bosons, and its couplings to down (up) type fermions are (inversely) proportional to $\tan\beta$. This is also the case for the couplings of the charged Higgs boson to fermions, which are admixtures of scalar and pseudoscalar currents and depend only on $\tan\beta$. For the CP-even Higgs bosons h and H, the couplings to down (up) type fermions are enhanced (suppressed) compared to the SM Higgs couplings for $\tan\beta > 1$. They share the SM Higgs couplings to vector bosons as they are suppressed by sin and $\cos(\beta - \alpha)$ factors,

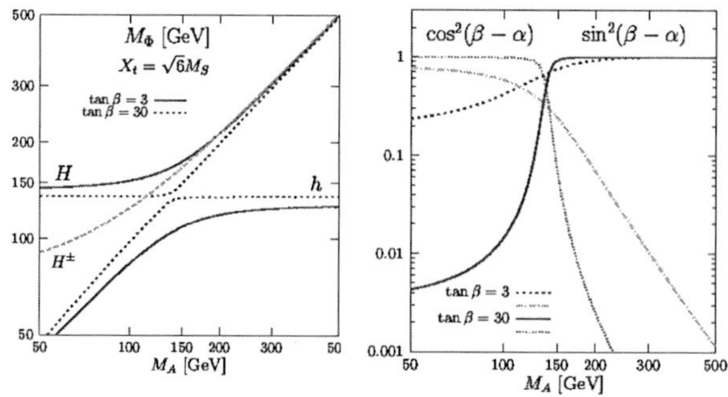

Figure 4. The masses and coupling of the Higgs bosons in the MSSM: as a function of M_A for $\tan\beta = 3, 30$ with $M_S = 2$ TeV and $X_t = \sqrt{6}M_S$

respectively for h and H. The Higgs couplings to the $W^{\pm}, Z$ bosons are displayed in the right panel of Fig. 4.

If the pseudoscalar mass is large, the h boson mass reaches its upper limit [which, depending on the value of $\tan\beta$ and stop mixing, is in the range 100–140 GeV] and its couplings to fermions and gauge bosons are SM-like; the heavier CP-even H and charged $H^{\pm}$ bosons become degenerate with the pseudoscalar A boson and have couplings to fermions and gauge bosons of the same intensity. In this *decoupling limit*, which can be already reached for masses $M_A \gtrsim 300$ GeV, it is very difficult to distinguish the Higgs sectors of the SM and MSSM if only the lighter h particle is observed.

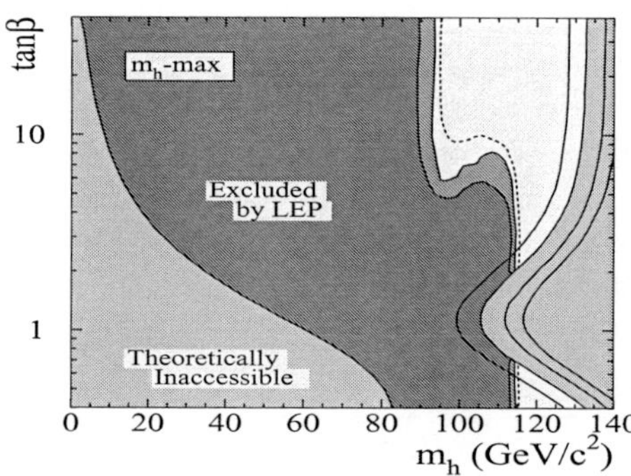

Figure 5. The MSSM exclusion contours from LEP at 95% (light green) and 99.7% c.l. (dark green) for the m_h–max scenario in the m_h–$\tan\beta$ plane [20]

Finally we note the experimental constraints on the MSSM Higgs masses, coming mainly from the negative LEP2 searches [20]. In the decoupling limit where the h boson is SM-like, the limit $M_h \gtrsim 114$ GeV from the Higgs–strahlung $e^+e^- \to hZ$ process holds; this constraint rules out $\tan\beta$ values smaller than $\tan\beta \sim 3$.

Combining all processes, the current limits in the CP conserving MSSM at 95% c.l., assuming no invisible decays are [8,20]: $M_h > 92.4$ GeV, $M_A > 93.4$ GeV for $\tan\beta > 0.4$ and $M_{H^\pm} > 79.3$ GeV. Figure 5 shows the current limits from LEP and Tevatron data on the MSSM Higgs sector.

Let us now turn to the decays of the MSSM Higgs particles (Fig. 6). The lightest h boson will decay mainly into fermion pairs since $M_h \lesssim 140$ GeV. This is, in general, also the dominant decay mode of the A and H bosons, since for $\tan\beta \gg 1$, they decay into $b\bar{b}$ and $\tau^+\tau^-$ pairs with BRs of the order of $\sim 90\%$ and 10%, respectively. For large masses, the top decay channels $H, A \to t\bar{t}$ open up, yet they are suppressed for large $\tan\beta$. The H boson can decay into gauge bosons or h boson pairs, and the A particle into hZ final states; however, these decays are strongly suppressed for $\tan\beta \gtrsim 5$. The $H^\pm$ particles decay into fermions pairs: mainly $t\bar{b}$ and $\tau\nu_\tau$ final states for $H^\pm$ masses, respectively, above and below the tb threshold. If allowed kinematically, they can also decay into hW final states for $\tan\beta \lesssim 5$. Adding up the various decays, the widths of all five Higgses remain rather narrow (Fig. 7).

Other possible decay channels for the heavy H, A and $H^\pm$ states are decays into light charginos and neutralinos, which could be important if not dominant; decays of the h boson into the invisible lightest neutralinos (LSP) can also be important, exceeding 50% in some parts of the parameter space and altering the searches at hadron colliders [21,22]. Decays into (third-generation) squarks and sleptons can also occur for the heavy Higgs states but are in general suppressed, in particular at high $\tan\beta$. See Ref. [18] for more details.

Note finally, that light SUSY particles can also affect the branching ratio of the loop-induced modes in a sizable way [23]. In particular, light stops can significantly affect the $h \to gg$ decay mode while light stops and charginos can alter the photonic Higgs decays.

2.3. Higgs Bosons in the CP-violating MSSM

A *quantitative* explanation of baryogenesis requires physics beyond the SM, one possibility being an additional source of CP violation beyond the one present in the SM via CKM mixing. CP-violation in the SUSY sector is one such source which allows explanation of baryogenesis at the electroweak scale. A general two Higgs doublet model seems to be able to generate adequate amount of baryon asymmetry in the universe (BAU) and be consistent with the current experimental constraints such as electric dipole moments [24]. In the MSSM, it may be possible to satisfy all the low energy constraints and still have sufficient CP viola-

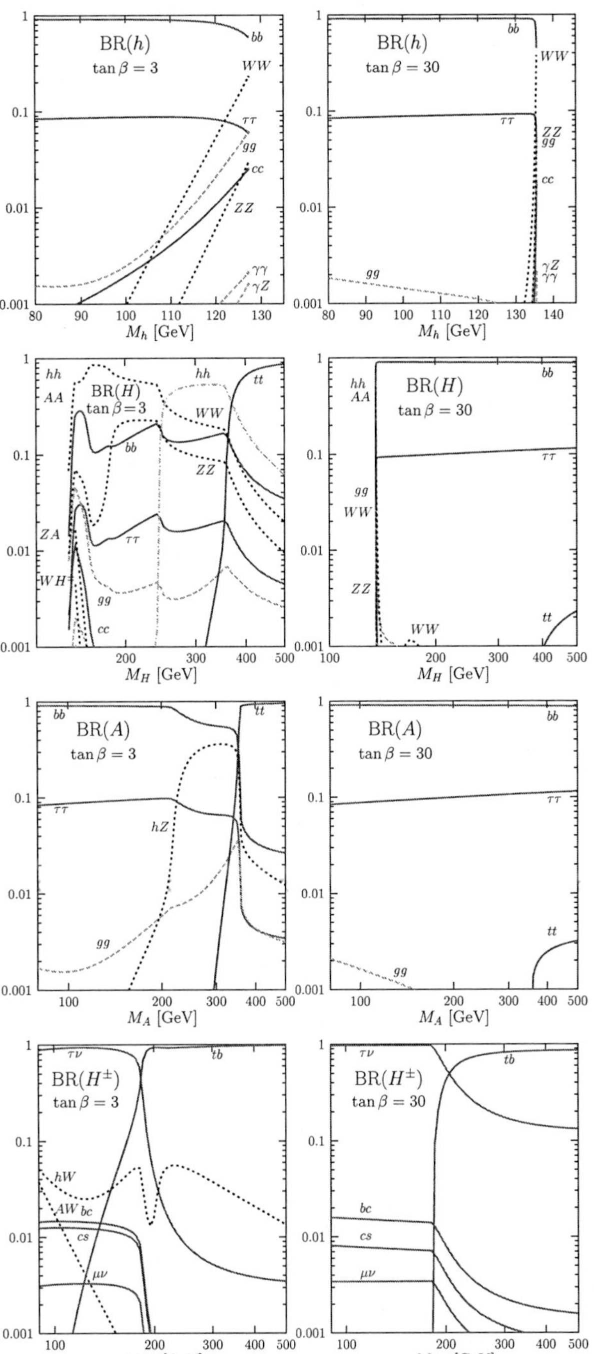

Figure 6. The decay branching ratios of the MSSM Higgs bosons as functions of their masses for $\tan\beta = 3$ and 30 as obtained with an update of `HDECAY` [16]; $m_t = 172$ GeV and the maximal mixing scenario $X_t = \sqrt{6}M_S$ with $M_S = 2$ TeV are assumed

tion in the theory to explain the BAU quantitatively (without requiring too much fine-tuning, one needs to go to non-minimal versions [25]). This further causes

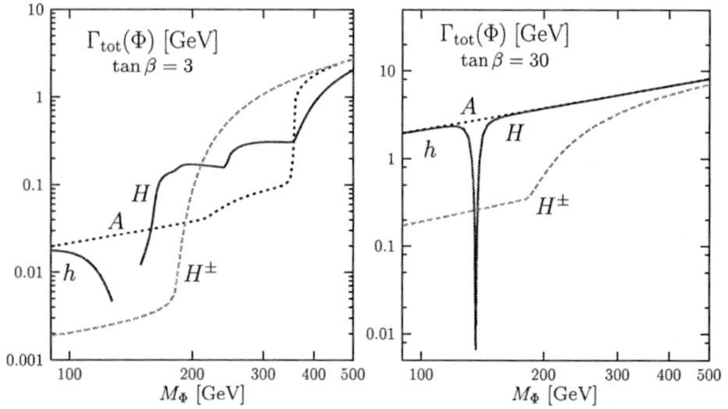

Figure 7. The total widths of the MSSM Higgs bosons as functions of their masses for the inputs of Fig. 6

new phases to enter the MSSM Higgs sector, which is CP–conserving at tree-level, through the large radiative corrections. These phases affect both the masses and the couplings of the neutral and charged Higgs particles, thus having very serious implications for the Higgs phenomenology at the LHC. This issue has received a lot of attention in the recent times [26–30].

Since CP is violated, the three mass eigenstates H_1, H_2, H_3 need no longer have definite CP quantum numbers and can be a mixture of the h, H, A states. The subscript i indicates the order of the mass m_{H_i} of the H_i boson in the spectrum, i.e. $m_{H_1} < m_{H_2} < m_{H_3}$. It is obvious that this will lead to significant modification of the properties of the various Higgs particles. Effect of this mixing on the couplings of the mixed CP states H_1, H_2, H_3 with a pair of gauge bosons/fermions i.e. $H_i f\bar{f}$, $H_i VV$, can change the Higgs phenomenology profoundly. For details, see Refs. [26–30].

In multi-Higgs doublet models, there exist in general sum rules which force the different H_i bosons to share among themselves the coupling of the SM Higgs boson to the massive gauge bosons [31], $\sum_i g^2_{H_i VV} = g^2_{H_{\rm SM}}$. However, it is only the CP-even component that is projected out. A CP violating MSSM is distinguished from a general CP violating two-Higgs doublet model by the fact that the former has a prediction for the mixing in terms of SUSY-breaking CP-violating phases of the MSSM. The possible dilution of the LEP limits on the Higgs masses due to CP violation had been discussed in a model independent formulation [32]. The specific feature of the CP-violating MSSM is the prediction for the mixing in terms of the SUSY parameters and CP breaking phases that they have.

As examples of new features in the CP violating MSSM, compared to the usual MSSM, we simply mention the possibility of a relatively light H_1 state with very weak couplings to the gauge bosons, and which could have escaped detection at LEP2 [20,30,33] and the possibility of resonant H/A mixing when the two Higgs particles are degenerate in mass [34]. An example of the Higgs mass spectrum in the so-called CPX scenario in which H_1 can be light is shown in Fig. 8 (left) as a function of the phase of the coupling A_t.

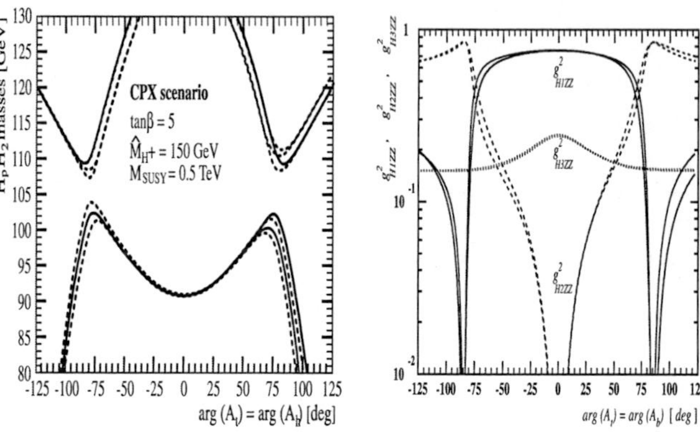

Figure 8. The mass spectrum of neutral Higgs particles and their couplings to the gauge bosons in the CP-violating MSSM CPX scenario (with $\text{Arg}(A_t) = \text{Arg}(A_b) = \text{Arg}(\mu)$ while $\text{Arg}(M_3) = 0$ or $\frac{\pi}{2}$, all the other parameters are indicated on the figure) [30]

Figure 8 (right) shows the result for the Higgs couplings to gauge bosons in the same CPX scenario for two different values of the gluino mass phase. In fact, the non-observation of a Higgs boson signal in the direct searches at the LEP now needs to be reinterpreted in the MSSM with CP violation. The analysis from LEP [20,33] shows that indeed there are holes in the excluded region at small $\tan\beta$ and m_{H_1} in the $\tan\beta$–M_{H_1} plane, consistent with the non-observations of a signal (Fig. 9). This corresponds to the H_1 decoupled from the Z boson as mentioned above.

Of course such features will have to be proved to be the result of CP-violation by, for instance, studying CP-odd observables and associated phenomenology.

2.4. The Higgs Sector in Non-minimal SUSY

The Higgs sector in SUSY models can be more complicated than previously discussed if some basic assumptions of the MSSM, such as the presence of only two Higgs doublet fields, or R-parity conservation are relaxed. A few examples are listed below.

2.4.1. Next-to-minimal SUSY extension

The next-to-minimal SUSY extension (NMSSM), in which the spectrum of the MSSM is extended by

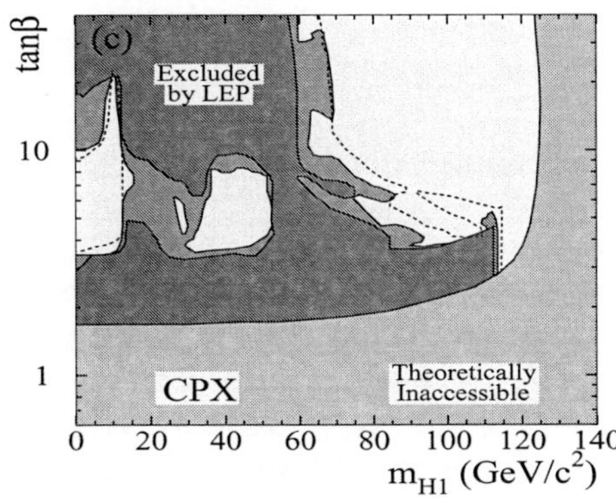

Figure 9. Regions in the $\tan\beta$–M_{H_1} plane disallowed theoretically or excluded by the current LEP searches [20]. The allowed 'hole' at the low $M_{H^+}, \tan\beta$ values can be seen very clearly

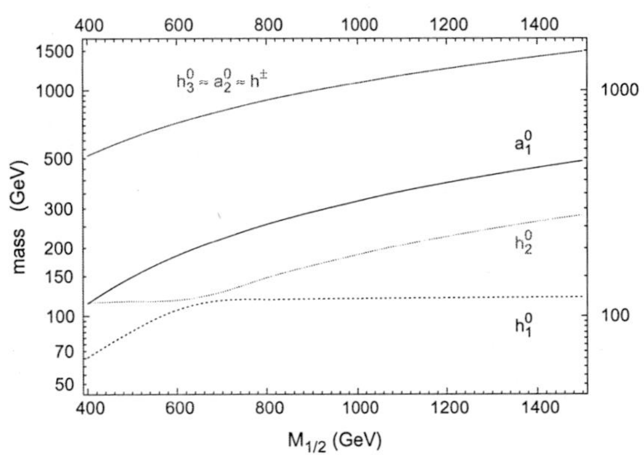

Figure 10. The Higgs masses as a function of the gaugino mass parameter $M_{1/2}$ in the cNMSSM [37]

one singlet superfield [35], was among the first SUSY models based on supergravity-induced SUSY-breaking terms. It has gained a renewed interest in the last decade, since it solves in a natural and elegant way the so-called μ problem [36] of the MSSM; in the NMSSM this parameter is linked to the vev of the singlet Higgs field, generating a μ value close to the SUSY-breaking scale. Furthermore, when the soft breaking terms are assumed to be universal at the GUT scale, the resulting constrained model (cNMSSM) is very constrained as one single parameter (e.g. the gaugino mass $M_{1/2}$) allows to fully describe its phenomenology [37].

The NMSSM leads to an interesting phenomenology [38,39] as the MSSM spectrum is extended to include an additional CP-even and CP-odd Higgs states as well as a fifth neutralino, the singlino. An example of the Higgs mass spectrum in the cNMSSM [37] is shown in Fig. 10 as a function of the gaugino mass parameter. As in the MSSM in the decoupling regime, the heaviest CP-even, CP-odd and charged Higgs states form a practically degenerate SU(2) multiplet with a common mass beyond 500 GeV; the lightest CP-even state is mostly SM-like, with a mass increasing slightly with $M_{1/2}$ from 115 GeV up to $\sim$ 120 GeV. The third CP-even state has a dominant singlet component: for small $M_{1/2}$ it is lighter than the SM-like Higgs boson, escaping LEP constraints due to the very small coupling to the Z boson. For increasing values of $M_{1/2}$, its mass increases until it becomes comparable and eventually exceeds the mass of SM-like CP-even Higgs state.

However, in the unconstrained NMSSM, the effect of the additional singlet to the scalar potential leads a relaxation of the upper bound on the mass of the lighter CP-even particle above that of the MSSM h boson [40]. As a result the constraints in the M_A–$\tan\beta$ plane, implied by the negative results of the LEP2 searches are less restrictive as compared to those in the MSSM [26,41]. In addition, there exists a small region not yet completely excluded, where the lightest CP-even Higgs boson might have escaped the LEP2 searches. Even more interestingly, there exist possibilities that one of the neutral Higgs particles, in particular the lightest pseudoscalar A_1, is very light with a mass of a few ten's of GeV. The light CP-even Higgs, which is SM-like in general, could then decay into pairs of A_1 bosons, $H_1 \to A_1 A_1 \to 4b, 4\tau$, with a large branching fraction.

2.4.2. Higgs bosons in GUT theories

A large variety of theories, string theories, grand unified theories, left–right symmetric models, etc. suggest an additional gauge symmetry which may be broken only at the TeV scale. This leads to an extended particle spectrum and, in particular, to additional Higgs fields beyond the minimal set of the MSSM. Especially common are new U(1)' symmetries broken by the vev of a singlet field (as in the NMSSM) which leads to the presence of a Z' boson and one additional CP-even Higgs particle compared to the MSSM; this is the case, for instance, in the exceptional MSSM [42] based on the string inspired E_6 symmetry. The secluded $SU(2) \times U(1) \times U(1)'$ model [43], in turn, includes four additional singlets that are charged under U(1)', leading to 6 CP-even and 4 CP-odd neutral Higgs states. Other exotic Higgs sectors in SUSY models [44] are, for instance, Higgs representations that transform as SU(2) triplets or bi-doublets under the $SU(2)_L$ and $SU(2)_R$ groups in left-right symmetric models, that are motivated by the seesaw approach to explain the small neutrino masses and which lead e.g. to a doubly charged Higgs boson H^{--}. These extensions, which also predict extra matter fields, would lead to a very interesting phenomenology and new collider signatures in the Higgs sector. We will not be discussing much about this subject in this review.

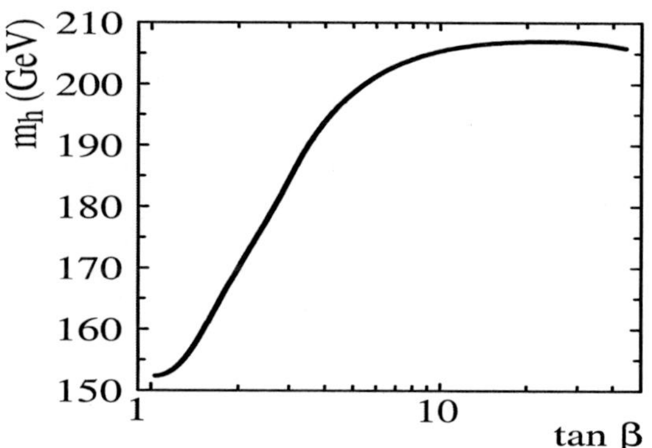

Figure 11. The upper bound on the lighter Higgs mass in a general SUSY model [45]

In a general SUSY model, with an arbitrary number of singlet and doublet scalar fields (as well as a matter content which allows for the unification of the gauge couplings), a linear combination of Higgs fields has to generate the W/Z masses and thus, from the triviality argument discussed earlier, a Higgs particle should have a mass below 200 GeV and significant couplings to gauge bosons [45]. The upper bound on the mass of the lightest Higgs boson in this most general SUSY model is displayed in Fig. 11 as a function of $\tan\beta$.

2.4.3. R–parity violating models

Models in which R–parity is spontaneously broken [and where one needs to either enlarge the SM symmetry or the spectrum to include additional gauge singlets], allow for an explanation of the light neutrino data [46]. Since $\not{R}_p$ entails the breaking of the total lepton number L, one of the CP-odd scalars, the Majoron J, remains massless being the Goldstone boson associated to $\not{L}$. In these models, the neutral Higgs particles have also reduced couplings to the gauge bosons. More importantly, the CP-even Higgs particles can decay into pairs of invisible Majorons, $H_i \to JJ$, while the CP-odd particle can decay into a CP-even Higgs and a Majoron, $A_i \to H_i J$ and three Majorons, $A \to JJJ$ [46].

2.5. Higgs Bosons in Alternative Scenarios

There are also many non supersymmetric extensions of the SM which might lead to a different Higgs phenomenology. In some cases, the Higgs sector would consist of one scalar doublet leading to a Higgs boson which would mimic the SM Higgs, but the new particles that are present in the models might alter some of its properties. In other cases, the Higgs sector is extended to contain additional scalar fields leading to the presence of new Higgs particles. Another possibility is a scenario with a composite and strongly interacting Higgs, or where no Higgs particle is present at all, leading to strong interactions of the W/Z bosons. Below we give a non-exhaustive list of various possible scenarios.

2.5.1. Scenarios with Higgs mixing

In warped extra-dimensional models [47] the fluctuations of the size of the extra dimension about its stabilised value manifest themselves as a single scalar field the radion. In the Randall-Sundrum model with a bulk scalar field, it is expected that the radion is the lightest state beyond the SM fields with a mass probably in the range between $\mathcal{O}(10\text{ GeV})$ and $\Lambda = \mathcal{O}(\text{TeV})$ [48–50]. The couplings of the radion are order of $1/\Lambda$ and are very similar to the couplings of the SM Higgs boson, except for one important difference: due to the trace anomaly, the radion directly couples to massless gauge bosons at one loop. Moreover, in the low energy four-dimensional effective theory, the radion can mix with the Higgs boson. This mixing can lead to important shifts in the Higgs couplings which become apparent in the Higgs decay widths and production cross sections (Fig. 12).

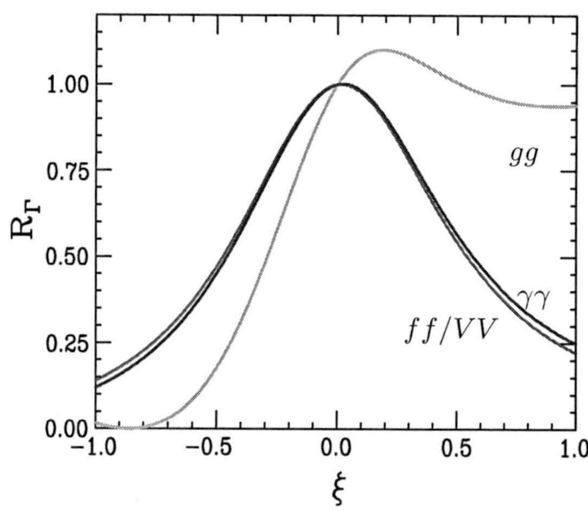

Figure 12. The ratio R_Γ of Higgs partial widths to their SM values, as a function of the Higgs-radion mixing parameter ξ with $M_H = 125$ GeV, $M_\phi = 300$ GeV and a scale $v/\Lambda = 0.2$ [48]

Another important consequence of radion mixing is the decays of the Higgs boson into a pair of radions. Indeed, if the radion is relatively light, the decays $H \to \phi\phi$ might be kinematically accessible and, for some mixing values the branching fractions might be substantial. This is exemplified in Fig. 13 where $\text{BR}(H \to \phi\phi)$ is displayed in a specific scenario.

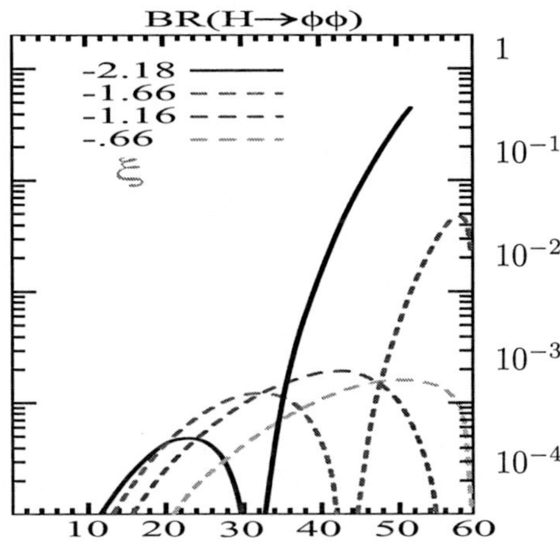

Figure 13. The branching fractions for the decays $H \to \phi\phi$ as a function of M_ϕ for different ξ values and $M_H = 120$ GeV, $\Lambda = 5$ TeV [50]

In large extra dimension models [51], mixing of the Higgs boson with graviscalars also occurs [52], leading to an invisible decay width. Mixing effects also occur if the SM is minimally extended in a renormalisable way to contain a singlet scalar field S that does not couple to the other SM particles; its main effect would be to alter the scalar potential and to mix with the SM Higgs field [53] and, in such a case, the Higgs could mainly decay into two invisible S particles.

2.5.2. Scenarios with extended Higgs/gauge/matter

Non-supersymmetric extensions of the Higgs sector with additional singlet, doublet and higher representation fields have also been advocated [44]. Examples are the minimal SM extension with a singlet discussed above, two–Higgs doublet models which potentially include CP-violation, triplet Higgs fields in models for light neutrino mass generation, etc. These extensions lead to a rich spectrum of Higgs particles which could be produced at the LHC. In other extensions of the SM, new gauge bosons and new matter particles are predicted and they can affect the properties of the SM-like Higgs boson. For instance the new fermions present in little Higgs and extra-dimensional models might contribute to the loop induced Higgs couplings, while new heavy gauge bosons could alter the Higgs couplings to W and Z bosons for instance. The anomalous ZZH and $t\bar{t}H$ couplings can be a good probe of the additional scalars and/or the novel features of the geometry in the extra dimensions [54].

2.5.3. Scenarios with a composite Higgs boson

In little Higgs models [55], the dynamical scale is around $\Lambda = 10$ TeV, unlike the traditional Technicolour model [56]. A light Higgs boson can be generated as a pseudo Goldstone boson and its mass of order 100 GeV is protected against large radiative corrections individually in the boson and the fermion sectors. The models predict a rich spectrum of new particles not only at the scale Λ but also at lower scales. Axion-type pseudoscalar bosons may be associated with the spontaneous breaking of U(1) factors in the extra global symmetries [57]. These particles have properties analogous to Higgs bosons and can be produced at the LHC; deviations in the production and decay rates of the SM-like Higgs boson can also be induced by these particles. Note that, recently, a model-independent description of a strongly interacting light Higgs has been given [58].

2.5.4. Higgsless models and strong W/Z interactions

Assuming the W/Z bosons to become strongly interacting at TeV energies, damping the rise of the elastic W/Z scattering amplitudes, is an alternative way to solve the problem of unitarity violation at high energies in the SM, without adding a relatively light Higgs boson. Naturally, the strong forces between the massive gauge bosons may be traced back to new fundamental interactions characterised by a scale of order 1 TeV [56]. Also in theories with extra space dimensions, EWSB can occur without introducing additional fundamental scalar fields, leading also to Higgsless theories [59]. Studying such difficult scenarios at the LHC will be possible with very high luminosity [60].

3. Higgs Production and Detection at the LHC

3.1. The SM Higgs Case

There are essentially four mechanisms for the single production of the SM Higgs boson at hadron colliders [61]; some Feynman diagrams are shown in Fig. 14.

The total cross sections, obtained with the programmes of Ref. [62], are displayed in Fig. 15 for the LHC with $\sqrt{s} = 14$ TeV as a function of the Higgs mass; the top quark mass is set to $m_t = 178$ GeV and

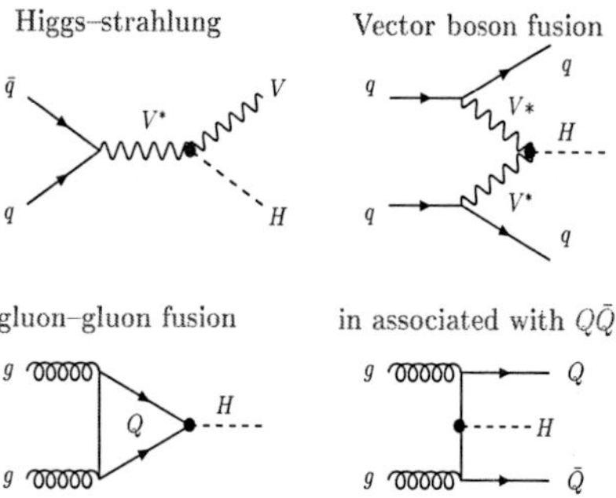

Figure 14. The production mechanisms for SM Higgs bosons at hadron colliders

the MRST parton distributions functions [63] have been adopted. The NLO and eventually NNLO, corrections have been implemented as will be summarised below, where we discuss the main features of each production channel.

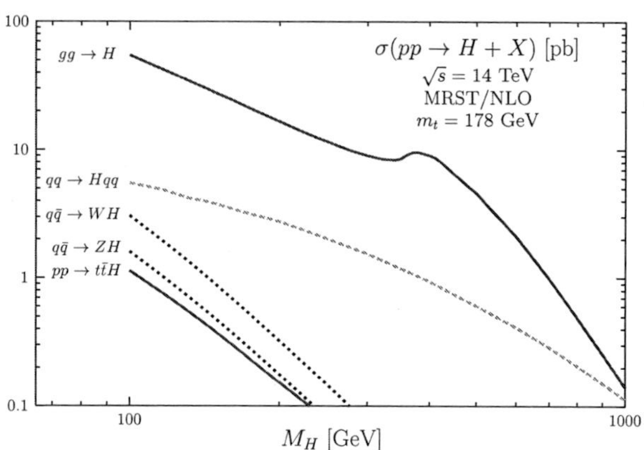

Figure 15. The production cross sections for the SM Higgs boson at the LHC in the main channels

3.1.1. $gg \to H$

This is by far the dominant production process at the LHC, up to masses $M_H \approx 1$ TeV. The most promising detection channels are [64] $H \to \gamma\gamma$ for $M_H \lesssim 130$ GeV and slightly above this mass value, $H \to ZZ^* \to 4\ell^\pm$ and $H \to WW^{(*)} \to \ell\ell\nu\nu$ with $\ell = e, \mu$ for masses below, respectively, $2M_W$ and $2M_Z$. For higher Higgs masses, $M_H \gtrsim 2M_Z$, it is the golden mode $H \to ZZ \to 4\ell^\pm$, which from $M_H \gtrsim 500$ GeV can be complemented by $H \to ZZ \to \nu\bar{\nu}\ell^+\ell^-$ and $H \to WW \to \nu\ell jj$ to increase the statistics [65–70].

The next-to-leading order (NLO) QCD corrections have been calculated in both the limit where the internal top quark has been integrated out [71], an approximation which should be valid in the Higgs mass range $M_H \lesssim 300$ GeV, and in the case where the full quark mass dependence has been taken into account [72]. The corrections lead to an increase of the cross sections by a factor of ~ 1.7. The "tour de force" of deriving the three-loop corrections has been preformed in the infinite top-quark mass limit; these NNLO corrections lead to the increase of the rate by an additional 30% [73] (see also Refs. [74,75]). This results in a nice convergence of the perturbative series and a strong reduction of the scale uncertainty, which is the measure of unknown higher order effects (Fig. 16). The resummation of the soft and collinear corrections, performed at next-to-next-to-leading logarithm accuracy, leads to another increase of the rate by $\sim 5\%$ and a decrease of the scale uncertainty [76]. The QCD corrections to the differential distributions, and in particular to the Higgs transverse momentum and rapidity distributions, have also been recently calculated at NLO [with a resummation for the former] and shown to be rather large [77]. The dominant components of the electroweak corrections, some of which have been derived very recently are comparatively very small [78].

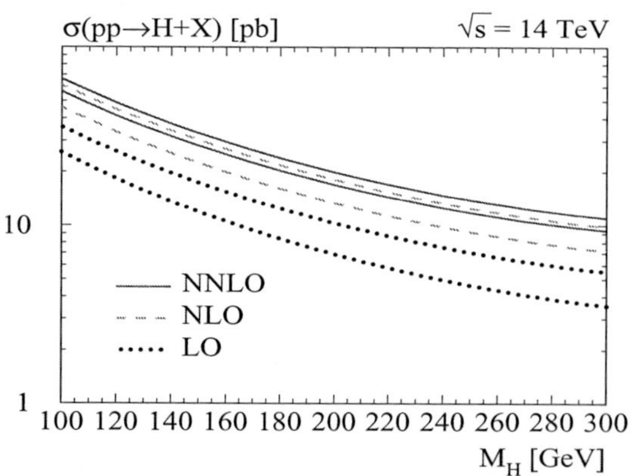

Figure 16. SM Higgs production cross sections in the gg fusion process at the LHC as a function of M_H at the three different orders with the upper (lower) curves are for the choice of the renormalisation and factorisation scales $\mu = \frac{1}{2}M_H$ ($2M_H$); from Harlander and Kilgore in Ref. [73]

3.1.2. $q\bar{q} \to HV$

The associated production with gauge bosons, with $H \to b\bar{b}$ and possibly $H \to WW^* \to \ell^+\nu jj$, is the most relevant mechanism at the Tevatron [7] [$gg \to H \to$

$WW \to \ell\nu\ell\nu$ being important for Higgs masses close to 160 GeV]. At the LHC, this process plays only a marginal role; however, the channels $HW \to \ell\nu\gamma\gamma$ and eventually $\ell\nu b\bar{b}$ could be useful for the measurement of Higgs couplings.

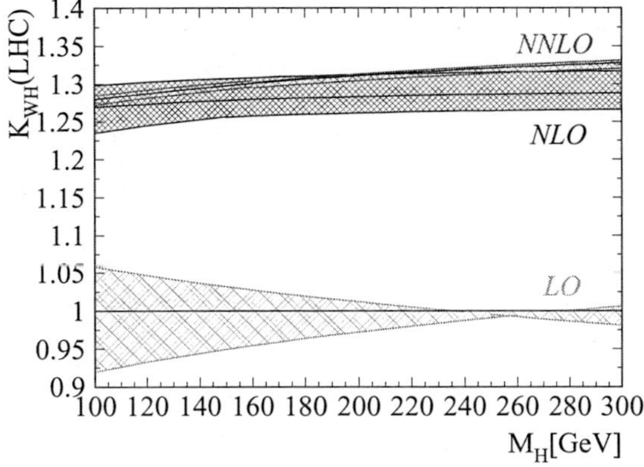

Figure 17. K-factors for $pp \to HW$ at the LHC as a function of M_H at LO, NLO and NNLO with the bands represent the spread of the cross section when the scales are varied in the range $\frac{1}{3}M_{HV} \leq \mu_R\,(\mu_F) \leq 3M_{HV}$ [81]

The QCD corrections, which at NLO [79,80], can be inferred from Drell–Yan production, have been calculated at NNLO [81]; they are $\sim 30\%$ in total (Fig. 17). The $\mathcal{O}(\alpha)$ electroweak corrections have been also derived recently [82] and decrease the rate by 5–10%. The remaining scale dependence is very small, making this process the theoretically cleanest of all Higgs production processes.

3.1.3. The WW/ZZ fusion mechanism

This process has the second largest cross section at the LHC. The QCD corrections, which can be obtained in the structure-function approach, are at the level of 10% and thus small [80,83] (the electroweak corrections [84] are at the level of a few percent). The corrections including cuts, and in particular corrections to the p_T and η distributions, have also been calculated and implemented into a parton-level Monte Carlo programme [85]. With the specific cuts to the process, the output for the production cross section is shown in Fig. 18 for a Higgs in the mass range 100–200 GeV.

For several reasons, the interest in this process has grown in recent years: it has a large enough cross section [a few picobarns for $M_H \lesssim 250$ GeV] and one can use cuts, forward-jet tagging, mini-jet veto for low luminosity as well as triggering on the central Higgs decay products [86], which render the backgrounds comparable to the signal, therefore allowing precision measurements. In the past, it has been shown that the decay $H \to \tau^+\tau^-$ and possibly $H \to \gamma\gamma, ZZ^*$ can be detected and could allow for coupling measurements [68,87,88]. In the last years, parton-level analyses have shown that various other channels can be possibly detected [89]: $H \to WW^*$ for $M_H \sim 125$–180 GeV, $H \to \mu^+\mu^-$ [for second-generation coupling measurements], $H \to b\bar{b}$ [for the $b\bar{b}H$ Yukawa coupling] and $H \to$ invisible (see later). Recent experimental simulations [67] have assessed more firmly the potential of this channel.

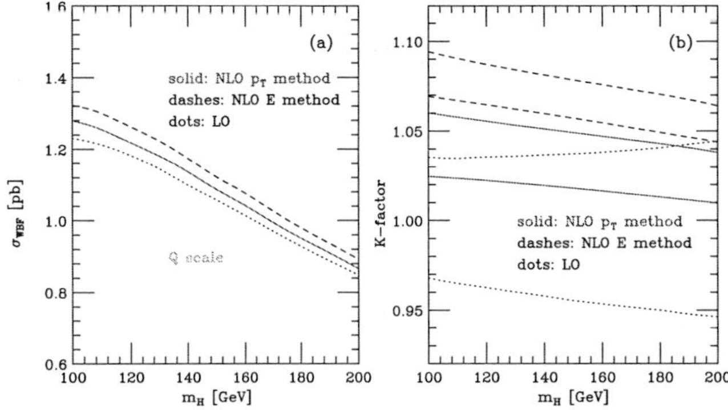

Figure 18. The $pp \to Hqq$ cross section after cuts as a function of M_H at LO (dotted line) and NLO with the tagging jets defined in the P_T and E_T methods (left) and the scale variation of the LO and NLO cross sections as a function of M_H (right) from Ref. [85]

3.1.4. $pp \to t\bar{t}H$

Finally, Higgs boson production in association with top quarks, with $H \to \gamma\gamma$ or $b\bar{b}$, can in principle be observed at the LHC and direct measurement of the top Yukawa coupling, as well as an unambiguous determination of the CP of the Higgs can be possible. (Recent analyses have however, shown that $pp \to t\bar{t}H \to t\bar{t}b\bar{b}$ might be subject to a too large jet background [66].) The cross section is rather involved at tree-level since it is a three-body process, and the calculation of the NLO corrections was a real challenge which was met a few years ago [90]. The K-factors turned out to be rather small, $K \sim 1.2$ at the LHC. However, the scale dependence is drastically reduced from a factor two at LO to the level of 10–20% at NLO (Fig. 19). Note that the NLO corrections to the $q\bar{q}/gg \to b\bar{b}H$ process, which is more relevant in the MSSM, have been also completed [91]: compared with the NLO rate for the $bg \to bH$ process where the initial b-quark is treated as a parton [92], the calculations agree within the scale uncertainties [93]. A similar situation occur for $H^\pm$ production in the gb process: the K–factor is moderate $\sim$1.2–1.5 if the cross section is evaluated at scales $\mu \sim \frac{1}{2}(m_t + M_{H^\pm})$ [94].

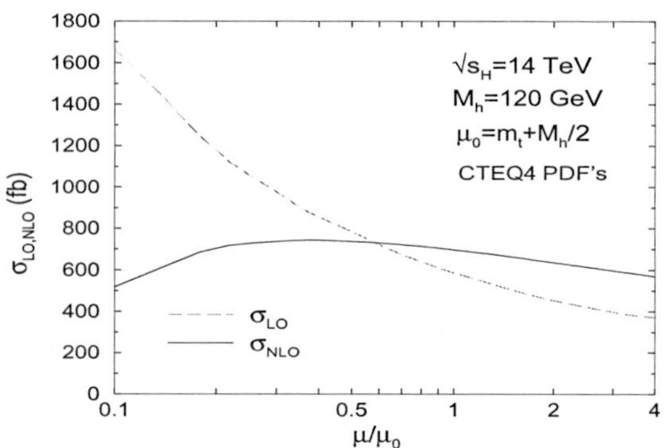

Figure 19. The Higgs production cross sections in the $t\bar{t}H$ process as a function of the renormalisation/factorisation scale μ; from Dawson et al. [90]

Note that the PDF uncertainties have also been estimated for the four production processes: at the LHC, the uncertainties range from 5–15% depending on the considered process and the Higgs mass [95].

All the various channels discussed above have been discussed in great detail over the past decades [65–70]. The significance for detecting the SM Higgs particle in the various production and decay channels is shown in Fig. 20, assuming a 100 fb^{-1} integrated luminosity.

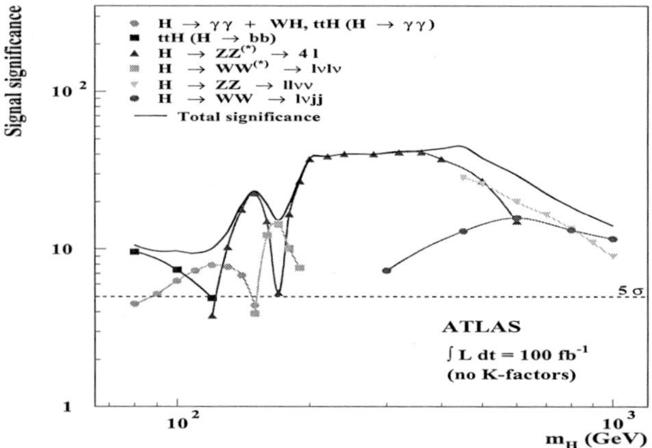

Figure 20. Significance for the experimental detection [65] of the SM Higgs boson at the LHC

3.2. The CP Conserving MSSM

In the CP conserving MSSM, the production processes for the CP-even h, H bosons are practically the same as for the SM Higgs and the ones depicted in Fig. 14 are all relevant. However, the b quark will play an important role for moderate to large $\tan\beta$ values as its Higgs couplings are enhanced. First, one has to take into account the b loop contribution in the $gg \to h, H$ process which becomes the dominant component in the MSSM [here the QCD corrections are available only at NLO where they have been calculated in the full massive case [72] and increase the rate by $\sim$ 1.5; SUSY-QCD corrections are discussed in Refs. [96,97]]. Moreover, in associated Higgs production with heavy quarks, $b\bar{b}$ final states must be considered, $pp \to b\bar{b}+h/H$, and this process for either h or H becomes the dominant one in the MSSM [here the QCD corrections are available in both the gg and $gb \to b\Phi, b\bar{b} \to \Phi$ pictures [91–93] depending on how many b-quarks are to be tagged, and which are equivalent if the renormalisation and factorisation scales are chosen to be small, $\mu \sim \frac{1}{4}M_\Phi$]. The rates for associated production with $t\bar{t}$ and W/Z pairs as well as for WW/ZZ fusion processes are suppressed for at least one of the particles as a result of coupling reduction.

Because of CP invariance which forbids AVV couplings, the A boson cannot be produced in the Higgs-strahlung and vector boson fusion processes; the rate for the $pp \to t\bar{t}A$ process is suppressed by the small $At\bar{t}$ couplings for $\tan\beta \gtrsim 3$. Hence, only the $gg \to A$ fusion with the b-quark loops included [where the QCD corrections are also available only at NLO and are approximately the same as for the CP-even Higgs boson with enhanced b-quark couplings] and associated production with $b\bar{b}$ pairs, $pp \to b\bar{b} + A$ [where the QCD corrections are the same as for one of the CP-even Higgs bosons as a result of chiral symmetry] provide large cross sections. The one-loop induced processes $gg \to AZ, gg \to Ag$ [which hold also for one of the CP-even Higgses] and associated production with other Higgs particles, $pp \to A + h/H/H^+$ are possible but the rates are much smaller in general, in particular for $M_A \gtrsim 200$ GeV [98].

For the charged Higgs boson, the dominant channel is the production from top quark decays, $t \to H^+b$, for masses not too close to $M_{H^\pm} = m_t - m_b$; this is particularly true at low or large $\tan\beta$ when the $t \to H^+b$ branching ratio is significant. For higher masses [99] the processes to be considered is the fusion process $gg \to H^\pm tb$ supplemented by $gb \to H^\pm t$. The two processes have to be properly combined and the NLO corrections for both processes have been derived [94] and are moderate, increasing the cross sections by 20–50% if they are evaluated at low scales, $\mu \sim \frac{1}{2}(m_t + M_{H^\pm})$. Additional sources [100] of $H^\pm$ states for masses below $M_{H^\pm} \approx 250$ GeV are provided by pair and associated production with neutral Higgs bosons in $q\bar{q}$ annihilation as well as H^+H^- pair and associated $H^\pm W^\mp$ production in gg and/or $b\bar{b}$ fusion but the cross sections are not as large, in particular for $M_{H^\pm} \gtrsim m_t$.

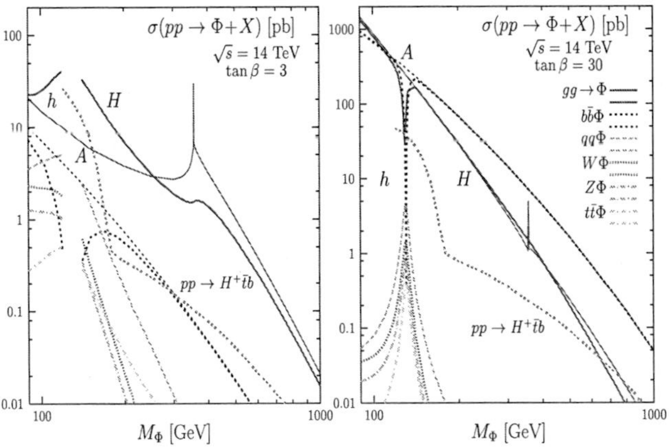

Figure 21. The cross section for the neutral and charged MSSM Higgs production in the main channels at the LHC as a function of their respective masses for $\tan\beta = 3$ and 30 in the maximal mixing scenario

The cross sections for the dominant production mechanisms are shown in Fig. 21, as a function of the Higgs masses for $\tan\beta = 3$ and 30 for the same set of input parameters as used for the widths and BR. The NLO QCD corrections are included, except for the $pp \to Q\bar{Q}$Higgs processes where, however the scales have been chosen as to approach the NLO results; the MRST NLO structure functions have been adopted. As can be seen, at high $\tan\beta$, the largest cross sections are by far those of the $gg \to \Phi_A/A$ and $q\bar{q}/gg \to b\bar{b}+\Phi_A/A$ processes, where $\Phi_A = H\,(h)$ in the (anti-)decoupling regimes $M_A > (<) M_h^{\max}$: the other processes involving these two Higgs bosons have cross sections that are several orders of magnitude smaller. The production cross sections for the other CP-even Higgs boson, that is $\Phi_H = h\,(H)$ in the (anti-)decoupling regime when $M_{\Phi_H} \simeq M_h^{\max}$ are similar to those of the SM Higgs boson with the same mass and are substantial in all the channels which have been displayed. At small $\tan\beta$, the gg fusion and $b\bar{b}$-Higgs cross sections are not strongly enhanced as before and all production channels, except for $b\bar{b}$-Higgs which is only slightly enhanced, have cross sections that are smaller than in the SM Higgs case, except for h in the decoupling regime.

The principal detection signals of the neutral Higgs bosons at the LHC, in the various regimes of the MSSM are as follows [18,65–68,101].

3.2.1. Decoupling regime

One of the most interesting region is the decoupling regime, i.e. when $M_h \simeq M_h^{\max}$, the lighter h boson is SM-like and has a mass smaller than ≈ 140 GeV. It can be detected in the $h \to \gamma\gamma$ decays [possibly supplemented with a lepton in associated Wh and $t\bar{t}h$ production], and eventually in $h \to ZZ^*, WW^*$ decays in the upper mass range, and if the vector boson fusion processes are used, also in the decays $h \to \tau^+\tau^-$ and eventually $h \to WW^*$ in the higher mass range $M_h \gtrsim 130$ GeV (Fig. 22). For relatively large values of $\tan\beta$ ($\tan\beta \gtrsim 10$), the heavier CP-even H boson which has enhanced couplings to down-type fermions, as well as the pseudoscalar Higgs particle, can be observed in the process $pp \to b\bar{b} + H/A$ where at least one b-jet is tagged and with the Higgs boson decaying into $\tau^+\tau^-$, and eventually, $\mu^+\mu^-$ pairs in the low mass range. With a luminosity of 30 fb^{-1} (and in some cases lower) a large part of the $[\tan\beta, M_A]$ space can be covered as can be seen from Fig. 23.

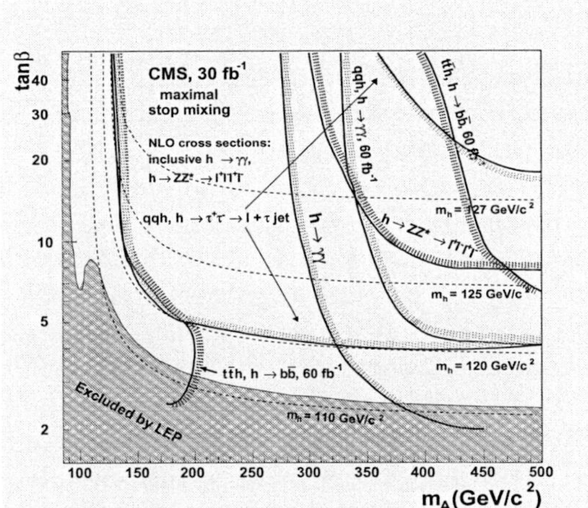

Figure 22. The areas in the $(M_A, \tan\beta)$ parameter space where the lighter MSSM Higgs boson can be discovered at the LHC with an integrated luminosity of 30 fb^{-1} in the standard production channels [66]

3.2.2. Anti-decoupling regime

In the anti-decoupling regime, i.e. when $M_A < M_h^{\max}$ and at high $\tan\beta$ ($\gtrsim 10$), it is the heavier H boson which will be SM-like and can be detected as above, while the h boson will behave like the pseudoscalar Higgs particle and can be observed in $pp \to b\bar{b} + h$ with $h \to \tau^+\tau^-$ or $\mu^+\mu^-$ provided its mass is not too close to M_Z not to be swamped by the background from Z production. The part of the $[\tan\beta, M_A]$ space which can be covered is also shown in Fig. 23 and corresponds to $M_A \lesssim 130$ GeV.

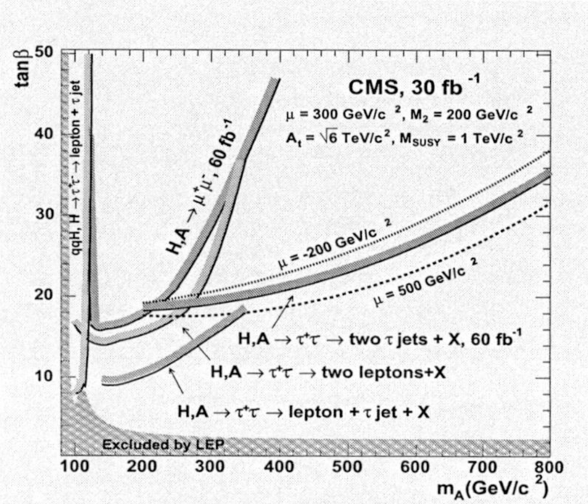

Figure 23. The areas in the $(M_A, \tan\beta)$ parameter space where the heavier MSSM neutral Higgs bosons can be discovered at the LHC with a luminosity of 30 fb^{-1} in the standard production channels [66]

3.2.3. Intermediate coupling regime

In the intermediate coupling region, that is for not too large M_A values and moderate $\tan\beta \lesssim 5$, the interesting decays $H \to hh$, $A \to hZ$ and even $H/A \to t\bar{t}$ [as well as the decays $H^\pm \to Wh$] still have sizable branching fractions and can be searched for. In particular, the $gg \to H \to hh \to b\bar{b}\gamma\gamma$ process (the $4b$ channel is more difficult as a result of the large background) is observable for $\tan\beta \lesssim 3$ and $M_A \lesssim 300$ GeV, and would allow to measure the trilinear Hhh coupling. These regions of parameter space may need to be reconsidered in the light of the new Tevatron value for the top mass.

3.2.4. Intense-coupling regime

In the intense-coupling region, that is for $M_A \sim M_h^{\max}$ and $\tan\beta \gg 1$, the three neutral Higgs bosons $\Phi = h, H, A$ have comparable masses and couple strongly to isospin $-\frac{1}{2}$ fermions leading to dominant decays into $b\bar{b}$ and $\tau\tau$ and large total decay widths [101,102]. The three Higgs bosons can only be produced in the channels $gg \to \Phi$ and $gg/q\bar{q} \to b\bar{b} + \Phi$ with $\Phi \to b\bar{b}, \tau^+\tau^-$ as the interesting $\gamma\gamma, ZZ^*$ and WW^* decays of the CP-even Higgses are suppressed. Because of background and resolution problems, it is very difficult to resolve between the three particles. A solution advocated in Ref. [101] (see also Ref. [103]), would be to search in the channel $gg/q\bar{q} \to b\bar{b} + \Phi$ with the subsequent decay $\Phi \to \mu^+\mu^-$ which has a small BR, $\sim 3 \times 10^{-4}$, but for which the better muon resolution, $\sim 1\%$, would allow to disentangle between at least two Higgs particles. The backgrounds are much larger for the $gg \to \Phi \to \mu^+\mu^-$ signals. The simultaneous discovery of the three Higgs particles is very difficult and in many cases impossible, as exemplified in Fig. 24 where one observes only one single peak corresponding to h and A production.

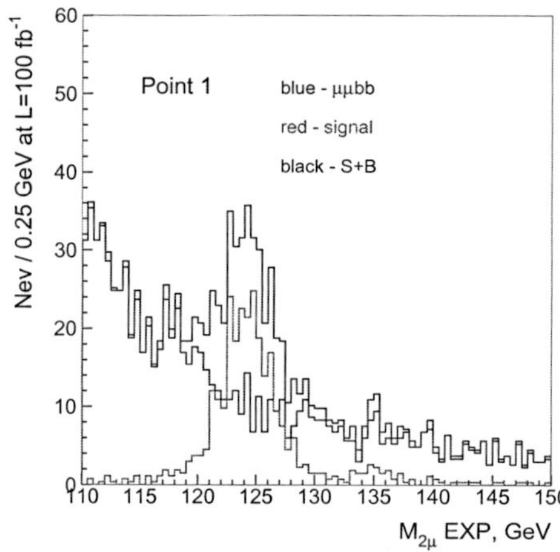

Figure 24. The $\mu^+\mu^-$ pair invariant mass distributions for the three Higgs signal peaks with $M_A = 125$ GeV and $\tan\beta = 30$ (leading to $M_h \sim 124$ GeV and $M_H \sim 134$ GeV) and backgrounds after detector resolution smearing; from Ref. [101]

Finally, as mentioned previously, light $H^\pm$ particles with masses below $M_{H^\pm} \sim m_t$ can be observed in the decays $t \to H^+ b$ with $H^- \to \tau\nu_\tau$ (Fig. 25). Heavier $H^\pm$ states can be probed for large enough $\tan\beta$, by considering the properly combined $gb \to tH^-$ and $gg \to t\bar{b}H^-$ processes using the decay $H^- \to \tau\nu_\tau$ and taking advantage of the τ polarisation to suppress the backgrounds, and eventually the decay $H^- \to \bar{t}b$ which however, seems more problematic as a result of the large QCD background. See Ref. [104] for more detailed discussions on $H^\pm$ production and search strategies at the LHC.

3.3. The SUSY Regime

The previous discussion on MSSM Higgs production and detection at the LHC might be significantly altered if some sparticles are relatively light. Some standard production processes can be affected, new processes

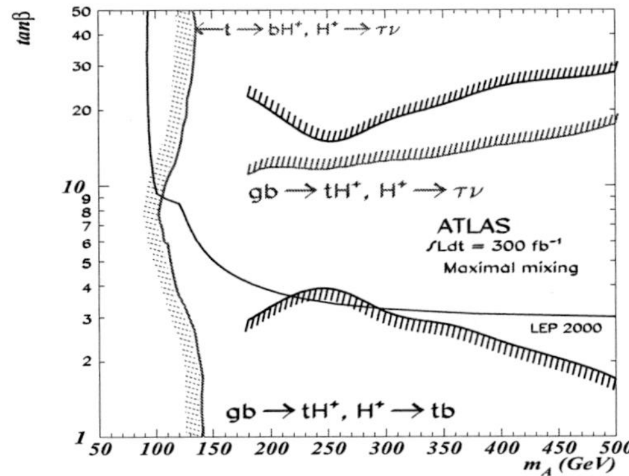

Figure 25. The coverage in the M_A–$\tan\beta$ plane in the search for the charged Higgs boson at the LHC in ATLAS simulations; from Ref. [65]

can occur and additional channels involving SUSY final states might drastically change the Higgs detection strategies. Let us comment on some possibilities.

The Hgg and hgg vertices in the MSSM are mediated not only by heavy t/b loops but also by loops involving squarks. If the top and bottom squarks are relatively light, the cross section for the dominant production mechanism of the lighter h boson in the decoupling regime, $gg \to h$, can be significantly altered by their contributions. In addition, in the $h \to \gamma\gamma$ decay which is one of the most promising detection channels, the same $\tilde{t}, \tilde{b}$ loops together with chargino loops, will affect the branching rate. The cross section times branching ratio $\sigma(gg \to h) \times \text{BR}(h \to \gamma\gamma)$ for the lighter h boson at the LHC can be thus very different from the SM, even in the decoupling limit in which the h boson is supposed to be SM-like [23]. The effects can be drastic and could lead to a strong suppression of $\sigma(gg \to h \to \gamma\gamma)$.

If one of the top squarks is light and its coupling to the h boson is enhanced, an additional process might provide a new source for Higgs particles in the MSSM: associated production with $\tilde{t}_1$ states, $pp \to gg/q\bar{q} \to \tilde{t}_1\tilde{t}_1 h$ [105]. This is similar to the standard $pp \to t\bar{t}h$ mechanism and in fact, for small masses and large mixing of the $\tilde{t}_1$ the cross section can be comparable. Final states with the heavier $H, A, H^\pm$ and/or other squark species than $\tilde{t}_1$ are less favoured by phase space.

Another possible source of MSSM Higgs bosons would be from the cascade decays of strongly interacting sparticles, which have large production rates at the LHC. In particular, the lighter h boson and the heavier A, H and $H^\pm$ particles with masses $\lesssim$ 200–300 GeV, can be produced from the decays of squarks and gluinos into the heavier charginos/neutralinos, which then decay into the lighter ones and Higgs bosons. This can occur either in "little cascades", $\chi_2^0, \chi_1^\pm \to \chi_1^0$+ Higgs, or in "big cascades" $\chi_{3,4}^0, \chi_2^\pm \to \chi_{1,2}^0, \chi_1^\pm$ + Higgs. The rates for ino decays into Higgs bosons can be dominant while decays of squarks/gluinos into the heavier inos are substantial. Detailed studies [106] have shown that these processes can be isolated in some areas of the SUSY parameter space and can be complementary to the direct production ones (Fig. 26). In particular, one can probe the region $M_A \sim 150$ GeV and $\tan\beta \sim 5$, where only h can be observed in standard searches.

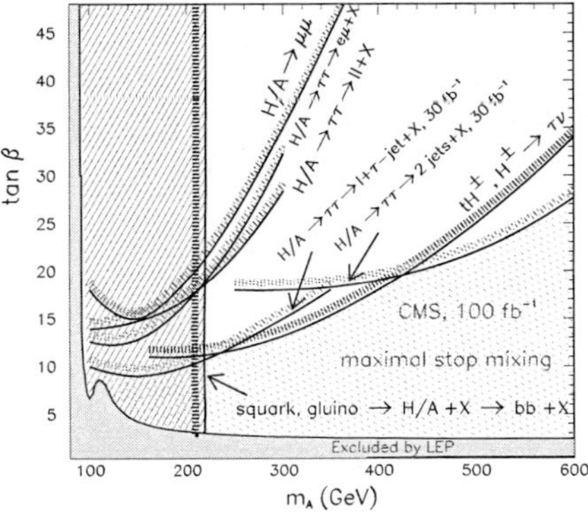

Figure 26. Areas in the $[M_A, \tan\beta]$ parameter space where the MSSM Higgs bosons can be discovered at the LHC with 100 fb^{-1} data in sparticle cascades [106]

The possibility of light charginos and neutralinos allows to search for the heavier H, A and $H^\pm$ states in regions of the parameter space in which they are not accessible in the standard channels [this is the case e.g. for $M_A \sim 200$ GeV and moderate $\tan\beta$ values]. There are situations in which the signals for Higgs decays into charginos and neutralinos are clean enough to be detected at the LHC. One of the possibilities is that the neutral H/A bosons decay into pairs of the second lightest neutralinos, $H/A \to \chi_2^0\chi_2^0$, with the subsequent decays of the latter into the LSP neutralinos and leptons, $\chi_2^0 \to \tilde{\ell}^*\ell \to \chi_1^0\ell\ell$ with $\ell^\pm = e^\pm, \mu^\pm$, through the exchange of relatively light sleptons. This leads to four charged leptons and missing energy in the final state. If the H/A bosons are produced in the gg-fusion processes, there will be little hadronic activity and the $4\ell^\pm$ final state is clean enough to be detected. Preliminary analyses show that the decays can be isolated from the large (SUSY) background (Fig. 27).

3.4. The CP Violating MSSM

There are two ways by which CP violation in the MSSM affects the Higgs production rates and detection at the

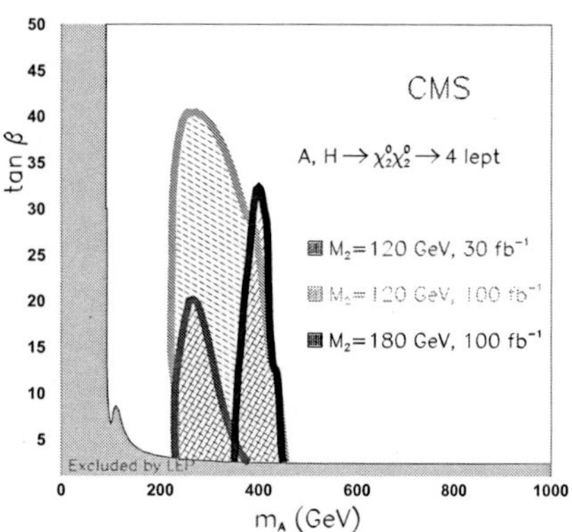

Figure 27. Areas in the $[M_A, \tan\beta]$ parameter space where the MSSM Higgs bosons can be discovered at the LHC with 100 fb^{-1} data in $A/H \to \chi_2^0\chi_2^0 \to 4\ell^\pm + X$ decays for a given set of the MSSM parameters [22]

LHC, through mixing in the Higgs states and/or modification of loop induced ggH_i and $\gamma\gamma H_i$ couplings due to CP violation in the squark couplings [26–30,107–110]. Ref. [107] for example, discusses the situation with no significant mixing between the h, H and A states and large effect on the ggH_1 coupling of CP violation in the squark-squark-Higgs vertex.

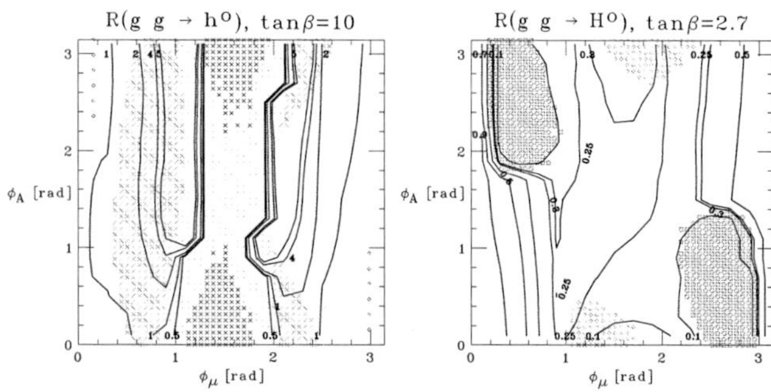

Figure 28. Contours of ratio of Higgs production to that expected in the CP conserving case, as a function of Φ_μ and Φ_A [107]. The left panel is for h and $\tan\beta = 10$ and the right panel is for H and for $\tan\beta = 2.7$. Also shown are the regions disallowed by the EDM constraints

In Fig. 28, the contours of ratios of h, H production rates in the CP violating MSSM to those without CP violation are shown. This corresponds to the case where the CP violation in the MSSM induces CPV $\tilde{q}\tilde{q}h(H)$ couplings. As expected from the sum rule we find that whereas the h production rate increase in the allowed region, the H production rate decreases. As can be seen from the Fig. 28 the effects can be considerable.

A more interesting aspect of CP violation in the Higgs sector is the vanishing of the coupling of the lightest Higgs scalar H_1 to the Z pair mentioned earlier, which in fact invalidates the lower limit on the mass of the lightest neutral at the LHC. Further, due to the reduced ggH_1 and $t\bar{t}H_1$ couplings in this case, one may miss this state at the LHC too. The region of the parameter space where this happens, generically seems to correspond to the case where all the three neutral scalars and the charged scalar are reasonably light. This region of Ref. [29] corresponds to $\tan\beta \sim 3.5$–5, $M_{H^+} \sim$125–140 GeV, $M_{H_1} \lesssim 50$ GeV and $\tan\beta \sim$2–3, $M_{H^+} \sim$ 105–130 GeV, $M_{H_1} \lesssim 40$ GeV, for $\Phi_{CP} = 90°$ and 60° respectively. (The details of the exact excluded region depend on the code used to compute the spectrum [111].) An analysis taking into account simulation of detector effects [112] confirms that there exists a region in the $\tan\beta$–M_{H^+} plane corresponding to $M_{H_1} < 50$ GeV, $100 < M_{H_2} < 110$ GeV and $130 < M_{H_3} < 180$ GeV [29], where LHC does not seem to have reach.

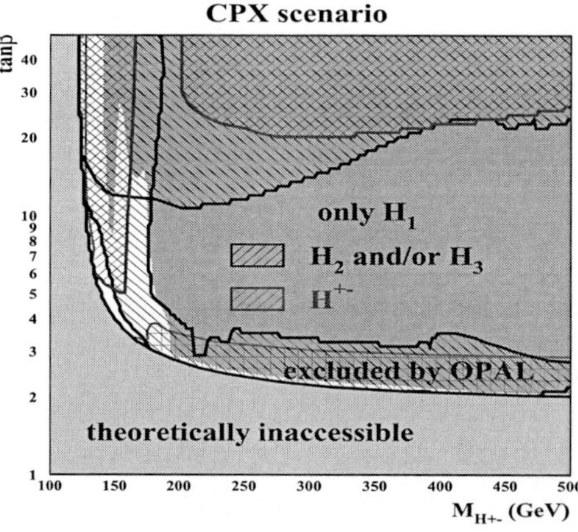

Figure 29. Reach of the LHC in $\tan\beta$–M_{H^+} plane in the CPX scenario [112]

This is shown in Fig. 29. In fact, the sum rules that the H_i couplings must satisfy, come to the rescue [110] in recovering the lost Higgs signal. This is a very generic example of how one can 'recover' the Higgs signal if the model parameters should contrive to make the usually investigated search channels ineffective. H_iVV and H_iH^+W couplings satisfy a sum rule given by: $g^2_{H_iVV} + |g_{H_iH^+W}|^2 = 1$. Further, there exists in the MSSM a correlation between the mass of the charged Higgs $M_{H^\pm}$ and that of the pseudo-scalar state. A suppressed H_1VV coupling implies a light pseudo-scalar state, which in turn implies a light charged Higgs, with $M_{H^+} < M_t$. Hence, a light H_1 which might have been missed at LEP also corresponds to an $H^\pm$ light enough to be produced in t decay, which in turn decays to $H_1W^\pm$, followed by H_1 decay to $b\bar{b}$. Due to the large $H^\pm \to H_1W^\pm$ branching ratio, the expected event rate for this final state is quite healthy ($\sim \mathcal{O}100$ fb), over the entire hole region (Fig. 30).

Thus one can look for the H_1 in final states containing $bW^+b\bar{b}\bar{b}W^-$ in the $t\bar{t}$ sample. The huge background ($\sim$ 8.5 pb) coming from QCD production of $t\bar{t}b\bar{b}$, can be reduced to $\sim$ 0.5 fb level and below, by demanding that one of the bW combination reconstructs to t mass and the $bbbW$ also to the t mass [69,110].

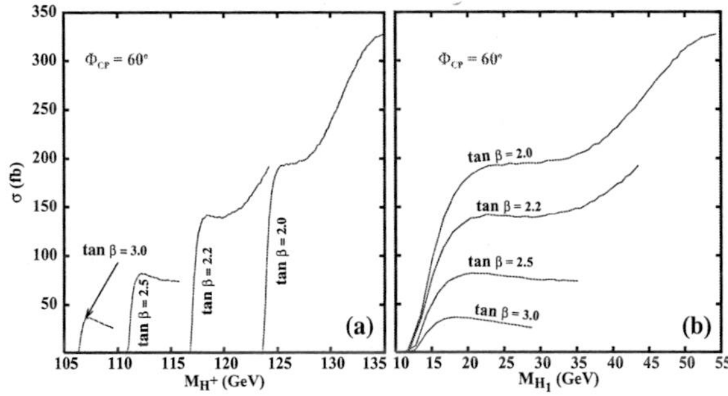

Figure 30. Variation of the expected cross-section with M_{H^+} for four values of $\tan\beta = 2, 2.2, 2.5$ and 3. The CP-violating phase $\Phi_{\rm CP}$ is 60° [110]

Figure 31 shows the clustering of the $b\bar{b}, b\bar{b}W$ and $b\bar{b}bW$ for the signal which can be used effectively to handle the background. These studies have thus opened up a new topology in which one should be looking for the lightest neutral Higgs in the decay of $H^\pm$ produced in t decay. A few points are worth noticing. Due to the rather small value of $\tan\beta$ the usual $\tau\nu_\tau$ decay mode for the H^+ is also not available for the H^+ search in this case. Thus in this region of the MSSM parameter space, the above process provides a search prospect not just for the light neutral state which might have been missed at LEP, but also the light charged Higgs H^+ in this parameter range (a similar situation attains in NMSSM as well [113]). A theorists analysis [114] indicates that it may be possible to look at $\bar{t}\bar{t}H_1$ production, which will be higher than in the corresponding CP conserving scenario due to lighter H_1, and have a signal for parameter values corresponding to the hole.

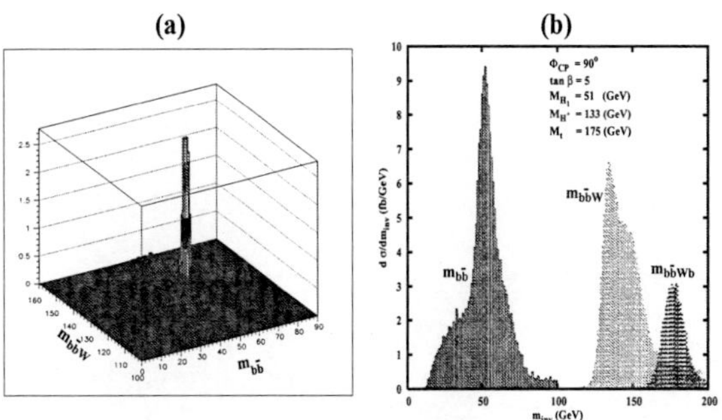

Figure 31. Clustering of the $b\bar{b}$, $b\bar{b}W$ and $b\bar{b}bW$ invariant masses. (a) three-dimensional plot for the correlation between $m_{b\bar{b}}$ and $m_{b\bar{b}W}$ distribution. (b) $m_{b\bar{b}}, m_{b\bar{b}W}$ and $m_{b\bar{b}Wb} = M_t$ distributions for $\Phi_{\rm CP} = 90°$. Appropriate M_t, M_W mass window cuts have been applied The other MSSM parameters are $\tan\beta = 5, M_{H^+} = 133$ GeV, corresponding to $M_{H_1} = 51$ GeV [110]

3.5. Extensions of the MSSM

In the NMSSM, where a complex iso-scalar field is introduced, leading to an additional pair of scalar and pseudoscalar Higgs particles, the axion–type or singlino character of the pseudoscalar A_1 boson makes it preferentially light and decaying into b quarks or τ leptons [26,39,69]. Therefore, in some areas of the NMSSM parameter space, the lightest CP-even Higgs boson may dominantly decay into a pair of light pseudoscalar A_1 bosons generating four b quarks or τ leptons in the final state, $H_1 \to A_1A_1 \to 4b, 2b2\tau, 4\tau$. In fact, it is also possible that H_1 is very light with small VV couplings, while H_2 is not too heavy and plays the role of the SM-like Higgs particle; the decays $H_2 \to H_1H_1$ can also be substantial and will give the same signature as above.

This situation, similar to the CPX scenario discussed above, is very challenging at the LHC. Indeed, all the production mechanisms of the light A_1 or H_1 singlino-like state will have small cross sections as both couplings to vector bosons and top quarks are tiny. The SM-like Higgs H_1 or H_2 will have reasonable production rates but the dominant decay channels into $4b, 2\tau 2b$ and 4τ will be swamped by the QCD background. Nevertheless, in the case of very light A_1 bosons with masses smaller than 10 GeV and, therefore decaying almost exclusively into $\tau^+\tau^-$ pairs, the $H_1 \to A_1A_1 \to 4\tau \to 4\mu + 4\nu_\mu + 4\nu_\tau$ final state with the H_1 boson dominantly produced in vector boson fusion can be isolated in some cases. This is exemplified in Fig. 32 where the result of a simulation of this process by members of the ATLAS collaboration is shown in the parameter space formed by the trilinear NMSSM couplings λ and κ. While there are regions in which the final state can be detected, there are other regions in which the light H_1 and A_1 states remain invisible even for the high luminosity which has been assumed.

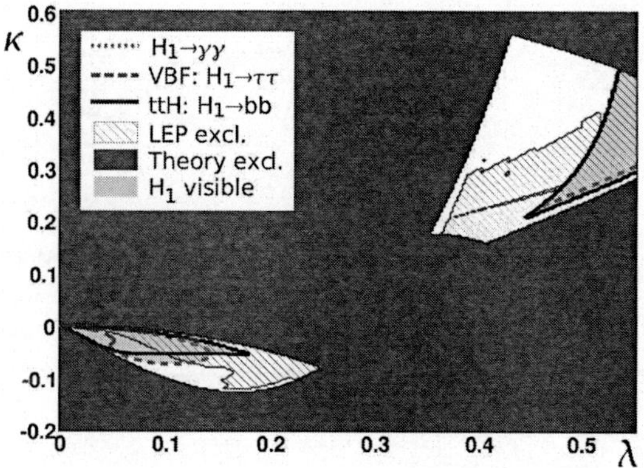

Figure 32. Regions of the NMSSM parameter space $[\lambda, \kappa]$ in which a light pseudoscalar Higgs boson can be detected in an ATLAS simulation [69]

In the most general SUSY model, with an arbitrary number of singlet and doublet fields and an extended matter content to allows for the unification of the gauge couplings, a Higgs boson should have a mass smaller than 200 GeV and significant couplings to gauge bosons and top quarks; this particle can be thus searched for in the gg and VV fusion channels with the signature $WW \to \ell\ell\nu\nu$ which would be hard to miss.

Furthermore, in scenarios with spontaneously broken R–parity, besides invisible decays of the h boson to be discussed later, decays of the pseudoscalar Higgs $A_i \to H_j Z \to Z$ and missing energy could be detected if the cross sections for A_i production are large enough.

Other SUSY scenarios can also be probed at the LHC [44,96]. In GUT theories which lead to the presence of an extra neutral gauge boson at low energies, the Z' boson decays $Z' \to Zh$ which occur via Z–Z' mixing could have non-negligible rates and would lead to a detectable $\ell\ell b\bar{b}$ signature; the Z' production cross section would be large enough for $M_{Z'} \lesssim 2$ TeV to compensate for the tiny mixing and hence, the small Z + Higgs branching ratio. If relatively light doubly charged Higgs bosons exist, they can be produced in the Drell-Yan process $q\bar{q} \to H^{++}H^{--}$ and, if their leptonic decays $H^{--} \to \ell\ell$ are not too suppressed, they would lead to a spectacular 4-lepton final state.

3.6. Alternative Scenarios and Invisible Higgs

Various beyond the SM physics options can in fact cause the Higgs to have large branching ratio in "invisible" final states: in the conventional MSSM, Higgs decays into LSP neutralinos $h \to \chi_1^0\chi_1^0$, in the MSSM with R-parity violation decays into escaping Majorons, $h \to JJ$, mixing with graviscalars in extra-dimensional model are few of the reasons. In some cases this impacts the branching ratio of the Higgs into the 'visible' final states such as $b\bar{b}$ or $\gamma\gamma$ severely. The issue of how to search for a Higgs which dominantly decays into invisible decay products, is therefore important from the point of view of recovering the lost reach as well as for measuring the invisible decay width.

There have been many parton level and detector level studies on this subject [115–120]. The most promising one is the the production of the h boson in the WW fusion process, $qq \to qqh$, which leads to two large rapidity jets with a rapidity gap [117] along with large missing momentum due to the invisible Higgs. Figure 33 shows that the distribution in the azimuthal angle between the two jets, clearly distinguishes between the dominant $Z + 2$ jets background and the signal. With 100 fb^{-1} luminosity this method is shown to be sensitive for invisible BR as low as 5% (12%) for Higgs mass 130 (400) GeV.

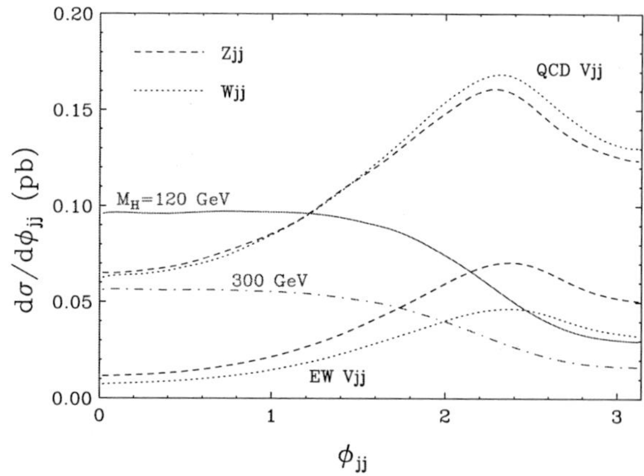

Figure 33. The azimuthal distribution between the jets for the signal for an invisibly decaying Higgs boson in the WW fusion process [117] and background

Alternatively, one can use the production of a h boson in association with Z boson followed by Z decaying into a large p_T lepton pair with missing transverse energy E_T [118]. Figure 34 shows the p_T distribution for the signal (dashed histogram) and the dominant ZZ background (solid histogram) in the case of the two-lepton signature. This is a result of judicious cuts on different kinematic variables exploiting the differences in the background and signal. In fact, the cuts significantly reduce the $Z + 2$ jets background while affecting the signal only slightly. However due to the small production rate, this process is useful only for large BR (40% or so) into the invisible channel and for Higgs masses in the lower mass range.

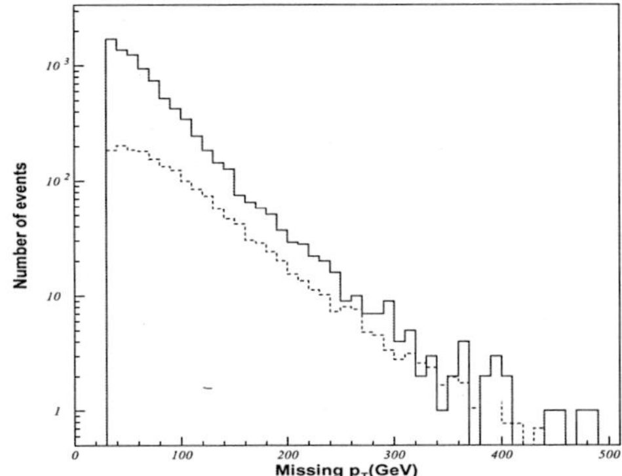

Figure 34. Comparison of the p_T distribution for the signal for the invisible Higgs in the $l^+l^- + E_T^{\rm miss}$ channel [118] and the irreducible ZZ background

Of course detection of an invisibly decaying Higgs boson will have to be followed by a study of attendant phenomenology predicted in each of the corresponding models. Decays into LSP can not give rise to a substantial invisible branching ratio in the simplest mSUGRA picture due to the current limits on chargino masses from LEP and the attendant lower limits on the neutralino mass that exist in these models. However, in the MSSM with non-universal U(1) and SU(2) gaugino masses M_1 and M_2, it is possible to have substantial invisible branching ratio corresponding to a light LSP and still be consistent with the LEP results [21]. Further, there still can exist regions of the parameter space where $\sigma(gg \to h \to \gamma\gamma)$, is suppressed below the value expected for the SM as previously discussed.

In this case, it is the consistency with the cosmological relic density of the LSP neutralino, which requires that the small value of the ratio $r = M_1/M_2$ be also accompanied by a light slepton (which in fact is preferred by the $(g-2)_\mu$ data), which constrains the allowed region of this version of the MSSM. In this case, the loss of the Higgs signal due to reduction in the useful $\gamma\gamma$ and $b\bar{b}$ channels is compensated by increased rate for production of h in the decays of heavier neutralinos and charginos caused by the fact that the LSP is a mixture of gaugino and higgsino in this case. Figure 35 shows this connection between the Higgs sector properties and the DM relic density in the universe. The usual signal for the light Higgs in the $\gamma\gamma$ final state is reduced here.

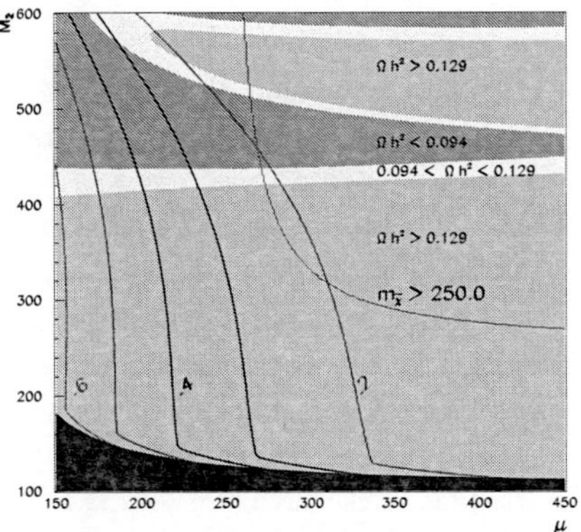

Figure 35. The invisible branching ratio for light Higgs in the M_2–μ plane overlaid with regions allowed by relic density constraints, for non-universal gaugino masses $M_1/M_2 = 0.2$ [21,70]

Invisible Higgs decays are also possible in non-SUSY models. In models with large extra dimensions [51], the interaction of the Higgs field and the Ricci scalar curvature of the induced four-dimensional metric also generates a mixing term with the closest Kaluza-Klein graviscalar fields [52]. This mixing results in an effective Higgs decay width, $\Gamma(H \to \text{graviscalar})$, which is invisible as the graviscalars are weakly interacting and mainly reside in the extra dimension while the Higgs is on the TeV brane. These invisible Higgs decays can be largely dominating. In addition, there is the possibility of Higgs decays into a pair of graviscalars, but the rates are smaller than the ones from mixing.

Finally, let us comment on suppressed Higgs couplings in alternative scenarios. As discussed previously, in Randall-Sundrum models [47], a scalar radion field is introduced to stabilise the distance between the SM and the gravity brane. Carrying the same quantum numbers, the Higgs and radion fields can mix and the properties of the Higgs boson will be altered [48,50] and can lead to important shifts in the Higgs couplings which become apparent in the various decay widths and production cross sections (Fig. 12). As can be seen, while the shifts in the $f\bar{f}/VV$ and $\gamma\gamma$ widths are rather similar, the shift in the $H \to gg$ partial decay width is different; the width can become close to zero for some values of the mixing. The impact of mixing in $f\bar{f}$ and VV final states is in general smaller and the BR will not be significantly affected as these decays are dominant. This implies that it will be imperative to perform a precise measurement of the Higgs total decay width in order to probe the mixing with radions.

Another important consequence of radion mixing is the decays of the Higgs boson into a pair of radions.

Indeed, if the radion is relatively light, the decays $H \to \phi\phi$ might be kinematically accessible and, for some mixing values, the branching fractions might be substantial. In some mass range, e.g. $M_\phi \lesssim 60$ GeV, the radion will mainly decay into $b\bar{b}$ and gg final states, while the $\gamma\gamma$ branching ratio is very small. Observing these final states will be rather difficult at the LHC.

The suppression of the Hgg loop induced coupling can occur in non-SUSY extensions of the SM as well. For instance, the $\mathrm{SU(2)_R}$ partner of the right-handed top quark in warped extra-dimensional models with an extended left-right symmetric structure will also contribute to the Hgg vertex and could interfere destructively with the top quark contribution, leading to a much smaller coupling [121]. In the strongly interacting light Higgs scenario proposed recently [58], the Higgs couplings to gluons, as well as the couplings to fermions and gauge bosons are also suppressed. The suppression of the Hgg coupling would lead to a decrease of the cross section for the dominant Higgs production mechanism, $gg \to H$ and would make the Higgs search more complicated at the LHC.

4. Measurements of the Higgs Properties

It is clear from the discussion so far that after seeing the Higgs signal at the LHC it will be essential to perform a measurement of the Higgs properties, to be able to establish the exact nature of EWSB and to achieve a more fundamental understanding of the issue. It is well known that a hadron collider can afford only a limited accuracy on measurements of most of the Higgs properties and that the next e^+e^- linear collider ILC will indeed be needed for a high precision measurement [122]. Nonetheless, since the LHC is the current collider, it is important to address the Higgs properties question when a large luminosity, ≈ 300 fb^{-1} has been collected. We summarise some of the information below.

4.1. Mass, Width and Couplings of the SM Higgs

The ease with which information can be obtained for the Higgs profile clearly depends on the mass (Fig. 36). The accuracy of the mass determination is driven by the $\gamma\gamma$ mode for a light Higgs and by the $H \to ZZ \to 4l$ mode for a heavier one and, in fact, is expected to be accurate at one part in 1000. For $M_H \gtrsim 500$ GeV, the precision deteriorates rising to about a percent level around $M_H \approx 800$ GeV, which is close to the theoretically expected upper limit, due to decreasing rates.

Using the same process, $H \to ZZ \to 4\ell^\pm$, the Higgs total decay width can be measured for $M_H \gtrsim 200$ GeV when it is large enough to be resolved experimentally.

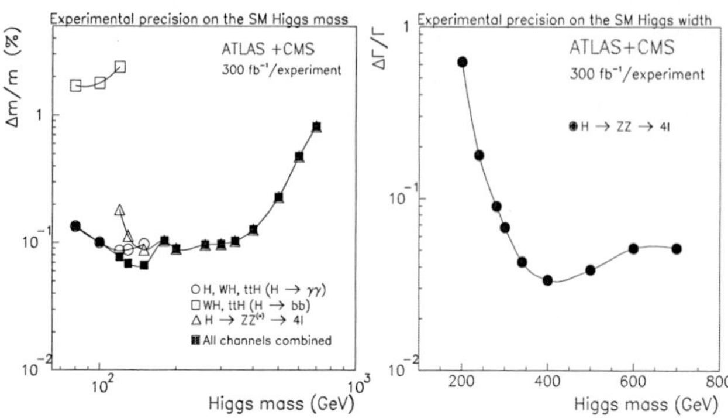

Figure 36. Precision possible for the mass (left) and total width (right) measurements for the SM Higgs for $\mathcal{L} = 300$ fb^{-1} combining ATLAS and CMS [65]

While the precision is rather poor near this mass value, it improves to reach the level of $\sim 5\%$ around $M_H \sim 400$ GeV and the precision stays almost constant up to masses of order $M_H \sim$ 700 GeV [65].

One would like to determine the couplings of the Higgs and test their proportionality to the masses of fermions/gauge bosons, which is absolutely essential for checking the Higgs mechanism of EWSB. Ratios of Higgs couplings squared can be determined by measuring ratios of production cross sections times decay BR and accuracies at the 10–50% can be obtained in some cases [87]. However, it has been shown in Ref. [88] that with some theoretical assumptions, which are valid in general for multi-Higgs doublet models, the extraction of absolute values of the couplings rather than just ratios of the couplings, is possible by performing a fit to the observed rates of Higgs production in different channels. For Higgs masses below 200 GeV they find accuracies of order 10–40% for the Higgs couplings after several years of the LHC running. Figure 37 shows the relative precision possible on fitted Higgs couplings-squared for $2 \times 300 + 2 \times 100$ fb^{-1} as explained on the figure. Thus at the LHC the various couplings can be determined with a relative precision of at most 30%. With just 30 fb^{-1} data per experiment this is perhaps only good to 50–60% level. Ref. [88] also discusses how one can carry out the programme, for example, for the MSSM or for other beyond SM models.

The trilinear Higgs boson self-coupling λ_{HHH} is too difficult to be measured at the LHC because of the smallness of the $gg \to HH$ [and, *a fortiori*, the $VV \to HH$ and $qq \to HHV$] cross sections and the very large backgrounds [123,124]. A parton level analysis has been recently performed in the channel

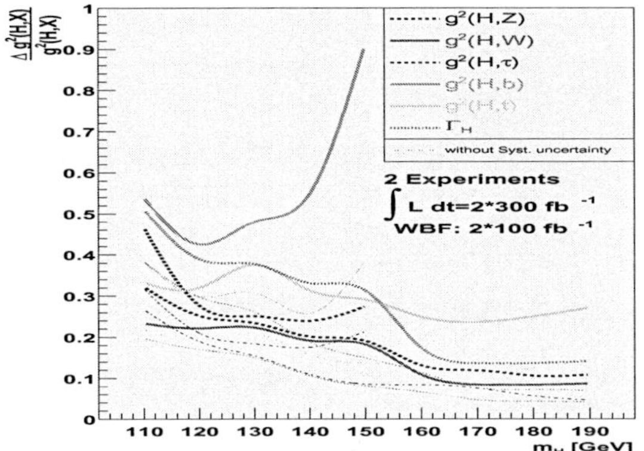

Figure 37. Relative precision of fitted Higgs couplings-squared as a function of the Higgs mass for the $2 \times 300 + 2 \times 100$ fb^{-1} luminosity scenarios. It is assumed that $g^2(H,V) < 1.05 \cdot g^2(H,V,SM)$ $(V = W,Z)$ but one allows for new particles in the loops for $H \to \gamma\gamma$ and $gg \to H$ and for unobservable decay modes [88]

$gg \to HH \to (W^+W^-)(W^+W^-) \to (jj\ell\nu)(jj\ell\nu)$ and $(jj\ell\nu)(\ell\ell\nu\nu)$ with same sign dileptons, including all the relevant large backgrounds [124]. The statistical significance of the signal is very small, even with an extremely high luminosity, and one can at most set rough limits on the magnitude of the Higgs self-coupling.

Thus for a very accurate and unambiguous determination of the Higgs couplings, clearly an e^+e^- Linear Collider [122] will be required.

4.2. Measurements in the MSSM

In the decoupling regime when $M_A \gg M_Z$, the measurements which can be performed for the SM Higgs boson with a mass $\lesssim 140$ GeV will also be possible for the h boson. Under some assumptions and with 300 fb^{-1} data, coupling measurements would allow to distinguish an MSSM from a SM Higgs particle at the 3σ level for A masses up to $M_A = 300$–400 GeV [87].

The heavier Higgs particles H, A and $H^\pm$ are accessible mainly in the $gg \to b\bar{b} + H/A$ and $gb \to H^\pm t$ production channels at large $\tan\beta$, with the decays $H/A \to \tau^+\tau^-$ and $H^+ \to \tau^+\nu$. The Higgs masses cannot be determined with a very good accuracy as a result of the poor resolution. However, for $M_A \lesssim 300$ GeV and with high luminosities, the H/A masses can be measured with a reasonable accuracy by considering the rare decays $H/A \to \mu^+\mu^-$ [66,101]. The discrimination between H and A is though difficult as the masses are close in general and the total decay widths large [101].

There is, however one very important measurement which can be performed in these channels. As the production cross sections above are all proportional to $\tan^2\beta$ and, since the ratios of the most important decays fractions are practically independent of $\tan\beta$ for large enough values [when higher-order effects are ignored], one has an almost direct access to this parameter. A detailed simulation shows that an accuracy of $\Delta\tan\beta/\tan\beta \sim 30\%$ for $M_A \sim 400$ GeV and $\tan\beta = 20$ can be achieved with 30 fb^{-1} data [125] (Fig. 38).

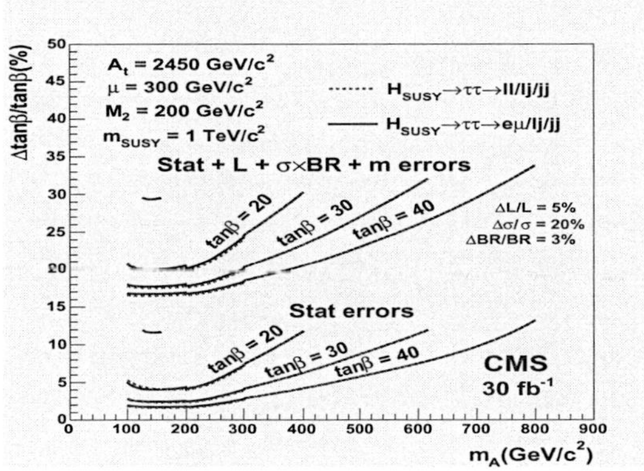

Figure 38. The uncertainty in the measurement of $\tan\beta$ in the channel $gg \to H/A + b\bar{b}$ with the combined $H/A \to \tau\tau$ decays at CMS with 30 fb^{-1} data. The three lower curves show the uncertainty when only statistical errors are taken into account, while the upper curves include the uncertainties from the mass (a few %) and luminosity (5%) measurements and the theoretical uncertainty (23%); from Ref. [125]

4.3. Determination of the Higgs Spin-parity

Apart from the mass, width and the couplings we also need to determine the spin of the Higgs and further establish that the Higgs is a CP-even particle. One can obtain information on these properties by studying various kinematical distributions such as the invariant mass distribution of the decay products and various angular correlations among them, which depend on the spin of the decaying object crucially, as well kinematical distribution of the production process. A large amount of work has been done on how to establish, at different colliders, that the Higgs boson is indeed $\mathrm{J^{PC}} = 0^{++}$ state [26,126]. Most of the analyses/suggestions for the LHC emanate by translating the strategies devised in the case of the ILC.

One example is to study the threshold behaviour of the M_{Z^*} spectrum in the $H \to ZZ^{(*)}$ decay for $M_H \lesssim 2M_Z$. Since the relative fraction of the longitudinally to transversely polarised Z varies with M_{Z^*}, this distribution is sensitive to both the spin and the CP property of the Higgs. This is seen in Figs. 39 and 40 where the behaviours for a CP-even and CP-odd states and for different spins are shown respectively.

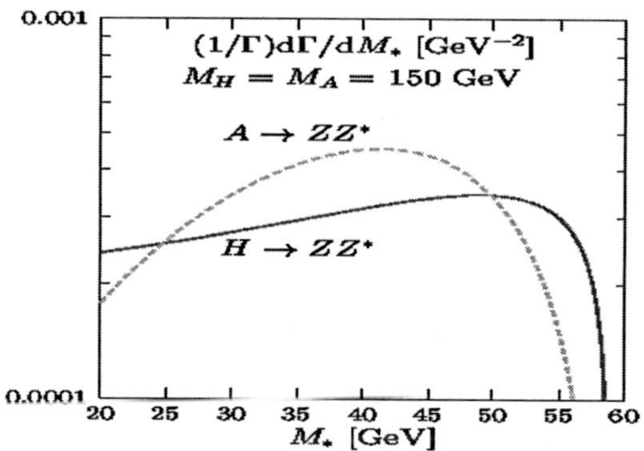

Figure 39. Dependence on the CP quantum number of the Higgs for the threshold behaviour of the distribution in M_{Z^*} for the $H \to ZZ^*$ decay [127]

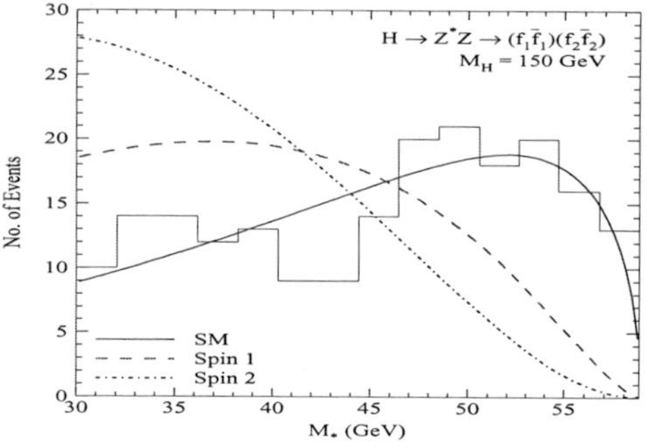

Figure 40. Spin determination of the Higgs boson via the threshold behaviour of the distribution in M_{Z^*} for the $H \to ZZ^*$ decay [128]

Another very useful diagnostic of the CP nature of the Higgs boson is the azimuthal distribution between the decay planes of the two lepton pairs arising from the $Z, Z^{(*)}$ bosons coming from the Higgs decay [26,127–131]. Alternatively, one can study the distribution in the azimuthal angle between the two jets produced in association with the Higgs produced in vector boson fusion [132–134] or in gluon fusion in Higgs plus jet events [135,136].

Figures 41 and 42, show the azimuthal angle distribution for the two jets produced in association with the Higgs, for the CP-even and CP-odd cases, for the vector boson and gluon fusion, and the gluon signal along with vector boson contribution and all the other backgrounds, respectively. One can see that with a high

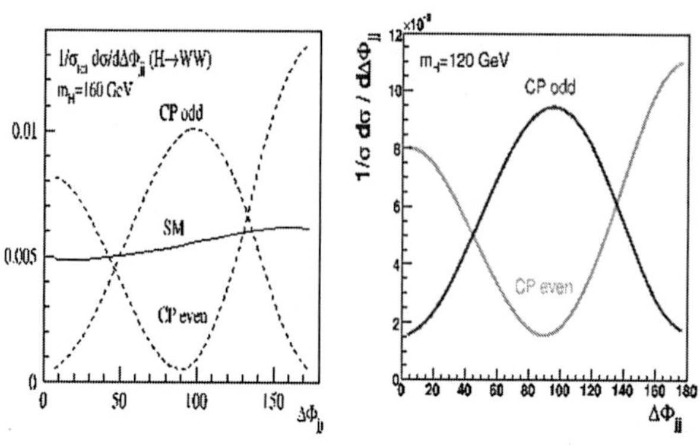

Figure 41. Azimuthal angle distribution for the two jets produced in association with the Higgs boson, for CP-even and odd cases. Left shows the vector boson fusion case, for $M_H = 160$ GeV and right, the gluon fusion for a mass $M_H = 120$ GeV [136]

luminosity of 300 fb^{-1}, it should be possible to use these processes quite effectively. Recall, however, that any determination of the CP property using a process which involves the coupling of the spin 0 particle to a pair of gauge bosons, is ambiguous as only the CP-even part of the coupling is projected out.

Couplings of a Higgs with heavy fermions offer therefore the best option. $t\bar{t}$ final states produced in the decay of an inclusively produced Higgs can be used to obtain information on the CP nature of the $t\bar{t}H$ coupling through spin-spin correlations [138,139]. Using optimal observable analyses, the associated $Ht\bar{t}$ production allows a determination of the CP-even and CP-odd part of the $t\bar{t}$ couplings with the Higgs boson separately [140], though it requires high luminosity. The use of τ polarisation in resonant $\tau^+\tau^-$ production at the LHC has also been recently investigated [34]. A novel approach [141,142], is to use double-diffractive processes with large rapidity gaps where only scalar Higgs production is selected.

In fact, recently, it was observed that the threshold rise of $\sigma(e^+e^- \to t\bar{t})+$ Higgs at the ILC offers a very clear and unambiguous determination of the CP nature of the $t\bar{t}$Higgs coupling [143]. The very different rise of the cross-section with the $t\bar{t}$Higgs invariant mass away from the threshold, $2M_t + M_H$, can be completely understood in terms of angular momentum and parity conservation. Interestingly, the same is found to also hold for $gg \to t\bar{t}$ + Higgs production as well (Fig. 43) [144].

Most of the suggested measurements should be able to verify the CP nature of a Higgs boson when the full luminosity of 300 fb^{-1} is collected at the LHC or even

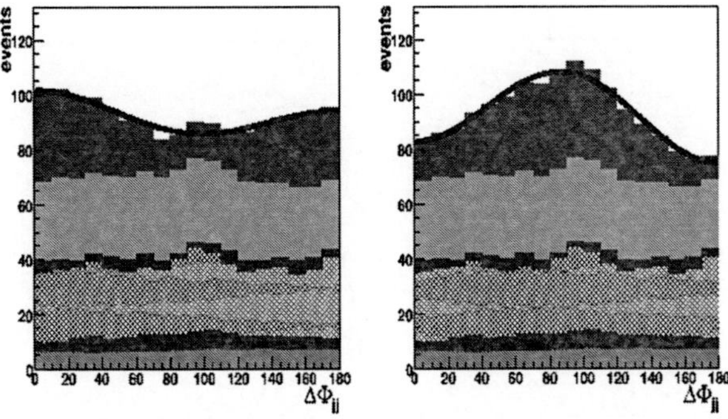

Figure 42. Azimuthal angle distribution for the two jets produced in association with a Higgs, for the CP even (left) and CP odd (right) cases, after selection cuts [136], for $M_H = 160$ GeV. Shown are the gluon signal and the other backgrounds from top to bottom

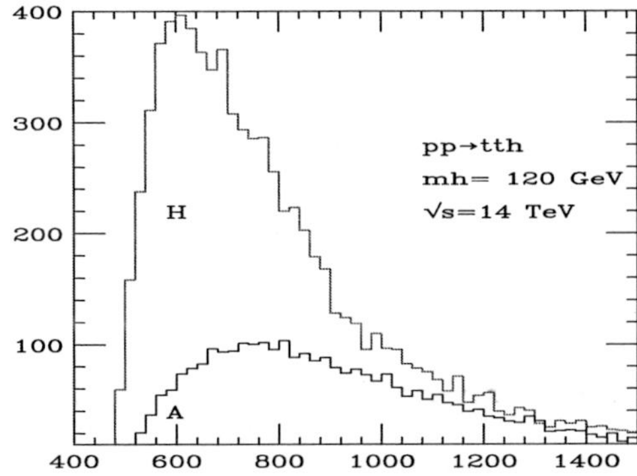

Figure 43. Distribution in the $t\bar{t}$Higgs invariant mass for $pp \to t\bar{t}$Higgs for scalar and a pseudoscalar bosons H and A of 120 GeV at the LHC [144]

before, provided the Higgs boson is a CP eigenstate. However, a measurement of the CP mixing is much more difficult, and a combination of several different observables will be essential.

The subject of probing CP mixing reduces more generally to the probing of the anomalous VVH and $t\bar{t}H$ couplings, the only two cases where such study can even be attempted at the LHC. Since CP-even and CP-odd Higgs bosons couple to the $t\bar{t}$ pair democratically where as the coupling to a VV pair is suppressed for the CP-odd case, the most unambiguous in this context will be the $t\bar{t}H$ process [140]. However, as already mentioned a CMS study shows that, at present, it is not clear whether it would be possible to detect the $t\bar{t}H$ signal above the $t\bar{t}b\bar{b}$ background. Hence, VVH is the only relevant case. In principle, the same studies which are used to determine the CP-even or CP-odd character of the Higgs boson mentioned above, can be used.

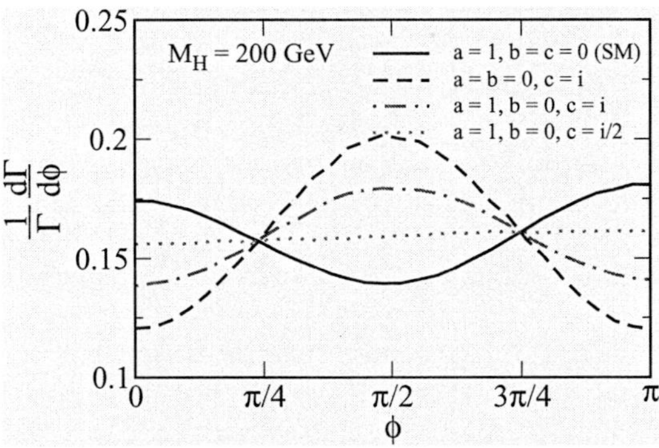

Figure 44. The normalised differential width for $H \to Z^{(*)}Z \to (f_1\bar{f}_1)(f_2\bar{f}_2)$ with respect to the azimuthal angle ϕ. The solid (black) curve shows the SM case ($a = 1$, $b = c = 0$) while the dashed (blue) curve is a pure CP-odd state ($a = b = 0$, $c = i$). The dot-dashed (red) curve and the dotted (green) curve are for states with CP violating couplings $a = 1$, $b = 0$ with $c = i$ and $c = i/2$, respectively [131]

As an example, we show in Fig. 44 the distribtuion in azimuthal angle ϕ, for $M_H = 200$ GeV, where H corresponds to a Higgs which may have indeterminate CP assignments. It should be kept in mind though, that this method cannot be applied for very large Higgs masses where this dependence is washed out. One must also beware of degenerate Higgs bosons of opposite CP; since the decay products are the same, they will both contribute to the rate and must be summed coherently, possibly mimicking the effect seen above. Also, in the context of the LHC with the QCD environment and modifications to shapes of distributions, it is useful to construct specific observables which may be directly proportional to the anomalous part of the coupling, as was done in the e^+e^- case recently [145].

Parameterising the anomalous vertex by

$$V^{\mu\nu}_{HZZ} = \frac{igm_Z}{\cos\theta_W}\left[a\, g_{\mu\nu} + b\,\frac{p_\mu p_\nu}{m_Z^2} + c\,\epsilon_{\mu\nu\alpha\beta}\,\frac{p^\alpha k^\beta}{m_Z^2}\right],$$

where $p = q_1 + q_2$ and $k = q_1 - q_2$, θ_W denotes the weak-mixing angle and $\epsilon_{\mu\nu\alpha\beta}$ is the totally antisymmetric tensor with $\epsilon_{0123} = 1$, one can develop a strategy to probe different parts of the anomalous couplings a, b and c directly. The general strategy is to construct different observables out of the available 4-momenta such that they have specific CP and $\tilde{\mathrm{T}}$ transformation properties, using partially integrated cross-sections, where

$\tilde{T}$ denotes naive time reversal. Then, the expectation value of the sign of this observable, which will correspond to partially integrated cross-sections will be directly proportional to the particular anomalous coupling (or product of these) which have the same transformation property. In the reasonable approximation of small anomalous couplings, these observables will then directly probe different anomalous couplings.

One example of such an observable is the cosine of the angle θ_1 made by the decay lepton with the Z direction in the rest frame of the Higgs boson. One can write

$$O_1 \equiv \cos\theta_1 = \frac{(\vec{p}_{\bar{f}_1} - \vec{p}_{f_1}) \cdot (\vec{p}_{\bar{f}_2} + \vec{p}_{f_2})}{|\vec{p}_{\bar{f}_1} - \vec{p}_{f_1}||\vec{p}_{\bar{f}_2} + \vec{p}_{f_2}|}$$

for the decay $H \to ZZ^{(*)} \to f_1\bar{f}_1 f_2\bar{f}_2$. The expectation value of the sign is

$$\mathcal{A}_1 = \frac{\Gamma(\cos\theta_1 > 0) - \Gamma(\cos\theta_1 < 0)}{\Gamma(\cos\theta_1 > 0) + \Gamma(\cos\theta_1 < 0)}.$$

This is $\propto \Im m(c)$ and is thus a direct probe of non-zero value for it and hence of CP violation. For example, for $M_H = 200$ GeV, values of A_1 of about 8% are possible. Figure 45 shows the sensitivity of such a measurement for ATLAS with an integrated luminosity of 300 fb^{-1} for a scalar of mass 200 GeV. One can, in fact, systematically construct observables, using this strategy, to probe the different parts of the anomalous couplings separately. Thus in principle, with high luminosities it will be possible to map the anomalous HZZ couplings at the LHC at the level of 40–50%. Of course, this precision is no comparison to what will achievable at the ILC, as can be seen, for example, from recent discussions in Refs. [122,145,146].

In short, all the discussions above indicate that while the LHC with 300 fb^{-1} data per experiment can perform measurements of different Higgs properties, it is really to the ILC [122] that we have to look to for precision information.

5. Conclusion

The LHC will tell!

Acknowledgments: We acknowledge support from the Indo-French IFCPAR for the project No. 3004-2 "SUSY, Higgs and CP at colliders and in astrophysics". A.D. acknowledges support from the Alexander von-Humbold Foundation (Bonn, Germany). R.G. would like to acknowledge the support from the Department of Science and Technology, India under the J.C. Bose Fellowship.

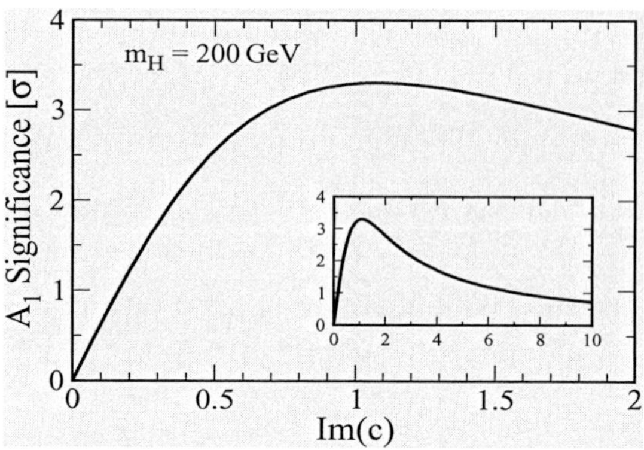

Figure 45. The significances corresponding to the asymmetry $\mathcal{A}_1$ as a function of $\Im m(c)$, for a Higgs boson of mass 200 GeV. We chose the CP-even coupling coefficient $a = 1$ and $b = 0$. The inserts show the same quantities for a larger range of $\Im m(c)$ [131]

REFERENCES

1. P.W. Higgs, Phys. Lett. 12 (1964) 132; Phys. Rev. Lett. 13 (1964) 506; F. Englert and R. Brout, Phys. Rev. Lett. 13 (1964) 321; G. Guralnik, C. Hagen and T. Kibble, Phys. Rev. Lett. 13 (1964) 585
2. Y. Nambu, Phys. Rev. Lett. 4 (1960) 380; Y. Nambu and G. Jona–Lasinio, Phys. Rev. 122 (1961) 345; ibid Phys. Rev. 124 (1961) 246; J. Goldstone, Nuov. Cim. 19 (1961) 154; J. Goldstone, A. Salam and S. Weinberg, Phys. Rev. 127 (1962) 965
3. For a review on the Higgs sector, see: J. Gunion, H. Haber, G. Kane and S. Dawson, "The Higgs Hunter's Guide", Addison–Wesley, Reading 1990
4. A. Djouadi, Phys. Rept. 457, 1 (2008) [arXiv:hep-ph/0503172]
5. The LEP collaborations and the LEP electroweak working group, hep-ex/0612034; http://lepewwg.web.cern.ch/LEPEWWG/
6. LEP working group for Higgs searches, R. Barate et al., Phys. Lett. B565, 61 (2003)
7. A. Duperrin, arXiv:0805.3624 [hep-ex]
8. C. Amsler et al. (Particle Data Group), Phys. Lett. B667, 1 (2008)
9. H. Flaecher et al., arXiv:0811.0009 [hep-ph]
10. B.W. Lee, C. Quigg and H.B. Thacker, Phys. Rev. D16, 1519 (1977)
11. C.H. Llewellyn Smith, Phys. Lett. B46 233 (1973); J. S. Bell, Nucl. Phys. B60 427 (1973) ; J. Cornwall et al, Phys. Rev. Lett. 30, 1268 (1973) Phys. Rev. D10, 1145 (1974) [Erratum-ibid. D 11, 972 (1975)]
12. N. Cabibbo et al, Nucl. Phys. B158 (1979) 295
13. M. Lüscher and P. Weisz, Phys. Lett. B212, 472 (1988); M. Göckeler et al, Nucl. Phys. B405, 555 (1993)
14. T. Hambye and K. Riesselmann, Phys. Rev. D55, 7255

(1997)
15. A. Djouadi, M. Spira and P. Zerwas, Z. Phys. C70 (1996) 427; A. Djouadi, J. Kalinowski and P. Zerwas, Z. Phys. C70 (1996) 435
16. A. Djouadi, J. Kalinowski and M. Spira, Comput. Phys. Commun. 108 (1998) 56
17. See e.g. M. Drees, R.M. Godbole and P. Roy, *Theory and phenomenology of sparticles*, World Scientific, 2005; H. Baer and X. Tata, *"Weak scale Supersymmetry: From superfields to scattering events," Cambridge, UK: Univ. Pr. (2006)*
18. A. Djouadi, Phys. Rept. 459, 1 (2008) [arXiv:hep-ph/0503173]
19. S. Heinemeyer, W. Hollik and G. Weiglein, Phys. Rept. 425, 265 (2006); B.C. Allanach et al., JHEP 0409 (2004) 044
20. The LEP Collaboration (ALEPH, DELPHI, L3, OPAL), Eur. Phys. J. C47, 547 (2006)
21. G. Bélanger et al., Nucl. Phys. B581 (2000) 3, [arXiv:hep-ph/0002039]; G. Belanger et al., Phys. Lett. B 519, 93 (2001), [arXiv:hep-ph/0106275]
22. F. Moortgat, hep-ph/0105081; F. Moortgat, S. Abdullin, D. Denegri, hep-ph/0112046
23. See, e.g.: J.I. Illana et al., Eur. Phys. J.C1 (1998) 149; A. Djouadi, Phys. Lett. B435 (1998) 101
24. For a recent status see: S. J. Huber, M. Pospelov and A. Ritz, arXiv:hep-ph/0610003
25. For a recent review and additional references see: J. M. Cline, arXiv:hep-ph/0609145
26. E. Accomando *et al.*, arXiv:hep-ph/0608079
27. R. M. Godbole, Pramana 67 (2006) 835
28. A. Pilaftsis and C.. Wagner, Nucl. Phys. B553, 3 (1999), Nucl. Phys. B586, 92 (2000), M. Carena, J. Ellis, A. Pilaftsis and C. Wagner, Nucl. Phys. B586, 92 (2000); S.Y. Choi, M. Drees and J.S. Lee, Phys. Lett. B481, 57 (2000); J. R. Ellis, J. S. Lee and A. Pilaftsis, Mod. Phys. Lett. A21, 1405 (2006)
29. M. Carena et al. Nucl. Phys. B659, 145 (2003)
30. M. Carena et al, Nucl. Phys. B625, 345 (2002)
31. A. Mendez and A. Pomarol, Phys. Lett. B272 313 (1991), J. Gunion, H. Haber and J. Wudka, Phys. Rev. D43 904 (1991), B.Grzadkowski, J.Gunion and J. Kalinowski, Phys. Rev. D60 075011 (1999)
32. J.F. Gunion, B. Grzadkowski, H.E. Haber and J. Kalinowski, Phys. Rev. Lett. 79, 982 (1997)
33. G. Abbiendi et al. [OPAL Collaboration], Eur. Phys. J. C37 (2004) 49
34. J. R. Ellis, J. S. Lee and A. Pilaftsis, Phys. Rev. D70, 075010 (2004); S.Y. Choi et al, Eur. Phys. J. C40, 55 (2005)
35. U. Ellwanger, M. Rausch de Traubenberg and C.A. Savoy, Phys. Lett. B **315** (1993) 331; Z. Phys. C67 (1995) 665 and Nucl. Phys. B **492** (1997) 307; T. Elliott, S.F. King and P. White, Phys. Lett. B351 (1995) 213; S.F. King and P. White, Phys. Rev. D **52** (1995) 4183
36. J. Kim and H. Nilles, Phys. Lett. B138 (1984) 150
37. A. Djouadi, U. Ellwanger and A. Teixeira, Phys. Rev. Lett. 101 (2008) 101802
38. For recent analyses, see: U. Ellwanger and C. Hugonie, Eur. Phys. J. C25, 297 (2002); U. Ellwanger et al., hep-ph/0111179 and hep-ph/0305109; U. Ellwanger, C. Hugonie and J. Gunion, JHEP 0502, 066 (2005); J. Gunion and R. Dermisek, Phys. Rev. Lett. 95, 041801 (2005); D.J. Miller, R. Nevzorov and P.M. Zerwas, Nucl. Phys. B681, 3 (2004); V. Barger et al., Phys. Rev. D73, 115010 (2006); C. Balazs and D. Carter, arXiv:0808.0770 [hep-ph]
39. For a recent discussion and more references, see A. Djouadi et al., JHEP 0807 (2008) 002 [arXiv:0801.4321 [hep-ph]]
40. See e.g., P.N. Pandita, Pramana 51, 169 (1998) and references therein
41. M. Drees et al., Phys. Lett. B433, 346 (1998)
42. See e.g. S. King, S. Moretti and R. Nevzorov, Phys. Rev. D73, 035009 (2006)
43. See e.g. T. Han, P. Langacker and B. McElrath, Phys. Rev. D70, 115006 (2004)
44. For a review, see: J. Gunion, hep-ph/0212150
45. See e.g., J. Espinosa and M. Quiros, Phys. Rev. Lett. 81, 516 (1998)
46. M. Hirsch, et al, Phys. Rev. D73, 055007 (2006); A. Villanova del Moral in Ref. [26]
47. L. Randall and R. Sundrum, Phys. Rev. Lett. 83, 3370 (1999)
48. J. L. Hewett and T. G. Rizzo, JHEP 08, 028 (2003)
49. M. Chaichian et al, Phys. Lett. B524, 161 (2002)
50. D. Dominici et al, Nucl. Phys. B671, 243 (2003)
51. N. Arkani-Hamed, S. Dimopoulos and G. Dvali, Phys. Lett. B429, 263 (1998) and Phys. Rev. D59, 086004 (1999); A. Antoniadis, N. Arkani-Hamed, S. Dimopoulos and G. Dvali, Phys. Lett. B436, 267 (1998)
52. G. Giudice, R. Rattazi and J. Wells, Nucl. Phys. 595, 250 (2001); M. Battaglia et al, hep-ph/0402062
53. J. van der Bij, Phys. Lett. B636, 56 (2006); S. Dilcher and J. van der Bij, Phys. Lett. B638, 234 (2006); see also, J. Kumar and J.D. Wells, Phys. Rev. D74, 115017 (2006)
54. See, e.g. D. Choudhury, A. Datta and K. Huitu, Nucl. Phys. B673, 385 (2003)
55. N. Arkani-Hamed et al., JHEP 08, 021 (2002); N. Arkani-Hamed et al, JHEP 07, 034 (2002)
56. S. Weinberg, Phys. Rev. D13, 974 (1979) and Phys. Rev. D19, 1277 (1979); L. Susskind, Phys. Rev. D20, 2619 (1979); For a review: C. Hill and E. Simmons, Phys. Rept. 381 (2003) 235
57. W. Kilian, D. Rainwater and J. Reuter, Phys. Rev. D74, 095003 (2006); ibid. D71, 015008 (2005)
58. G. Giudice et al, hep-ph/0703164
59. C. Csaki et al., Phys. Rev. D69, 055006 (2004); C. Csaki et al, Phys. Rev. Lett. 92, 101802 (2004)
60. See e.g., A. Birkedal, K. Matchev and M. Perelstein,

Phys. Rev. Lett. 94, 191803 (2005)

61. H. Georgi et al., Phys. Rev. Lett. 40 (1978) 692; S.L. Glashow, D.V. Nanopoulos and A. Yildiz, Phys. Rev. D18 (1978) 1724; R.N. Cahn and S. Dawson, Phys. Lett. B136 (1984) 196; K. Hikasa, Phys. Lett. B164 (1985) 341; G. Altarelli, B. Mele and F. Pitolli, Nucl. Phys. B287 (1987) 205; Z. Kunszt, Nucl. Phys. B247 (1984) 339; J. Gunion, Phys. Lett. B253 (1991) 269
62. M. Spira, `http://mspira.home.cern.ch/mspira/proglist.html`
63. A.D. Martin et al. [MRST Collaboration], Eur. Phys. J. C28 (2003) 455
64. See e.g., J.F. Gunion et al., Phys. Rev. D34 (1986) 101; J. Gunion, G. Kane and J. Wudka, Nucl. Phys. B299 (1988) 231; M. Dittmar and H. Dreiner, Phys. Rev. D55 (1997) 167
65. ATLAS Collaboration, Technical Design Report, CERN/LHCC/99-15 (1999)
66. CMS Collaboration, Physics TDR, CERN/LHCC/2006-021, June 2006
67. E. Richter–Was et al., Int. J. Mod. Phys. A13 (1998), ATLAS Note PHYS–No–074. D. Denegri et al. hep-ph/0112045 and CMS–Note 2003/033; G. Branson et al. (CMS and ATLAS Collaborations), Eur. Phys. J. direct C4 (2002) N1; S. Asai et al., Eur. Phys. J. C32S2 (2004) 19; V. Büscher and K. Jakobs, Int. J. Mod. Phys. A20 (2005) 2523
68. Proceedings of the Les Houches Workshops on "Physics at TeV Colliders": A. Djouadi et al. (1999), hep-ph/0002258 ; D. Cavalli et al. (2001), hep-ph/0203056; K.A. Assamagan et al. (2003), hep-ph/0406152; C. Buttar et al., hep-ph/0604120 (2005)
69. Proceedings of the Les Houches Workshop 2007, S. Dawson et al., arXiv:0803.1154 [hep-ph]
70. G. Weiglein et al., "LHC/LC Study Group", Phys. Rept. 426 (2006) 47
71. A. Djouadi, M. Spira and P. Zerwas, Phys. Lett. B264 (1991) 440; S. Dawson, Nucl. Phys. B359 (1991) 283
72. M. Spira et al., Nucl. Phys. B453 (1995) 17; Phys. Lett. B318 (1993) 347
73. R.V. Harlander and W. Kilgore, Phys. Rev. Lett. 88 (2002) 201801; C. Anastasiou and K. Melnikov, Nucl. Phys. B646 (2002) 220; V. Ravindran, J. Smith and W.L. Van Neerven, Nucl. Phys. B665 (2003) 325
74. S. Moch and A. Vogt, Phys. Lett. B631 (2005) 48; S. Marzani et al., Nucl. Phys. B800 (2008) 127 and arXiv:0809.4934 [hep-ph]; V. Ahrens et al., arXiv:0808.3008 [hep-ph] and arXiv:0809.4283 [hep-ph]
75. C. Anastasiou, K. Melnikov and F. Petriello, Nucl. Phys. B274 (2005) 197; C. Anastasiou, G. Dissertori and F. Stockli, JHEP 0709 (2007) 018; S. Catani and M. Grazzini, Phys. Rev. Lett. 98 (2007) 222002; M. Grazzini, JHEP 0802 (2008) 043
76. S. Catani et al, JHEP 0307 (2003) 028
77. C.R. Schmidt, Phys. Lett. B413 (1997) 391; D. de Florian, M. Grazzini and Z. Kunszt, Phys. Rev. Lett. 82 (1999) 5209. C. Balazs and C. P. Yuan, Phys. Lett. B478 (2000) 192; C. J. Glosser and C. R. Schmidt, JHEP 0212 (2002) 016; E. L. Berger and J. W. Qiu, Phys. Rev. D67 (2003) 034026; A. Kulesza, G. Sterman, W. Vogelsang, Phys. Rev. D69 (2004) 014012; G. Bozzi et al, Phys. Lett. B564 (2003) 65; V. Ravindran, J. Smith and W.L. Van Neerven, Mod. Phys. Lett. A18 (2003) 1721-1734; C. Anastasiou, L. Dixon and K. Melnikov, hep-ph/0211141; C. Anastasiou K. Melnikov and F. Petriello, Phys. Rev. Lett. 93 (2004) 262002; J. Campbell, R. Ellis and G. Zanderighi, JHEP 10 (2006) 028; J. Butterworth et al., Phys. Rev. Lett. 100 (2008) 2420001
78. A. Djouadi and P. Gambino, Phys. Rev. Lett. 73 (1994) 2528; U. Aglietti et al. Phys. Lett. B595 (2004) 432; G. Degrassi and F. Maltoni, Phys. Lett. B600 (2004) 255; S. Actis et al., arXiv:0809.3667 [hep-ph]
79. G. Altarelli, R.K. Ellis and G. Martinelli, Nuc. Phys. B157 (1979) 461; J. Kubar–André and F. Paige, Phys. Rev. D19 (1979) 221; T. Han and S. Willenbrock, Phys. Lett. B273 (1991) 167; J. Ohnemus and W. J. Stirling, Phys. Rev. D47 (1993) 2722
80. For a review of NLO QCD corrections, see: M. Spira, Fortschr. Phys. 46 (1998) 203; see also: hep-ph/9711394 and hep-ph/9810289; A. Djouadi and M. Spira, Phys. Rev. D62 (2000) 014004
81. O. Brein, A. Djouadi and R. Harlander, Phys. Lett. B579 (2004) 149
82. M. L. Ciccolini, S. Dittmaier and M. Krämer, Phys. Rev. D68 (2003) 073003
83. T. Han, G. Valencia and S. Willenbrock, Phys. Rev. Lett. 69 (1992) 3274
84. M. Ciccolini, A. Denner and S. Dittmaier, Phys. Rev. Lett. 99 (2007) 161803 and Phys. Rev. D77 (2008) 013002
85. T. Figy, C. Oleari and D. Zeppenfeld. Phys. Rev. D68 (2003) 073005
86. V. Barger et al., Phys. Rev. D44 (1991) 1426; V. Barger, R. Phillips, D. Zeppenfeld, Phys. Lett. B346 (1995) 106; D. Rainwater and D. Zeppenfeld JHEP 9712 (1997) 5
87. D. Zeppenfeld et al, Phys. Rev. D62 (2000) 013009 and in the first report of Ref. [68]
88. M. Dührssen et al., Phys. Rev. D70 (2004) 11300
89. T. Plehn, D. Rainwater and D. Zeppenfeld, Phys. Rev. D61 (2000) 093005; O. Eboli and D. Zeppenfeld, Phys. Lett. B495 (2000) 147; N. Kauer, T. Plehn, Rainwater and D. Zeppenfeld, Phys. Lett. B503 (2001) 113; T. Plehn and D. Rainwater, Phys. Lett. B520 (2001) 108; M.L. Mangano et al., Phys. Lett. B556 (2003) 50
90. W. Beenakker et al., Phys. Rev. Lett. 87 (2001) 201805; Nucl. Phys. B653 (2003) 151; S. Dawson et al., Phys. Rev. Lett. 87 (2001) 201804 and Phys. Rev. D67 (2003) 071503
91. S. Dittmaier, M. Krämer and M. Spira, Phys. Rev. D70 (2004) 074010; S. Dawson et al, Phys. Rev. D69 (2004)

074027
92. J. Campbell et al, Phys. Rev. D67 (2003) 095002; R. Harlander and W. Kilgore, Phys. Rev. D68 (2003) 013001; F. Maltoni, Z. Sullivan and S. Willenbrock Phys. Rev. D67 (2003) 093005; E. Boos and T. Plehn, Phys. Rev. D69 (2004) 094005; T. Plehn, Phys. Rev. D67 (2003) 014018
93. J. Campbell et al. in Ref. [68] 2001
94. T. Plehn, Phys. Rev. D67 (2003) 014018
95. A. Djouadi and S. Ferrag, Phys. Lett. B586 (2004) 345
96. A. Djouadi, arXiv:0810.2439 [hep-ph]
97. For recent reviews and a summary of recent developments, see: R. Harlander, J.Phys.G35 (2008) 033001; M. Mühlleitner, arXiv:0810.1158
98. C. Kao, Phys. Rev. D46 (1992) 4907 and Phys. Lett. B328 (1994) 420; Y. Jun et al., hep-ph/0209279; C. Kao, G. Lovelace and L.H. Orr, Phys. Lett. B567 (2003) 259; O. Brein and W. Hollik, Phys. Rev. D68 (2003) 095006 (2003); B. Field et al., Phys. Lett. B551 (2003) 137; B. Field, S. Dawson and J. Smith, Phys. Rev. D69 (2004) 074013; Q.H. Cao, S. Kanemura and C.P. Yuan, Phys. Rev. D69 (2004) 075008; U. Langenegger, JHEP 0606 (2006) 035
99. A. Bawa, C. Kim and A. Martin, Z. Phys. C47 (1990) 75; V. Barger, R. Phillips and D.P. Roy, Phys. Lett. B324 (1994) 236; S. Moretti and K. Odagiri, Phys. Rev. D55 (1997) 5627; J. Gunion, Phys. Lett. B322 (1994) 125; F. Borzumati, J.L. Kneur and N. Polonsky, Phys. Rev. D60 (1999) 115011; D. Miller et al, Phys. Rev. D61 (2000) 055011; D.P. Roy, Phys. Lett. B459 (1999) 607
100. For a review, see S. Moretti, Pramana 60 (2003) 369 and hep-ph/0205104
101. E. Boos, A. Djouadi and A. Nikitenko, Phys. Lett. B578 (2004) 384
102. E. Boos et al., Phys. Rev. D66 (2002) 055004; E. Boos et al., Phys. Lett. B622 (2005) 311; A. Djouadi and Y. Mambrini, JHEP 0612 (2006) 001
103. V. Barger and C. Kao, Phys. Lett. B424 (1998) 69; Tao Han and B. McElrath, Phys. Lett. B528 (2002) 81; S. Dawson, D. Dicus and C. Kao, Phys. Lett. B545 (2002) 132
104. D.P. Roy, Mod. Phys. Lett. A19 (2004) 1813
105. A. Djouadi, J.L. Kneur, G. Moultaka, Phys. Rev. Lett. 80 (1998) 1830 and Nucl. Phys. B569 (2000) 53; G. Bélanger et al., Eur. Phys. J. C9 (1999) 511; A. Dedes and S. Moretti, Eur. Phys. J. C10 (1999) 515 and Phys. Rev. D60 (1999) 015007
106. Asesh Datta et al., Phys. Rev. D65 (2002) 015007; Asesh Datta et al., Nucl. Phys. B681 (2004) 31
107. A. Dedes and S. Moretti, Phys. Rev. Lett. 84, 22 (2000); Nucl. Phys. B576, 29 (2000)
108. S. Moretti, S. Munir and P. Poulose, Phys. Lett. B649, 206 (2007); S. Hesselbach et al, Eur. Phys. J. C54, 129 (2008)
109. S.Y. Choi, K. Hagiwara and J.S. Lee, Phys. Rev. D64, 032004 (2001); Phys. Lett. B529, 212 (2002); A. G. Akeroyd and A. Arhrib, Phys. Rev. D64, 095018 (2001); A. Arhrib, D. K. Ghosh and O. C. W. Kong, Phys. Lett. B537, 217 (2002); S. Y. Choi et al, Eur. Phys. J. C25, 307 (2002)
110. D. K. Ghosh, R. M. Godbole and D. P. Roy, Phys. Lett. B628, 131 (2005) [arXiv:hep-ph/0412193]
111. J. S. Lee et al. Comput. Phys. Commun. 156 (2004) 283; S. Heinemeyer, W. Hollik and G. Weiglein, Comp. Phys. Commun. 124 (2000) 76
112. M. Schumacher, hep-ph/0410112.
113. R. M. Godbole and D. P. Roy, in Ref. [26]
114. P. Bandyopadhyay et al., Phys. Rev. D78, 015017 (2008)
115. J. Gunion, Phys. Rev. Lett. 72 (1994) 199
116. D. Choudhury and D.P. Roy, Phys. Lett. B322 (1994) 368
117. O. Eboli and D. Zeppenfeld, Phys. Lett. B495 (2000) 147
118. R. M. Godbole et al., Phys. Lett. B571, 184 (2003), [arXiv:hep-ph/0304137]
119. H. Davoudiasl, T. Han and H. Logan, Phys. Rev. D71 (2005) 115007
120. B. Di Girolamo et al., in Ref. [68] (2001), G. Bélanger et al. in [68] (2001), S. Balatenychev et al., in [68] (2001)
121. A. Djouadi and G. Moreau, Phys. Lett. B660 (2008) 67 [arXiv:0707.3800 [hep-ph]]
122. A. Djouadi *et al.* [ILC Collaboration], arXiv:0709.1893 [hep-ph]
123. E. Glover and J. van der Bij, Nucl. Phys. B309 (1988) 282; T. Plehn, M. Spira and P. Zerwas, Nucl. Phys. B479 (1996) 46; M. Muhlleitner et al., Eur. Phys. J. C10 (1999) 45
124. U. Baur, T. Plehn and D. L. Rainwater, Phys. Rev. Lett. 89 (2002) 151801; Phys. Rev. D67 (2003) 033003; Phys. Rev. D69 (2004) 053004
125. R. Kinnunen, S. Lehti, F. Moortgat, A. Nikitenko and M. Spira in Ref. [68] (2003)
126. R. M. Godbole et al., arXiv:hep-ph/0404024
127. V. Barger et al., Phys. Rev. D49 (1994) 79
128. S. Y. Choi et al, Phys. Lett. B553 (2003) 61
129. C. P. Buszello et al., Eur. Phys. J. C32 (2004) 209; C. P. Buszello and P. Marquard, in Ref. [26]; C. P. Buszello, P. Marquard and J. J. van der Bij, arXiv:hep-ph/0406181
130. R. M. Godbole et al, Pramana 67 (2006) 617; *ibid.* in Ref. [26]
131. R. M. Godbole, D. J. Miller and M. M. Muhlleitner, JHEP 0712, 031 (2007); [arXiv:0708.0458 [hep-ph]]
132. T. Plehn, D. Rainwater and D. Zeppenfeld, Phys. Rev. Lett. 88 (2002) 051801
133. B. Zhang et al, Phys. Rev. D67 (2003) 114024
134. C. P. Buszello and P. Marquard, arXiv:hep-ph/0603209
135. V. Del Duca et al., arXiv:hep-ph/0109147
136. V. Hankele, G. Klamke and D. Zeppenfeld, arXiv:hep-ph/0605117; V. Hankele et al, Phys. Rev. D74 (2006)

095001
137. K. Odagiri, JHEP 0303 (2003) 009
138. W. Bernreuther, M. Flesch and P. Haberl, Phys. Rev. D58 (1998) 114031. W. Bernreuther, A. Brandenburg and M. Flesch, arXiv:hep-ph/9812387
139. W. Khater and P. Osland, Nucl. Phys. B661 (2003) 209
140. J. F. Gunion and X. G. He, Phys. Rev. Lett. 76 (1996) 4468; J. Albert et al., in Ref. [26]; B. Field, Phys. Rev. D 66 (2002) 114007
141. V. A. Khoze et al, Eur. Phys. J. C 23 (2002) 311; A. De Roeck et al, Eur. Phys. J. C 25 (2002) 391
142. J. R. Ellis et al, Phys. Rev. D 71 (2005) 075007
143. P. S. Bhupal Dev et al., Phys. Rev. Lett. 100, 051801 (2008) [arXiv:0707.2878 [hep-ph]]
144. A. Djouadi et al., in preparation
145. S. S. Biswal et al, Phys. Rev. D73, 035001 (2006) [Erratum-ibid. D74, 039904 (2006)] [arXiv:hep-ph/0509070]; S. S. Biswal et al, arXiv:0809.0202 [hep-ph]
146. S. Dutta et al, arXiv:0808.0477 [hep-ph]

Prospects of Searches for the Higgs Boson at the LHC

Bruce Mellado

Physics Department, University of Wisconsin - Madison, Madison, Wisconsin 53706, USA

One of the most exciting prospects for the Large Hadron Collider is to observe the Higgs boson. The most important experimental characteristics of the two general-purpose detectors, the CMS and ATLAS experiments are summarised. The sensitivity for the CMS and ATLAS experiments at the LHC to discover a Standard Model Higgs boson with relatively low integrated luminosity per experiment is outlined. The most relevant discovery modes are covered. A brief discussion about the expected performance from these experiments in searches for one or more of the Higgs bosons from the minimal version of the supersymmetric theories is also included.

1. Introduction

In the Standard Model (SM) of electroweak and strong interactions, there are four types of gauge vector bosons (gluon, photon, W and Z) and twelve types of fermions (six quarks and six leptons) [1–4]. These particles have been observed experimentally. At present, all the data obtained from the many experiments in particle physics are in agreement with the SM. In the SM, there is one particle the Higgs boson, that is responsible for giving masses to all the particles [5–10]. In this sense, the Higgs particle occupies a unique position.

Prior to the end of the year 2000, the Higgs particle was not observed experimentally. After the centre-of-mass energy at the LEP[1] accelerator of CERN reached 205 GeV/c^2 in 2000, excess candidates began to show up in the SM Higgs boson analysis in the ALEPH experiment, consistent with a Higgs boson mass, M_H, around 115 GeV/c^2 [11, 12]. One of the most exciting prospects for the Large Hadron Collider (LHC) [13] is confirming or rejecting the first possible experimental evidence for the Higgs particle at a mass around 115 GeV/c^2.

The Higgs boson mass is not predicated by theory and, to date, direct experimental searches for the Higgs boson have put a lower limit on its mass at $M_H >$ 114.4 GeV/c^2 @ 95% confidence level (CL) [12, 14].[2] Figure 1 shows the $\Delta\chi^2$ curve derived from high-Q^2 precision electroweak measurements, performed at LEP and by SLD, CDF and D0, as a function of the Higgs boson mass, assuming the SM. While this is not a proof that the SM Higgs boson actually exists, it does serve as a guideline in what mass range to look for it. A preferred value for the Higgs boson mass, derived by fitting precision electroweak data [15] is currently $M_H = 84^{+34}_{-26}$ GeV/c^2 with an upper bound of 154 GeV/c^2 @ 95% CL.

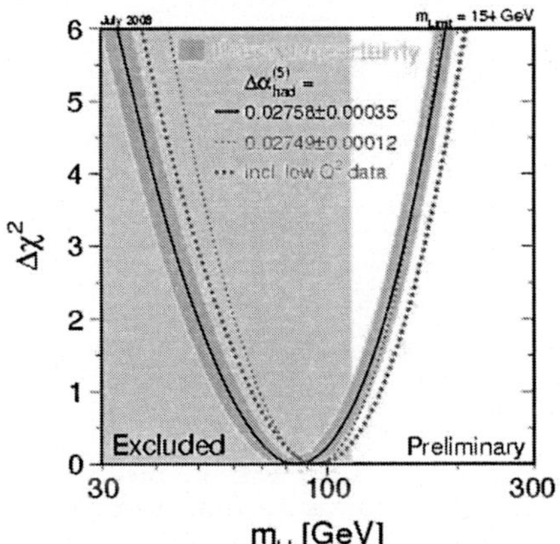

Figure 1. $\Delta\chi^2$ curve derived from high-Q^2 precision electroweak measurements, performed at LEP and by SLD, CDF and D0, as a function of the Higgs boson mass

Both the CMS and ATLAS experiments at the LHC, scheduled for proton-proton collision data-taking in 2009, have been designed to search for the Higgs boson over a wide mass range [16, 17]. The CMS and ATLAS detectors were designed for the search for the Higgs boson using, for the most part, inclusive signatures. In recent years significant progress has been made, both in the phenomenology and experimentally, in demonstrating the feasibility of observing a low mass SM Higgs boson in association with jets using different decay modes [16, 18–25].[3] Here we summarise the

[1] LEP stands for Large Electron-Positron Collider.

[2] This is usually referred to as the LEP limit.

[3] The search for a heavy SM Higgs boson in association with high transverse momentum hadronic jets had been considered earlier [26].

sensitivity for each experiment to discover a SM Higgs boson with relatively low integrated luminosity per experiment (1–30 fb^{-1}) as well as recent developments that have enhanced this sensitivity.

In this paper we briefly review the experimental apparatus, covering the most relevant parameters of the LHC accelerator facility (Section 2.1) together with the CMS and ATLAS detectors (Sections 2.2 and 2.3). In section 3 we'll touch upon the production mechanisms, of the Higgs boson in proton-proton collisions with the most important discovery modes and related experimental aspects. Section 4 summarises the discovery potential of the CMS and ATLAS detectors.

2. The Experimental Apparatus

2.1. The Large Hadron Collider

The LHC [13] at the European Organisation for Nuclear Research (CERN) is a two-ring-superconducting hadron accelerator collider constructed in the already existing 26.7 km tunnel that was used for the LEP machine. The LHC displays high luminosity experiments, CMS [27] and ATLAS [28], both aiming at registering proton-proton collisions with a peak instantaneous luminosity of $L = 10^{34}$ $\text{cm}^{-2}\text{s}^{-1}$. There are also two low luminosity experiments: LHCb [29][4] for B-physics, aiming at a peak instantaneous luminosity of $L = 10^{34}$ $\text{cm}^{-2}\text{s}^{-1}$, and TOTEM [30] for the detection of protons from elastic scattering at small angles, aiming at a peak instantaneous luminosity of $L = 10^{29}$ $\text{cm}^{-2}\text{s}^{-1}$. In addition to the proton beams, the LHC will also be operated with ion beams. The LHC has one dedicated ion experiment, ALICE [31],[5] aiming at a peak instantaneous luminosity of $L = 10^{27}$ $\text{cm}^{-2}\text{s}^{-1}$ for nominal lead-lead ion operation.

To collide two counter-rotating proton beams requires opposite magnetic dipole fields in both rings. The LHC is designed as a proton-proton collider with separate magnet fields and vacuum chambers in the main trajectories and with common sections only at the insertion regions where the detectors are located. The two beams share an approximately 130 m long beam-pipe along the interacting regions. According to the design, a large number of bunches (2808 for each proton beam) will be provided with a nominal bunch spacing of 25 ns. The long common beam pipe implies 34 parasitic collision points at each insertion region where the detectors will observe collisions. The LHC uses twin bore magnets since there is not enough room for two separate rings. These consist of two sets of coils and

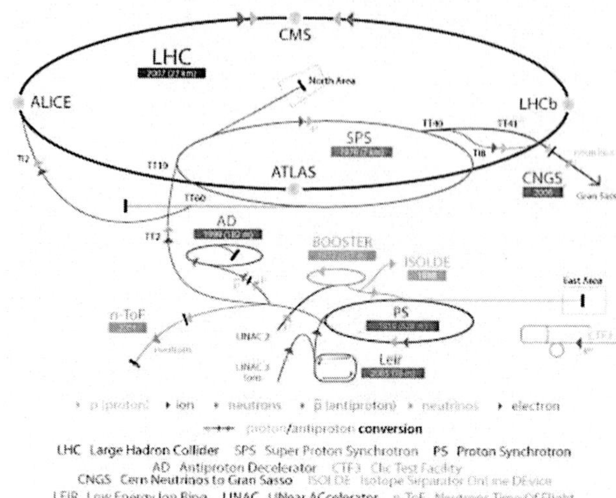

Figure 2. The LHC accelerator complex. CERN's accelerator complex is a succession of particle accelerators that can reach increasingly higher energies. Each accelerator boosts the speed of a beam of particles, before injecting it into the next one in the sequence

beam channels within the same mechanical structure and cryostat. The peak beam energy depends on the integrated dipole field around the storage ring. This implies a peak dipole field of 8.33 T for the 7 TeV in the LHC machine and the use of superconducting magnets.

CERN's accelerator complex is a succession of particle accelerators that can reach increasingly higher energies (Fig. 2). Each accelerator boosts the speed of particles clustered in bunches, before injecting it into the next one in the sequence. Protons are produced by ripping off electrons from hydrogen atoms. They are injected from the linear accelerator (LINAC2) into the PS booster, then into the Proton Synchrotron (PS), followed by the Super Proton Synchrotron (SPS). Finally, the protons reach the LHC. Protons will circulate in the LHC for 20 minutes before reaching the maximum speed and energy.

2.2. The CMS Experiment

The Compact Muon Solenoid (CMS) is one of two general-purpose experiments at the LHC designed to explore the physics at the energy frontier [27]. The main volume of the CMS detector (Fig. 3) is a multi-layered cylinder, about 21.3 m long and 16 m in diameter, weighing more than 13,000 tons. The innermost

[4]LHCb stands for Large Hadron Collider Beauty Experiment.
[5]ALICE stands for A Large Ion Collider Experiment.

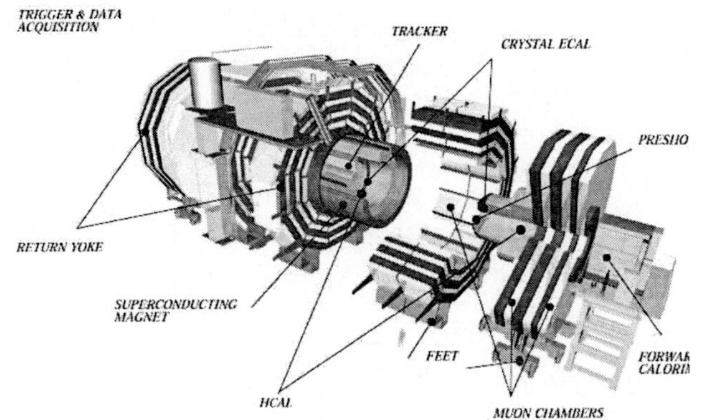

Figure 3. The CMS detector

layer is a silicon-based particle tracker, surrounded by electro-magnetic and hadronic calorimeters for measuring particle energies. These detector devices are placed inside a central superconducting solenoid magnet, 13 m long and 6 m in diameter. The magnetic field produced by the superconducting solenoid is used to measure the momentum of particles. Outer muon detectors surround the central magnet. The CMS detector is a hermetic detector in the range $|\eta| < 5$.

2.2.1. The Inner Tracker

The inner tracking system of CMS is designed to give a precise measurement of the trajectories of charged particles coming out from the LHC collisions [27]. Secondary vertexes will be reconstructed precisely, as well. The Inner Tracker surrounds the interaction point and has a length of 5.8 m and a diameter of 2.5 m. The CMS solenoid provides a homogeneous magnetic field of 4 T over the full volume of the Inner Tracker. The tracker is entirely based on silicon detectors due to the requirements of granularity, occupancy, speed and radiation hardness. The CMS Inner Tracker is composed of a pixel detector with three barrel layers at radii between 4.4 cm and 10.2 cm and a silicon strip tracker with 10 barrel detection layers to a radius of 1.1 m. Each system is covered on backward and forward sides by end-caps which consist of two disks in the pixel detector and three plus nine disks in the strip tracker on each side of the barrel, extending the acceptance of the tracker up to a pseudorapidity of $|\eta| < 2.5$.[6] The CMS Inner Tracker, with about 200 m^2 of active silicon are, is the largest silicon tracker ever assembled. In addition, CMS is the first experiment using silicon detectors in this outer tracker region. The construction of the CMS tracker, composed of 1440 pixel and 15,148 strip detector modules, required the development of production methods and quality control procedures that are new to the field of particle physics detectors.

In order to keep the occupancy at or below 1%. pixelated detectors with radii below 10 cm need to be constructed. The pixel size is 100×150 μmm in $-\phi$ and z, respectively. This is driven by the impact parameter resolution. In this configuration the occupancy is of the order 10^{-4} per pixel and the LHC bunch crossing. At intermediate radii (20 cm < < 55 cm) the reduced particle flux allows the use of silicon micro-strip detectors with a typical cell size of 10 cm $\times$ 80 μm. This results in an expected occupancy of up to $2 \div 3\%$ per strip and the LHC bunch crossing.

In order to maintain a good signal to noise ratio well above 10, CMS uses thicker silicon sensors for the outer tracker region (500 mm thickness as opposed to the 320 mm in the Inner Tracker) with correspondingly higher signal. Cell sizes up to about 25 cm $\times$ 180 μm can be used in the outer region of the tracker, with an occupancy of about 1%.

2.2.2. The Calorimeter

One of the driving criteria in the design was the capability to detect two photons coming from the decay of a Higgs boson (Section 3). This capability is enhanced by the good energy resolution provided by a homogeneous crystal calorimeter. The use of high density crystals has allowed the design of a calorimeter which is fast, has fine granularity and is radiation resistant, all important characteristics in the LHC environment.

The electro-magnetic calorimeter of CMS (ECAL) is a hermetic homogeneous calorimeter consisting of 61,200 lead tungstate ($PbWO_4$) crystals mounted in the central barrel part [27]. The barrel part is closed by 7324 crystals in each of the two end-caps. Avalanche photo-diodes are used as photo-detectors in the barrel and vacuum photo-triodes in the end-caps. A preshower detector is placed in front of the end-cap crystals.

The barrel part of the ECAL (EB) covers the pseudorapidity range $|\eta| < 1.479$. The barrel granularity is 360-fold in ϕ and 2×85-fold in η, resulting in a total of 61,200 crystals. The crystals have a tapered shape, slightly varying with position in η. They are mounted in a quasi-projective geometry to avoid cracks aligned with particle trajectories, so that their axes make a small angle (3°) with respect to the vector from the nominal interaction vertex, in both the ϕ and η projec-

[6]Pseudorapidity, η, is defined as $\eta = -\log(\tan(\theta/2))$, where, θ is the polar angle.

tions. The crystal cross-section corresponds to approximately 0.0174×0.0174 in $\eta \times \phi$ or 22×22 mm^2 at the front face of crystal, and 26×26 mm^2 at the rear face. The crystal length is 230 mm corresponding to $25.8\,X_0$.

The end-caps (EE) cover the range $1.479 < |\eta| < 3.0$ [27]. The distance between the interaction point and the end-cap envelope is 315.4 cm. The end-cap consists of identically shaped crystals grouped in mechanical units of 5×5 crystals (super-crystals, or SCs) consisting of a carbon-fiber alveolar structure. The crystals and SCs are arranged in a rectangular x–y grid, with the crystals pointing at a focus 1300 mm beyond the interaction point, giving off-pointing angles ranging from 2–8 degrees. The crystals have a rear face cross section 30×30 mm^2, a front face cross section of 28.62×28.62 mm^2 and a length of 220 mm ($24.7\,X_0$).

The hadronic barrel calorimeter (HB) is a sampling calorimeter covering the range $|\eta| < 1.3$ consisting of 36 identical azimuthal wedges arranged in two half-barrels [27]. The wedges are constructed out of flat brass absorber plates aligned parallel to the beam axis. The plastic scintillator is divided into 16 η sectors, resulting in a segmentation $\Delta\eta \times \Delta\phi = 0.087 \times 0.087$. The total absorber thickness at 90° is 5.82 interaction lengths, λ. The HB effective thickness increases with polar angle (θ) as $1/\sin\theta$, resulting in 10.6 λat $|\eta| < 1.479$. The electro-magnetic crystal calorimeter in front of HB adds about 1.1 λ of material. In the central pseudorapidity region, the combined stopping power of EB plus HB does not provide sufficient containment for hadron showers. To ensure adequate sampling depth for $|\eta| < 1.3$, the hadron calorimeter is extended outside the solenoid with a tail catcher called the outer calorimeter.

The hadron calorimeter end-caps (HE) cover $1.3 < |\eta| < 3$ and it is a sampling calorimeter [27]. The total length of the calorimeter, including electro-magnetic crystals, is about $10\,\lambda$. Quartz fibers were chosen as the active medium for the forward calorimeter, extending the calorimeter coverage to $|\eta| < 5$ [27]. The calorimeter consists of a steel absorber structure that is composed of 5 mm thick grooved plates and fibers are inserted in these grooves.

2.2.3. The Muon System

CMS uses three different types of gaseous particle detectors for muon identification [27]. In the barrel region the neutron-induced background and the muon rate are low. Here the $\int Bdl$ is uniform and mostly contained in the steel yoke. Standard rectangular drift chambers are used, which are outside the cryostat. The barrel drift tube (DT) chambers cover the range $|\eta| < 1.2$.

In the two end-cap regions the muon rates and background levels are high and the magnetic field is large and non-uniform. Here the muon system consists of cathode strip chambers (CSC). With their fast response time, fine segmentation and radiation resistance, the CSCs identify muons in the range $0.9 < |\eta| < 2.4$. A complementary, dedicated trigger system consisting of resistive plate chambers (RPC) was added. The RPCs are assembled in both the barrel and end-cap regions. The RPCs provide a fast, independent and highly segmented trigger with a sharp transverse momentum threshold over a large portion of the rapidity range $|\eta| < 1.6$.

2.2.4. The Superconducting Magnet

The superconducting magnet for CMS was designed to produce a 4-T magnetic field in a free bore of 6 m diameter and 12.5 m length with a total stored energy of 2.6 GJ at full current [27]. The return flux occurs through a 10000-t yoke comprising 5 wheels and 2 end-caps, composed of three disks each. The distinctive feature of the 220-tonne cold mass is the 4-layer winding made from a stabilised reinforced NbTi conductor. The ratio between stored energy and cold mass is high (11.6 KJ/kg), causing a large mechanical deformation (0.15%) during energising. This is well beyond the values of previous solenoidal detector magnets. The magnet was designed to be assembled and tested in a surface hall (SX5), prior to being lowered 90 m below ground to its final position in the experimental cavern. After provisional connection to its ancillaries, the CMS magnet has been fully and successfully tested and commissioned in SX5 during the autumn 2006.

2.2.5. The Trigger and Data Acquisition Systems

In order to reduce the GHz proton-proton collision rate expected at design instantaneous luminosity, the CMS detector pursues a two-level trigger system: Level-1 (L1) trigger and High-Level Trigger (HLT), respectively [27]. The L1 trigger consists of custom-designed, largely programmable electronics, whereas the HLT is a software system implemented in a filter farm of about 1000 commercial processors. The rate reduction capability is designed to be at least a factor of 10^6 for the combined L1 trigger and HLT. The design output rate limit of the L1 trigger is 100 kHz. The L1 trigger uses coarsely segmented data from the calorimeters and the muon system, while holding the high resolution data in pipelined memories in the front-end electronics. The HLT has access to the complete read-out data and can therefore perform complex calculations similar to those

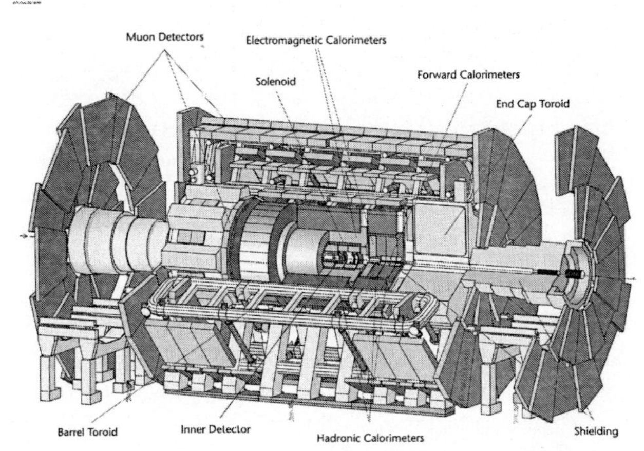

Figure 4. The ATLAS detector

made by the analysis off-line software if required for specially interesting events. The Global Calorimeter and Global Muon Triggers determine the highest-rank calorimeter and muon objects across the entire experiment and transfer them to the Global Trigger. The latter takes the decision to reject an event or to accept it for further evaluation by the HLT. The output rate of the HLT is 200 Hz. The CMS Data Acquisition System is designed to handle the data flow driven by the trigger chain.

2.3. The ATLAS Experiment

ATLAS (A Toroidal LHC ApparatuS) is one of the two general-purpose experiments at the LHC designed to explore the physics at the energy frontier [28]. The ATLAS detector consists of four major components, the Inner Tracker which measures the momentum of each charged particle, the calorimeter which measures the energies carried by the particles, the muon spectrometer which identifies and measures muons and the magnet system that bends charged particles for momentum measurement (Fig. 4). The detector is a cylinder with a total length of 42 m and a radius of 11 m and weighs approximately 7000 tonnes. The ATLAS detector is hermetic in the range $|\eta| < 4.9$.

2.3.1. The Inner Tracker

The ATLAS Inner Tracker provides precise measurement of the trajectories of charged particles that pass through the region in the range $|\eta| < 2.5$ [28]. The innermost layer is a precision tracker comprised of three layers of silicon pixel detectors, arranged in a cylindrical geometry, with an inner radius of about 50 mm, an outer radius of about 150 mm and a total length of about 1.3 m. It provides position measurements with an accuracy of about 10 μm in R–ϕ and about 115 μm in z in the barrel, with similar precision for the disks at either end.

The middle layer, the Semiconductor Tracker (SCT), uses four stereo layers of silicon strip detectors with one set of strips in each layer offset by a small angle with respect to the other, to provide additional position measurements with a precision of 17 μm in R–ϕ and about 580 μm in z, again with similar precision for the disks at either end. The SCT has an inner radius of about 250 mm, an outer radius of about 600 mm, and a length of about 5.4 m.

The outer layer, the Transition Radiation Tracker (TRT), is a straw tube tracker that provides a large number (about 36 per track) of R–ϕ measurements in the region with radius, R, from about 550 mm to about 1100 mm. The straws are interleaved with polypropylene (fibers in the barrel region, foil in the end-caps) to provide transition radiation to help with electron identification. The total length of the TRT is about 5.4 m; this corresponds to the range $|\eta| < 2.5$. The precision of the R–ϕ measurement is 130 μm per straw.

The material budget of the ATLAS Inner Detector ranges from < 0.5 radiation lengths and 0.2 interaction lengths near $\eta = 0$ to almost 2.5 X_0 and 0.7 λ in the region around $|\eta| = 1.6$–1.7. This large material budget can lead to degraded performance for the reconstruction of physics objects.

2.3.2. The Calorimeter

The innermost layer, comprised of the Liquid Argon (LAr) barrel and end-cap electromagnetic calorimeters, is a lead-LAr detector with accordion-shaped absorbers and electrodes [28]. This unusual geometry is intended to improve the readout time of the detector by avoiding current loops. It has a thickness $> 22\ X_0$ in the barrel region and a thickness $> 24\ X_0$ in the end-cap. The electromagnetic calorimeters are arranged in three layers, plus a pre-sampler in the region $|\eta| < 1.8$. The granularity of the pre-sampler in $\eta \times \phi$ is 0.025×0.1. The first sampling is segmented into strips of $\eta \times \phi$ is $0.025/8 \times 0.1$ in the range $|\eta| < 1.4$; the granularity decreases in the forward regions. Similarly, the second sampling has a granularity of 0.025×0.025 in $|\eta| < 1.4$, with similar or lower granularity in other regions. Finally, the third sampling is segmented into cells of 0.05×0.025 in $|\eta| < 1.35$ and in the range $1.5 < |\eta| < 2.5$.

The Tile Calorimeter (TileCal), placed directly outside the LAr calorimeter, is a hadronic sampling calorimeter using steel as the absorber and scintillating tiles as the active material [28]. It is separated into two regions, the barrel region with $|\eta| < 1$ and the extended barrel region with $0.8 < |\eta| < 1.7$. Like the electromagnetic calorimeter, the tile calorimeter is segmented longitudinally into three layers. In the barrel, the first layer has a thickness of about 1.5 λ, the second about 4.1 λ, and the thickness of the third sampling is 1.8 λ. In the extended barrel, these figures are 1.5 λ, 2.6 λ and 3.3 λ, respectively. The granularity of the first two layers of the tile calorimeter is 0.1×0.1; the granularity of the last layer is 0.1×0.2.

In the region $|\eta| > 1.5$ liquid argon is used as the active medium for the hadronic calorimetry. The Hadronic End Cap (HEC) uses copper plates as the absorber and covers the range up to $|\eta| < 3.2$ to a depth of about 10 λ [28]. It has three layers, with a granularity of 0.1×0.1 in the range $|\eta| < 2.5$ and 0.2×0.2 in the range $|\eta| > 2.5$.

The forward calorimeter (FCal) covers the region with $3.1 < |\eta| < 4.9$; like the other calorimeters, it is also divided into three longitudinal layers [28]. The FCal is split into one electromagnetic calorimeter and two hadronic calorimeters, each about 45 cm deep. The absorber for the electromagnetic FCal is copper; for the hadronic FCal modules, it is tungsten. The granularity is dependent on η and it is slightly different for the three different FCal modules, but in general it is lower for the FCal than for the other calorimeters.

2.3.3. The Muon System

The muon spectrometer consists of a large air-core toroidal magnet system instrumented with a variety of detectors that provide position measurements for muon tracks passing through its volume. The toroidal magnetic field geometry was chosen so that the momentum resolution will not be degraded significantly in the forward directions. Muon trajectories are measured by three concentric cylindrical shells of detectors with radii of about 5, 7.5 and 10 m in the barrel region, and by four end-cap "wheels" located at $|z| \approx 7.4$, 10.8, 14 and 21.5 m. The Monitored Drift Tubes (MDTs) are aluminum tubes with a diameter of about 30 mm; they are filled with a gas mixture of 93% argon and 7% of CO_2, with a tungsten-rhenium wire running down the centre [28]. Each MDT chamber provides a measurement of the bending direction, z (in the barrel) or R (in the end-cap), with precision of approximately 34 μm. The MDTs cover the full range of $|\eta| < 2.7$, except for the first layer which only covers $|\eta| < 2$.

Cathode Strip Chambers (CSCs) consist of thin anode wires suspended halfway between two parallel plates covered with cathode strips, with one layer oriented parallel to the wires and the other perpendicular; position measurements (in both the bending and azimuthal directions) are taken from these cathode strips and not the anode wires [28]. Four wire planes per chamber provide position measurements with a precision of about 60 μm per CSC plane in the bending direction and about 0.5 cm in the azimuthal direction

In addition to the MDTs and the CSCs ATLAS displays two muon subsystems used for triggering on muons candidates. Resistive Plate Chambers (RPCs) cover the region with $|\eta| < 1.05$ [28]. RPCs are parallel-plate electrodes filled with a gaseous mixture. Thin Gap Chambers (TGCs) cover the region with $1.05 < |\eta| < 2.4$ [28]. These are multi-wire proportional chambers similar to the CSCs.

2.3.4. The Magnet System

The central ATLAS solenoid has a length of 5.3 m with a bore of 2.4 m. The conductor is a composite that consists of a flat superconducting cable located in the centre of an aluminum stabiliser with rectangular cross-section. It is designed to provide a field of 2 T in the central tracking volume with a peak magnetic field of 2.6 T. To reduce the material build-up the solenoid shares the cryostat with the Liquid Argon calorimeter.

The ATLAS toroid magnet system consists of eight barrel coils housed in separate cryostats and two End-Cap cryostats housing eight coils each. The End-Cap coils systems are rotated by 22.5^o with respect to the barrel toroids in order to provide radial overlap and to optimize the bending power in the interface regions of both coil systems.

2.3.5. The Trigger and Data Acquisition Systems

The ATLAS trigger consists of a three-level system: the first-level trigger or Level-1, the second-level trigger or Level-2 and the third-level trigger or the Event Filter (EF) [32, 33]. Event selection is performed at Level-2 and EF with software algorithms. Both triggers have strong similarities in terms of software frameworks and architecture. They are commonly referred to as the High Level Trigger (HLT). The Level-1 trigger [32] reduces the initial 40 MHz to < 75 kHz in < 2.5 μs, the maximum output rate and latency the trigger hardware can tolerate. In the HLT [33], where the boundary between the two trigger steps is purposefully kept flexible, the Level-2 trigger will reduce the rate to O(2) kHz and the EF further to O(200) Hz. The available average latency of the two steps is substantially different, with ~ 10 ms for the Level-2 trigger and ~ 1 s for the EF.

The ATLAS Data Acquisition System is designed to handle the data flow driven by the trigger chain.

3. The Search for the Higgs Boson

In this section we deal with the most relevant aspects that define the different experimental analyses. In section 3.1 a brief overview of the production mechanisms and decay mode is given. In section 3.2 we give a brief overview of the experimental reconstruction of physics objects relevant to Higgs boson searches. In section 3.3 we give the specifics of the most relevant analysis for the observation of the Higgs boson.

3.1. Production Mechanisms and Decay Modes

When it comes to defining the experimental strategy for searches of the Higgs boson several theoretical inputs are necessary: the Higgs boson width, decay products and production mechanisms. The CMS and ATLAS detectors have been designed to a significant degree according to these theoretical inputs and within the context of inclusive analyses. A very important theoretical input is the expected natural width of the Higgs boson. Figure 5 shows the total width of the SM Higgs boson as a function of the mass [34]. In the mass range that seems to be preferred by the electroweak constraints (Section 1) the SM Higgs boson width is well below a GeV/c^2. For masses close to the LEP limit, the SM Higgs boson width is expected to be of the order of (1) MeV/c^2, which is much smaller than the resolutions that can be achieved experimentally. As a result, for practical purposes the width of the low mass SM Higgs boson is neglected in experimental searches. The SM Higgs boson natural width becomes comparable to the experimental resolution in some of the discovery channels (Section 3.3) for $M_H > 250 \div 300$ GeV/c^2. The SM Higgs boson width grows rapidly with the mass and at some point reaches (10^2) GeV/c^2, making experimental observation more challenging. On the other hand, when at the SM Higgs boson mass approaches TeV/c^2 perturbation theory breaks down and some of the theoretical inputs used here are not valid anymore.

Figure 6 shows the branching ratios of the SM Higgs boson to different decay products [34]. Branching ratios in Fig. 6 are given as a function of the Higgs boson mass. The final states most suitable for discovery at the LHC vary depending on the branching ratios, which are a function of the Higgs boson mass, and the relevant backgrounds. For $M_H < 2M_W$ the dominant decay mode is through $b\bar{b}$. However, due to the enormous QCD background, this channel is only considered in

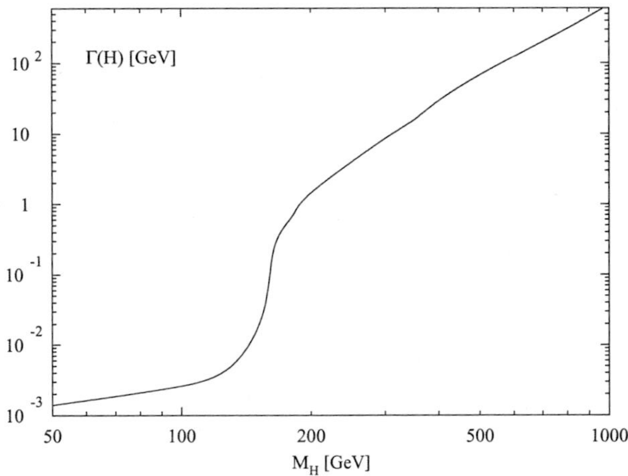

Figure 5. Total width of the SM Higgs boson to different decay products as a function of the mass

the $t\bar{t}H$ final state where handles exist for the rejection of this background.[7] The $\gamma\gamma$ final state, which appears when the Higgs boson decays via bottom, top and W loops, has a small branching fraction but excellent γ/jet separation and γ resolution help to make this a very significant channel. The $H \rightarrow \tau\tau$ decay is accessible if the Higgs boson is produced in association with jets (Section 3.3).

If the Higgs boson mass is large enough to make the WW and ZZ modes kinematically accessible, the $H \rightarrow WW^{(*)}$ final-states are powerful over a very large mass range (WW accounts for ~95% of the branching ratio at M_H ~160GeV/c^2), as is the $H \rightarrow ZZ^{(*)} \rightarrow 4l$ final state–the latter of which is commonly referred to as the "Golden Mode" as with four leptons in the final state the signal is easy to trigger on and allows for full reconstruction of the Higgs boson mass.

For SM Higgs boson masses close to $2m_{top}$, the channel $H \rightarrow t\bar{t}$ opens up, thus reducing the branching ratio of $H \rightarrow ZZ, WW$. Due to the very large cross-section for the production of non-resonant $t\bar{t}$ pairs, the inclusive search for the SM Higgs boson with $H \rightarrow t\bar{t}$ is not considered feasible.

The SM Higgs boson will be produced at the LHC via several mechanisms. The Higgs boson will be predominantly produced via gluon-gluon fusion [35] (see the left diagram in Fig. 7). For Higgs boson masses, such that $M_H > 100$ GeV/c^2, the second dominant process is the Vector Boson fusion (VBF) [36, 37] (see right diagram

[7]The observation of a Higgs boson with the $b\bar{b}$ decay in association with $t\bar{t}$ is not considered here as a discovery channel for the SM Higgs boson.

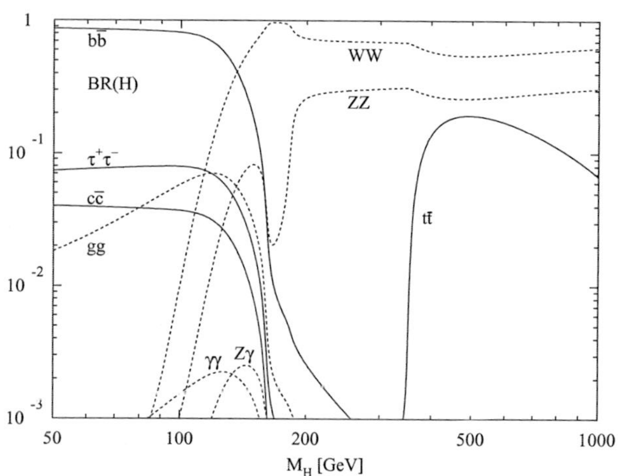

Figure 6. Branching ratio of the SM Higgs boson to different decay products as a function of the mass

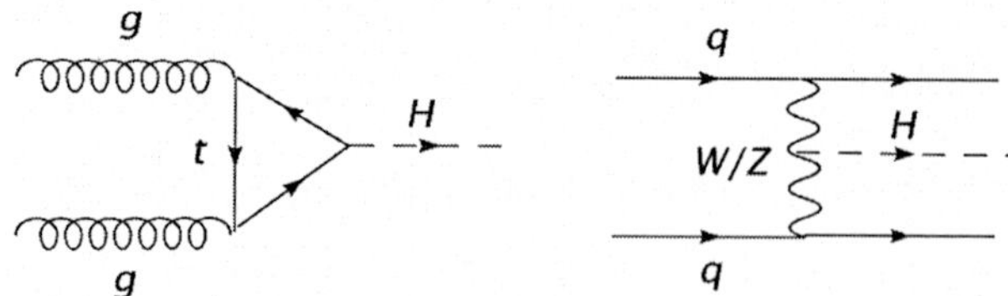

Figure 7. Leading order diagrams for the dominant processes involving the production of a SM Higgs boson at the LHC: gluon-gluon fusion (left) and VBF (right)

in Fig. 7).

In addition to inclusive searches, the CMS and ATLAS collaboration have investigated the feasibility of observing the Higgs boson in association with at least one or two high transverse momentum, p_T, hadronic jets. In the case of gluon-gluon fusion the Higgs boson can be produced alone. However, when one of the gluons or top quarks emits a gluon, the Higgs boson is produced with the gluon, which is seen in the detector as a hadronic jet. When the Higgs boson has a significant transverse momentum the associated jet tends to be back-to-back with the Higgs boson in the transverse plane (for reasons of transverse momentum balance). In the case of VBF the Higgs boson is produced with at least two jets. In both cases, jets produced in association with the Higgs boson are most useful in the identification of the Higgs boson, suppressing significantly the QCD backgrounds. The presence of one or two high p_T hadronic jets in association with the Higgs boson produces two distinct type of topologies. The hadronic jet in final states with one jet tends to be more forward and the invariant mass of the Higgs boson system and the jet tend to be larger than that of the backgrounds [20,24,25]. These and other particular properties are used to suppress backgrounds.

In events with two high p_T jets produced by the VBF process display different peculiarities. The VBF process gives final state in which the two struck quarks appear as high-p_T jets in the forward (high-η) and opposite regions of the detectors (backward-forward). In addition, due to the lack of QCD colour flow between the scattered quarks, low hadronic activity is expected to arise, giving rise to a final state with a depleted hadronic activity. The application of a central jet veto, in addition to the presence of two forward high p_T jets in opposite hemispheres and a cut on the invariant mass of the di-jet system reduces QCD backgrounds very strongly and enhances significantly the sensitivity of searches for a low mass Higgs boson at the LHC [18–22].

Early analyses performed at the parton level associated with one or two high transverse momentum hadronic jets indicated that this final state can be a powerful observation mode for a low mass Higgs. The ATLAS and CMS collaborations have performed feasibility studies for SM Higgs boson searches in association with two jets including more detailed detector description and the implementation of initial state and final state parton showers, hadronisation and multiple interactions, which has confirmed the strong potential of these final states [16,23].

The SM Higgs boson production cross-sections at the LHC to QCD Next-to-Leading-Order (NLO) [38–42],[8] as a function of Higgs boson mass are shown in Fig. 8[9] fusion mode. The Higgs boson cross-section with this mechanism reaches over 30 pb for masses around 115 GeV/c^2. The VBF process is the second most dominant production mode at the LHC. It typically takes up $\approx 10\%$ of the total Higgs boson cross-section for low masses and up to $\approx 50\%$ for a very heavy Higgs boson. Associated production modes, where the Higgs boson is produced via $^{-\prime} \to HW$, $^{-} \to HZ$ and gg, $^{-} \to t\bar{t}H$, have smaller cross-sections. The presence of a W, Z or

[8]The references given here correspond to the first computations of the QCD NL corrections to the gluon-gluon fusion and the VBF processes (Section 3). Great progress has been made in recent years in understanding QCD Next-to-Next-to-Leading- rder effects, QCD NL corrections to other production mechanisms and EW corrections. We do not deem it appropriate to review this question here, instead we refer the reader to a comprehensive review [43].

[9]Cross-sections were computed with [34].

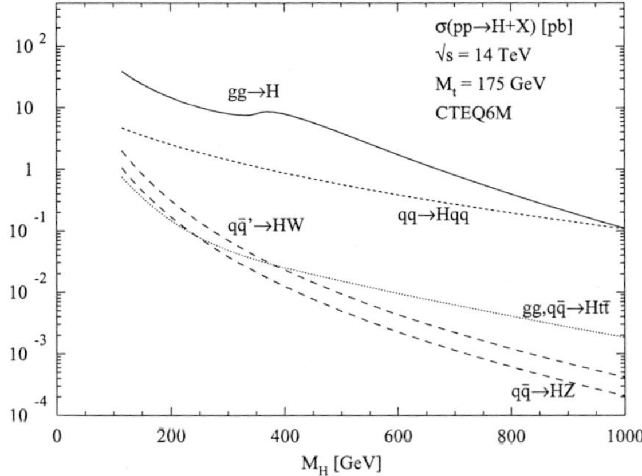

Figure 8. Production cross-section for the SM Higgs boson in proton proton collisions with the centre of mass of the LHC. Results are given for different production mechanisms as a function of the mass (see text)

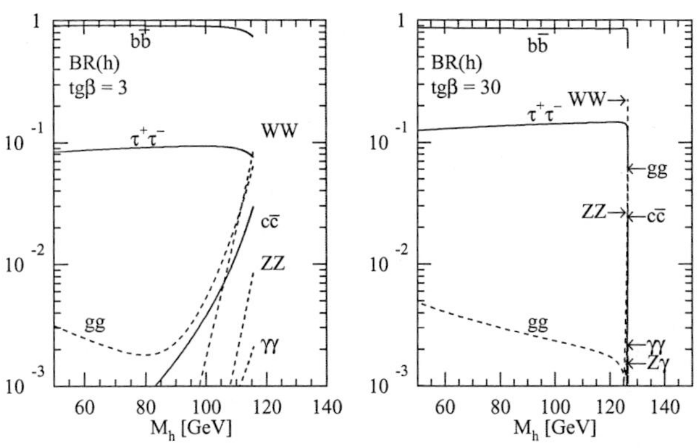

Figure 9. Branching ratio of the MSSM Higgs boson h to different decay products as a function of the mass. Results are given for two regions in the MSSM parameter space, $\tan\beta = 3, 30$

top-quark alongside the Higgs, or high-p_T high-η jets from VBF, allow for triggering on events with Higgs boson in invisible final states.

Efforts have been developed to understand the sensitivity of the detectors for the minimal expression of the Higgs boson sector, a single Higgs boson doublet. With the extension of the Higgs boson sector by the addition of a second Higgs boson doublet, the situation becomes more complex, as the multiplicity of relevant final states is enhanced. In the latter case, within the context of the minimal supersymmetric (MSSM) extension, the Higgs boson sector contains two charged ($H^{\pm}$) and three neutral (h, H, A) physical states.[10] All Higgs boson masses and couplings are expressed in terms of two parameters: the mass of the CP-odd boson, m_A, and the ratio of the vacuum expectation values of the Higgs boson doublets, $\tan\beta$. Here we consider the case when the mass of supersymmetric particles are large enough so that they do not play an important role in the phenomenology. A review of the impact of supersymmetric particles can be found in [44].

Whereas Figures 9-11 show the branching ratio of the MSSM Higgs bosons $h, A, H^{\pm}$ as a function of the mass, Fig. 12 refers to production cross-section of the MSSM Higgs boson in proton-proton collision. Results are given for two representative points in the MSSM parameter space.

The production processes shown in Fig. 7 for the SM Higgs can also be computed for the CP-even h, H bosons. However, the production mechanism involving the b-quark will play an important role for moderate to large $\tan\beta$ values as the couplings to the Higgs boson are enhanced. Firstly, one has to take into account the b-quark loop contribution in the $gg \to h, H$ mechanism. Secondly, associated Higgs boson production with heavy quarks in the final states must be considered, $pp \to b\bar{b}h, H$ and this process for either h or H becomes the dominant one in the MSSM. The cross sections for the associated production with $t\bar{t}$ pairs and with W/Z bosons as well as the WW/ZZ fusion processes, are suppressed for at least one of the particles as a result of the VV coupling reduction.

Because of CP invariance, which forbids AVV couplings, the A boson cannot be produced in the Higgs-strahlung and VBF processes; the rate for the $pp \to t\bar{t}A$ mechanism is suppressed due to the small $At\bar{t}$ coupling for $\tan\beta > 3$. The processes $gg \to A$, including the b-quark loop and the production in association with $b\bar{b}$ pairs yield significant cross-sections. The one-loop induced diagrams $gg \to AZ, Ag$ and the associated production with other Higgs bosons are possible but their cross-sections are typically small.

Concerning the production of the charged Higgs boson, it is important to consider the mass. For Higgs boson masses lighter than the top-quark mass, the dominant channel is the production from top-quark decays. For higher masses, one needs to consider the processes

[10] It is relevant to note that the notation for a SM Higgs boson is H, where the notation for the lightest and heaviest neutral CP-even MSSM Higgs boson is h, H, respectively.

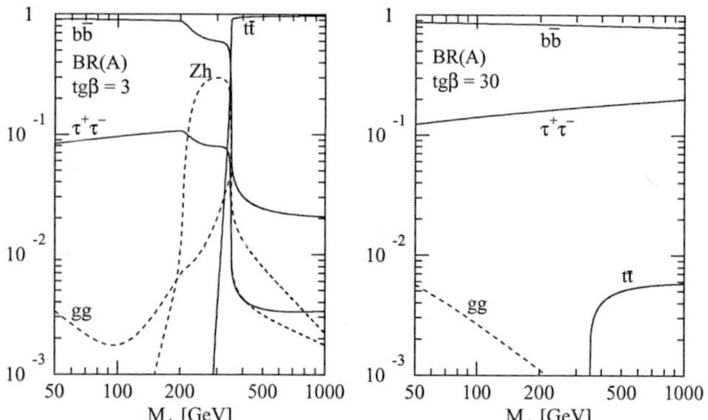

Figure 10. Branching ratio of the MSSM Higgs boson A to different decay products as a function of the mass. Results are given for two regions in the MSSM parameter space, $\tan\beta = 3, 30$

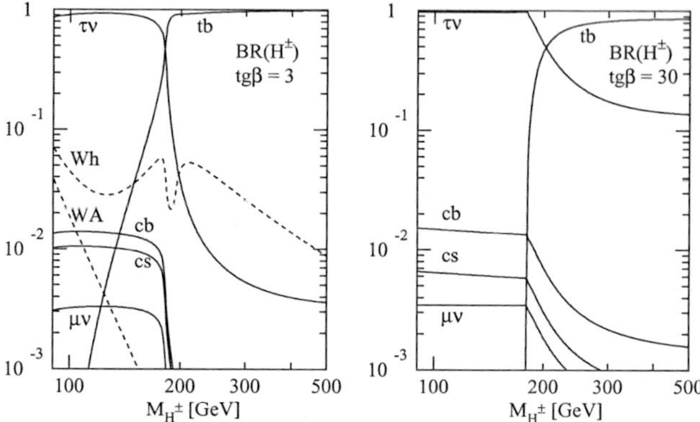

Figure 11. Branching ratio of the MSSM Higgs boson $H^{\pm}$ to different decay products as a function of the mass. Results are given for two regions in the MSSM parameter space, $\tan\beta = 3, 30$

$gg \rightarrow H^{\pm}tb$ and $gb \rightarrow H^{\pm}t$, which have to be properly combined in order to compute the physical cross-section.

A comprehensive review of the properties and the production mechanism of the MSSM Higgs bosons can be fond in [44]. A brief discussion on the sensitivity for these experiments to discover one or more of the Higgs bosons from the minimal version of the supersymmetric theories is also included (Section 4).

3.2. Experimental Reconstruction of Physics Objects

Here we briefly review the experimental reconstruction of physics objects that are relevant to the search of the Higgs boson with the CMS and ATLAS detectors. As seen in section 3.1 the search for the Higgs boson will involved a variety of final states. This will include the reconstruction of photons, electrons, muons, hadronic τ decays, hadronic jet reconstruction, b-jet reconstruction and the computation of missing transverse momentum, $\not{p}_T$, carried by neutrinos.

Electrons are reconstructed as objects that have a track in the Inner Tracker and an electromagnetic cluster in the electromagnetic section of the calorimeter. In order to separate isolated electrons originating from interesting events, from QCD background (hadrons, jets and photons) with similar topology, several of their characteristics are exploited. The electro-magnetic cluster in the calorimeter is required to match with a track in the Inner Tracker and the ratio of its energy over its momentum measured by the tracker (E/p) to be that of an electron. Cuts on the longitudinal (and lateral) shape of the shower are applied, and minimal energy is allowed to be deposited in the hadronic calorimeter. Rejections of $10^4 \div 10^6$ against hadronic jets can be achieved, depending on the definition of the electron identification, transverse momentum and angle.

Photons are also reconstructed as electromagnetic clusters with a different treatment of the information given by the inner detector. For identification purposes photon candidates are classified into unconverted (no high p_T tracks are associated) and converted (one or two high p_T tracks are associated). Due to the large amount of inactive material in front of the face of the calorimeter, a significant fraction of photons will convert into a e^+e^- pair, leading to the observation of one or two tracks, depending on the circumstances of the conversion. Sophisticated tracking algorithms are implemented in order to identify a vertex produced by a e^+e^- pair with a large distance to the beam-line. In addition, single tracks are searched for such that no hits

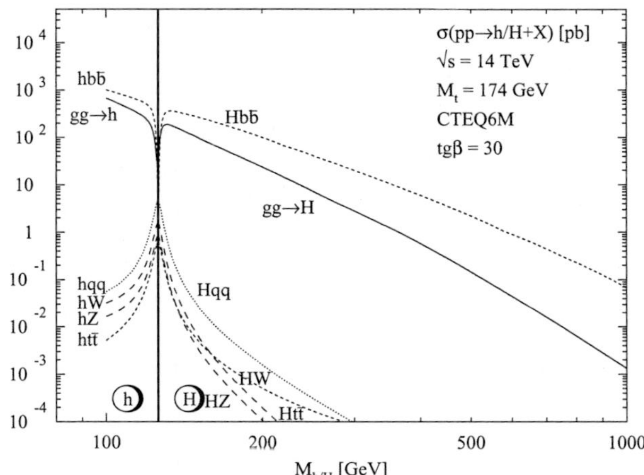

Figure 12. Production cross-section of the MSSM Higgs boson h, H in proton-proton collisions with the centre of mass of the LHC, as a function of the mass. The results are presented for $\tan\beta = 30$

are found in the layer of the silicon detectors closest to the beam-pipe. Rejections of $10^3 \div 10^4$ against hadronic jets can be achieved, depending on the definition of the photon identification, transverse momentum and angle.

The calorimeter is designed to contain almost all of the energy of a high p_T (TeV range) electron/photon, and has an energy resolution of $2 \div 10\%/\sqrt{E[\text{ GeV}]}$, depending on the experiment. The energy resolution for electron and photon candidates is similar.

Muons are reconstructed as objects that have a track in the muon spectrometer and a corresponding ("matched") track in the Inner Tracker. In the case of ATLAS, the good resolution of the muon spectrometer provides the possibility to trigger and reconstruct muons in "stand alone" mode (no matching with the inner detector involved). The momentum resolution is maintained high for both experiments. For low momenta the muon momentum resolution is dominated by the resolution of the Inner Tracker. The Inner Tracker has an intrinsic p_T resolution of a few times $10^{-4}p_T(\text{ TeV}/c)$, which is limited by early bremsstrahlung in its material. For muon p_T in the TeV$/c$ range the resolution is limited by detector alignment in the case of ATLAS and can be kept at $\sigma/p_T \approx 10\%$, whereas in the case of CMS it is limited by energy losses in the iron yoke, and it varies between $15 \div 30\%$.

For the reconstruction of hadronic jets, a seeded fixed-cone reconstruction algorithm with a cone size $\Delta R = \sqrt{\Delta\phi^2 + \Delta\eta^2} = 0.4$ is presently used for search studies the Higgs boson. For future studies also the SISCone (Seedless Infrared Safe Cone) jet algorithm and the fast K_T algorithm are considered. If one neglects the noise term, the jet energy resolution varies between $50 \div 100\%/\sqrt{E[\text{ GeV}]}$, depending on the angle and the detector. Both experiments have strong capabilities for the identification of b-jets and τ-jets in wide range of transverse momentum for $|\eta| < 2.5$. For a b-jet tagging efficiency of 60% and transverse momentum of $20 < p_T < 100$ GeV$/c$ a rejection above 100 and about 10 may be achieved against light an c-jets, respectively, with degradation of performance for $p_T > 100$ GeV$/c$. For a τ-jet efficiency of 50%, the rejection against hadronic jets improves with p_T, reaching rejection values of $(10^2) \div (10^3)$.

The missing transverse momentum carried by neutrinos is primarily reconstructed from the energy deposits in the calorimeter and the reconstructed muon tracks. Apart from the hard scattering process of interest, many other sources, such as the underlying event, multiple interactions, pileup and electronic noise lead to energy deposits and/or fake muon tracks. Classifying these energy deposits into various types (e.g. electrons, taus or jets) and calibrating them accordingly, is the essential key for optimal $\not{p}_T$ measurement. In addition, the loss of energy in regions of inactive material and dead detector channels make the $\not{p}_T$ measurement a real challenge.

The $\not{p}_T$ reconstruction algorithm starts from the energy deposits in calorimeter cells or clusters of cells ("raw $\not{p}_T$"). The raw $\not{p}_T$ is then cleaned up from a number of sources of fake $\not{p}_T$: hot cells, overlay of beam-halo, cosmics, detector malfunctions, detector hermiticity. Overall, the reconstruction of $\not{p}_T$ is a challenging task and it requires a good understanding of the calorimeter response and the topology of different signatures. The $\not{p}_T$ resolution roughly scales with $\sqrt{\sum E_T}$, where $\sum E_T$ is the scalar sum of the energies of the particles in the final state, for $\sqrt{\sum E_T} <$ 1.5 TeV.

Both CMS and ATLAS have conducted extensive fully-simulated GEANT-based [45] Monte Carlo studies to determine the experimental viability of all of these channels. A few of these signatures are highlighted below. A more comprehensive and complete account can be found elsewhere [16, 17, 23].

3.3. Discovery Channels

In this section we succinctly put in perspective the relevant theoretical inputs (Section 3.1) and corresponding experimental parameters (Section 3.2) needed to define the strategy for the observation of the Higgs boson.

Analysis	$\gamma\gamma$	$\tau\tau$	WW	ZZ
Inclusive	Yes	NA	Yes	Yes
≥ 1 jet	Yes	Yes	Yes	NA
≥ 2 jet	Yes	Yes	Yes	NA
Z/WH	Yes	NA	Yes	NA
$t\bar{t}H$	Yes	NA	Yes	NA

Table 1. Final states with the potential to give a significant contribution to the discovery of the SM Higgs boson at the LHC. The those final states labeled as NA correspond to channels that are not feasible or for which feasibility studies are not available.

Table 1 summarises the final states with the potential to give a significant contribution to the discovery of the SM Higgs boson at the LHC. Those final states labeled as NA correspond to channels that are not feasible or for which feasibility studies (for discovery purposes) are not available. Below we give a summary of the most relevant aspects of the experimental analysis involving $H \rightarrow \gamma\gamma, \tau\tau, WW, ZZ$ decays.

The feasibility of channels intrinsic to the observation of MSSM Higgs bosons will be also briefly discussed in section 4. This includes the search for di-μ and di-τ pairs. The feasibility of these searches is usually performed when the di-μ and di-τ pairs are produced in association with at least one b-jet.[11] The potential of the decay chains, such as $Wh \rightarrow l\nu b\bar{b}$ $(l = e, \mu)$, $A/H \rightarrow t\bar{t}$, $A \rightarrow Zh \rightarrow llb\bar{b}$ $(l = e, \mu)$ and $H \rightarrow hh \rightarrow b\bar{b}\gamma\gamma$ has been evaluated. The decay mode $H^{\pm} \rightarrow \tau^{\pm}\nu$ is considered, as well.

3.3.1. $\mathbf{H} \rightarrow \gamma\gamma$

Despite the small branching ratio, $H \rightarrow \gamma\gamma$ remains a very attractive channel for $115 < M_H < 140$ GeV/c^2. The feasibility of this channel heavily relies on excellent photon resolution due to the small signal to ratio expected for the inclusive analysis. In addition to the required calorimeter resolution, the reconstruction of the di-photon vertex plays a significant role. This is achieved by using the main vertex of the event produced by the multiplicity of low p_T tracks (the underlying event) and hadronic jets.

The backgrounds to this channel are usually divided into two types: irreducible and reducible. Photon pairs from $^{-} \rightarrow \gamma\gamma$, $gg \rightarrow \gamma\gamma$ and quasi-collinear quark bremsstrahlung comprise the irreducible background, while jet-jet and γ-jet events, where one or more jets are misidentified as photons (mostly from the production of energetic π^0s), take up the bulk of the reducible background. $Z \rightarrow e^+e^-$ events, with both electrons misidentified as photons, can be reduced using electron/photon separation techniques. Excellent photon identification is required in order to suppress the reducible backgrounds, and, in particular, to separate π^0s leading to two collimated photons from a single photon. The high-granularity Liquid Argon calorimeter of ATLAS is capable of resolving single photons from π^0s,[12] while CMS has a superior energy resolution.[13]

Studies conducted of the inclusive analysis by both experiments consider the signal and backgrounds to QCD NLO [46–52]. Both experiments have looked beyond a simple cut-based analysis. The discriminating power of the di-photon transverse momentum and the photon decay angle in the Higgs boson rest frame with respect to the Higgs boson lab flight direction, $|\cos\theta^*|$, are evaluated in conjunction with the di-photon invariant mass.

Efforts have been made to evaluate the feasibility of Higgs boson searches in association with at least one or two high p_T hadronic jets. The sensitivity of the Higgs boson production in association with Z, W and $t\bar{t}$ has also been evaluated.

The sensitivity of this channel is similar for both experiments. For $M_H = 130$ GeV/c^2, and an integrated luminosity of 30 fb^{-1} of integrated luminosity more than a $5\,\sigma$ effect may be achieved with one experiment, as illustrated in Fig. 13. The sensitivity of this channel can increase significantly with a more sophisticated treatment of discriminating variables. This is illustrated in Fig. 13, where the sensitivity of the cut-based analysis is compared to that of a more complete analysis.

3.3.2. $\mathbf{H} \rightarrow \tau\tau$

As pointed out in Table 1, the search for a SM Higgs boson with the $H \rightarrow \tau\tau$ decay requires the presence of at least one high p_T jet.[14] The final state with at least two high p_T jets is the best studied by the CMS and ATLAS collaborations. The distinct experimental signature of Higgs boson production via VBF, with jets from the "struck quarks" at high-η and Higgs boson

[11] The feasibility of purely inclusive searches using $A/H \rightarrow \mu\mu, \tau\tau$ at the LHC is being evaluated.

[12] The ATLAS calorimeter is able to suppress π^0s by a factor of $2.5 \div 3$ in the range of transverse momentum relevant to the SM Higgs boson search [28].

[13] Detailed comparisons of the sensitivity of the two experiments are underway.

[14] The inclusive search for excess of events di-τ pairs carries sensitivity to the MSSM light h for large values of $\tan\beta$.

decay products in the central region is a great asset for channels like $H \to \tau\tau$. CMS and ATLAS now both consider three final states, thus covering all combinations of leptonically- and hadronically-decaying taus.[15] Triggering on the fully hadronic mode by using combinations of low-p_T tau and other triggers (e.g. missing transverse momentum or forward jets) are currently under investigation. Despite the presence of multiple neutrinos in the final-state, mass reconstruction can typically be done via the collinear approximation where the tau decay daughters are assumed to be in the same direction as their parent. The reconstruction of the di-τ invariant mass using the collinear approximation is feasible when the τs are not back-to-back, which is the case considered here. The resolution on the reconstructed mass (~10 GeV/c^2 for M_H = 120 GeV/c^2) is mainly affected by the missing energy resolution. Other experimental issues, such as the identification of hadronic taus and low p_T electrons and muons, together with the reconstruction of forward jets and the understanding of the central jet veto are central to the analysis, as well. Data-driven methods for understanding the dominant backgrounds (Z + jets, QCD and $t\bar{t}$) have been investigated. With the combination of all the decay modes, one experiment may achieve 5 σ significance for 30 fb^{-1} of integrated luminosity (Fig. 13).

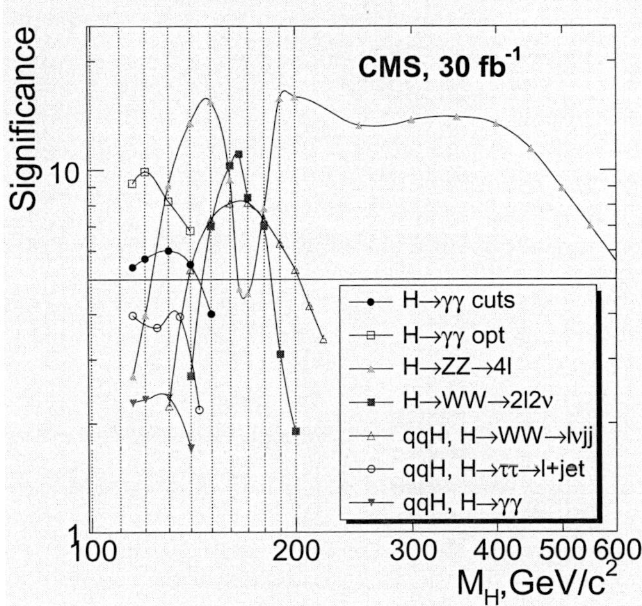

Figure 13. The discovery potential at CMS for SM Higgs boson searches, as obtained using NLO cross-sections, for 30 fb^{-1} of integrated luminosity

3.3.3. H $\to ZZ^{(*)} \to 4l$ (4e, 4μ, 2e2μ)

At M_H > 130 GeV/c^2, the 4-lepton channels gain in importance on account of the energy reconstruction and identification of both ATLAS and CMS for electrons and muons. This analysis is performed inclusively due to the small cross-section of the expected SM backgrounds giving four isolated leptons (e, μ).[16] The dominant backgrounds for these channels are $ZZ^{(*)}$, $Zb\bar{b}$ and $t\bar{t}$ production. Through the use of impact parameter and lepton isolation requirements the latter two can be significantly reduced. The $^{-}$ component of the $ZZ^{(*)}$ background is known at NLO [54–57], however due to the lack of a Monte Carlo generator for $gg \to ZZ^{(*)}$, typically the contribution from this process is added as 30% of the LO $^{-} \to ZZ^{(*)}$.[17] After the imposition of an appropriate event selection the contribution of the reducible $Zb\bar{b}$ and $t\bar{t}$ backgrounds becomes negligible for searches of a heavy Higgs boson and become significantly smaller than the contribution of reducible backgrounds for low mass Higgs boson searches. Collectively, the significance for these channels is more than 5σ for 30 fb^{-1} of integrated luminosity in a wide range of Higgs boson masses (Fig. 13).[18]

3.3.4. H $\to WW^{(*)} \to l\nu l\nu$ ($l = e, \mu$)

As the branching ratio for a SM Higgs boson decaying to WW is more than 95% at ~ 160 GeV/c^2, this is the most significant channel at that mass point. Unlike other channels, in the $H \to WW \to l\nu l\nu$ final state full mass reconstruction is not possible and the analysis is essentially reduced to a counting experiment; therefore an accurate background estimate is critical. The dominant backgrounds for this analysis are W^+W^- and $t\bar{t}$ production. The former can be suppressed by exploiting spin correlations between the two leptons while the latter has been shown to be suppressed significantly by a jet veto. The CMS and ATLAS collaborations have extensively studied two regimes: an analysis based on

[15]The search for a low mass SM Higgs boson when both τs decay hadronically has been investigated rendering the analysis challenging. Searches for di-τ pairs are feasible when dealing with heavy MSSM Higgs bosons.

[16]The impact of the search for ZZ pairs in association with jets is still being assessed by the collaborations. The isolation of ZZ pairs in conjunction with two high p_T and well separated jets has been addressed within the context of Higgs boson coupling measurements [53].

[17]Currently, the **gg2ZZ** generator is being used in order to properly take into account the $gg \to ZZ^{(*)}$ process, including Z/γ interference effects [58].

[18]The sensitivity of this channel is very strong with the exemption of a narrow window 160 < M_H < 175 GeV/c^2, where branching ratio of $H \to ZZ$ is diminished (Fig. 6).

the application of a full jet veto[19] and with the tagging of two high p_T hadronic jets.[20] The feasibility of search for WW pairs in association with Z/W and $t\bar{t}$ has been evaluated.

Using NLO [55–57, 59, 60] cross-sections and conservative estimates for the effect of systematic uncertainties, a significance of around 5σ for $M_H = 165$ GeV/c^2 using an integrated luminosity of $\sim$1 fb^{-1} is estimated. The sensitivity of this channel is also evaluated for other Higgs boson masses showing strong sensitivity for the mass range $140 < M_H < 190$ GeV/c^2, as illustrated in Fig. 13.

4. Summary of Higgs Boson Discovery Potential

The expected significance in 30 fb^{-1} of integrated luminosity, for various final states as a function of SM Higgs boson mass is summarised in Fig. 13 for the CMS experiment. The discovery potential at CMS and ATLAS is quite similar. When combining all the most important channels one experiment is expected to reach a $5\,\sigma$ effect in the entire SM Higgs boson mass range of interest with about 10 fb^{-1} of integrated luminosity.

Discovery prospects for the detection of MSSM Higgses (A, h, H and $H^{\pm}$) have also been evaluated [16,17]. At tree-level, all Higgs boson masses and couplings can be expressed in terms of m_A and $\tan\beta$. The complete region of the m_A–$\tan\beta$ parameter space (m_A = 50–500 GeV/c^2 and $\tan\beta$ = 1–50) should be accessible to the LHC experiments. The sensitivity for the discovery of MSSM Higgses, in the minimal mixing scenario for 30 fb^{-1} of data, is summarised in Fig. 14 for the ATLAS experiment. With the further increase of the integrated luminosity the sensitivity the ability to observe more than one Higgs boson for intermediate $\tan\beta$ and large M_A is enhanced. However, there wil always remain a region at intermediate $\tan\beta$ and $M_A > 200\div250$ GeV/c^2 for which the observation of more than one Higgs boson will not be feasible. As in the case of the SM Higgs boson, the discovery potential at CMS and ATLAS is quite similar.

Acknowledgments: I would like to whole heartedly thank Prof. Biswarup Mukhopadhyaya and Prof. Amitava Raychaudhuri for the encouragement and strong support. I would also like to thank Prof. Sau Lan Wu, Prof. W. Smith, Prof. S. Dasu, Dr. D. Loveless, Dr. M. Spira, Dr. A. Nikitenko, Dr. W. Quayle, Dr. T. Vickey, Dr. Y. Fang and Dr. X. Chen for their invaluable help in the preparation of the manuscript.

This work was supported in part by the Department of Energy through Grant No. DE-FG0295-ER40896 and the University of Wisconsin Alumni Research Foundation.

[19]Events are rejected in which a hadronic jet is found with $p_T >$ 20 GeV/c (hadron level) in the pseudorapidity range $|\eta| < 4.9$. With this selection the leading background is the non-resonant production of W^+W^-.

[20]This analysis is designed to isolate the VBF signal. The leading background is the $t\bar{t}$ production.

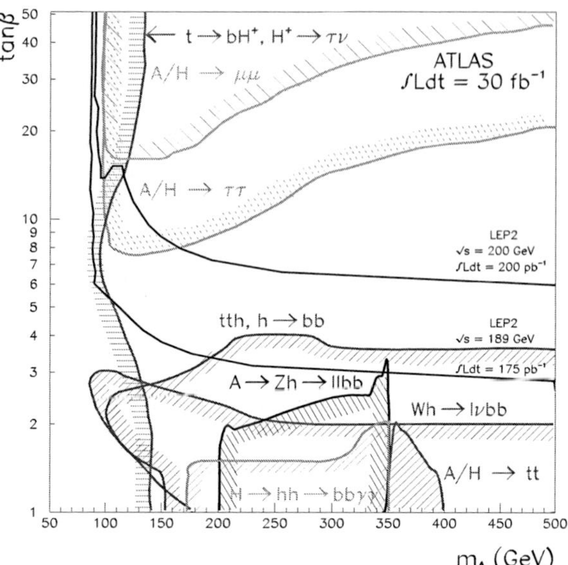

Figure 14. The discovery potential at ATLAS for MSSM Higgs boson searches, for 30 fb^{-1} of integrated luminosity

REFERENCES

1. S. L. Glashow, Nucl. Phys. **22** 1961 579 .
2. S. Weinberg, Phys. Rev. Lett. **19** 1967 1264.
3. A. Salam, Proceedings to the Eigth Nobel Symposium, May 1968, ed N. Svartholm Wiley, 1968 357.
4. S.L. Glashow, J. Iliopoulos and L. Maiani, Phys. Rev. **D2** 1970 1285.
5. F. Englert, R. rout, Phys. Rev. Lett. **13** 1964 321.
6. P. W. Higgs, Phys. Lett. **12** 1964 132.
7. P. W. Higgs, Phys. Rev. Lett. **13** 1964 508.
8. P. W. Higgs, Phys. Rev. **145** 1966 1156.
9. G. S. Guralnik, C.R. Hagen and T.W. . ibble, Phys. Rev. Lett. **13** 1964 585.
10. T.W. . ibble, Phys. Rev. **155** 1967 1554.
11. ALEPH Collaboration, R. arate *et al.*, Phys. Lett. **B495** 2000 1.
12. P. McNamara and Sau Lan Wu, Higgs Particle in the Standard Model Experimental Results from LEP, Reports on Progress in Physics **65** 2002 465.
13. L. Evans and P. ryant editors , LHC Machine, 2008 JINST **3** S08001 2008 .

14. The LEP Higgs Working Group, Phys. Lett. **B565** -2003- 61.
15. The LEP Electroweak Working Group, http-//lepewwg.web.cern.ch/.
16. CMS Collaboration, CMS PTDR - .2- Physics Performance, CERN/LHCC 2006-021.
17. ATLAS Collaboration, Detector and Physics Performance Technical Design Report, CERN-LHCC/99-14/15 -1999-.
18. - . Hagiwara, D.L. Rainwater and D. -eppenfeld, Phys. Rev. **D59** -1999- 014037.
19. T. Plehn, D.L. Rainwater and D. -eppenfeld, Phys. Rev. **D61** -2000- 093005.
20. S. Abdullin *et al.*, Phys. Lett. **B431** -1998- 410.
21. D.L. Rainwater and D. -eppenfeld, Phys. Rev. **D60** -1999- 113004.
22. N. - auer, T. Plehn, D.L. Rainwater and D. -eppenfeld, Phys. Lett. **B503** -2001- 113.
23. S. Asai *et al.*, Eur. Phys. J. C 32 -2004- s19-s54.
24. -. Mellado, W. - uayle and Sau Lan Wu, Phys. Lett. **B611** -2005- 60.
25. -. Mellado, W. - uayle and Sau Lan Wu, Phys. Rev. **D76** -2007- 093007.
26. R.N. Cahn, S.D. Ellis, R. - leiss and W.J. Stirling, Phys. Rev. **D35** -1987- 1626.
27. The CMS Collaboration, S. Chatrchyan *et al*, The CMS experiment at the CERN LHC, 2008 JINST **3** S08004.
28. The ATLAS Collaboration, G. Aad *et al*, The ATLAS experiment at the CERN Large Hadron Collider, 2008 JINST **3** S08003.
29. The LHCb Collaboration, A.A. Alves Jr *et al*, The LHCb Detector at the LHC, 2008 JINST **3** S08005.
30. The TOTEM Collaboration, G. Anelli *et al*, The TOTEM Experiment at the CERN Large Hadron Collider, 2008 JINST **3** S08007.
31. The ALICE Collaboration, - . Aamodt *et al*, The ALICE experiment at the CERN LHC, 2008 JINST **3** S08002.
32. ATLAS Collaboration, ATLAS Level-1 Trigger- Technical Design Report, CERN-LHCC-98-014, ATLAS-TDR-12 -1998-.
33. ATLAS Collaboration, High-Level Trigger Data Ac-uisition and Controls, CERN-LHCC-03-016, ATLAS-TDR16 -2003-.
34. A. Djouadi, J. - alinowski and M. Spira, HDECA- - a Program for Higgs - oson Decays in the Standard Model and its Supersymmetric Extention, Comp. Phys. Comm. **108** -1998- 56.
35. H.M. Georgi, S.L. Glashow, M.E. Machacek and D.- . Nanopoulos, Phys. Rev. Lett. **40** -1978- 692.
36. R. Cahn and S. Dawson, Phys. Lett. **B136** -1984- 196.
37. G. - ane, W. Repko and W. Rolnick, Phys. Lett. **B148** -1984- 367.
38. S. Dawson, Nucl. Phys. **B359** -1991- 283.
39. A. Djouadi, M. Spira and P.M. -erwas, Phys. Lett. **B264** -1991- 440.
40. D. Grauden-, M. Spira and P.M. -erwas, Phys. Rev. Lett. **70** -1993- 1372.
41. M. Spira, A. Djouadi, D. Grauden- and P.M. -erwas, Nucl. Phys. **B453** -1995- 17.
42. T. Han, G. - alencia and S. Willenbrock, Phys. Rev. Lett. **69** -1992- 3274.
43. A. Djouadi, Phys. Rep. **457** -2008- 1.
44. A. Djouadi, Phys. Rep. **459** -2008- 1.
45. S. Agostinelli et al., NIM A **506** -2003- 250–303.
46. T. - inoth *et al.*, E. Phys. J. **C16** -2000- 311.
47. C. - ala-s, E. - erger, S. Mrenna and C.-P. - uan, Phys. Rev. **D57** -1998- 6934.
48. C. - ala-s, E. - erger, P. Nadolsky, C. Schmidt and C.-P. - uan, Phys. Lett. **B489** -2000- 157.
49. C. - ala-s, E. - erger, P. Nadolsky and C.-P. - uan, Phys. Lett. **B637** -2006- 235.
50. S. Catani, M. Fontanna-, J.P. Guillet, and E. Pilon, JHEP **05** -2002- 028.
51. -. Nagy, Phys. Rev. Lett. **88** -2002- 122003.
52. -. Nagy, Phys. Rev. **D68** -2003- 094002.
53. M. Duhrssen *et al.*, Phys. Rev. **D70** -2004- 113009.
54. -. Mele, P. Nason and G. Ridolfi, Nucl. Phys. **B357** -1991- 409.
55. J. Ohnemus, Phys. Rev. **B50** -1994- 1931.
56. J. Campbell, R.- . Ellis, Phys. Rev. **D60** -1999- 113006.
57. S. Frixione and - .R. Webber, JHEP **0206** -2002- 029.
58. T. - inoth, N. - auer and P. Mertsch, Gluon-induced - CD corrections to $pp \rightarrow ZZ \rightarrow l\bar{l}l'\bar{l'}$, To appear in the proceedings of 16th International Workshop on Deep Inelastic Scattering and Related Subjects -DIS 2008-, London, England, 7-11 Apr 2008. e-Print- ar- iv-0807.0024 -hep-ph-.
59. J. Ohnemus, Phys. Rev. **B44** -1991- 1403.
60. S. Frixione, Nucl. Phys. **B410** -1993- 280.

QCD at the Large Hadron Collider—Higgs Searches and Some Non-SUSY Extensions Beyond the SM

Prakash Mathews[a] *and V. Ravindran*[b]

[a]*Saha Institute of Nuclear Physics, 1/AF Bidhan Nagar, Kolkata-700064, India*

[b]*Regional Centre for Accelerator-based Particle Physics, Harish-Chandra Research Institute, Chhatnag Road, Jhunsi, Allahabad-211019, India*

We present a brief overview of the physics potential of the Large Hadron Collider LHC and the role of uantum chromodynamics CD in predicting various observables at the LHC with unprecedented accuracy. We have studied the production of Standard Model SM Higgs boson through gluon fusion channel and various signals of physics beyond the Standard Model SM restricted to non-supersymmetric scenarios. These are models with large extra-dimensions such as ADD and Randall-Sundrum models and also physics senario resulting from scale/conformal invariant sector, namely unparticle physics. We have presented CD effects to several of the observables in these models through higher order perturbative CD corrections and parton distribution functions. We have demonstrated how the these corrections reduce the scale ambiguities coming from renormalisation and factorisation. Our study shows that the precise and unambiguous predictions are possible for various SM studies at the LHC.

1. Large Hadron Collider

The Large Hadron Collider (LHC) is a gigantic machine situated at CERN [1], Geneva, spanning the border between Switzerland and France about 100 m underground. This is a particle accelerator specially designed to study the smallest fundamental building blocks that constitute our universe. This is aimed at to shed light on our understanding of the nature at a very fundamental level. It consists of six distinct experiments, characterised by their unique particle detector and the purpose. Out of which, ATLAS and CMS are two large experiments, independently designed for discovery purpose and will serve to study wide range of phenomena at sub atomic level. Two medium-size experiments, ALICE and LHCb are designed to study specific phenomena. The remaining experiments, namely TOTEM and LHCf will focus on forward particles that just brush past each other in the collision region.

At the LHC, two beams of protons will be first accelerated in opposite directions in the underground accelerators of 27 km in circumference, and then collided head-on at close to the speed of light, after they gain very high energy which is around 14 TeV. During the collision, some of the energy will be converted into mass and already known as well as new particles that will unravel the physics which is never explored at the fundamental level. The particles produced due to these collisions will be recorded by specially designed detectors for further study. Since the experiment is going to recreate the environment of our early universe, it will improve our understanding of the basic forces that govern the nature and the mysteries such as the origin of mass, whether there exists extra-dimensions of space not seen so far, evidence for dark matter candidates and microscopic black holes and scale or conformal invariant sectors.

The detectors consist of layers of detector materials to detect and measure the energy or momentum of different particles produced. Particles produced in the collision will pass through the tracking system, made of silicon pixels and silicon strip detectors which record their positions. The momenta of charged particles can be obtained by the curvature of their paths resulting due to the magnets present. The calorimeters in the next layers will record the energies of the particles. They are of two types, the first one records the energies of the electrons and photons called electromagnetic calorimeter (ECAL) and the second one records the those of hadrons (also jets) called hadronic calorimeter (HCAL). Muons that escape these detectors can be detected in the detectors outside these calorimeters from their tracks in the presence of magnets. Neutrons that can not be trapped can be studied from the missing energy and momentum of each collision.

There are several physics issues which the LHC is expected to provide partial or full answers. We discuss here few of them. The most likely explanation of the origin of mass may be found in the Higgs boson predicted by the SM of particle physics [2]. This particle is yet to be found and the LHC has sufficient energy to discover it. The universe we live is made up of ordinary particles forming only 4%, the rest called dark matter and dark energy forming 96% which are difficult to detect. The supersymmetric particles [3] which may explain the existence of dark matter can be detected at the LHC. The other mystery of our universe is why only tiny fraction of matter survived after Big Bang leaving hardly any antimatter. The LHCb experiment is designed to study the differences between matter and antimatter to find the answer. In the very early universe just after the Big Bang, the temperature would have been very high to overcome the binding strong interaction force between quarks and gluons to form nucleon making new phase of matter called plasma, a very hot and dense mixture of quarks and gluons. The ALICE experiment will recreate conditions similar to those just after the Big Bang to analyse the properties of the quark-gluon plasma. We know that the gravity appears weak as compared to other three interactions of nature. This is called hierarchy problem. Theories with hidden dimensions of space may explain this. String theory implies that there are additional spatial dimensions yet to be observed. Motivated by this idea, models such as large, universal extra-dimensional models [4, 5] provide the explanation for the smallness of gravitational interaction. These models may be testable at very high energies and the LHC detectors are also specially designed to look for the signs of extra dimensions. Recently, the idea of detecting a sector which is scale or conformal invariant has been proposed, under the name unparticle physics [6]. This can also be tested at the LHC because these effects can influence various observables significantly.

2. Quantum Chromodynamics

At the LHC the particles that collide are protons which interact through strong interaction force. The theory of strong interaction [7], namely quantum chromodynamics (QCD) plays an important role for all the physics studies to be undertaken at the LHC. QCD describes the dynamics behind the strong force between quarks, anti-quarks and gluons that constitute all the hadrons that we know of.

The theory of strong interaction force is described by the QCD Lagrangian given by

$$-_{Q\ \ D} = -\frac{1}{2}T\ [F^a_{\mu\nu}T^aF^{\mu\nu b}T^b]$$

$$+\overline{\psi}[i(\mathbb{I}\partial\!\!\!/ - ig_s A\!\!\!/^a T^a) - m\mathbb{I}]\psi\ . \quad (1)$$

The above Lagrangian is called the Yang-Mills (YM) Lagrangian. It is invariant under local non-abelian group $SU(3)$. The $SU(3)$ gauge symmetry is called non-abelian gauge symmetry and the gauge fields are called non-abelian gauge fields. The matrix T^a is the generator of the gauge group which in the fundamental representation is a 3×3 matrix. There are 8 such matrices denoted by the index $a = 1, ..., 8$. The parameters g_s and m are the gauge coupling and quark mass matrix respectively. The fermionic field given by ψ is a three component vector with each component carrying a specific index, called colour index. In other words, each quark or anti-quark comes in three different colour quantum numbers. There are 6 such quarks and 6 anti-quarks, namely up(u), down(d), charm(c), strange(s), top(t), bottom(b) and their charge conjugate partners, called anti-quarks. The kinetic energy part of the $SU(3)$ gauge fields contains second rank tensor fields given by

$$F^a_{\mu\nu} \quad = \quad \partial_\mu A^a_\nu - \partial_\nu A^a_\mu + g_s f^{abc} A^b_\mu A^c_\nu\ . \qquad (2)$$

The interaction of gauge fields among themselves comes from $F^a_{\mu\nu}(x)$ in the action Eq. (1) which contains a term $g_s f^{abc} A^b_\mu(x) A^c_\nu(x)$ (Eq. (2)). Notice that the above action describes not only the interaction of 6 quarks (also 6 anti-quarks) with 8 gauge fields, called gluons but also describes the interaction of gauge fields among themselves. This feature is characteristic of theories with non-abelian gauge symmetry. Since the theory of electrons and electromagnetic gauge fields is invariant abelian symmetry i.e. $U(1)$, the electromagnetic gauge fields do not interact with each other. This theory is further quantised and the standard gauge fixing procedure introduces the ghost fields that interact with gauge fields. Hence one has to take into account the ghost field contributions which enter through quantum loops. This theory is also renormalisable. Hence the strong coupling constant g_s that appear in the Eqs. (1, 2) depend on a scale where the theory is renormalised. This scale is called renormalisation scale denoted by μ_R. The standard renormalisation programme can be applied to this theory to find out how the coupling constant changes with the renormalisation scale. One finds that the scale dependence is governed by the renormalisation group equation (RGE), defining

$$a_s = \frac{g_s^2}{16\pi^2}\,, \qquad (3)$$

the RGE reads,

$$\mu_R^2 \frac{da_s}{d\mu_R^2} = \beta(a_s(\mu_R^2)) = -\sum_{i=0}^{\infty} a_s^{i+2}(\mu_R^2)\beta_i\,, \qquad (4)$$

where $\beta(a_s(\mu_R^2))$ is the beta function. The one-loop contribution to it is given by β_0

$$\beta_0 = \frac{11}{3}C_A - \frac{2}{3}n_f T_f \,, \qquad (5)$$

where $C_A = 3, T_f = 1/2$. This is positive because $n_f = 6$. The results for β_1, β_2 and β_3 are available in the literature. The solution to Eq. (4) at one-loop level is given by

$$a_s(Q^2) = \frac{a_s(\mu_0^2)}{1 + \beta_0 a_s(\mu_0^2)\ln(Q^2/\mu_0^2)} + \mathcal{O}(a_s^2). \qquad (6)$$

Notice that the strong coupling constant decreases as the scale Q increases. It vanishes when $Q \to \infty$. This behaviour is called asymptotic freedom. In other words, the non-abelian theory which describes strong interaction force is a free theory at very high energies. This means that the quarks, anti-quarks and gluons do not interact with other at infinitely high energies. On the contrary, the coupling becomes very large at low energies. Hence at moderately high energies, the coupling constant is not only finite but also a small parameter which can allow us to apply the standard perturbative methods to compute observables involving quark, anti-quarks and gluons. Hence, for the LHC study where the energy scale is large, we can safely apply perturbative methods to compute observables using QCD. Similar renormalisation group analysis for the masses implies that at high energies the masses approach to zero. In addition the energy scales involved at the LHC is so high the effects coming from the finite mass of the quarks except for top quark can be safely ignored.

2.1. QCD Factorisation Theorem

At the LHC, due to large energy available, the high energetic scattering of proton beams will break the protons into their constituents such as quarks, anti-quarks and gluons and the scattering of protons essentially means the scattering of these particles. We call them partons. These partons are so energetic that they can interact and produce the particles we know of through the dynamics summarised by SM of particles. It is capable of producing the Higgs boson which is responsible for symmetry breaking mechanism and for the mass generation of all the particles within the SM. Also we can unravel physics beyond the Standard Model (BSM). That is, the LHC can produce new particles and demonstrate existence of new dynamics which we have not seen in the past. This is possible at the LHC because of the high energy available in each collision but also the high luminosity of the proton beams. These factors will boost the production rates through their large cross sections and luminosity. In addition, the specially designed detectors are capable of recording most important events for the physics study and discovery purposes. Huge statistics and high level detector technologies will make the measurement to an unprecedented accuracy. Hence it is important to have theoretical predictions for various observables at the LHC with least uncertainties so that we will be able not only to test the SM to very good accuracy but also understand any deviations from the SM in terms of phenomena resulting from the physics BSM. The theoretical uncertainties resulting from computations can be estimated and further reduced using the tools developed with aid of various algebraic and numerical techniques. Often the leading order (LO) results are highly sensitive to theory uncertainties. Sometimes, the standard perturbative methods are not suitable for the computation of kinematical distributions in certain phase space regions due large logarithm appearing at every order in perturbative expansion. Analytical computation of the distributions at next to leading order (NLO) involves technical complications which can be solved only by seminumerical approach and it turns out they are most suitable for direct comparison with the experimental results. They are called Monte Carlo approach.

The high energy hadronic cross section involving two protons in the initial state and a specific set of final state particles denoted by X having momenta given by P_X can be written in terms of renormalised partonic cross sections using QCD factorisation theorem [8] as follows:

$$\begin{aligned} d\sigma_{\mathcal{S}}(P_1, P_2, \{P_X\}) &= \sum_{a,b=q,\bar{q},g} \sum_{\{\ _I\}} \int_0^1 dx_1 \int_0^1 dx_2 \\ &\times f_{a/\ _1}(x_1, \mu_F^2) f_{b/\ _2}(x_2, \mu_F^2) \\ &\times \Delta(x_1, x_2, \{P_I\}, \{P_X\}, \mu_F^2) \\ &\times \mathcal{S}(x_1, x_2, \{P_I\}, \{P_X\}) \,, \qquad (7) \end{aligned}$$

where $f_{a/\ }(x, \mu_F^2)$ is the probability of finding a parton of type a inside the proton with a momentum fraction x of the proton. They are called parton distribution functions (PDFs), $\Delta(x_1, x_2, \{P_I\}, \{P_X\})$ is the collinear finite partonic cross section. The incoming partonic states (a, b) with momentum fractions x_1, x_2 and intermediate states (I) with momenta $\{P_I\}$ are summed over. Notice that the massless partons produce divergent cross sections which are mass factorised in a process independent way and then suitably absorbed into bare PDFs leaving the resultant $\Delta(x_1, x_2, \{P_I\}, \{P_X\})$

collinear finite. The scale at which this is done is called factorisation scale, denoted by μ_F. Hence both PDFs and the collinear finite cross sections are individually dependent on μ_F but the observable $d\sigma(P_1, P_2, \{P_X\})$ does not. The function $\mathcal{S}(x_1, x_2, \{P_I\}, \{P_X\})$ take care various constraints that we can impose on the final states. Even though this function uniquely defines the observable we are interested in, it is chosen in such a way that it does not affect the factorisation of collinear singularities.

The PDFs are not computable within perturbative approach because they carry long distance (low energy) part of the cross section. They are usually fitted from the experiment for various values of x and μ_F. The renormalisation group invariance of the hadronic cross section with respect to the factorisation scale controls the scale dependence of the PDFs as well the collinear finite cross sections through perturbatively calculable anomalous dimensions/splitting functions denoted by $P_{ab}(x, \mu_F^2)$

$$\frac{df_{a/}(x,\mu_F^2)}{\mu_F^{-2} d\mu_F^2} = \sum_b \int_x^1 \frac{dy}{y} P_{ab}(y,\mu_F^2) f_{b/}\left(\frac{x}{y},\mu_F^2\right) . \quad (8)$$

This equation is called DGLAP evolution equation. $P_{ab}(x, \mu_F^2)$ is expanded in powers of a_s

$$P_{ab}\left(x,\mu_F^2\right) = \sum_{=1}^{\infty} a_s(\mu_F^2) P_{ab}^{(-1)}(x) . \quad (9)$$

The result for P_{ab} upto order a_s^3 is available in the literature [9]. The solution to DGLAP evolution equation determines the scale dependence of these PDFs. One parameterises these initial distributions for the partons at an experimental scale, say Q_{expt} and wide range of x available and then they are compared against the data to fix the parameters of the fit. The fit depends on various factors namely, the range of x and Q_{expt} that the experiments can provide and the accurate theoretical expressions. Fixed target as well as collider experiments such as deep inelastic scattering experiments and proton anti-proton hadronic machines have provided huge amount of rich data to fit these PDFs to a very good accuracy for a wide range of x and Q.

The collinear finite cross sections $\Delta_{ab}(\mu_F^2)$ are computable in perturbative QCD as a series expansion in the strong coupling constant a_s

$$\Delta_{ab}(\mu_F^2) = \sum_{=0}^{\infty} a_s(\mu_R^2) \Delta_{ab}^{( )}(\mu_F^2, \mu_R^2) , \quad (10)$$

where all the argument except μ_F is suppressed for brevity. The scale μ_R is called the ultraviolet (UV) renormalisation scale at which the theory is renormalised. The computation involves careful treatment of soft and collinear singularities that arise due to soft gluons and massless partons respectively. KLN theorem ensures the cancellation of soft singularities and factorisation theorem, the collinear finiteness. Note that the scales μ_F and μ_R are artifacts of factorisation and renormalisation and hence they should not affect the physical cross sections which are renormalisation group invariants with respect to these scales independently. On the other hand, the truncated perturbative series is not RG invariant. This introduces a scale uncertainty in any fixed order perturbative results. As is clear, the dependence goes down as the order of perturbation increases. The computation of these partonic cross sections beyond LO is feasible only for few process with definite final states due to technical complexities. The results beyond the LO are available only for few processes, for example Drell-Yan, Z, $W^\pm$, Higgs, prompt photon, di-photon, di-boson and jet productions, etc. are known to NLO level. In fact, NNLO QCD results for total as well as several kinematical distributions are available only for Drell-Yan, Z, $W^\pm$ and Higgs productions. In addition, NLO QCD results resulting from physics BSM are also available for certain processes making the predictions stable under perturbation. The Z and $W^\pm$ productions at the LHC at the initial stages of the operation, will be very important to constrain PDF uncertainties in the (x, Q) range not measured by previous experiments. These processes are said to serve as a standard candles. This will help to make unambiguous predictions for other SM processes for precision study as well as for new physics discoveries. A careful analysis indicate that the PDF uncertainties for the rapidity distributions of Z and $W^\pm$ at the LHC using the available data from various experiments have reached a precision of around 8% and this is good for using these processes as luminosity monitors. The analytical approach is often not possible if one is interested in the observables involving definite set of final states with well defined kinematic configurations. For example, the choice of the function $\mathcal{S}$ given in Eq. (7) can often make the analytical computation not feasible. The alternate approach is to use numerical approach called parton level generators. Here, one generates parton level events which are weighted with appropriate probabilities given by the partonic scattering matrix elements to predict the cross sections. This way, it is now possible to compute cross sections involving several jets, photons, Z and $W^\pm$ etc. Shower Monte Carlo event generators incorporate the hadronisation of final state partons using theoretically motivated models.

Even though these numerical approaches provide theoretical estimates for many realistic observables measurable in the experiments, they are often known at LO level due to technical difficulties involved in implementing higher order matrix elements. NLO improved monte carlo codes for certain observables are already available in the literature for physics study. The fixed order perturbative QCD predictions have limitations in applicability due to the appearance of large logarithms in some kinematical regions of the phase space. In such regions, the applicability of fixed order perturbative results becomes questionable due to the missing higher order corrections that are hard to compute. The alternate approach is to resum these logarithms in a closed form. Such an approach of resuming a class of large logarithms supplemented with fixed order results can almost cover the entire kinematic region of the phase space. In addition, these threshold corrections are further enhanced when the flux of the incoming partons become large in those regions. In the case of Higgs production through gluon fusion, the gluon flux at small partonic energies becomes large improving the role of threshold corrections.

3. Higgs Production Through Gluon Fusion at the LHC

In the SM, the Higgs mechanism is responsible for mass generation and its prediction is existence of the Higgs boson, a neutral scalar particle which couples to fermions and gauge boson. The only unknown parameter here is its mass. In the minimal supersymmetric extension of the standard model (MSSM), one needs two Higgs doublets to preserve supersymmetry and to give mass to the fermions. After symmetry breaking, we are left with two CP even (h, H), one CP odd (A) and two charged higgs $(H^{\pm})$. To lowest order, mass m_A of A and $\tan\beta$, the ratio of vacuum expectation values $v_i, i = 1, 2$ are the unknown parameters. Experiments which are going on for last two decades have set bounds on the mass of the Higgs bosons. The limit coming from the direct searches of SM Higgs boson is 114.4 GeV (at 95% CL) while it is <180 GeV from precision measurements. In the MSSM, at tree level, the lightest neutral Higgs boson has an upper bound of M_Z and it goes up to 130–140 GeV if we include higher order radiative corrections. At the LHC, these Higgs bosons will be produced and the properties can be measured. For example the SM Higgs boson mass can be measured at the per-mille level and its coupling at 5–20% level. If we restrict ourselves to SM Higgs boson and MSSM pseudo scalar A, they are produced largely through gluon-gluon fusion at the parton level through a top quark loop [10]. This LO process is proportional to square of the strong coupling constant $a_s(\mu_R^2)$ and the gluon flux, $\Phi_{gg}(\mu_F^2)$

$$\Phi_{gg}(x, \mu_F^2) = \int_x^1 \frac{dy}{y} f_{g/} \ (y, \mu_F^2) f_{g/} \ \left(\frac{x}{y}, \mu_F^2\right) . \quad (11)$$

They bring in un compensated renormalisation and factorisation scale dependencies make the predictions unreliable. The NLO computation [11] that reduces this problem results in large correction by about 80–100% of the LO result making the perturbative predictions questionable. Such a computation involves evaluation of technically complicated two loop integrals with top quarks in the loops and three body phase space integrals. Going beyond NLO is a difficult task and hence a computation at NNLO in the large top quark mass limit has been accomplished to stabilise the perturbative results as well as to reduce the scale uncertainties. Here the top quark degrees of freedom can integrated out which results in an effective Lagrangian containing new tree level interactions of Higgs boson with the gluons as well as photons. The effective Lagrangian for the Higgs boson coupling to gluons reads

$$_{-eff} = G_H(a_s(\mu_R^2))\phi(x)\mathcal{O}(x) , \quad (12)$$

where $\phi(x)$ is Higgs scalar field and $\mathcal{O}(x) = -1/4 \, F^a_{\mu\nu}(x)F^{a,\mu\nu}(x)$. The constant G_H is Wilson coefficient which is also function of top quark mass m_t, Fermi constant G_F and the Higgs boson mass m_H. Such an approach is justifiable if the mass of the Higgs boson is in the intermediate mass range, that is of the order of a few hundred GeV. In fact, this approach works very well if the Higgs boson is less than twice of top quark mass. This makes the computation of total cross sections and several distributions at NNLO level possible [12]. At NLO level, this approximation leads to only 1–4% error when the mass of the Higgs boson is less than 220 GeV. The NNLO result computed using the effective theory increases the cross section by about 10–15% and the scale uncertainties go down significantly, which is of the order of 10%. The resummation of soft gluons [13] can be systematically carried out beyond NNLO level. We have used renormalisation group (RG) invariance, collinear (mass) factorisation and Sudakov resummation of QCD amplitudes as guiding principles to perform the resummation in this region. Using the resummed results in $z = m_H^2/\hat{s}$ ($\hat{s}$-parton centre of mass) space we can now predict the soft-plus-virtual parts (also called threshold corrections) of the dominant partonic differential cross sections beyond N^{2}LO due to the recent results on three loop splitting functions [9] and the form factors [14]. The threshold corrections dominate when the partonic scaling variable z approaches its kinematic limit which is 1. They manifest in terms of the distributions $\delta(1-z)$ and $\mathcal{D}_i$, where

$$\mathcal{D}_i = \left[\frac{\ln^i(1-z)}{(1-z)}\right]_+ \qquad i = 0, 1, \cdots \tag{13}$$

The impact of partial soft plus virtual parts of N^3L and N^4L contributions to Higgs production through gluon fusion at the LHC is presented in Fig. 1. The left panel in Fig. 1 shows the total cross for producing Higgs at the LHC against its mass and the right panel contains the dependence on the scale variation. For LO, NLO and NNLO we used the exact results which contain both soft plus virtual as well as regular hard contributions. For N^iL $(i = 3, 4)$, we use only soft plus virtual results extracted from the resummed formula. Here we have set $\mu_F = \mu_R = m_H$. We find that the inclusion of N^iL $(i = 3, 4)$ does not change the cross section much confirming the reliability of the perturbative approach. Defining

$$R^I\left(\mu_R^2\right) = \left(d\sigma^I\left(\mu_R^2 = \;^2\right)\right)^{-1} d\sigma^I\left(\mu_R^2\right) \tag{14}$$

in the right panel of Fig. 1, we plot R as a function of $\mu/\mu_0 = \mu_R/m_H$, where we have fixed $\mu_F^2 = m_H^2$. It shows that the inclusion of the higher order terms reduces the sensitivity to the choice of the scale. The

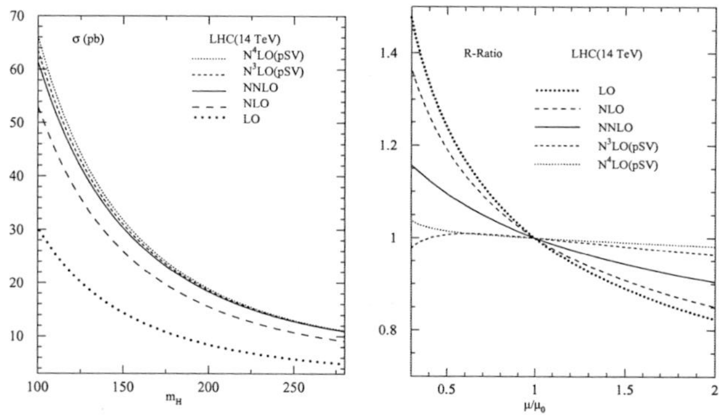

Figure 1. Total cross section for Higgs production through gluon fusion at the LHC, and its scale dependence for the Higgs boson of mass $m_H = 150$ GeV with $\mu = m_R, \mu_0 = \mu_H$. The abbreviation "pSV" means partial soft plus virtual

approach can be systematically extended to to resum the dominant soft gluon contributions to rapidity y distribution of Higgs production [15]. The resummation is done in the $z_i (i = 1, 2)$ space of the kinematic variables, which are the scaling variables that enter the rapidity distribution of the partonic cross sections. The threshold region corresponds to $z_i \to 1$ and in this region all the partonic cross sections are symmetric in $z_1 \leftrightarrow z_2$. Using RG invariance, mass factorisation and Sudakov resummation of QCD amplitudes as guiding principles we have performed the resummation in this region

$$\Delta^{\rm sv}_{gg}(z_1, z_2) = \mathcal{C} \exp\left(\Psi^H_{gg}(z_1, z_2, \varepsilon)\right)\Big|_{\varepsilon=0}, \tag{15}$$

where the symbol $\mathcal{C}$ means that all the normal products resulting from expansion of exponential are replaced by convolutions, sv means soft plus virtual and ε, regulator in the dimensional regularisation. The function Ψ^H_{gg} depends on the gluon form factor F^g, DGLAP kernel Γ_{gg}, the operator renormalisation constant Z_g and the soft distribution function Φ^H_S. Using the above resummed result given in z_i space the predictions are made for the soft-plus-virtual parts of the dominant partonic differential cross sections beyond N^{2}LO. The

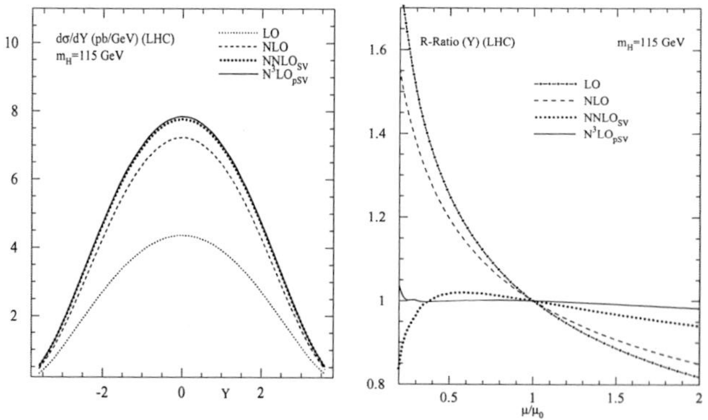

Figure 2. Rapidity distributions for Higgs production through gluon fusion at the LHC, and their μ_R scale dependence for the Higgs boson of mass $m_H = 115$ GeV with $\mu_F = m_H$

impact of soft-plus-virtual parts N^{2}LO and the partial soft-plus-virtual N^{3}LO contributions to Higgs production through gluon fusion at the LHC is presented in left panel of Fig. 2. We see that the inclusion of the higher order terms does not make any appreciable change in the magnitude of the rapidity distribution. Again this confirms the reliability of the perturbation series. The plot in the right panel of Fig. 2 shows R as a function of $\mu/\mu_0 = \mu_R/m_H$, where again we choose $\mu_F^2 = m_H^2$. We find the inclusion of higher order effect reduces the uncertainty coming from the scale.

4. Non-SUSY Extensions Beyond the SM

The gauge hierarchy problem has been one of the main motivations for physics beyond the SM. The hierarchy between the electroweak scale and the Planck scale is a long standing problem and has been mainly addressed in the past by modifying the particle content of the theory—Supersymmetry (SUSY) and Technicolour belong to this category. A paradigm shift in this approach was proposed by Arkani-Hamed, Dimopoulos and Dvali (ADD) [4] in 1998, wherein they modified the gravity sector instead of additional structure in the particle physics. Though the idea of extra dimension existed since the 1900s, all models assumed that gravity together with all other known interactions could live in the full extra dimensions. Consistency with experimental observations, demands that these extra dimensions be very small. The ADD scenario had explored the possibility of allowing only gravity to probe the extra dimensions and studied the constraints on the size of the extra dimensions. It turns out that for more then one extra dimension, its size could be large without contradicting any known experimental observation and consequently explain the weakness of gravity in 4-dimensions. An alternate solution of the hierarchy problem was suggested by Randall and Sundrum [5] in 1999 with a single extra dimension in a Anti-de-Sitter (AdS_5) metric. The ADD and RS models used the geometry of the extra dimensions to account for the hierarchy between the electroweak scale and the Planck scale.

The idea of extra dimensions has a rich history dating back to 1914 from the initial work of Gunnar Nordström who attempted to unify the electromagnetism and scalar gravity (the only forces then known), by invoking one extra spacial dimension. Later Kaluza (1921) and Klein (1926) tried to unify the electromagnetism and Einstein's general theory of relativity using the same geometrical approach. In the recent times string theory has been the main motivation for extra dimensions as quantised strings are not consistent unless embedded in a ten-dimensional spacetime. In order that these models or scenarios are consistent with present experimental length scale probed at colliders, its essential that these extra dimensions remain hidden. The various extra dimensional models have used different physical mechanism to hide the extra dimensions. The extra dimension could be small and compact, wherein all SM fields are allowed to propagate or alternatively the brane world scenarios wherein the SM particles are confined to the brane and hence can not probe the extra dimensions. In the brane world scenarios the extra dimensions could be of macroscopic size without contradiction with present experiments.

ADD and RS are both brane-world models where gravity exists and only gravity can probe the extra dimensions. Another possibility is the universal extra dimensions (UED) [16] wherein all the SM fields can probe the extra dimension while gravity coupling remains suppressed. Though the UED model has nothing to do with the solution of the hierarchy problem, it could possibly account for dark matter. Various combinations of these scenarios have also been considered. At the LHC it would be possible to test these non-SUSY extensions beyond the SM. In what follows we review some of these non-SUSY extensions beyond the SM and look at its experimental signatures at the LHC. The two main detectors which would be used for this purpose are the general purpose detectors ATLAS and CMS.

4.1. Large Extra Dimensions

In the ADD case the compactified extra dimensions are flat, of equal size and could be large. The large volume of the compactified extra spacial dimension would account for the dilution of gravity in 4-dimensions and hence the hierarchy. This was the first extra dimension model in which the compactified dimensions could be of macroscopic size and consistent with present experiments. In this model, new physics can appear at a mass scale of the order of a TeV. A viable mechanism to hide the extra spacial dimension, is to introduce a 3-brane with negligible tension and localise the SM particles on it. Only the graviton is allowed to propagate the full $4 + d$ dimensional spacetime. As a consequence of these assumptions, it follows from Gauss Law that the effective Planck scale M in 4-dimension is related to the $4 + d$ dimensional fundamental scale M_S through the volume of the compactified extra dimensions [4]. The extra dimensions are compactified on a torus of common circumference R. The number of extra spacial dimension possible is $d \geq 2$ from current experimental limits on deviation from inverse square law [17]. The spacetime is factorisable and the 4-dimensional spectrum consists of the SM confined to 4-dimensions and a tower of Kaluza-Klein (KK) modes of the graviton propagating the full $4 + d$ dimensional spacetime.

The interaction of the KK modes $h^{(\vec{n})}_{\mu\nu}$ with the SM fields localised on the 3-brane is given by

$$-_{int} \sim -\frac{1}{M} \sum_{\vec{n}=0}^{\infty} T^{\mu\nu}(x)\ h^{(\vec{n})}_{\mu\nu}(x)\,, \qquad (16)$$

where $T^{\mu\nu}$ is the energy-momentum tensor of the SM fields on the 3-brane. The zero mode corresponds to the usual 4-dimensional massless graviton. The KK modes are all M suppressed but the high multiplicity could lead to observable effects at the LHC. The Feynman rules are given in [18,19]. The ADD scenario raises the exciting possibility of observing quantum gravity at the LHC.

4.2. Warped Extra Dimensions

In the RS model there is only one extra spacial dimension and the extra dimension is compactified to a circle of circumference $2L$ and further orbifolded by identifying points related by $y \to -y$. Two brane are placed at orbifold fixed points, $y = 0$ with positive tension called the Planck brane and a second brane at $y = L$ with negative tension called the TeV brane. For a special choice of parameters, it turns out that the 5-dimensional Einstein equations have a warped solution for $0 < y < L$ with metric $g_{\mu\nu}(x^\rho, y) = \exp(-2\eta y)\ \eta_{\mu\nu}$, $g_{\mu y} = 0$ and $g_{yy} = 1$. This space is not factorisable and has a constant negative curvature—AdS_5 spacetime. η is the curvature of the AdS_5 spacetime and $\eta_{\mu\nu}$ is the usual 4-dimensional flat Minkowski metric. In this model the mass scales vary with y according to the exponential warp factor. If gravity originates on the brane at $y = 0$, TeV scales can be generated on the brane at $y = L$ for $\eta L \sim 10$. The apparent hierarchy is generated by the exponential warp factor and no additional large hierarchies appear. The size of the extra dimension is of the order of M^{-1}. Further it has been showed that [20] the value of ηL can be stabilised without fine tuning by minimising the potential for the modulus field which describes the relative motion of the two branes. In the RS model graviton and the modulus field can propagate the full 5-dimensional spacetime while the SM is confined to the TeV brane. The 4-dimensional spectrum contains the KK modes, the zero mode is M suppressed while the excited modes are massive and are only TeV suppressed. The mass gap of the KK modes is determined by the difference of the successive zeros of the Bessel function $J_1(x)$ and the scale $m_0 = \eta\ e^{-\pi k L}$. As in the ADD case the phenomenology of the RS model concerns the effect of massive KK modes of the graviton, though the spectrum of the KK mode is quite different.

In the RS model the massive KK modes $h^{(n)}_{\mu\nu}(x)$ interacts with the SM fields

$$-_{int} \sim -\frac{1}{M} T^{\mu\nu}\ h^{(0)}_{\mu\nu} - \frac{e^{\pi k L}}{M} \sum_{n=1}^{\infty} T^{\mu\nu}\ h^{(n)}_{\mu\nu}\,, \quad (17)$$

where $T^{\mu\nu}$ is the energy-momentum tensor of the SM fields on the 3-brane at $y = L$. The masses of $h^{(n)}_{\mu\nu}$ are given by $M_n = x_n\ \eta\ e^{-\pi k L}$, where x_n are the zeros of the Bessel function $J_1(x)$. In the RS model there are two parameters which are $c_0 = \eta/M$, the effective coupling and M_1 the mass of the first KK mode. Expect for an overall warp factor the Feynman rule of RS is the same as those of the ADD model.

5. Extra Dimension Searches at Hadron Colliders to NLO-QCD

In this section we will consider the brane-world models ADD and RS and study how the KK spectrum would modify the dilepton production at the LHC, which would be produced through the $P_1(p_1) + P_2(p_2) \to \ell^+(l_1) + \ell^-(l_2) + X(P_x)$. The final hadronic state is denoted by X and it carries a momentum P_X. We will also look at how precisely the various observables can be measured subject to the various uncertainties.

At hadron colliders, it is important to have a precise knowledge of the PDFs to predict production cross sections of both signals and backgrounds. These universal PDFs are non-perturbative inputs that are extracted from global fits to available data on DIS, DY and other hadronic processes. Parametrisation of PDFs to a particular order in QCD would involve various theoretical and experimental uncertainties. Recently there has been a series of papers [21–23] which for the first time have calculated the NLO-QCD corrections to various distributions of the DY process for both ADD and RS model. These NLO results would certainly reduce one aspect of the theoretical uncertainties as results prior to this calculation were only to LO in QCD for process involving gravity.

The cross-section $d\sigma/dQ$ as a function of Q to NLO is presented in Fig. 3a for the ADD model [21] and Fig. 4a for RS model [22] at the LHC. We have chosen the representative values of the RS model parameters: $M_1 = 1.5$ TeV the first RS resonance mass and the coupling constant $c_0 = 0.01$. The width of the resonance is related to c_0 and hence a smaller c_0 corresponds to a narrow resonance. The subsequent resonance are determined by m_0 and x_n. To LO the dilepton case has been presented in [24]. To see the effect of the NLO effect we study the K-factor for the Q and Y distribution. The K-factor for the invariant lepton pair mass

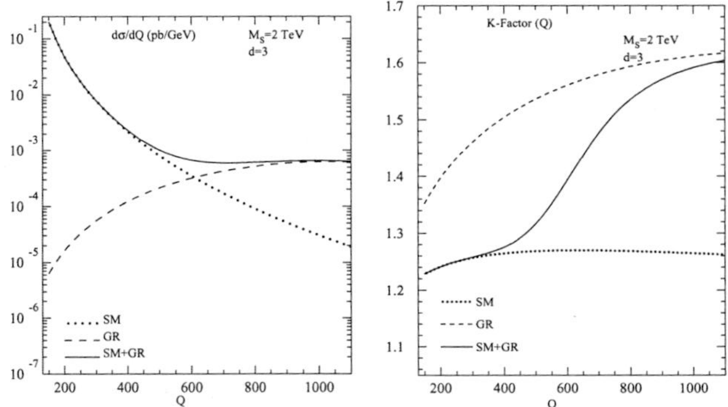

Figure 3. (a) Invariant mass distribution of the dilepton pair production $d\sigma/dQ$ for $M_s = 2$ TeV and $d = 3$ at the LHC to NLO in QCD (left panel). (b) The K-factor for the invariant mass distribution at the LHC (right panel)

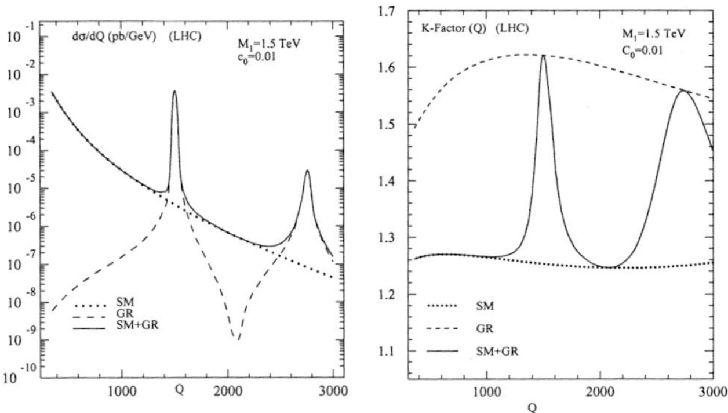

Figure 4. (a) The cross section is plotted as a function of invariant mass Q of the lepton pair for $M_1 = 1.5$ TeV at the LHC (left panel). (b) The corresponding K-factor for Q distribution SM, gravity and SM plus gravity (right panel)

distribution defined by

$$K^I = \left[\frac{d\sigma^I_{LO}(Q)}{dQ}\right]^{-1} \left[\frac{d\sigma^I_{NLO}(Q)}{dQ}\right], \tag{18}$$

where $I = SM$, $I = SM + GR$ for both SM and gravity combined and $I = GR$ for only gravity. It is possible to define K^{GR} for the invariant lepton pair mass distribution, as there is no interference with SM [21]. The results are presented in Fig. 4b. The parameters chosen are the same as in Fig. 4a.

The spectrum of the KK modes in the ADD and RS models are very distinct. In the ADD case the tower of KK modes leads to an enhancement of the tail of the invariant mass distribution Fig. 3a. In the RS case the KK spectrum consists of a few a few resonance and hence can be seen as resonant enhancement Fig. 4a over the SM.

5.1. Theoretical Uncertainties

In the QCD improved parton model the hadronic cross section can be expressed in terms of the partonic cross section convoluted with appropriate PDF. The subprocess cross section is a perturbative expansion in the strong coupling constant $\alpha_s(\mu_R)$. The partonic flux is non-perturbative and is given in terms of the PDF $f_a(z, \mu_F)$. In perturbative QCD, the unknown higher order corrections and the scale uncertainties are strongly correlated. The factorisation of mass singularities from the perturbatively calculable partonic cross sections leads to the introduction of factorisation scale μ_F in both non-perturbative parton densities $f_a(x, \mu_F)$ as well as the finite partonic cross sections $d\hat{\sigma}_{ab}(x, \mu_F)$. The value of the scale is arbitrary and one demands that the physical cross sections be independent of them. In addition to the factorisation scale, the partonic cross sections are dependent on the renormalisation scale μ_R. This is the scale at which the bare parameters of the theory become finite renormalised ones. The choice of the scale is again arbitrary. Gravity couples to the SM fields via its energy momentum tensor, and the calculations are done in the high energy limits where masses of the SM particles are ignored. Only parameter that requires UV renormalisation is the strong coupling constant. The scale uncertainties come about from the truncation of the perturbative series.

5.2. PDF Uncertainty

Unlike the perturbatively calculable partonic cross sections, the PDFs being non-perturbative in nature are extracted from various experiments. These are fitted at a scale of the experiments and then evolved according to the Altarelli-Parisi evolution equations to any other relevant scale. They are not only sensitive to experimental errors but also to theoretical uncertainties that enter through the partonic cross section calculations and the splitting functions that are known only to certain orders in strong coupling constant in perturbative QCD. There are various groups that are involved in parameterising these PDFs taking the uncertainties into account. These groups use not only different experiments but also different methods to parametrise these PDFs. Here, we mainly concentrate on the uncertainties coming from the various PDFs, *viz.* Alekhin [25], CTEQ [26] and MRST [27], in detail and quantify their impact on the new physics searches in extra-dimensional models. Differences among the various PDFs would translate as uncertainties on the physical observable. To NLO in QCD for various PDFs, we consider the following differential distributions [28]

$$\frac{d\sigma}{dQ}, \qquad \frac{d^2\sigma}{dQ\, dY}, \qquad \frac{d^2\sigma}{dQ\, d\cos\theta^*}, \tag{19}$$

where Q is the invariant mass, Y the rapidity and the θ^* is the angle between the final state lepton momenta and the initial state hadron momenta in the $c.o.m$ frame of the lepton pair (Figs. 5a, b). The corresponding K-factor which is the ratio of NLO to LO of the above distributions are also plotted for the various PDFs. The K-factor is as large as about 1.6, this clearly shows the need to go beyond LO in QCD in these models [21–23, 28].

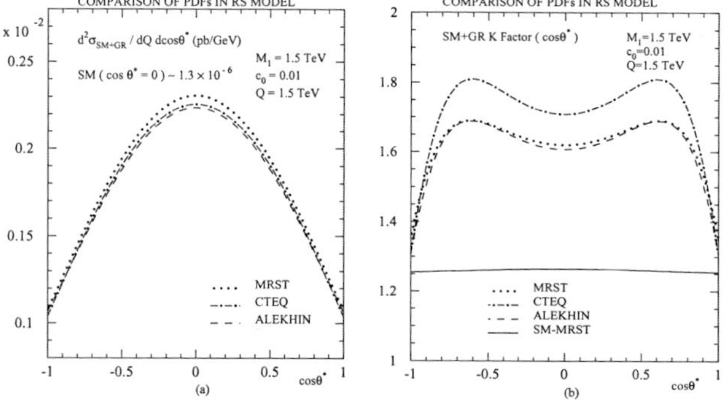

Figure 5. (a) In the region of first RS resonance, the double differential with respect to invariant mass and angular distribution of the lepton is plotted for the various PDFs at the LHC (left panel). (b) The corresponding K-factor for the various PDFs (right panel)

5.3. Renormalisation/Factorisation Scale Uncertainties

The μ_F variation is studied by varying μ_F in the range $0.5\ Q \le \mu_F \le 1.5\ Q$. We see that for both the ADD and RS model in going from LO to NLO in QCD, the uncertainties due to μ_F variation considerably get reduced [28]. The spread of K-factor with μ_F is much smaller for the SM as compared to $SM+GR$. This certainly indicates need to go beyond NLO. In Table 1 we tabulate the percentage spread of the factorisation scale μ_F dependence of various distributions at the LHC and Tevatron. On the average at the LHC and Tevatron, the percentage spread of the scale variation get reduced by about 2.88 times in going from LO to NLO. The dependence of cross section on μ_R comes from the strong coupling constant at NLO and at LO there is no μ_R dependence. At NLO μ_R dependence for the Y distribution is plotted for the μ_R range $0.5\ Q \le \mu_R \le 1.5\ Q$. The μ_R spread is largest in the central rapidity region and would only reduce at the NNLO level when the μ_R dependences would be compensated for by the dependence coming from the coefficient functions. We have studied the K-factor for SM and $SM+GR$ and see its dependence on μ_R. The uncertainties due to μ_R are much larger when the gravity is included. The percentage spread is of the order of 3.5% which is comparable to the μ_F spread at NLO.

	Distributions	Tevatron		LHC	
		LO	NLO	LO	NLO
ADD	$d^2\sigma/dQdY$	22.8	7.4	9.5	3.5
	$d^2\sigma/dQdcos\theta$	24.2	8.2	10.9	3.8
RS	$d^2\sigma/dQdY$	23.2	7.7	18.7	6.9
	$d^2\sigma/dQdcos\theta$	24.2	8.0	18.4	6.8

Table 1. Percentage spread as a result of factorisation scale variation in the range $0.5Q \le \mu_F \le 1.5Q$. For the ADD case $Q = 0.7$ TeV. For the RS first resonance region $Q = 1.5$ TeV for the LHC and $Q = 0.7$ TeV for Tevatron.

5.4. Experimental Uncertainties

In addition to the theoretical uncertainties that we have described in the previous section, there are uncertainties due to errors on the data. Various groups have studied the experimental errors and have estimates of the uncertainties on the PDFs within NLO QCD framework [29,30]. Now that NLO QCD results are also available for extra dimension searches [21] for the dilepton production, we consider some of the distributions and estimate the uncertainties due to the experimental error. We have plotted the error band for the MRST 2001 PDF [30] in the ADD model for the dilepton invariant mass distribution at the LHC. This error band is comparable to the spread associated with the different set of PDFs [28]. At $Q = 1$ TeV the percentage of experimental error is 7.5% for $SM + GR$ while the pure SM error is about 3.3%. For the RS case at the LHC in the first resonance region at $Q = 1.5$ TeV the experimental error is about 12.8%. At Tevatron the ADD model experimental error is 7.4% at $Q = 1$ TeV. In Fig. 6a we have plotted PDF comparison plots for the double differential distribution with respect to invariant mass and rapidity at a fixed value of $Q = 0.7$ TeV. The experimental error for this distribution for the central rapidity region is about 3.5% and is indicated in the Fig. 6a by the coloured lines.

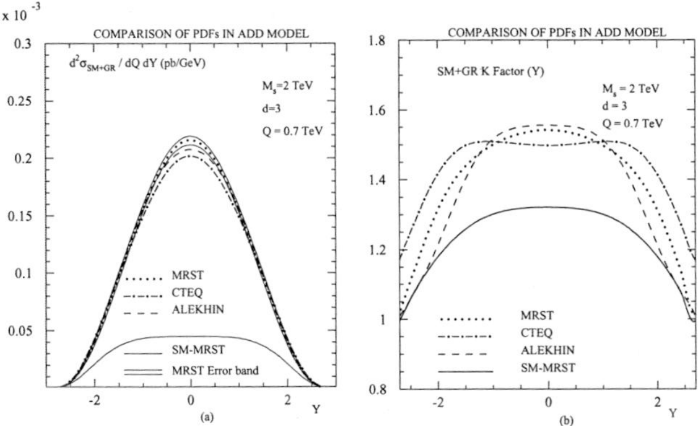

Figure 6. The comparison plots for the various PDFs sets for $Q = 0.7$ TeV at the LHC. (a) The double differential cross section with respect to invariant mass and rapidity as a function of rapidity. The coloured thin line gives the range of experimental error on the MRST PDF (left panel). (b) The corresponding K-factor as a function of rapidity (right panel)

6. Unparticle Physics

In a recent paper, Georgi [6,31] considered a scenario wherein the SM is weakly coupled to a sector which at an IR scale $\Lambda_{\mathcal{U}}$ has a non-trivial fixed point. At low energies this sector would hence be scale invariant and consequently conventional particle interpretation in this new sector fails—which Georgi termed as unparticle stuff. Essential ingredients of this scenario are (a) the coupling of these unparticle fields to the SM, which makes it possible to look for phenomenological consequences and (b) a mechanism to couple the SM to the unparticle fields which is accomplished by the transition in the Banks-Zaks ($\mathcal{BZ}$) theory [32] from which unparticle physics emerges at low energies.

Banks and Zaks studied the infrared stable fixed point of the $SU_c(3)$ gauge theory with massless fermions. Its well known that the asymptotic freedom of QCD disappears if the number of fermion species n_f or the dimension of their colour representation n_c is too large. QCD β function has been calculated up to 4-loops [33]. It is interesting to study the behaviour of the various coefficients of the β function as a function of n_f for $n_c = 3$. β_3 is positive for all positive values of n_f. As n_f is increased from zero the coefficients of the beta function changes sign as follows: $\beta_2(n_f = 5.844)$, $\beta_1(n_f = 8.05)$ and $\beta_0(n_f = 16.5)$. This indicates the presence of a non-trivial zero of the β function. Now if we denote $n_f = 16.5 = N^*$ and $n_f = 8.05 = N'$ and study the behaviour of β function in the range $N' < n_f < N^*$, we note that as $n_f \to N^*$, the zeros of the β function occurs at lower values of α_s, which implies that the zero of the β function is in the perturbative region. Hence the two loop result is sufficient to extract the IR fixed point. This study of Banks and Zaks showed that for n_f in the range $N' < n_f < N^*$ and close to N^* the theory has an exact scale invariant sector and has no particle interpretation.

Motivated by Banks and Zaks, Georgi [6] proposed the following scheme: Theory at very high energy contains the fields of the SM and fields of a sector called $\mathcal{BZ}$ sector, with a nontrivial IR fixed point. These two sectors interact through exchange of particles with a large mass scale $M_{\mathcal{U}}$. Below $M_{\mathcal{U}}$ the couplings have generic form

$$\frac{1}{M_{\mathcal{U}}^k}\eta_{\,SM}\eta_{\,\mathcal{BZ}}\,, \tag{20}$$

where $\eta_{\,SM}$ and $\eta_{\,\mathcal{BZ}}$ are operators built out of the SM and the $\mathcal{BZ}$ fields respectively. Scale invariance in the $\mathcal{BZ}$ sector emerges at energy scale $\Lambda_{\mathcal{U}}$. In the effective theory below $\Lambda_{\mathcal{U}}$, the interaction of Eq. (20) matches onto

$$C_{\mathcal{U}}\frac{\Lambda_{\mathcal{U}}^{d_{\mathcal{BZ}}-d_{\mathcal{U}}}}{M_{\mathcal{U}}^k}\eta_{\,SM}\eta_{\,\mathcal{U}}\,, \tag{21}$$

where $d_{\mathcal{U}}$ is the scaling dimension of the unparticle operator $\eta_{\,\mathcal{U}}$ and $C_{\mathcal{U}}$ is a coefficient in the low energy effective theory. $M_{\mathcal{U}}$ should be large enough that its coupling to SM be sufficiently weak. Few of the generic operators that can describe the interaction of unparticle fields with those of the SM are found to be

$$\frac{\lambda_s}{\Lambda_{\mathcal{U}}^{d_{\mathcal{U}}}}T^\mu_\mu\;\eta_{\,\mathcal{U}},\quad \frac{\lambda_v}{\Lambda_{\mathcal{U}}^{d_{\mathcal{U}}-1}}\bar\psi\gamma_\mu\psi\;\eta^{\,\mu}_{\,\mathcal{U}},\quad \frac{\lambda_t}{\Lambda_{\mathcal{U}}^{d_{\mathcal{U}}}}T_{\mu\nu}\;\eta^{\,\mu\nu}_{\,\mathcal{U}}\,. \tag{22}$$

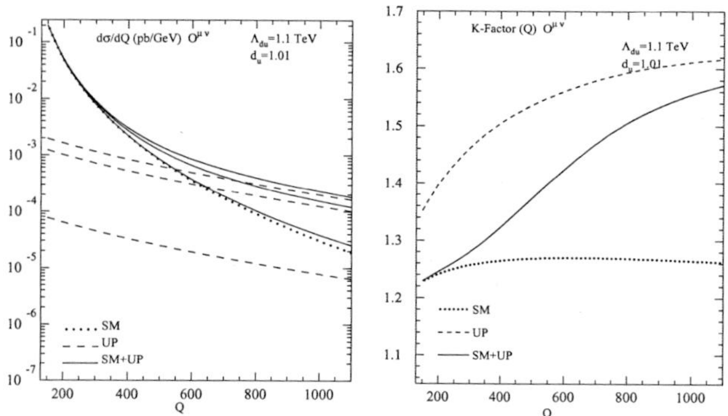

Figure 7. The tensor unparticle coupling to quarks and gluons in the initial state to produce a dilepton pair. (a) Left panel corresponds to the invariant mass distribution of the dilepton Q. The lowest set of lines corresponds to $\lambda_t = 0.4$, middle $\lambda_t = 0.8$ and upper $\lambda_t = 0.9$. (b) Right panel corresponds to the K-factor for the invariant mass distribution

The dimensionless coupling λ_κ corresponds to the unparticle operator $\eta^{\kappa}_{\mathcal{U}}$, where $\kappa = s, v, t$ refers to the scalar, vector and tensor operators respectively. $T_{\mu\nu}$ is the energy momentum tensor of the SM. These operators are Hermitian and transverse.

The unparticle propagator is given by

$$\int d^4x\, e^{i\ \ x} \langle 0|T\eta^{\kappa}_{\mathcal{U}}(x)\eta^{\kappa}_{\mathcal{U}}(0)|0\rangle = \frac{iA_{d_\mathcal{U}}}{2\sin(d_\mathcal{U}\pi)} \frac{B_\kappa}{(-P^2 - i\epsilon)^{2-d_\mathcal{U}}}, \quad (23)$$

where B_κ depends on the Lorentz structure of the operator $\eta_\mathcal{U}$ as given below

$\eta_\mathcal{U}$	1
$\eta^{\rho}_{\mathcal{U}}$	$\eta_{\mu\nu}(P) = -g_{\mu\nu} + \frac{\mu\ \nu}{\alpha^2}$
$\eta^{\rho\sigma}_{\mathcal{U}}$	$B_{\mu\nu\alpha\beta} = \frac{1}{2}\left(\eta_{\mu\alpha}\eta_{\nu\beta} + \eta_{\mu\beta}\eta_{\nu\alpha} - \frac{2}{3}\eta_{\mu\nu}\eta_{\alpha\beta}\right)$.

The constant $A_{d_\mathcal{U}}$ is given by

$$A_{d_\mathcal{U}} = \frac{16\pi^{5/2}}{(2\pi)^{2d_\mathcal{U}}} \frac{\Gamma(d_\mathcal{U} + 1/2)}{\Gamma(d_\mathcal{U} - 1)\Gamma(2d_\mathcal{U})}, \quad (24)$$

where $1 < d_\mathcal{U} < 2$.

The couplings of the unparticle operators to the SM fields are given by

$$\lambda_\kappa = C^{\kappa}_{\mathcal{U}} \left(\frac{\Lambda_\mathcal{U}}{M_\mathcal{U}}\right)^{d_{\mathcal{BZ}}} \frac{1}{M_\mathcal{U}^{d_{\rm SM}-4}} . \quad (25)$$

A priori we have no information on any of the parameters in the above equation. For our numerical analysis we have taken λ_κ in the range $0.4 \leq \lambda_\kappa < 1$, so that the unparticle effects are treated as perturbation. The other parameter that appears in this model is $\Lambda_\mathcal{U}$ which we choose to be 1 TeV.

The unparticles could be interpreted as fractional number of massless particles [6] or a collection of particles with a particular mass distribution [34] as a result of deconstruction. It would hence be interesting to look for phenomenological consequences of unparticle phase space or propagator. The coupling of unparticles to the SM Higgs could lead to a constraint on the unparticle sector due to breaking of the scale invariance when the Higgs gets a vacuum expectation value [35]. As a consequence unparticle physics may be relevant at the high energy colliders which have been actively investigated [6, 31, 36–38]. For the dilepton invariant mass distribution we plot the tensor unparticle contribution for various values of λ_t in Fig. 7a and the corresponding K factor in Fig. 7b.

7. Summary

In this article, we have summarised in detail the role of perturbative QCD at the LHC in predicting various SM as well as non-susy observables with least theoretical uncertainties. We have presented a short introduction to the LHC experiment and the basics of perturbative QCD relevant for our theoretical predictions. Higgs boson will be one of the important particles that the LHC will search for. We have studied the impact of NNLO corrections to the Higgs production through gluon fusion in the large top quark mass limit. The results for its total cross section as well as rapidity distribution beyond NNLO are presented using the formalism of soft gluon resummation. The numerical impact these results are presented for a wide range of Higgs masses and rapidities. The results show stability of these observables against higher order perturbation and scale variations. Next we have studied the role played by QCD in predicting various observables resulting from physics BSM such as large extra-dimensional, Randall-Sundrum and unparticle models. The impact of various parton density sets at NLO in QCD on the Drell-Yan production of dileptons at hadron colliders such as the LHC is studied. This process can probe the physics beyond SM through exchange of new particles that these theories predict. At hadron colliders, the precise measurement of DY production cross sections is possible. In this context, we have studied the theories of extra dimensions such as ADD and RS which attempt to explain gauge hierarchy problem in SM. We have discussed various theoretical uncertainties that enter through renormalisation, factorisation scales and the parton density sets. We have quantified the uncertainties coming from various parton density sets using the recent results on NLO QCD corrections to parton level cross sections and recent PDF sets that take into account various theoretical and experimental errors. Our entire analysis is model independent thanks to the factorisation of QCD radiative corrections from the model dependent contributions. We find that the K-factor for various observable depends on the choice of PDFs.

Acknowledgement: Wc thank M.C. Kumar and A. Tripathi for discussion. VR thanks J. Smith for discussion. The work of VR has been partially supported by funds made available to the Regional Centre for Accelerator-based Particle Physics (RECAPP) by the Department of Atomic Energy, Govt. of India.

REFERENCES

1. International Europhysics Conference On High Energy Physics fiEPS-HEP2007fi 19-25 Jul 2007, Manchester, England
2. A. Djouadi, Phys. Rept. **457** fi2008fi 1 farfi ivfhep-ph/0503172ffi R. Harlander, J. Phys. G **35** fi2008fi 033001.
3. S. P. Martin, arfi ivfhep-ph/9709356fiA. Djouadi, Phys. Rept. **459** fi2008fi 1 farfi ivfhep-ph/0503173ffi
4. I. Antoniadis, N. Arkani-Hamed, S. Dimopoulos and G. R. Dvali, Phys. Lett. fi **436** fi1998fi 257 farfi ivfhep-ph/9804398ffi
5. L. Randall and R. Sundrum, Phys. Rev. Lett. **83** fi1999fi 3370 farfi ivfhep-ph/9905221ffi W.D. Goldberger and M.fi. Wise, Phys. Rev. Lett. **83** fi1999fi 4922.
6. H. Georgi, Phys. Rev. Lett. **98** fi2007fi 221601 farfi ivfhep-ph/0703260fi
7. T. Muta, Foundations of fi uantum Chromodynamicsfi An Introduction to Perturbative Methods, World Scientific Publishing Company, August 1998.
8. G. Sterman, An Introduction to fi uantum Field Theory, Cambridge fi niversity Press, September 1993.
9. S. Moch, J. A. M. fiermaseren and A. fiogt, Nucl. Phys. fi **688** fi2004fi 101 farfi ivfhep-ph/0403192ffi A. fiogt, S. Moch and J. A. M. fiermaseren, Nucl. Phys. fi **691** fi2004fi 129 farfi ivfhep-ph/0404111fi
10. F. Wilcfiek, Phys. Rev. Lett. **39**, 1304 fi1977ffi H. M. Georgi, S. L. Glashow, M. E. Machacek and D. fi. Nanopoulos, Phys. Rev. Lett. **40**, 692 fi1978ffi J. R. Ellis, M. fi. Gaillard, D. fi. Nanopoulos and C. T. Sachrajda, Phys. Lett. fi **83**, 339 fi1979fi
11. S. Dawson, Nucl. Phys. fi **359**, 283 fi1991ffiA. Djouadi, M. Spira and P. M. fierwas, Phys. Lett. fi **264**, 440 fi1991ffiD. Graudenfi, M. Spira and P. M. fierwas, Phys. Rev. Lett. **70**, 1372 fi1993ffi M. Spira, A. Djouadi, D. Graudenfi and P. M. fierwas, Nucl. Phys. fi **453**, 17 fi1995fi farfi ivfhep-ph/9504378fi
12. R. fi. Harlander and W. fi. fiilgore, Phys. Rev. Lett. **88**, 201801 fi2002fi farfi ivfhep-ph/0201206ffi C. Anastasiou and fi. Melnikov, Nucl. Phys. fi **646**, 220 fi2002fi farfi ivfhep-ph/0207004ffi fi. Ravindran, J. Smith and W. L. van Neerven, Nucl. Phys. fi **665**, 325 fi2003fi farfi ivfhep-ph/0302135fi
13. fi. Ravindran, Nucl. Phys. fi **752**, 173 fi2006fi farfi ivfhep-ph/0603041fi
14. S. Moch, J. A. M. fiermaseren and A. fiogt, Phys. Lett. fi **625** fi2005fi 245 farfi ivfhep-ph/0508055fi
15. fi. Ravindran, J. Smith and W. L. van Neerven, Nucl. Phys. fi **767** fi2007fi 100 farfi ivfhep-ph/0608308fi
16. T. Appelfiuist, H. C. Cheng and fi. A. Dobrescu, Phys. Rev. **D64** fi2001fi 035002.
17. C. D. Hoyle *et. al*, *Phys. Rev.* **D70** fi2004fi 042004.
18. G. F. Giudice, R. Rattaffi and J. D. Wells, *Nucl. Phys.* **B544** fi1999fi 3.
19. T. Han, J. D. Lykken and R-J. fihang, *Phys. Rev.* **D59** fi1999fi 105006.
20. W.D. Goldberger and M.fi. Wise, *Phys. Rev. Lett.* **83** fi1999fi 4922fi *Phys.Lett.* **B475** fi2000fi 275.
21. Prakash Mathews, fi. Ravindran, fi. Sridhar and W.L. van Neerven Nucl. Phys. **B713** fi2005fi 333.
22. Prakash Mathews, fi. Ravindran and fi. Sridhar, JHEP **0510** fi2005fi 031.
23. Prakash Mathews, fi. Ravindran, Nucl. Phys. **B753** fi2006fi 1.
24. H. Davoudiasl, J.L. Hewett and T.G. Rififio, *Phys. Rev. Lett* **84** fi2000fi 2080fi *ibid. Phys. Rev.* **D63** fi2001fi 075004.
25. S. Alekhin Phys. Rev. D68 fi2003fi 014002.
26. J. Pumplin *et. al.*, JHEP 0207 fi2002fi 012.
27. A.D. Martin *et. al.*, Eur. Phys. J. C23 fi2002fi 73.
28. M. C. fi umar, Prakash Mathews and fi. Ravindran, Eur. Phys. J. C49 fi2007fi 599.
29. S. I. Alekhin, Phys. Rev. D 63 fi2001fi 094022fi CTEfi Collaborationfi J. Pumplin et al., JHEP 0207 fi2002fi 012.
30. A.D. Martin *et. al.*, Eur. Phys. J. C28 fi2003fi 455.
31. H. Georgi, Phys. Lett. fi **650** fi2007fi 275.
32. T. fianks and A. fiaks, Nucl. Phys. fi **196**, fi1982fi 189.
33. T. van Ritbergen, J. A. M. fiermaseren, S. A. Larin Phys. Lett. fi400 fi1997fi 379.
34. M. A. Stephanov, Phys. Rev. D **76** fi2007fi 035008.
35. P. J. Fox *et al.*, Phys. Rev. D **76** fi2007fi 075004fiA. Delgado *et al.*, JHEP **0710** fi2007fi 094fiM. fiander *et al.*, Phys. Rev. D **76** fi2007fi 115002.
36. fi. Cheung, W. fi. fieung and T. C. fiuan, Phys. Rev. Lett. **99** fi2007fi 051803.
37. P. Mathews and fi. Ravindran, Phys. Lett. fi **657** fi2007fi 198.
38. M. C. fi umar *et al.*, Phys. Rev. D **77** fi2008fi 055013.

Heavy Ions at the LHC: A Quest for Quark-gluon Plasma

Rajeev S. Bhalerao and Rajiv V. Gavai

Department of Theoretical Physics, Tata Institute of Fundamental Research, Homi Bhabha Road, Mumbai-400005, India

Quantum chromodynamics (QCD), the theory of strong interactions, predicts a transition of the usual matter to a new phase of matter, called Quark-gluon Plasma (QGP), at sufficiently high temperatures. The non-perturbative technique of defining a theory on a spacetime lattice has been used to obtain this and other predictions about the nature of QGP. Heavy ion collisions at the Large Hadron Collider in CERN can potentially test these predictions and thereby test our theoretical understanding of confinement. This brief review aims at providing a glimpse of both these aspects of QGP.

1. Introduction

There are two very commonly quoted motivations for the upcoming Large Hadron Collider (LHC) at CERN in Geneva, the centre of attraction for the articles in this volume. Perhaps the primary one is that the LHC will provide us a key to understand the origin of the visible mass of our universe. This alludes to that fact that our Standard Model (SM) of particle interactions has to start with matter in the form of massless quarks and leptons. The famous Higgs mechanism [1] of spontaneous breaking of gauge symmetries provides masses to them, and the carriers of the weak force, namely $W^{\pm}$, Z. The LHC is widely expected to discover the Higgs boson which is tied with this mechanism. The other motivation rests on the fact that the SM has been well understood due to the many impressive precision tests carried out in many experiments, including those at the Large Electron Positron (LEP) at CERN and the Tevatron at the Fermilab in the USA. However new physics beyond the Standard Model (BSM) has to exist [2] since SM contains many, at least 19, arbitrary parameters and thus cannot be the final theory. Indeed, it is even hoped that the LHC may provide us not only a glimpse of the BSM physics, but it will hopefully also explain the origin of the mass of the dark matter in the universe.

While these motivations are largely correct, there are certain oversimplifications in them, leading to a few misconceptions, especially in the popular media. First of all, even if the expected Higgs particle is actually discovered, the origin of the mass of up/down (u/d) quarks can be claimed to be understood only after it is also established that the Higgs particle couples to them with a strength of $\sim 10^{-6}$, not an easily achievable goal at the LHC. Indeed, one may as well need an electron-positron collider to establish this in the post-LHC era. Moreover the protons and neutrons, which make up most of the visible mass in our universe, have each a much larger, almost a factor of 100 larger, mass than the sum of the masses of their constituent u/d quarks. Therefore, the understanding of the visible mass of the universe will emerge from the efforts to figure out why protons/neutrons have such large binding energies. Starting from molecules to atoms and nuclei, we are accustomed to the idea that the interactions which bind the respective constituents give rise to binding energies much smaller, less than even a per cent. This has given rise to the very successful idea of treating these interactions perturbatively as an expansion in the strength of the interaction. As we shall see below, one needs new suitable techniques to investigate these large binding energies, in Quantum chromodynamics (QCD), the theory of interactions of quarks with gluons, the carriers of the strong force.

As may be seen from the articles in this volume itself, QCD is an integral part of our SM of particle and their interactions. From various experiments in the past, it is well known that quarks carry both flavour quantum numbers such as, electric charge or strangeness, as well as colour: they transform as a triplet under the colour $SU(3)$ group. As in the case of electric charge, the colour charge is also mediated by massless vector particles, gluons. Structurally, the theory of quark-gluon interactions, QCD, looks very similar to that of electron-photon interactions, QED. A key difference though is that there are eight gluons which themselves carry colour charge, transforming as an octet under $SU(3)$-colour group. Consequently, gluons can interact amongst themselves. Furthermore the QED coupling is rather small at the scales we probe, being $1/137$, whereas the smallest measured QCD coupling, α_s, is about 0.12. In fact, more often, one has to deal with $\alpha_s = 0.3$ or so and it is $\gtrsim 1$ in the bound states like proton or neutron. QCD exhibits a much richer structure and a variety of phenomena as a result of this large α_s. Quark confinement and dynamical chiral symmetry breaking can be named as typical examples. A lack of observation of quarks in experiments led to the hypothesis that quarks are permanently confined in the hadrons, i.e. protons or pions whereas the lightness of pions compared to protons is expected to be understood as the phenomenon of dynamical breaking of the chiral symmetry by the vacuum. QCD as the theory of strong interactions has to explain these phenomena. Since, QCD is too complex, simple models based on underlying symmetries are often employed to account for its non-perturbative aspects. Indeed, most if not all of the "precision tests" are either performed experimentally only at small coupling, α_s, corresponding to rather rare events, or employ the simple QCD-based models. The latter are in many cases possible weak links in the precision tests of the SM: physics beyond SM may even show up in non-perturbative QCD beyond these models. We need to look for it and rule out such a mundane possibility for BSM-physics in order to be sure that other exotic possibilities are indeed worth looking for. Thus non-perturbative techniques are needed for real precision tests of QCD. As a glaring example, let me mention that the easiest precise measurement at the LHC will perhaps be the total proton-proton cross section at 14 TeV. The current best theoretical prediction for it is [3] $\sigma^{\text{tot}} = 125 \pm 25$ mb ! As explained in [3], one uses the so-called Regge models to arrive at it, and one such model can even explain the currently observed Q^2-variation of the structure function of proton, F_2, as well. Recall that a key cornerstone for establishing QCD as the theory of strong interaction is this Q^2 variation.

While obtaining a reliable prediction for the above cross section from QCD still seems far away, a non-perturbative technique does exist today to obtain other quantities, such as the decay constants or the weak matrix elements, from QCD using first principles and these could still provide non-perturbative precision tests of the SM. QCD defined on a spacetime lattice is such a tool. Not only does it explain many of the above mentioned phenomena but it provides quantitative estimates of many physical observables. Furthermore, the *same* techniques of lattice QCD lead to spectacular predictions for the behaviour of matter under extreme conditions. Thus, lattice QCD predicts the existence of a new phase, called Quark-gluon Plasma (QGP) at sufficiently high temperature, and a phase transition of the strongly interacting matter of protons, neutrons and pions to the new phase QGP at high enough temperature. The dynamically broken chiral symmetry of QCD at low temperatures in our world is expected to be restored in the QGP phase, 'melting' away the constituent mass of the light quarks acquired due to interactions.

Our universe ought to have existed in such a phase a few microseconds after the Big Bang, and about 20 microseconds later the phase transition to the normal hadrons like protons, neutrons and pions ought to have taken place in it. Whether there are any imprints of this phase transition on the astronomical objects observed today depends on the nature of the phase transition. There have been speculations of stars with strange matter, consisting of neutral baryons made from an up, down and a strange quark each. Similarly attempts have been made to study the influence of such a phase transition on the Big Bang nucleosynthesis. More excitingly, the LHC itself will provide us with an opportunity to create these early universe-like conditions of high energy densities, or equivalently high temperature, in the laboratory in its proposed heavy ion collisions of Lead on Lead at 5.5 TeV colliding energy. Heavy ion collisions at relativistically high energy have had an illustrious past, and even more impactful present. Early such experiments were made at the SPS collider in CERN, Geneva at a colliding energy of 17 GeV per nucleon in the centre of mass (cm) frame. The relativistic heavy ion collider (RHIC) has been operative in BNL, New York, since a last few years and has produced heavy ion collision data for a variety of ions, deuterium (D), copper (Cu) and gold (Au), at a spectrum of energies, 62–200 GeV per nucleon in the cm frame. Experiments at the LHC will thus see a further jump in the colliding energy by a factor of about 30. It is hoped that this will offer us cleanest environment yet for investigating the physics of QGP.

In this short review, we shall attempt to provide a glimpse of how lattice QCD leads to QGP and predicts many of its properties as well as those of the corresponding phase transition and how the heavy ion collision experiments amazingly provide us an opportunity to produce QGP in a laboratory, including the expectations of what we may observe at the LHC.

2. QGP from Lattice QCD

In order to understand and appreciate the fundamental importance of attempts to discover QGP at the LHC, let us first review the basics of lattice QCD and why it facilitates a truly reliable treatment of non-perturbative physics. In the process, we shall also see why essentially the *same* tested technique for obtaining, say, the hadron masses, comes into play for predicting new phases or phase transitions.

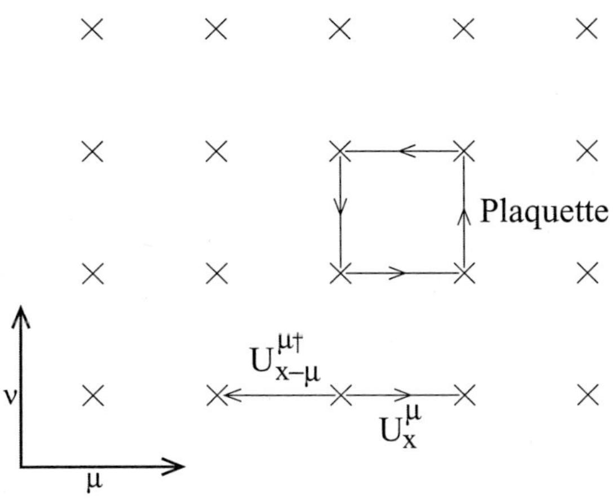

Figure 1. Quark and gluon fields on a spacetime lattice

2.1. Basic Lattice QCD

Lattice field theory is defined by discretising the space-time. The (inverse of the) lattice spacing a acts as the ultra-violet cut-off needed to tame the divergences in a quantum field theory. One places the anti-commuting quark fields $\psi(x)$ and $\bar{\psi}(x)$ on lattice sites whereas the gluon fields reside on the links, as shown in Fig. 1. A directed link from site x in the positive direction $\hat{\mu}$ is associated with the gluon field U^{μ}_{x}, while the link to the site $x - \hat{\mu}$ in the opposite direction is $U^{\mu}_{x-\hat{\mu}}$. A gauge transformation $V_x \in SU(3)$ rotates the quark field in the colour space: $\psi'(x) = V_x \psi(x)$. Demanding that the gluon field at the link x in the direction $\hat{\mu}$, $U_{\mu}(x)$, change to $U'_{\mu}(x) = V_x U_{\mu}(x) V^{-1}_{x+\hat{\mu}}$, ensures that the (discrete) kinetic energy term of quarks remains invariant under such a gauge transformation. Constructing gauge actions from closed Wilson loops of the links, like e.g. the smallest square loop, called plaquette and displayed in Fig. 1, ensures their gauge invariance.

It turns out that a straightforward discretisation of the derivative, given by $[a \cdot \partial_{\mu}\psi(x) = \psi(x + a\hat{\mu}) - \psi(x - a\hat{\mu})]$, can be made gauge invariant as shown in the Fig. 1, where the links end on respective quark fields ψ at the sites. Thus a sum over all independent terms of both types shown in Fig. 1 yields the QCD action on the lattice. However, it leads to the so-called fermion doubling problem: each lattice fermion corresponds to $2^d = 16$ flavours in the continuum limit of $a \to 0$. Various lattice fermion actions, referred to as the staggered, Wilson, domain wall or overlap fermions, have been proposed to alleviate this problem. In view of their simplicity and an exact chiral symmetry even on the lattice, the staggered fermions have dominated the field of interest for this article, namely lattice QCD at finite temperature and density. Briefly these are single component Grassmann variables on each site, with the γ-matrices replaced by suitably defined sign factors. They have a $U(1) \times U(1)$ chiral symmetry and 4 flavours in continuum limit. An oft-discussed problem of the staggered fermions, though, is that two or three light flavours are not simple to define, and the currently used methods may miss out on important physics aspects related to anomalies. It is often argued that for the bulk thermodynamic properties these issues are likely to be unimportant.

Typically, for any lattice computation one needs to evaluate the expectation value of an observable Θ

$$\langle \Theta(m_v) \rangle = \frac{\int DU \exp(-S_G)\Theta(m_v) \text{ Det } M(m_s)}{\int DU \exp(-S_G) \text{ Det } M(m_s)} , \tag{1}$$

where M is the Dirac matrix in x, colour, spin, flavour space for sea quarks of mass m_s, S_G is the gluonic action, and the observable Θ may contain fermion propagators of mass m_v. $S_G \sim 6 \sum trU_{plaq}/g_0^2$, with g_0 the bare coupling and U_{plaq} the product links along a plaquette as shown in Fig. 1. Amongst the many methods of evaluation of Eq. (1), numerical simulations stand out due to the ability to achieve the goal of removing the lattice scaffolding, i.e. taking the continuum limit $a \to 0$. Using the two-loop β-function, it is easy to show that

$$M \cdot a = \frac{M}{\Lambda} (g_0^2 b_0)^{-b_1/2b_0^2} e^{-\frac{1}{2b_0 g_0^2}} (1 + 0(g_0^2) \tag{2}$$

defines the way a mass scales $M \cdot a$ on the lattice changes as the bare coupling $g_0 (= \sqrt{6/\beta})$ is changed. Here b_0 and b_1 are the universal coefficients of the β-function. Typically, one needs larger and larger lattice sizes as $a \to 0$ in order to keep physical volume fixed.

Numerically, the $\langle \Theta \rangle$ is computed by averaging over a set of configurations $\{U_{\mu}(x)\}$ which occur with probability $\propto \exp(-S_G) \cdot \text{Det } M$. Thus the main problem is to generate the ensembles of such configurations with the desired probability distribution. Complexity of evaluation of Det M has lead to various levels of approximations in the process of generation of configurations: the quenched approximation consists of sea quark mass, $m_s = \infty$ limit whereas the full theory should have low sea quark masses: $m_u = m_d$ with a moderately heavy strange quark. The computer time required to obtain results at the same precision increases as the sea quark mass is lowered.

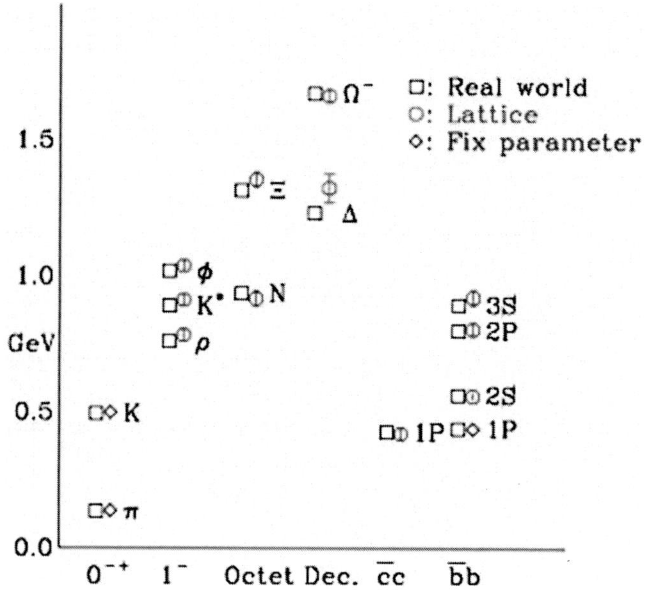

Figure 2. Comparison of experimental hadron spectra with lattice results [4]

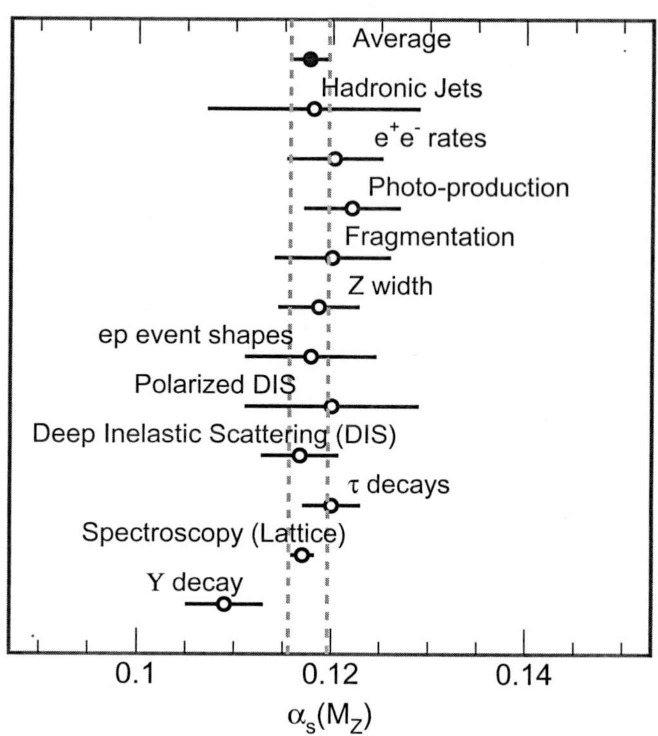

Figure 3. Various determinations of α_s. From [5]

2.2. Some Results from Lattice QCD

A variety of qualitative and quantitative results have been obtained using the lattice techniques. It will be both impractical and unnecessary to review all of them here. However, in order to appreciate the power of these techniques, we limit ourselves to providing a glimpse of them for the staggered fermions; similar sometimes better in quality/precision, results have been obtained with the Wilson fermions as well. Figure 2 shows [4] the results of the MILC and HPQCD collaborations for the light as well as heavy hadrons obtained with light sea quarks. Using the pion and kaon masses to fix the scales of the corresponding quark masses, most other particle masses are found to be in good agreement with the experiment. Furthermore the spontaneous breaking of the chiral symmetry has been demonstrated by many groups since the early days of the lattice QCD, showing a non-vanishing chiral condensate, $\langle\bar{\psi}\psi\rangle \neq 0$. Moreover, the goldstone nature of the pion has also been verified by checking that $m_\pi^2 \propto m_u$. Figure 3 displays a comparison [5] of the lattice determination of the strong coupling, $\alpha_s(M_Z)$, with other perturbative determinations from experimental data. While these results verify that QCD is indeed the correct theory of the strong interactions, and the lattice technique is the most reliable quantitative tool to extract its non-perturbative properties, making new predictions for the experiments is where the real challenges and excitement lies. It is very heartening to note that the decay constants of pseudo-scalar mesons containing a heavy quark were first obtained using lattice techniques: $f_{D^+} = 201 \pm 3 \pm 17$ MeV and $f_{D_s} = 249 \pm 3 \pm 16$ MeV [6]. These have since been measured experimentally to be $f_{D^+} = 223 \pm 16 \pm 7$ MeV [7] and $f_{D_s} = 283 \pm 17 \pm 14$ MeV [8] in excellent agreement with the lattice QCD predictions.

2.3. Lattice QCD at Non-zero Temperature and Density

Investigations of QCD under extreme conditions, such as high temperatures and/or densities, provide a solid platform for its most spectacular non-perturbative tests. Since the results from hadron spectroscopy fix the quark masses as well as the scale Λ_{QCD}, these tests are even completely free of any arbitrary parameters. Based on simple models, which build in the crucial properties of confinement or chiral symmetry breaking and allow asymptotically for the free quark gluon gas, one expects phase transitions to new phases such the QGP or the colour superconductors. As we shall see in the next section, the experimental possibilities of creating the required temperature, and thus the new QGP phase, exist in the heavy ion collisions at high energies in BNL, New York and CERN, Geneva. Considering the scale of the entire experimental enterprise, both in man-years invested and money spent, it seems absolutely necessary to have a better theoretical foundation for these results compared to merely relying on simple models. Fortunately, one can use the canonical Euclidean field theory formalism for equilibrium thermodynamics to look for the new phases, and the phase transitions in *ab initio* calculations from the underlying field theory, i.e. QCD. Indeed, properties of the QGP phase can be predicted theoretically using the lattice QCD approach and tested in the experiments at BNL and CERN. As a first principles based and parameter-free approach, Lattice QCD is an ideal reliable tool to establish the QCD phase diagram and the properties of its many phases. While most other basic features of the lattice formalism required for such an exercise remain the same as in section **2.1**, a key difference for simulations at finite temperature is the need of an $N_s^3 \times N_t$ lattice with the spatial lattice size, $N_s \gg N_t$, the temporal lattice size for the thermodynamic limit of $V = N_s^3 a^3 \to \infty$. The temperature $T = 1/(N_t \cdot a)$ provides the scale to define the continuum limit: Fixing the transition temperature in physical (MeV) units and using Eq. (2), the continuum limit is obtained by sending $N_t \to \infty$.

The lattice QCD approach has provided information on the transition temperature, the order of the phase transition, and the equation of state of QCD matter. One exploits the symmetries of the theory to construct order parameters which are then studied as a function of temperature to look for phase transitions, if any. QCD has two different symmetries in opposite limits of the quark mass m_q. For N_f flavours of massless quarks, QCD has $SU(N_f) \times SU(N_f)$ chiral symmetry while for $m_q \to \infty$, it has a global $Z(3)$ symmetry. Such symmetries usually imply zero expectation values for observables which transform non-trivially under it unless the symmetry is broken spontaneously due to dynamical reasons and the vacuum transforms non-trivially under it. Lattice techniques enabled us to establish that the chiral symmetry is broken spontaneously at low temperatures, as indicated by its non-vanishing order parameter, the chiral condensate $\langle \bar{\psi}\psi \rangle \neq 0$. Its abrupt restoration to zero at high temperature will be a signal of a chiral symmetry restoring phase transition. Since the chiral condensate can be regarded as an effective mass of a quark, arising due to QCD interactions, the chiral transition can be interpreted as thermal effects 'melting' this mass. Similarly, the global $Z(3)$ symmetry breaking can be shown to be equivalent to a single quark having a finite free energy, i.e. the existence of a free quark. A non-zero expectation value for its order parameter, the Polyakov loop $\langle L \rangle$, is the a signal for deconfinement. Of course, in our world with two light and one moderately heavy flavours, neither symmetry is exact but these order parameters may still act as beacons for transitions, depending on how mildly or strongly broken they are.

2.4. Results from Lattice QCD at $T \neq 0$

The transition temperature T_c can be determined by locating the point of discontinuity or sudden change in the order parameter as a function of the temperature (or other external parameter such as density). Since numerical results are necessarily obtained on finite lattices, there is an inevitable rounding which makes the determination of T_c a little tricky. A lot of work has been done on this question in the statistical mechanics area and standard finite size scaling techniques exist to pin down T_c as well as the order of the transition. Since the early days, numerical simulations of lattice QCD have progressively tried to approach the real world of light quarks with vanishing effects from the lattice cut-off. The efforts began from the quenched approximation, i.e. QCD without dynamical quarks, where the deconfinement order parameter $\langle L \rangle$ on small N_t-lattices was used to establish a first order deconfinement phase transition. Later QCD with three or more light dynamical quarks was also shown to have a first order chiral transition. Recent work on simulations for QCD with a realistic quark spectrum seems [9] to rule out a first order chiral transition or a second order transition with the expected $O(4)$-exponents, but suggests a rapid cross over. Determination of T_c, now the point of sharpest change, is even more tricky as a result. The current range for it can be summarised to be 170–190 MeV. A value on the lower end of the range was obtained [10] by using larger N_t-lattices while a value at the upper end was obtained [11] using improved action but smaller N_t. There are other technical differences, such as the physical observable used to set the scale of lattice QCD, as well. Since the energy density is proportional to T^4, the current uncertainty in the value of T_c translates to a $\sim 60\%$ difference in the corresponding energy density estimates at T_c. In view of the tremendous impact it has on the requirements of heavy ion collision experiments, it is hoped that a narrowing of the range takes place as a result of future lattice QCD work.

Quantities of thermodynamic interest such as the energy density, or the pressure or various quark number susceptibilities can be obtained by using the canonical relations from statistical mechanics. Thus

$$\epsilon = \frac{T^2}{V} \frac{\partial \ln \mathcal{Z}}{\partial T}\bigg|_{V,\mu} \quad \text{or} \quad \chi_B = \frac{T}{V} \frac{\partial^2 \ln \mathcal{Z}}{\partial \mu_B^2}\bigg|_{V,T}, \text{ etc.} \qquad (3)$$

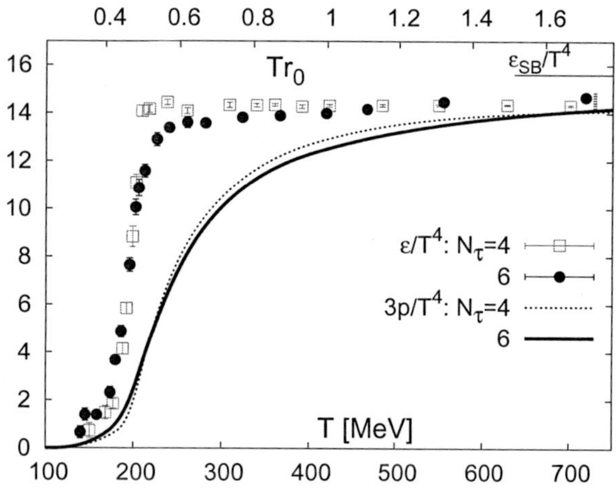

Figure 4. Energy density and pressure from lattice QCD. Taken from [12]

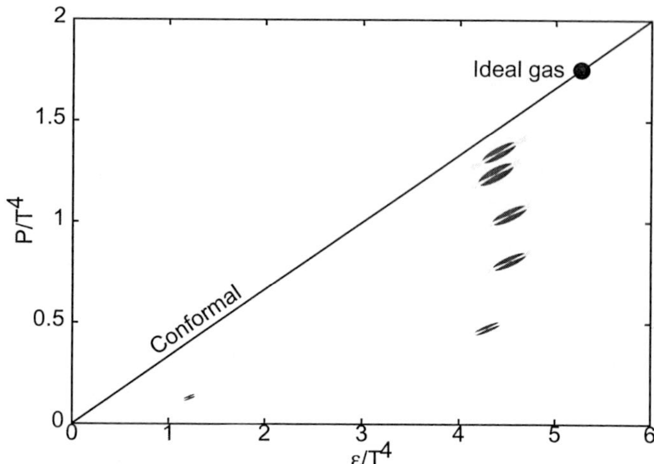

Figure 6. Equation of state for (quenched) lattice QCD. Taken from [13]

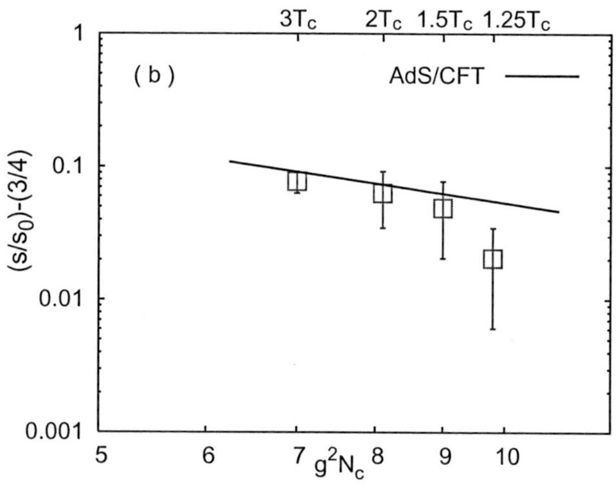

Figure 5. Entropy density s (in units of ideal gas entropy s_0) as a function of 't Hooft coupling. From [13]

Early results in the quenched QCD showed the existence of a QGP phase which has energy density of about 85% of the corresponding ideal gas. The progress since then has been in employing large N_t and inclusion of light quark loops. Figure 4 displays recent results from such efforts. Obtained on two different lattice sizes, $N_t = 4$ and 6 with nearly realistic u, d and s masses, these results also exhibit similar kind of, $\sim 15\%$, deviations from the ideal gas and do seem to hint towards the lattice cut-off effects to be small. The spatial volumes are perhaps not large enough to ensure that the thermodynamic limit is reached. However this question is likely be addressed in near future soon. The results also suggest at most a continuous transition or even a rapid cross over; a strong first order phase transition assumed/constructed in many phenomenological models seems clearly ruled out. This has implications for the hydrodynamical models used to analyse the experimental data: possible mixed state of QGP and hadronic gas must be short lived, if at all it exists.

From a theoretical perspective investigation of equation of state offers hints of developing analytic or semi-analytic approaches. Thus conformal invariant theories are known to yield a variety of predictions for the thermodynamic quantities using the famous AdS-CFT correspondence. Figure 5 shows an attempt to confront the entropy density [13] for the quenched QCD in terms of the entropy of the ideal gas with the prediction of $N = 4$ SYM [14]. The agreement is impressive, considering the differences of the underlying theories. On the other hand, it is really in the stronger coupling region that it is not as good. Moreover, resummed weak coupling perturbation theory approaches seem to perform equally well at the lower couplings. Figure 6 shows the results [13] for the equation of state to highlight how conformal QCD really is. The ellipses denote 66% error bounds on the measured EOS. The wedges piercing the ellipses have average slope c_s^2, the speed of sound and the opening half-angle of these wedges indicate the error in c_s^2. Conformal invariance is indeed violated significantly in the region close to the transition, with least violation at the same temperatures where in AdS-CFT prediction does well in Fig. 5.

Viscosities of the QGP both the shear (η) and bulk (ζ), can also be determined using the lattice approach

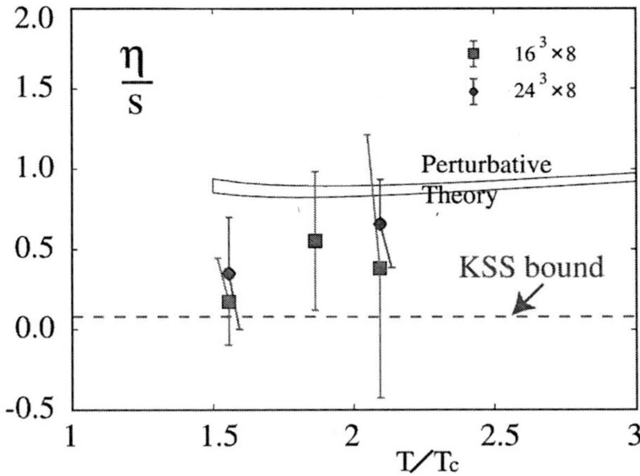

Figure 7. Ratio of shear viscosity to entropy in (quenched) QCD vs temperature. Taken from [15]

although unlike the equation of state these determinations need extra ansätze some of which are not universally accepted. Kubo's linear response theory lays down the framework to obtain such transport coefficients from certain equilibrium correlation functions. In particular, one obtains correlation functions of energy-momentum tensor using the lattice approach above. These are, of course, defined at discrete Matsubara frequencies. Recall that the simulations at $T \neq 0$ need lattices with (a) periodic boundary conditions and (b) small N_t compared to N_s. The correlation function is thus defined at few discrete points only. One then continues it analytically to get the so-called retarded propagators in real time from which the the η and ζ are obtained in the zero frequency limit. Figure 7 shows the results [15] in the quenched approximation. Close to T_c, rather small values are obtained for the ratio of η to the entropy density s. These are seen to be consistent with the famous bound [16] from AdS-CFT. As shown in the Figure, perturbation theory suggests rather large values for this ratio. These results have since been refined [17] and made more precise but the general picture remains the same, as do the various theoretical uncertainties which plague these determinations. Larger lattices and inclusion of dynamical quarks will surely reduce some of these in near future. What is needed though for a more convincing demonstration of the fact the shear viscosity is indeed as small as hinted by the experimental data (see the next section) is a better control over the systematic errors in the analytic continuation.

Analogous to the baryon number susceptibility, defined in Eq. (3), various quark number susceptibilities can be defined by taking derivatives with the appropriate chemical potential. These determine the fluctuations in the given conserved quantum number, say, strangeness. It has been argued [18] that under certain assumptions, testable experimentally, the strange susceptibility can be related to the Wróblewski parameter λ_s extracted from the data of heavy ion collisions. Interestingly, lattice QCD computation in both quenched approximation and full QCD yield a $\lambda_s(T_c) \simeq 0.4 - 0.5$, whereas various experimental results [19] lead to a value 0.47 ± 0.04. Taking derivatives with two different chemical potentials in Eq. (3), one obtains off-diagonal susceptibilities. These have the information on flavour correlations. Such a baryon-strangeness [20] or electric charge-strangeness [18] correlation has been proposed as a signature for identifying the nature of the high temperature phase as that of the quark-gluon phase.

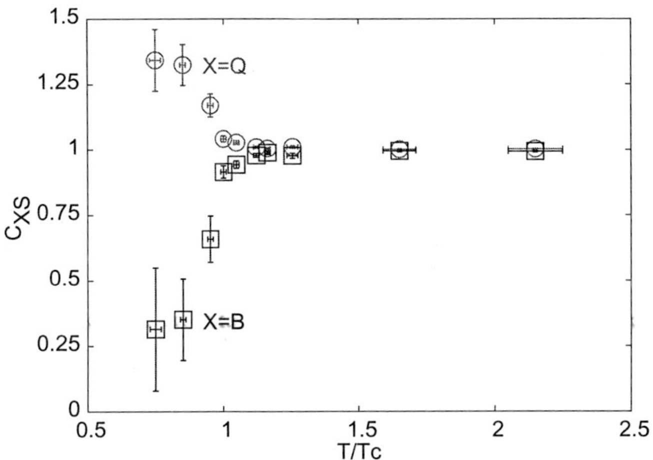

Figure 8. Baryon-Strangeness and electric charge-strangeness correlation vs temperature [18]

Figure 8 shows the lattice results for QCD with 2 light dynamical quarks for both these correlations. They have been so normalised that a value unity, as seen in most of the high temperature phase in Fig. 8, characterises the existence of quark degrees of freedom with the appropriately fractional baryon number or charge. It has been shown that the correlation in the low temperature phase are consistent with the hadronic degrees of freedom. Indeed, any lack of the expected transition should lead to much milder temperature dependence as well as a value different from unity for these correlation functions. Being ratios of the quark number susceptibilities, these correlations are robust, both theoretically and experimentally. Systematic errors due to lattice cut-off or dynamical quark masses are therefore very small as are the systematic errors from experimental sources.

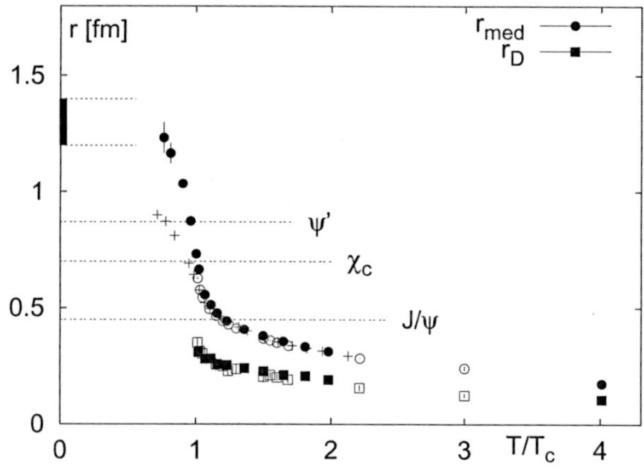

Figure 9. Debye radii for charmonia vs temperature [21]

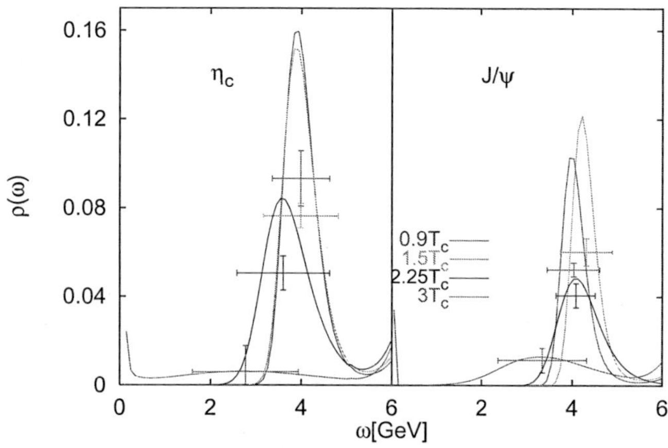

Figure 10. Spectral function of η_c and J/ψ. From [22]

Debye screening of coloured heavy quarks in the deconfined phase had long been recognised [23] as a possible signal of formation of QGP, detectable in the suppression of heavy quarkonia in the heavy ion collisions. In view of the impressive data from CERN at lower SPS energies, and the expectations from the upcoming LHC experiments, a critical assessment of the original theoretical argument seems prudent. Lattice QCD has contributed handsomely in finite temperature investigations of both the heavy quark-antiquark potential, which can be used in the Schrödinger equation to look for the melting of heavy quarkonia, and directly in the spectral function at finite temperature. Figure 9 displays the results [21] for the screening radii estimated from the inverse non-perturbative Debye mass m_D in quenched (open squares) and full (filled squares) QCD. For $r < r_{med}$, the medium effects are suppressed, leading to the same heavy quark potential as at $T = 0$. The horizontal lines correspond to the mean squared charge radii of J/ψ, χ_c and $\psi\prime$ charmonia, and are thus the averaged separations r entering the effective potential in potential model calculations. Figure 9 therefore suggests that the χ_c and $\psi\prime$ states would melt just above the transition while J/ψ may need higher temperatures to be so affected. Direct spectral function calculations [22] provide a strong support for such a qualitative picture. Such computations have been made feasible by the recognition of the maximum entropy method (MEM) technique as a tool to extract spectral functions from the temporal correlators computed on the Euclidean lattice. However as in the case of shear viscosity above, the data for such temporal correlators are sparse, making the extraction more of an art. Nevertheless, large lattices, $48^3 \times 12$ to $64^3 \times 24$ have been used in this case to avoid such criticisms. Figure 10 shows typical results for the J/ψ and η_c mesons in the quenched approximation. The vertical error bars denote the possible uncertainties on the area under the peak as defined by the horizontal error bar. The peaks in both spectral functions appear to persist up to 2.25 T_c, i.e. have non-zero area within the computed error-band, and are gone by 3 T_c unlike the χ_c which has no peak already by 1.1 T_c. Further technical improvements, such as the inclusion of light dynamical quarks, are clearly desirable. Another important issue is that of the huge widths of the peak compared to their known zero temperature values. If real, they could hint at rather loosely bound states which could be dissociated by thermal scatterings.

2.5. QCD Phase Diagram

The QGP phase and the corresponding quark-hadron transition which we discussed so far is a special case of the conditions that could be created in the heavy ion collisions. Indeed, the lattice QCD thermodynamics that we considered was for the case of zero net baryon density and an almost baryon-free region can be produced in the heavy ion collisions in the so-called central rapidity region, as we explain in the next section. It also pervaded our universe a few microseconds after the Big Bang. In general, of course, one should expect hot regions with some baryon number since the colliding nuclei themselves carry substantial baryon number. Massive stars could also have regions of huge baryon densities in the core which could even be at rather low temperatures. It is natural to ask what these generalised extreme conditions lead us to. One could have new phases and different natures of phase transitions which may even have astrophysical consequences. The vast research area of QCD phase diagram in the plane of temperature T and the baryonic chemical potential μ_B deals with these and several other interesting issues. While the current theoretical expectations suggest such physics at non-trivial baryon densities to be better accessible to the colliders at lower energies, such at the RHIC in New York or the forthcoming FAIR facility at GSI, Darmstadt, we feel that the physics may be interesting in its own right to be included in this article dedicated to the LHC; with some luck the LHC experiments may have important contributions to this area as well.

Using simple effective QCD models, such as the Nambu-Jana Lasinio model at finite temperature and densities [24], several speculations have been made about how the QCD phase diagram in the T-μ_B plane should be. At asymptotically high densities, one expects quarks to be effectively free, and therefore to exhibit various colour superconducting phases [25]. In the limit of large number of colours N_c for quarks, it has also been argued that a "quarkyonic" phase may exist [26] at low enough temperatures. A crucial question, especially in the context of either the massive stars, or heavy ion collisions, is the quantitative reliability of the predicted regions in the T-μ_B space. Alternatively, it is unclear how low can the asymptotic predictions be trusted. Nevertheless most model considerations seem to converge [25] on the idea of the existence of a critical point in the T-μ_B plane for the realistic case of 2 light flavours ($m_u = m_d$) of dynamical quarks with a moderately heavy strange quark. Establishing it theoretically and/or experimentally would have huge profound consequences in our (non-perturbative) understanding of QCD.

Extending the lattice approach to the case of QCD at finite density has turned out to be a challenging task at both conceptual and computational level. In principle, it really is straightforward. One just has to add a term $\mu_B N_B = \mu_B \bar{\psi}\gamma_0\psi$ term to the fermionic part of the action, hence the Dirac matrix M, in Eq. (1). In order to eliminate certain spurious divergences, even in the free case, some care is needed [27] and the naive form above has to be modified. A big conceptual block has however turned up in form of our inability to define exact chiral invariance in the presence of the chemical potential [28]: both the overlap and the domain wall fermions lose their exact chiral invariance for any non-zero μ. The staggered fermions do preserve the chiral invariance for non-zero μ. Furthermore, they are simpler to handle numerically. Again most of the numerical work has therefore employed the staggered fermions, although they are plagued with the difficulties of precise definition of flavour and spin as mentioned earlier. Indeed, the existence of the critical point depends [25] crucially on how many flavours of light quarks the theory has. Proceeding none the less with the staggered quarks, another tough problem arises in form of the fact that the Det $M(\mu \neq 0)$ in Eq. (1) is complex whereas the numerical methods of evaluation, employed to obtain the results in the sections above, work only if the determinant is positive definite. This is akin to the sign problem well known to the statistical physicists and is largely unsolved in its full generality.

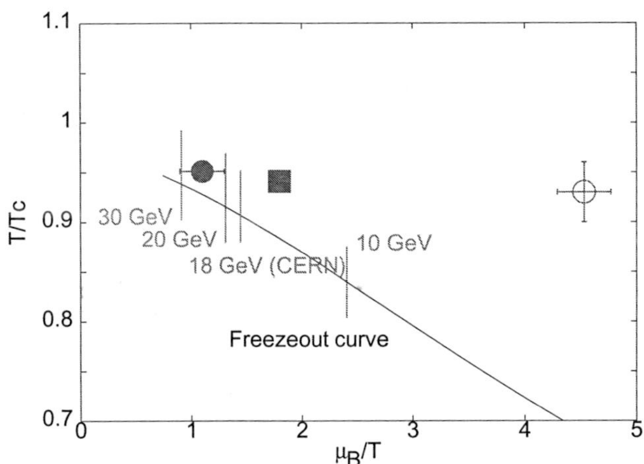

Figure 11. QCD phase diagram for 2 light flavours of quarks. The circles [29,31] and the square [32] denote the location of the critical point on lattices with $1/4T$ and $1/6T$ cut-offs respectively. Taken from [31], where more details can be found

A bold breakthrough was achieved [29] by applying the method of re-weighting in the vicinity of the finite temperature transition at $\mu = 0$. A flurry of activity saw many new methods emerge [30], such as analytic continuation of computations at imaginary chemical potential and Taylor series expansions of the free energy. These have been employed to get a glimpse of whether a critical point does exist, and if yes, what its location may be. The field is really in its infancy and unfortunately at present no consensus amongst the results obtained so far has emerged. Figure 11 exhibits the results obtained for the critical point for the case of two flavours of light quarks with a pion mass $m_\pi/m_\rho = 0.31 \pm 0.01$, compared to 0.18 in the real world. The results [29,31] denoted by circles in the Fig. 11 are for a lattice cut-off $a = 1/4T$ whereas the square [32] denotes the first attempt towards the continuum limit by lowering a to $1/6T$. Large finite volumes have been observed. The shift in the location of the open circle in the Fig. 11 was shown [31] to be due to the use of a 10 times larger volume than the open circle [29]. In order to be brief, we prefer to close this section by noting that different results have been claimed in the literature for larger pion masses and for a different number of flavours. It is hoped that a clear and solid picture will emerge in the near future.

3. Relativistic Heavy-Ion Collisions

At energies of a few GeV/N to a few 10's of GeV/N, colliding nuclei tend to stop each other thereby forming a dense, baryon-rich matter. At higher energies, they nearly pass through each other forming a dense, nearly baryon-number-free matter in the mid-rapidity region. This is evident in the shapes of rapidity distributions (dN/dy vs y) of the net proton (i.e. proton−antiproton) production observed at various beam energies. This apparent transparency of nuclear matter at ultra-relativistic energies can be understood in the spacetime picture of the collision, proposed by Bjorken [33,34].

3.1. Bjorken Picture

Consider, for simplicity, a central (i.e. head-on or zero impact parameter) collision of two identical spherical nuclei in their CM frame. Coordinate axes are chosen such that the two nuclei approach each other along the z-axis and collide at the origin at time $t = 0$. Deep inelastic scattering experiments have revealed the parton structure of hadrons: In the proton, e.g. the valence quark distributions $xu_v(x)$, $xd_v(x)$ peak around $x \sim 0.2$ and vanish as $x \to 0/1$. (x is the Bjorken scaling variable.) The gluon and sea quark distributions, $xg(x)$, $xu_s(x)$, $xd_s(x)$, on the other hand, shoot up as $x \to 0$. These numerous low-momentum partons are called *wee partons*. As a result of the Lorentz contraction, the longitudinal (i.e. parallel to the beam axis) spread of the valence quark wave function is reduced to $\sim 2R/\gamma$ where R is the nuclear radius and γ its Lorentz factor. However, no matter how high the beam energy (or γ), the incoming nuclei always have in them wee partons with typical momenta $p \sim \Lambda_{QCD}$ and hence longitudinal spread ~ 1 fm [33]. The wee partons prevent the nucleus from shrinking below ~ 1 fm in the z-direction. If $2R/\gamma < 1$ fm, they play an important role in the collision dynamics.

As a result of the collision of two nuclei, or rather two clouds of wee partons, a highly excited matter with a large number of virtual quanta is created in the mid-rapidity region. (In the modern parlance one talks about coherent "glasma" formed by a collision of two sheets of "colour glass condensates (CGC)" [35].) Hereinafter we discuss only the mid-rapidity region. The virtual quanta need a finite time ($\tau_{\rm dec}$) to decohere and turn into real quarks and gluons. Here $\tau_{\rm dec}$ refers to the rest frame of an individual parton. In the overall CM frame, the relevant time is $\gamma\tau_{\rm dec}$ due to the time dilation, γ being the Lorentz factor of the parton. It is now clear that "slow" partons decohere earlier and hence near the origin, than the "fast" ones which emerge later at points farther away from the origin. (This is known as the inside-outside cascade.) In other words, the large-x part of each nuclear wave function continues to move along its light-cone trajectory leaving the small-x part behind. Thus in the limit of high beam energy, the time dilation effect causes the near transparency of nuclei, referred to earlier.

Figure 12 shows this schematically in $1+1$ dimension for simplicity. The curves are hyperbolas of constant proper time $\tau = \sqrt{t^2 - z^2}$. All points on a given hyperbola are at the same stage of evolution. In particular, let the hyperbola labelled '1' refer to $\tau = \tau_{\rm dec}$. Parton at z undergoes decoherence at time $t = \sqrt{\tau_{\rm dec}^2 + z^2}$. The larger the z, the larger the time t and higher the parton velocity $v_z = z/t$ [34].

If the partons thus formed interact amongst themselves a multiple number of times, the system approaches local thermal equilibrium. Thermalisation time $\tau_{\rm th}$ ($> \tau_{\rm dec}$) is estimated to be of the order of 1 fm.

Figure 12 indicates a possible scenario. $1, ..., 5$ are the hyperbolas with proper times $\tau_1, ..., \tau_5$.

$t = 0 = z$: the instant of collision
$0 < \tau < \tau_1$: formation of quark-gluon matter
$\tau_1 < \tau < \tau_2$: (local) equilibration of quark-gluon matter, i.e. formation of QGP
$\tau_2 < \tau < \tau_3$: hydrodynamic evolution of QGP (partonic EOS)
$\tau = \tau_3$: hadronisation
$\tau_3 < \tau < \tau_4$: hydrodynamic evolution (hadronic EOS)
$\tau_4 < \tau < \tau_5$: transport theoretic evolution of hadrons
$\tau = \tau_5$: freezeout
$\tau > \tau_5$: free-streaming to detectors.

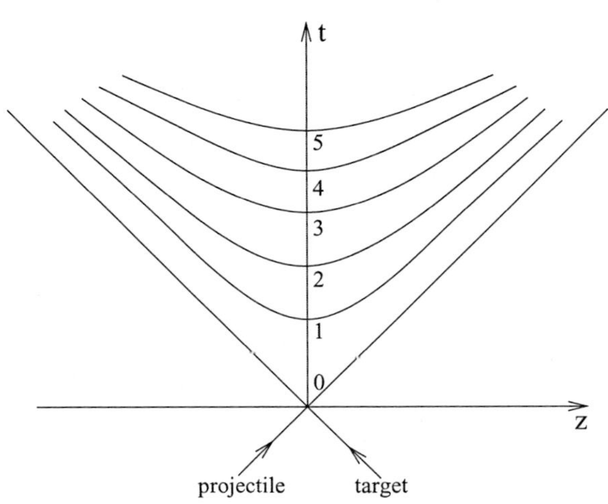

Figure 12. Spacetime picture of an ultra-relativistic nucleus-nucleus collision in 1 + 1 D for simplicity

The above is a rather simple-minded picture: in reality, there are no such "water-tight compartments". The framework of hydrodynamics is applicable, if at all, only when the system is at or near (local) thermal equilibrium. If the matter formed in ultrarelativistic heavy-ion collisions is fully thermalised, one may use the framework of relativistic ideal fluid dynamics to study its evolution. If it is only partially thermalised, one could use relativistic dissipative fluid dynamics. In any case, the covariant transport theory provides a more general framework for this purpose.

Bjorken [34] presented the following formula to estimate the energy density attained in the mid-rapidity region

$$\varepsilon_0 = \frac{1}{\pi R^2 \tau_f} \frac{dE_T}{dy}\bigg|_{y=0}, \tag{4}$$

where R is the nuclear radius, $\tau_f \sim 1$ fm/c is the formation time of QGP and E_T is the transverse energy.

It is clear that even if QGP is formed, its lifetime will be of the order of a few fm/c or $\mathcal{O}(10^{-23})$ seconds, and what experimentalists detect in their detectors are not quarks or gluons, but the standard hadrons, leptons, photons, etc. It is a highly non-trivial task to deduce the formation of QGP from the properties of the detected particles. This is analogous to the situation in cosmology where one tries to deduce the information on the early epochs after the Big Bang by studying the cosmic microwave background radiation and its anisotropy.

Actually the analogy between the Big Bang and the "Little Bang" is quite striking. In both the cases the initial conditions are not accurately known, but there are plausible scenarios. In the former case, there is inflation occurring at $\sim 10^{-35}$ sec, with the inflaton energy converting into matter and radiation, leading to a thermal era. In the latter case, one talks about a highly excited but coherent glasma converting, on the time scale of $\sim 10^{-24}$ sec, into quarks and gluons which may thermalise to form QGP. In both the cases the "fireball" expands, cools and undergoes one or more (phase) transitions. Decoupling or freezeout follows—of photons in the former case and of hadrons in the latter. The unknown initial conditions are parameterised and one tries to learn about them by working one's way backwards, starting with the detected particles. As we shall see shortly, the anisotropy of the detected particles plays a crucial role in the diagnostics of the Little Bang too.

Definition: The STAR collaboration at RHIC has defined the QGP as "a (locally) thermally equilibrated state of matter in which quarks and gluons are deconfined from hadrons, so that colour degrees of freedom become manifest over *nuclear*, rather than merely nucleonic, volumes" [36]. The two essential ingredients of this definition are (a) local equilibration of matter and (b) deconfinement of colour over nuclear volumes. Recent claims of the discovery of QGP at RHIC [37] were based on two observations which, for the first time, provided a good evidence that each of these two requirements has been fulfilled. We discuss them one by one in the next two subsections (**3.2, 3.3**). That will be followed by brief descriptions of a few other signals of QGP in subsections **3.4, 3.5**.

3.2. Anisotropic Flow

Consider now a non-central (or non-zero impact parameter) collision of two identical (spherical) nuclei travelling in opposite directions. Choose x, y axes as shown in Fig. 13. The collision or beam axis is perpendicular to the plane of the figure. Length of the line AB connecting the centres of the two nuclei is the impact parameter b. Plane xy is the azimuthal or transverse plane. Plane xz is the reaction plane. It is determined by the impact parameter vector $\mathbf{b}$ and the collision axis. (Obviously the reaction plane cannot be defined for a central collision.) $\phi = \tan^{-1}(p_y/p_x)$ is the azimuthal angle of an outgoing particle. The almond-shaped shaded area is the overlap zone. In a real experiment, Fig. 14 the x, y axes need not coincide with the lab-fixed X, Y axes. Indeed the reaction plane subtends an arbitrary angle ϕ_R with the X axis. ϕ_R varies from event to event. It is a priori unknown and special experimental techniques are needed for its determination.

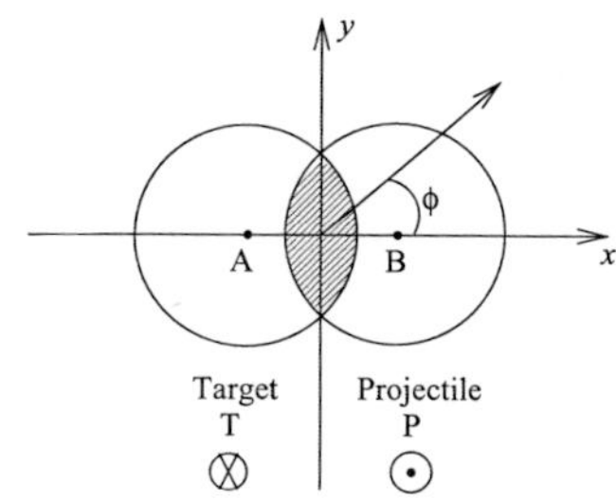

Figure 13. Non-central collision

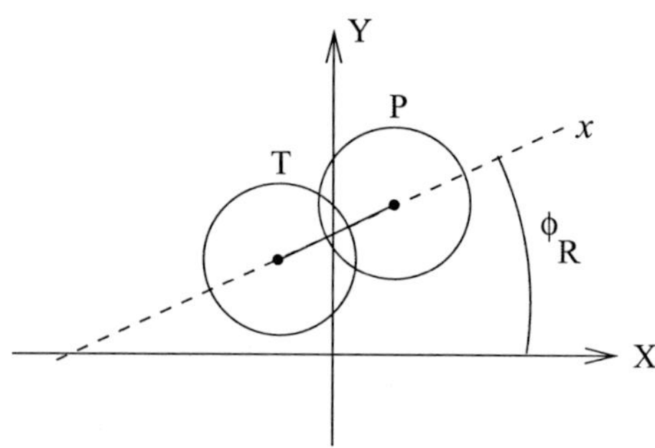

Figure 14. Non-central collision. XY are lab-fixed axes

The triple differential invariant distribution of particles emitted in the final state of a heavy-ion collision is a periodic even function of ϕ, and can be Fourier decomposed as

$$E\frac{d^3N}{d^3p} = \frac{d^3N}{p_T dp_T dy d\phi} = \frac{d^2N}{p_T dp_T dy}\frac{1}{2\pi}\left[1+\sum_1^\infty 2v_n\cos(n\phi)\right],$$

where y is the rapidity and ϕ is measured with respect to the reaction plane. The leading term in the square brackets in the above expression represents the azimuthally symmetric *radial flow.* v_1 is called the *directed flow* and v_2 the *elliptic flow.* $v_n \equiv \langle\cos(n\phi)\rangle$ is actually a function of p_T and y. Here the average is taken with a weight equal to the triple differential distribution of particles in the (p_T, y) bin under consideration. v_2 can also be written as $\langle(p_x^2 - p_y^2)/(p_x^2 + p_y^2)\rangle$. For a central collision the distribution is azimuthally isotropic and hence $v_n = 0$ for $n = 1, 2, ...$ In other words, only the radial flow survives.

Measurement of the radial flow: Radial flow gives a radially outward kick to the emerging hadrons thereby depleting the low-p_T population and making their p_T spectra flatter. The heavier the hadron, the stronger the momentum kick it receives. By measuring the slopes of the p_T spectra of various hadrons, the radial flow velocity can be extracted. At RHIC it turns out to be a sizeable fraction ($\sim 50\%$) of the speed of light. Thus the flow is compressible.

Measurement of the anisotropic flow v_n: There are several methods. (a) The most obvious one is based on the definition $v_n \equiv \langle\cos n(\phi - \phi_R)\rangle$ where both ϕ and ϕ_R are measured with respect to a lab-fixed frame of reference. This, however, requires the knowledge of ϕ_R which varies from event to event and is not easy to determine. (b) Two-particle correlation method: This gives $v_n^2 = \langle\cos n(\phi_1 - \phi_2)\rangle$, where ϕ_1 and ϕ_2 are azimuthal angles of two outgoing particles. This method has an advantage that the reaction plane need not be known. However v_n is determined only up to the sign. There are several other methods such as the cumulant method [38], mixed-harmonic method [39], Lee-Yang zeroes method [40], etc. For a recent review, see [41].

Importance of the anisotropic flow v_n: Consider a non-central collision (Fig. 13). Thus the initial state is characterised by a spatial anisotropy in the azimuthal plane. Consider particles in the almond-shaped overlap zone. Their initial momenta are predominantly longitudinal. Transverse momenta, if any are distributed isotropically. Hence $v_n(\text{initial}) = 0$. Now if these particles do not interact with each other, the final (azimuthal) distribution too will be isotropic. Hence $v_n(\text{final}) = 0$.

On the other hand, if these particles interact with each other a multiple number of times, then the (local) thermal equilibrium is likely to be reached. Once that happens, the system can be described in terms of thermodynamic quantities such as temperature, pressure, etc. The *spatial anisotropy* of the almond-shaped overlap zone ensures anisotropic pressure gradients in the transverse plane. This leads to a final state characterised by a *momentum anisotropy* in the $p_x p_y$ plane or equivalently[1] to an anisotropic distribution of particles in the transverse (xy) plane, and hence a non-

[1]Since $\phi = \tan^{-1}(p_y/p_x)$.

vanishing v_n. *Thus v_n is a measure of the degree of thermalisation of the matter produced in a non-central heavy-ion collision.*

To sum up, if either of the two ingredients, namely initial spatial anisotropy and adequate rescatterings, is missing, there is no anisotropic flow (v_n).

Sensitivity of v_n to properties of matter at *early* times ($\sim$ fm/c): We saw above that the spatial anisotropy of the initial state (together with multiple rescatterings) leads to more matter being transported in the directions of the steepest pressure gradients, and thus to a non-zero v_n. That in turn results in the reduction in spatial anisotropy ("self-quenching"). In other words expansion of the source gradually diminishes its spatial anisotropy. Thus v_n builds up early (i.e. when the spatial anisotropy is significant) and tends to saturate as the spatial anisotropy continues to decrease. (This is unlike the radial flow which continues to grow until freeze-out and is sensitive to early- as well as late-time history of the matter.) Thus v_n is a measure of the degree of thermalisation of the matter produced *early* in the collision. In other words, v_n is a signature of pressure at *early* times.

Hydrodynamic calculations of v_n involve the equation of state of QGP. Thus one hopes to learn about the material properties of the medium, such as the speed of sound, sheer and bulk viscosities, relaxation times, etc.

Flow may also be affected by the dynamics of the hadronic phase. Study of the flow would provide constraints on the properties of hadronic matter too. (It is expected that at the LHC, the relative contribution of the QGP phase to v_n would be larger than that at SPS and RHIC. This would reduce the effect of the uncertainties in the hadronic phase.)

It should, however, be kept in mind that the initial conditions for the hydrodynamic evolution are not known with certainty. Hence the task of unravelling the properties of medium is not as easy as it may appear.

Figure 15 shows the impressive agreement between RHIC data on $v_2(p_T)$ and ideal hydro calculations for p_T up to $\sim$ 1.5 GeV/c. In particular note the mass ordering: the heavier the hadron, the smaller the $v_2(p_T)$. This can be understood heuristically as follows.

Mass ordering of $v_2(p_T)$: Recall that the radial flow depletes the population of low-p_T hadrons (by shifting them to larger values of p_T). This effect is more pronounced for larger flow velocities and for heavier hadrons. Suppose v_2 is positive as at RHIC, which means more hadrons emerge in-plane (x-direction) than out-of-plane (y-direction). Now due to higher pressure gradients in the x-direction, hadrons which emerge in-plane experience a larger flow velocity than those which emerge out-of-plane. So the depletion is greater for the hadrons emerging in-plane than out-of-plane. This tends to reduce the anisotropy and hence v_2 of all hadron species. For a heavier hadron species this reduction is more pronounced. The net result is $v_2^{light\ hadron}(p_T) > v_2^{heavy\ hadron}(p_T)$. Mass-ordering signifies a common radial velocity field.

Hydrodynamic model calculations predicted mass ordering of $v_2(p_T)$. The broad agreement between the RHIC data and the predictions of ideal hydro (Fig. 15) led to the claims of thermalisation of matter and discovery of a perfect fluid — more perfect than any seen before.

In order to claim the discovery of a new state of matter, namely QGP one needs to demonstrate unambiguously that (local) equilibrium is attained. There are indications that the equilibrium attained at RHIC is incomplete [42].

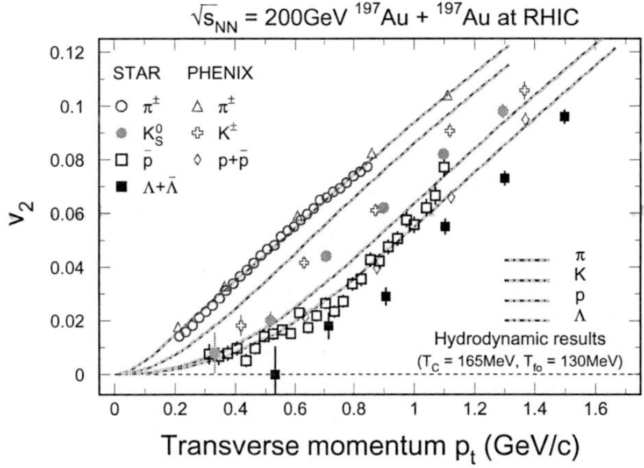

Figure 15. Minimum-bias data. Curves represent ideal hydro results with a first-order QGP-hadron phase transition. Figure taken from [43]

3.2.1. Constituent Quark Scaling

For $p_T \gtrsim 2$ GeV/c, ideal hydro results are in gross disagreement with the $v_2(p_T)$ data: calculated $v_2(p_T)$ continues to rise with p_T, while the data tend to saturate and the mass ordering is reversed. In the intermediate momentum range (2 GeV/c $\lesssim p_T \lesssim$ 5 GeV/c), it is observed that the v_2/n_q vs p_T/n_q (or KE_T/n_q)

data fall on a nearly universal curve (Fig. 16). Here n_q is the number of constituent quarks and KE_T is the transverse kinetic energy. This is called the constituent quark scaling. It shows that the flow is developed at the quark level, and that the hadronisation occurs by quark recombination.

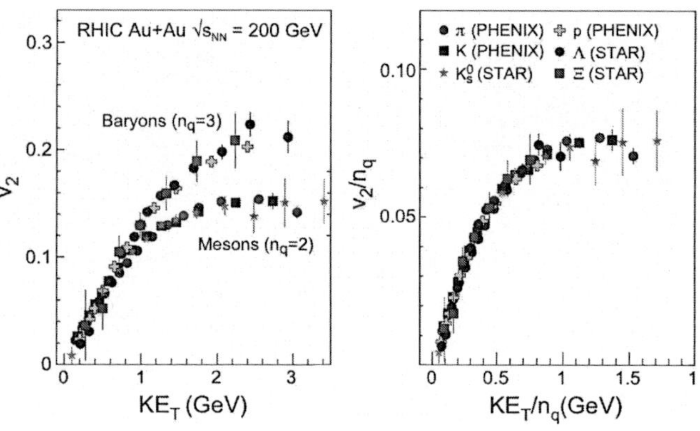

Figure 16. Left: Note the two distinct branches. Right: Universal curve. Figure taken from [44]

3.3. Jet Quenching

A variety of signatures of QGP have been proposed. Some of the more popular ones are excess strangeness production, thermal dileptons and photons, jet quenching, J/ψ-suppression and event-by-event fluctuations. A common theme underlying all of these is the idea of exploiting the consequences of those properties of QGP which distinguish it from alternatives like a hot hadron gas. Since QGP is expected to form and exist predominantly in the early phase of the collision, the so-called hard probes are potentially the cleaner direct probes of this early phase. It is experimentally known that rare but highly energetic scatterings produce jets of particles: $g+g \to g+g$, where energetic gluons from the colliding hadrons produce two gluons at large transverse momenta, which fragment and emerge as jets of showering particles. Their typical production time scale is $t \sim 1/Q$, where $Q = p_T$, the transverse momentum of the jet, is the hard scale of production. Thus jets at large transverse momenta are produced very early and by traversing through the produced medium carry its memory while emerging out. QGP or any medium in general, interacts with the jet, causing it to lose energy. This phenomenon goes by the name of jet quenching.

Using the well-known factorisation property of perturbative QCD [45], which allows a separation between the hard and soft scales, a typical cross section at hard scale, say that of hadron h at large transverse momenta in the process $A+B \to h+X$, can be symbolically written as

$$\sigma^{AB\to h} = f_A(x_1, Q^2) \otimes f_B(x_2, Q^2) \otimes \sigma(x_1, x_2, Q^2) \otimes D_{i\to h}(z, Q^2) \ . \quad (5)$$

Here f_A, f_B are parton distribution functions of the colliding hadrons A and B at scale Q^2, $\sigma(x_1, x_2, Q^2)$ is the elementary pQCD cross section for partons of momentum fractions x_1 and x_2 to produce a parton i with the hard scale $Q = p_T$ for jet production, and $D_{i\to h}(z, Q^2)$ is its fragmentation function to hadron h with momentum fraction z. Various convolution integrations are denoted symbolically by $\otimes$. Clearly, there are many more details which are not spelt out here for brevity, such as the kinematic integration region or the summation over all allowed many parton level processes, such quark-quark or gluon-quark etc. These can be found in textbooks [45].

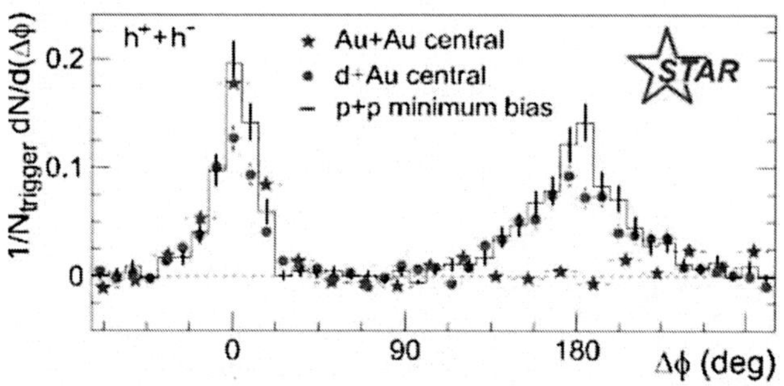

Figure 17. Comparison of the various dihadron angular correlations. Taken from [47]

In presence of a medium, of hot hadron gas or quark-gluon plasma, the function D above will get modified by the interactions with medium. The medium provides scattering centres for the fast moving seed particle of the jet which typically impart a transverse momentum kick to it. The medium induced transverse momentum squared per unit path length, $\hat{q}$, characterises the quenching weight function $P(\Delta E)$ [46] which is the probability that a hard parton loses an additional energy ΔE due to its interactions with the medium. In hot matter with a temperature of about $T = 250$ MeV, a perturbative estimate [49] for $\hat{q}$ is about 0.5 GeV^2/fm. It is typically a lot smaller in the cold nuclear matter. In terms of the quenching weight, one can write

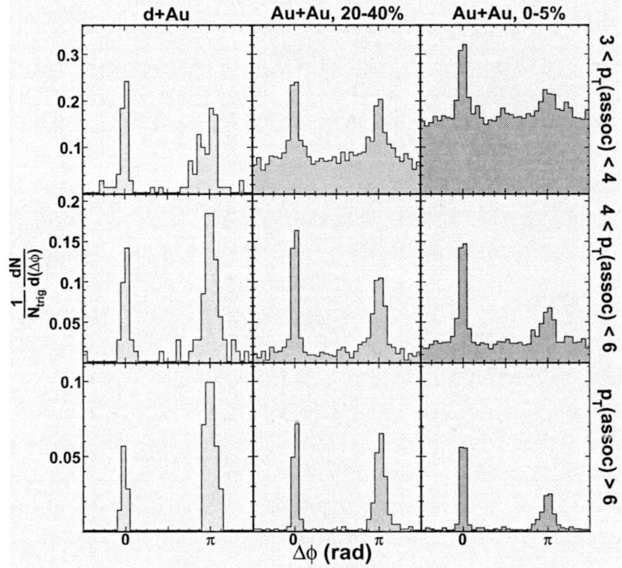

Figure 18. Comparison of the dihadron azimuthal correlations as a function of the associated p_T for $8 < p_T^{trigg} < 15$ GeV. Taken from [48]

down [46] a medium modified fragmentation function for a jet passing through a medium as

$$D^{\rm med}_{i\to h}(x,Q^2) = \int_0^1 d\epsilon \frac{P_E(\epsilon)}{1-\epsilon} D_{i\to h}(\frac{x}{1-\epsilon},Q^2) \ . \qquad (6)$$

For a heavy quarkonium like J/ψ, the analogue of D, is the wave function of a heavy quark-antiquark pair ($c\bar{c}$), and it will be presumably flatter in a hot medium, corresponding to "its melting".

RHIC experiments have cleverly exploited their capabilities to perform tests which have an on-off nature and are therefore rather convincing about the qualitative existence of the jet quenching phenomenon in heavy ion collisions. In the case of the elementary $g + g \to g + g$ hard process, one expects back-to-back jets, i.e. a well-determined azimuthal correlation between the fast particles. As jets are hard to identify in the complex multi-particle environment at RHIC, the STAR collaboration constructed the angular correlation of hadrons, using a high transverse momentum p_T^{trigg} particle as the trigger, and studying the azimuthal distributions of the associated particles ($p_T^{assoc} < p_T^{trigg}$). Figure 17 compares the results for gold-gold central collisions, where one expects formation of a hot medium, with the proton-proton or deuterium-gold collisions, where one expects to have turned off the medium effects. The expected correlation, signalling a lack of any quenching/medium, is clearly visible in the two peaks separated by 180° for the d-Au and pp collisions. Remarkably the gold-gold central collision data show only the peak at zero degree or the near-side. A hint of the creation of some medium is given by the vanishing of the away-side jet, at 180° degrees, which appears to have been fully quenched by the medium. For high enough trigger p_T, one can do the same comparison as a function of range of the associated p_T. Clearly, as the p_T^{assoc} increases, one ought to see the away-side re-emerge. This is beautifully seen in the Fig. 18. It shows the azimuthal correlations for $8 < p_T^{trigg} < 15$ GeV for d-Au and Au-Au collisions in two centrality bins, with the data for most central collisions displayed in the last column. The p_T of the associated particle is restricted to ranges marked on the right side, and increases as one goes from top to the bottom. All panels show comparable strengths for the near-side peak. As the p_T^{assoc} grows above 6 GeV, the away-side peaks in all the three systems also show comparable strengths whereas for lower p_T^{assoc} ranges one has diminishing away-side peaks, characteristic of jet-quenching. The same phenomena can also be studied by varying the p_T^{trigg} and the away-side peak is seen clearly to emerge as p_T^{trigg} increases.

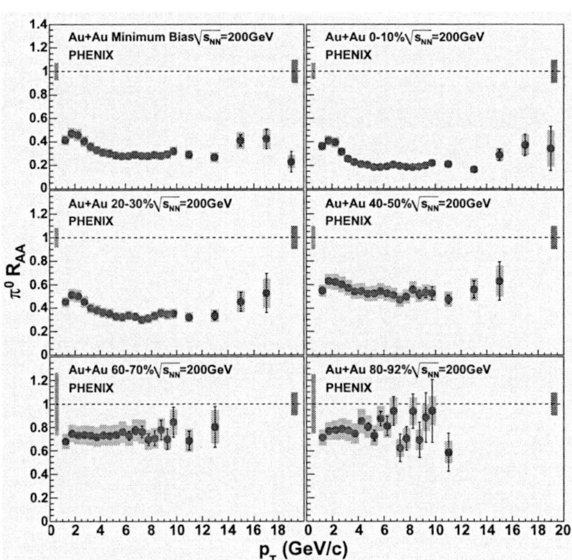

Figure 19. Nuclear modification factor, R_{AA}, for neutral pions as a function of transverse momentum for different centralities. Taken from [51]

A more quantitative investigation of the jet quenching phenomena needs to extract the transport coefficient $\hat{q}$, and establish the presence of the hot matter by comparing it with the corresponding theoretical estimates, directly from QCD. Many such attempts have been made. Recently, the PHENIX experiment [51] reported their measurement of neutral pion production in Au-Au collisions at 200 GeV at the RHIC collider in BNL. They define the now-famous nuclear suppression factor R_{AA} as the weighted ratio of the nuclear differential distribution in rapidity y and transverse momentum p_T and their own earlier measurements for the same quantity in proton-proton

$$R_{\mathrm{AA}} = \frac{1/N_{\mathrm{evt}} dN/dy dp_T}{\langle T_{\mathrm{AB}} \rangle d\sigma_{pp}/dy dp_T} , \tag{7}$$

where further details of determinations of various factors above are given in [51]. Their results for R_{AA} are displayed in Fig. 19. While the first panel shows the results for their entire data set, the other panels exhibit data for increasing peripherality of the collisions (indicated by the increasing range of the percentage label of each panel) or decreasing centrality. The error bars indicate the statistical errors, whereas various systematic errors are shown by the boxes. Note that if the nucleus-nucleus collisions were merely scaled proton-proton ones, one expects $R_{AA} = 1$. What the data in Fig. 19 indicate, however is a five-fold suppression that is essentially constant for $5< p_T <20$ GeV for the most central bin of 0–10%. The qualitative pattern is the same in all centralities, although the magnitude of suppression comes down. The highest centrality bin was used to determine the transport coefficient in the the parton quenching model [50] to obtain $\hat{q} = 13.2^{+2.1}_{-3.2}$ GeV^2/fm. Typically, fits with varying model assumptions do tend to yield a $\hat{q}$ of 5–15 GeV^2/fm. This order of magnitude or so higher value of the transport coefficient compared to the expectations from perturbative QCD, ~ 0.5, as mentioned above is an unresolved puzzle. Nevertheless the value hints at a hot medium, presumably even stronger interacting than the pQCD picture, as the cold matter expectations for $\hat{q}$ are even more in disagreement with the experimental determination. Clearly a lot more needs to be understood from the data by further delving into the detail predictions of the models and confronting them with data, as [51] attempts to do, in order to establish the nature of hot medium produced as that of QGP.

Having discussed the two main observations, anisotropic flow and jet quenching, which lend support to the claims of discovery of QGP at RHIC, we now discuss some corroborative evidences which strengthen these claims. There are also surprises in the RHIC data when compared with the expectations from the earlier lower energy heavy ion collisions at SPS in CERN. We discuss some with the aim to prepare ourselves for the expectation at yet higher energy in the LHC.

3.4. Anomalous J/ψ Suppression

Amongst the many signatures proposed to look for QGP experimentally, the idea of J/ψ-suppression has attracted the most attention as the likely "gold-plated" signal. Soon after the pioneering work of Matsui and Satz [23], arguing that (a) as a hard QCD process, the heavy charm pair production takes place very early, (b) the Debye screening of the QGP prevents formation of a J/ψ state in heavy ion collisions and (c) the low temperatures at the hadronisation do not permit production of charm-anticharm pair kinematically, it was further proposed that the suppression pattern ought to have a characteristic [52] transverse momentum dependence. Recognising that the gluon and quark distribution functions depend on the atomic number A, known by the famous EMC-effect, it was shown in a perturbative QCD calculation that the suppression signal [53] itself as well as its p_T-dependence [54] can be mimicked by the mundane nuclear shadowing. Thus it became clear since the early days that a detailed quantitative analysis is necessary to disentangle the effects of the Debye screening in QGP. It has since been recognised that other effects, notably the absorption [55] of the produced J/ψ in the nucleus, causes suppression of J/ψ in all pA and AB-collisions. Thus one has to first account for this expected or normal suppression and then look for additional or anomalous J/ψ-suppression as the possible signal of QGP. Considering the general wisdom that J/ψ-production can be computed in pQCD, it ought to be a straightforward task to compute this normal suppression. Unfortunately, it is not so. One reason is that the gluon distribution function, and the nuclear shadowing effects are not well known. Another perhaps much more important reason, is that the hadroproduction of J/ψ needs to tackle the vexing issue of its formation from the perturbatively produced charm-anticharm pair. One usually depends [56] on models, such as the colour evaporation or the colour octet model, hoping that the effective theory descriptions are valid. It turns out to be true for large p_T charmonium production but not for the total cross sections of interest for the QGP signal.

The preferred phenomenological method [55] has been to parametrise the ratio of J/ψ-cross sections, total or appropriate differential cross sections in its transverse momentum p_T or forward momentum fraction x_F etc. in pA and pp collisions at the same colliding en-

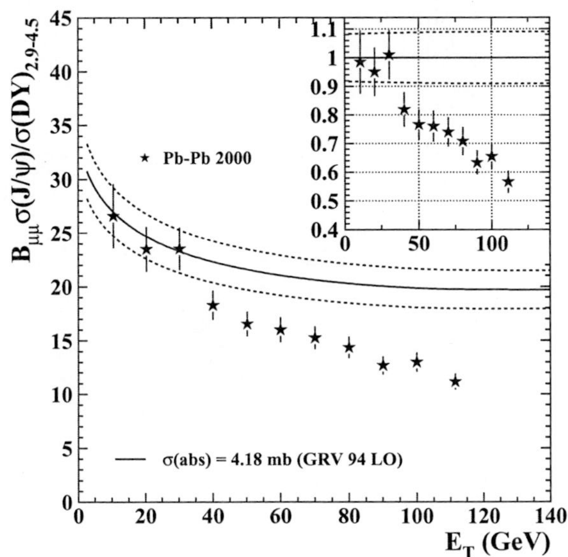

Figure 20. J/ψ-suppression in Pb-Pb collisions at SPS as a function of transverse energy E_T. Figure taken from [57]

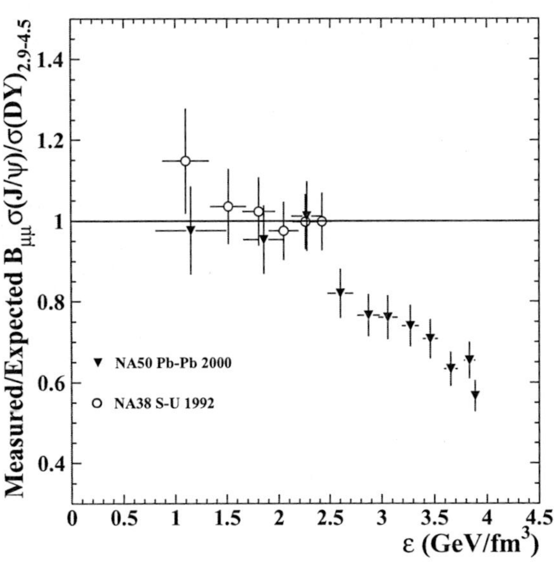

Figure 21. J/ψ-suppression in Pb-Pb collisions at SPS as a function of the energy density ϵ. Figure taken from [57]

ergy, $\sqrt{s}$, as $\exp(-\sigma_{\text{abs}}(J/\psi)\rho_0 L)$, where L is the mean length of the trajectory of the produced $c\bar{c}$ pair in nuclear matter and ρ_0 is the nuclear density. The parameter, $\sigma_{\text{abs}}(J/\psi)$ is obtained by fitting the data. Defining a mean free path $\lambda = 1/\sigma_{\text{abs}}(J/\psi)\rho_0$, one then extends this idea to the heavy-ion collisions to define the normal or expected J/ψ suppression due to the traversing of the $c\bar{c}$-pair in the nuclear matter as $\exp(-(L_A + L_B)/\lambda)$. Here L_A and L_B are the lengths for the trajectories of the $c\bar{c}$ in the projectile (A) and target (B) respectively. They are calculated from collision geometry by using the oft-used relations between mean transverse energy of the bin, E_T and the average impact parameter b.

Figure 20 exhibits [57] the results of the NA50 collaboration on J/ψ cross section as a function of the transverse energy E_T in Pb-Pb collisions at $\sqrt{s} \simeq 17$ GeV. It is normalised to the Drell-Yan cross section in the mass range shown and $B_{\mu\mu}$ is the branching fraction of J/ψ in the dimuon channel. The full curve depicts the expected normal suppression as a function of E_T, computed as explained above using the fitted J/ψ cross section of 4.18 mb obtained from the NA50's own pA data. The dashed lines show the computed error bars on the expected suppression, and the inset shows the ratio of measured to the expected suppression. Using the Bjorken formula in Eq. (4), one obtains this ratio of the measured to the expected cross section ratio of the J/ψ and the Drell-Yan as a function of the energy density in GeV/fm^3 units, as shown in Fig. 21, taken from [57]. One sees that the anomalous suppression, i.e. depletion of the measured cross section from that expected, sets in at an energy density of about 2.5 GeV/fm^3, comparable to the expectations from lattice QCD, as seen in Fig. 4. A natural explanation of the anomalous suppression was, therefore the formation of QGP. Since the J/ψ-production takes place both directly and through other charmonium states like χ_c, the slow fall-off with the energy density in Fig. 21 could be interpreted as gradual progress towards the full suppression. However, one could also explain the anomalous suppression in alternative ways, using hadronic [58] or thermal [59] models. Since one expects the higher collision energy at RHIC to produce higher temperatures/energy densities, one expected a further stronger suppression at RHIC. Indeed, this seems to be true both in the QGP models as well as the alternatives, the difference between them being quantitative in nature.

The RHIC results [60], however brought a big surprise by being different from any of those expectations. Analogous to the case of jet quenching in the previous section, the PHENIX collaboration at RHIC constructs the ratio R_{AA} of the J/ψ (differential) produc-

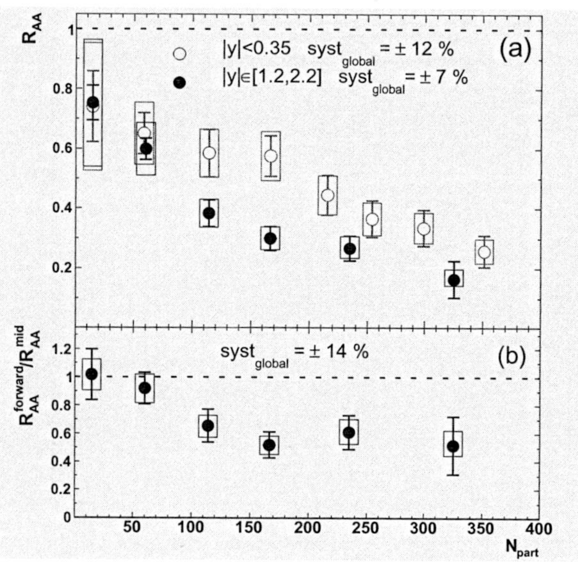

Figure 22. J/ψ-suppression in Au-Au collisions at PHENIX, BNL as a function of number of participants. Figure taken from [60]

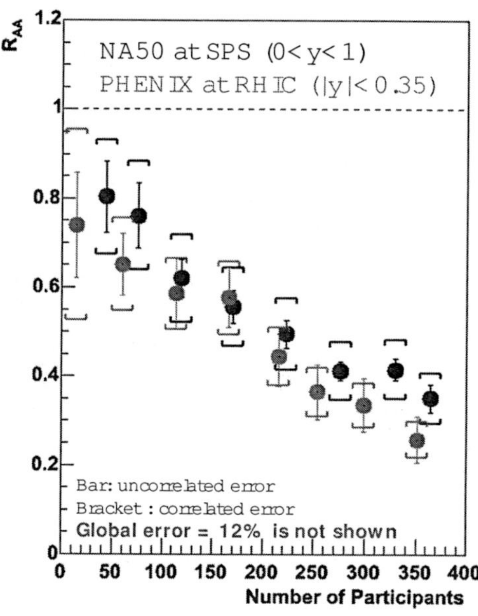

Figure 23. Comparison of NA50 and PHENIX results on J/ψ-suppression as a function of number of participants. Figure taken from [61]

tion cross section in AA collisions and the corresponding pp cross section weighted by the number of binary collisions. Figure 22 displays their results for R_{AA} in Au-Au collisions at $\sqrt{s}$ = 200 GeV. They show *more* suppression in the forward region ($|y| \in [1.2, 2.2]$, filled circles in the top panel), than the central ($|y| < 0.35$, open circles in the top panel) for number of participants > 100 (alternatively for large enough transverse energy E_T). More importantly, a direct comparison citemjt in Fig. 23 clearly demonstrates that the PHENIX data in the central rapidity region are in very good agreement with the CERN NA50 results [57]. The trends for both the central region of the CERN and RHIC experiments, as seen in Fig. 23 and the ratio of forward to the central rapidity region, as seen in the bottom panel of Fig. 22 are against [61] the predictions of the models which successfully accounted for the NA50 data.

There have been some attempts to solve this J/ψ-puzzle. As we saw in the Fig. 10 of section **2.4**, the lattice QCD results suggest melting of the J/ψ takes place at higher temperatures ($> 2T_c$) than predicted by simple models. A way to understand the results in Fig. 23 could then suggest itself if the temperature reached at both the SPS and RHIC energy is $\lesssim 2T_c$. In that case, only χ_c and ψ' would have melted [62], suppressing the corresponding decay J/ψ's, and giving similar results for CERN and RHIC experiments. Since the temperature reached at the LHC is expected to cross $2T_c$, a clear prediction of such a scenario would then be much more suppression for the LHC than that in Fig. 23. However there are other scenarios, including thermal enhancement [63] arising due to recombination of the large number of thermal produced charm-anticharm quarks. These would predict an overall enhancement. In any case, J/ψ-suppression could provide a lot of excitement again at the LHC.

3.5. Particle Ratios and Bulk Properties

A variety of hadrons are produced in an ultra-relativistic heavy-ion collision. They are identified and their relative yields measured (Fig. 24). These hadron abundance ratios can be calculated in a simple statistical model [64]: It is assumed that these particles emerge from a chemically equilibrated hadron gas characterised by a chemical potential (μ_i) for each hadron species and a common temperature (T). The number density n_i of hadron of type i is then given by the standard Fermi-Dirac (+) or Bose-Einstein (−) formulas

$$n_i = d_i \int \frac{d^3p}{(2\pi)^3} \frac{1}{\exp\left[(E_i - \mu_i)/T\right] \pm 1},$$

where d_i is the spin degeneracy. At chemical equilibrium, the chemical potential μ_i can be written as $\mu_i = \mu_B B_i - \mu_S S_i - \mu_I I_i^{(3)}$ where B_i, S_i and $I_i^{(3)}$ stand for the baryon number, the strangeness and the third component of the isospin quantum numbers, respectively, of the hadron of type i. The two unknown parameters T and μ_B are fitted to the data. This simple model has been quite successful in explaining the SPS and RHIC data; see Fig. 24 for SPS and a similar figure in [65] for RHIC. Note that even the multistrange particles seem to be consistent with the model. This suggests that they are produced in a partonic environment rather than in a hadronic one. $T \equiv T_{\rm ch}$ is the chemical freezeout temperature. The fitted values are

$$\begin{aligned} T_{\rm ch} &= 170\ \text{MeV},\ \mu_B = 270\ \text{MeV},\ \text{(SPS)}, \\ T_{\rm ch} &= 176\ \text{MeV},\ \mu_B = 41\ \text{MeV},\ \text{(RHIC 130 GeV)}, \\ T_{\rm ch} &= 177\ \text{MeV},\ \mu_B = 29\ \text{MeV},\ \text{(RHIC 200 GeV)}. \end{aligned}$$

Note the trend of the chemical freezeout point to approach the temperature axis of the QCD phase diagram as the collision energy is increased. Data obtained at the AGS and SIS energies are also consistent with this trend; see Fig. 1.3 in [66]. For more recent fits to the statistical model, see [67].

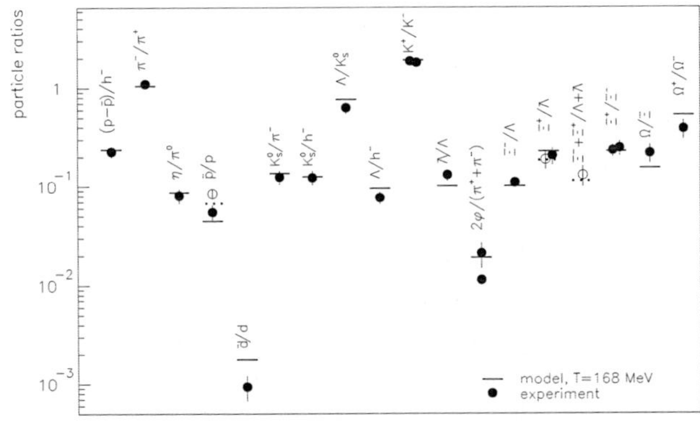

Figure 24. Comparison between the statistical model (horizontal bars) and experimental particle ratios (filled circles) measured at SPS CERN. From Braun-Munzinger et al. [64]

4. Hydrodynamics

Hydro plays a central role in modelling relativistic heavy-ion collisions: It is first used for the calculation of the p_T spectra and the elliptic flow v_2. The resultant energy density or temperature profiles are then used in the calculations of jet quenching, J/ψ melting, thermal photon and dilepton production, etc.

Hydrodynamic framework consists of a set of coupled partial differential equations for energy density, number density, pressure, hydrodynamic four-velocity, etc. In addition, these equations also contain various transport coefficients and relaxation times.

Hydro is a very powerful technique because given the initial conditions and the EOS it predicts the evolution of the matter. Its limitation is that it is applicable at or near (local) thermodynamic equilibrium only.

4.1. A Perfect Fluid?

How robust is the claim of discovery of a perfect fluid at RHIC, or is there any need of the viscous hydrodynamics for RHIC? A closer scrutiny shows that the claim is not really robust, and it is necessary to do viscous hydro calculations:

- Agreement between data and ideal hydro is far from perfect. (Ideal) "hydro models seem to work for minimum-bias data but not for centrality-selected π and $\bar{p}$ data" [68].
- Initial (and final) conditions for the hydrodynamic regime are uncertain. It is entirely possible that the ideal hydro mimics viscous hydro if the initial (and/or final) conditions are suitably tuned. Most ideal hydro calculations so far have been done with Glauber-type initial conditions. It has recently been realised that the CGC-type initial conditions yield higher eccentricity of the overlap zone [69] and hence higher v_2. To push these results down to agree with data, some viscous corrections are needed. The same is true with fluctuations in the initial conditions [70]. Event-to-event fluctuations in nucleon positions result in higher eccentricity and hence higher v_2 [71].
- Some v_2 may build up during the pre-equilibrium (i.e. pre-hydro) regime. Success of ideal hydro may be due to the neglect of this contribution to v_2 in most calculations [72].
- For realistic light quark masses, the deconfinement transition is known to be a smooth crossover. However it seems that the ideal hydro calculations need a first-order transition for a best fit to the data [73].
- The shear viscosity to entropy density ratio (η/s) may be small in the transition region. But there

are indications that the bulk viscosity to entropy density ratio (ζ/s) may be rising dramatically near T_c [74]. If this result holds, QGP discovered at RHIC cannot be called a perfect fluid.

- It is known that for helium, water, nitrogen, η/s at constant pressure plotted as a function of temperature, exhibits a minimum with a cusp-like behaviour at the critical point (Fig. 25). There are indications that the QCD matter too shows similar trends. Viscous hydro calculations of the QCD matter would allow us to extract η/s from data and might help us pinpoint the location of the QCD critical point [75].

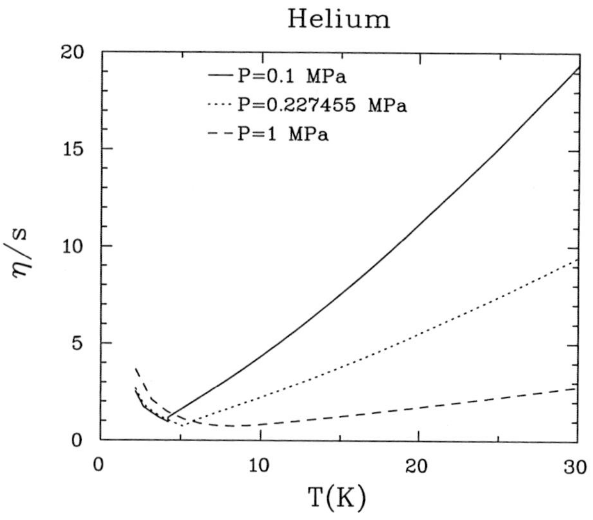

Figure 25. Each curve is at a fixed pressure. Solid: below the critical pressure P_c, dotted: at P_c, dashed: above P_c. From [75]

- If the inequality $\eta/s > 1/4\pi$ obtained [16] from the AdS/CFT duality is applicable to QCD, then also viscous hydro calculations become necessary.
- Assume a quasiparticle picture. Quantum mechanical uncertainty principle tells us that the mean-free path (λ) cannot be less than the inverse of the typical momentum of the quanta. It also makes no sense to have a mean-free path smaller than the interparticle spacing [76]. Since $\eta \propto \lambda$, η cannot vanish.
- Finally, to claim success for ideal hydro, one should calculate viscous corrections and show explicitly that they are indeed small.

4.2. Relativistic Dissipative Hydro–a Brief History

Relativistic version of the Navier-Stokes equation was obtained by Eckart [77] and by Landau and Lifshitz [78]. This is called the standard or the first-order formalism because terms only up to first order in dissipative quantities are retained in the entropy four-current. (The Euler's equation constitutes the zeroth-order formalism.) However, it was soon realised that this formalism suffers from the following problems:

- Acausality: Equations are parabolic and they result in superluminal propagation of signals [79, 80].
- Instability: Equilibrium states are unstable under small perturbations for a moving fluid [81]. This makes it difficult to perform controlled numerical simulations.
- Lack of relativistic covariance: This problem is related to the previous one. First-order theories look covariant, but they are not.

A causal dissipative formalism was developed by Müller [79] and Israel and Stewart [80], in the non-relativistic and relativistic sectors, respectively. It is also called a second-order formalism because the entropy four-current now contains terms up to second order in dissipative quantities. The resulting hydrodynamic equations are hyperbolic. Application of causal dissipative hydro to relativistic heavy-ion collisions was pioneered by Muronga [82]. Since then many others have contributed to this effort. We shall describe some of them in subsection **4.4**.

Recent years have witnessed intense activity in the area of causal hydro of gauge theory plasmas from AdS/CFT duality; for reviews see [83].

4.3. Basic Idea of Causal Dissipative Hydro

Before we discuss hydrodynamics, let us first consider a simpler example of diffusion. Consider a fluid in equilibrium with a uniform density ρ. If the fluid is perturbed such that the density is no longer uniform, it responds by setting up currents which tend to restore the equilibrium. In the linear response theory, the induced current J_i is simply proportional to the gradient of ρ (Fick's law)

$$J_i = -D\partial_i\rho, \tag{8}$$

where D is the diffusion coefficient. D is an example of a transport coefficient. Transport coefficients play an important role in the study of relaxation phenomena in non-equilibrium statistical mechanics or fluid dynamics. Equation (8) connects the applied force $(-\partial_i\rho)$ with the flux (J_i). Such equations are called constitutive equations because they describe a physical property of the material. (The familiar Ohm's law $\mathbf{J} = \sigma\mathbf{E}$ is another example of this.) In addition to Eq. (8), we also have the usual current conservation equation

$$\partial_\mu J^\mu = 0. \tag{9}$$

If D is constant, elimination of J_i gives

$$\partial_0\rho - D\partial_i^2\rho = 0.$$

This is the diffusion equation. It is parabolic. Its solution is

$$\rho \sim \exp(-x^2/4Dt)/\sqrt{4\pi Dt}.$$

It is easy to see that the solution violates causality: Initially (i.e. in the limit $t \to 0$), this is the Dirac delta function. But at any finite time, howsoever small, it is non-zero everywhere, even outside the lightcone. Now Eq. (9) cannot be wrong. So to restore causality the constitutive Eq. (8) which anyway was a hypothesis, is replaced by

$$\tau_J\partial_0 J_i + J_i = -D\partial_i\rho, \tag{10}$$

where τ_J is a parameter with dimensions of time. In Eq. (8), if the force vanishes, the flux vanishes instantaneously without any time lag. In contrast, in Eq. (10) the flux relaxes to zero exponentially. τ_J is called the relaxation time. The new diffusion equation is

$$\tau_J\partial_0^2\rho + \partial_0\rho - D\partial_i^2\rho = 0.$$

This equation is hyperbolic and is called the Telegraphist's equation [84]. If $v^2 \equiv D/\tau_J < 1$, causality is restored.

Now consider hydrodynamics. The conservation and constitutive equations are

$$\begin{aligned}\partial_\mu T^{\mu\nu} &= 0,\\ T_{ij} &= P\delta_{ij} - \eta(\partial_i u_j + \partial_j u_i - \frac{2}{3}\delta_{ij}\partial_k u_k)\\ &- \zeta\delta_{ij}\partial_k u_k.\end{aligned}$$

Here $T^{\mu\nu}$ is the energy-momentum or stress-energy tensor, P is the equilibrium pressure, and η and ζ are the coefficients of shear and bulk viscosity, respectively. Tensor decomposition is now more complicated. But the basic idea remains the same. Causality is restored by introducing higher-order terms in the gradient expansion. This forces introduction of a new set of transport coefficients, e.g. τ_π and τ_Π which are relaxation times corresponding to shear and bulk viscosities. They are important at early times or for a rapidly evolving fluid. For details, see e.g. [82].

4.4. Recent Results from Causal Viscous Hydro

The Israel-Stewart formulation [80] of the causal dissipative hydro is commonly used for numerical applications. However, it is not the only causal formulation available. There are others such as Müller's theory [79], Carter's theory [85], Öttinger-Grmela formulation [86], memory function method of Koide et al. [87], etc.

We have already mentioned the early work by Muronga [82]. Since then several authors have studied various aspects of the causal viscous hydro. We now describe briefly only a few of the most recent of these papers. This will also give the reader a feel for the complexities of these calculations and the uncertainties therein. (Other very recent papers which we shall not describe are listed in [88].)

Romatschke and Romatschke [89] used the Israel-Stewart theory. They assumed longitudinal boost invariance and used Glauber-type initial conditions. The initial shear pressure tensor $\pi^{\mu\nu}$ was assumed to be zero. η/s was treated as a fixed number independent of temperature. The bulk viscosity was ignored. For the EOS they used the semirealistic result of Laine and Schroder [90], and calculated the elliptic flow v_2. Their conclusion was that p_T-integrated v_2 is consistent with η/s up to 0.16 (Fig. 26). However, the minimum-bias $v_2(p_T)$ favoured $\eta/s < 1/4\pi$ violating the KSS bound [16] (Fig. 27).

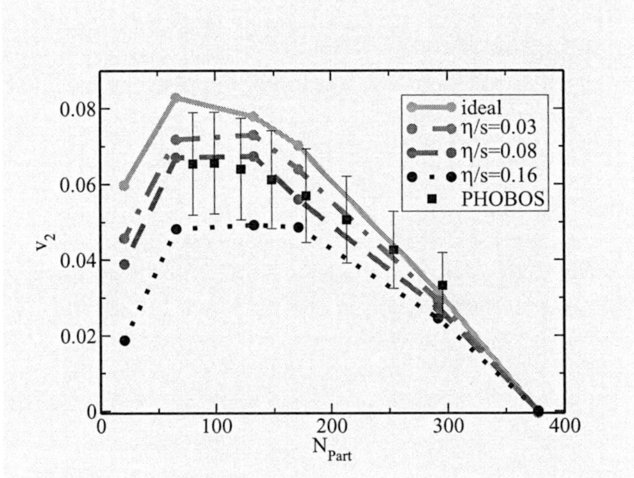

Figure 26. Au-Au, 200 GeV, p_T-integrated v_2 for charged particles vs number of participant nucleons. PHOBOS: 90% confidence level systematic errors. From [89]

Dusling and Teaney [91] used the Öttinger-Grmela formalism of causal viscous hydro. They assumed lon-

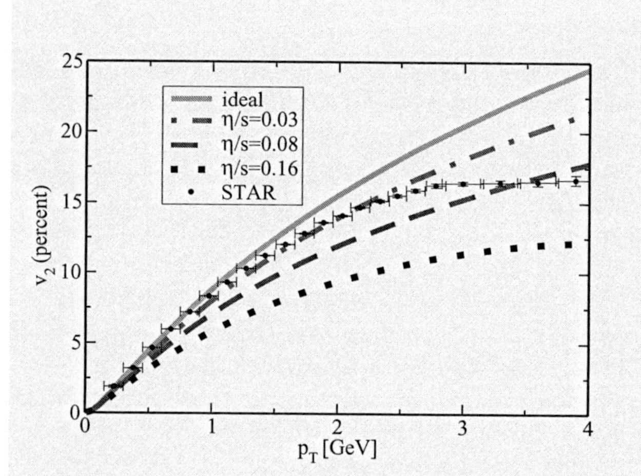

Figure 27. Au-Au, 200 GeV, minimum-bias $v_2(p_T)$ for charged particles. STAR: only statistical errors. From [89]

gitudinal boost invariance and used Glauber-type initial conditions. The initial shear pressure tensor π^{ij} was taken to be $\eta \left\langle \partial^i u^j \right\rangle$ as in the Navier-Stokes theory. η/s was treated as a fixed number independent of temperature. The bulk viscosity was ignored. The EOS used by them was simply $p = \epsilon/3$ without any phase transition. Their conclusion was that if the effects of viscosity are included in the evolution equations but not in the freezeout, then the v_2 is affected only modestly. If, however they are included at both the places, then v_2 is significantly reduced at large p_T.

Why does the shear viscosity suppress $v_2(p_T)$? Shear viscosity represents a frictional force proportional to velocity. For an in-plane elliptic flow, the in-plane flow velocity is higher than that out of plane. So the in-plane frictional force is stronger. This tends to reduce the flow anisotropy and hence $v_2(p_T)$.

Calculations described above include the shear viscosity in some approximation, but ignore the bulk viscosity completely. What do we know about the bulk viscosity of the strongly interacting matter? In the high-temperature limit, pQCD calculations [92] give the following results for the shear and bulk viscosity coefficients

$$\eta \sim \frac{T^3}{\alpha_s^2 \ln \alpha_s^{-1}} \quad \text{and} \quad \zeta \sim \frac{\alpha_s^2 T^3}{\ln \alpha_s^{-1}}.$$

As T increases, both η and ζ increase. However, the ratio ζ/η decreases showing the reduced importance of the bulk viscosity at high T. Also note that the entropy density $s \sim T^3$, and hence η/s increases with T, whereas ζ/s decreases with T. This is easy to understand because QCD becomes conformally symmetric at high temperatures.

In the deconfinement transition region the conformal symmetry is badly broken, and there is no reason to expect the bulk viscosity to be negligible. Extracting ζ for temperatures in this region from lattice QCD is difficult; see section **2.4**. However, some preliminary results are now available, and they indicate a dramatic rise of ζ/s as $T \to T_c$ [74] (Fig. 28).

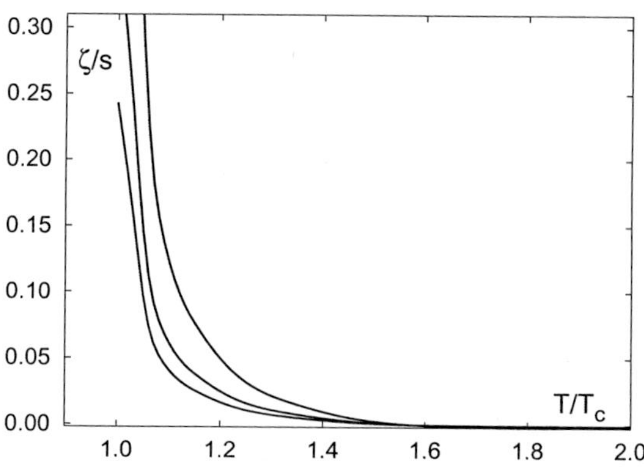

Figure 28. Bulk viscosity based on lattice data. $\omega_0 = 0.5, 1, 1.5$ GeV (top to bottom) is the scale at which pQCD is applicable. From [74]

Taking these results at their face value, Fries et al. [93] have studied the effect of inclusion of the bulk viscosity in the hydro equations. They studied 1D expansion of the fluid assuming longitudinal boost invariance. η/s was held fixed at $1/4\pi$. A realistic EOS based on the lattice results of Cheng et al. [12] was used. Various initial conditions were tried. They concluded that (a) Large bulk viscosities around T_c lead to sizeable deviations from equilibrium throughout the entire lifetime of QGP. (b) Bulk viscosities just slightly larger than currently favoured could easily lead to breakdown of hydro around T_c. (c) The decreased pressure should slow down the expansion and increase the time spent by the fluid in the vicinity of the phase transition. (d) The amount of entropy produced through *bulk stress* around T_c is smaller than that produced by *shear stress* at earlier times. Hence no large increase of the final particle multiplicity is expected.

4.5. What Remains to be Done?

- Bulk as well as shear viscosity (together with temperature dependence of ζ/s and η/s) needs to be incorporated.

- Can causal viscous hydro with CGC-type initial conditions reproduce dN/dy, $\langle p_T \rangle$ and v_2 data? If so, what are the extracted ζ/s, η/s?

- Causal viscous hydro + hadronic cascade is not done yet.

- There are issues related to the hydro formalism itself. For example, Baier et al. [94] have recently shown that the Müller and Israel-Stewart theories do not contain all allowed second-order terms.

- Present uncertainties in the hydro calculations limit the accuracy with which conclusions can be drawn. A coherent, sustained collaboration of experts in all stages of heavy-ion collisions is needed for a detailed, quantitative analysis of experimental data and theoretical models. Various numerical codes need to be compared with each other. To that end a new Theory-Experiment Collaboration for Hot QCD Matter (TECHQM) has been initiated. For details, see [95].

5. Predictions for the LHC

Pb-Pb collisions at $\sqrt{s_{\rm NN}} = 5.5$ TeV is an important part of the LHC experimental programme. 5.5 TeV represents about 30-fold increase in the CM energy compared to the maximum energy explored at RHIC which in turn was about 10 times higher than that at SPS. Measurements on pp collisions as well as collisions of p, d, light ions with Pb will provide important benchmarks.

Among the experiments at the LHC, CMS and ATLAS are primarily particle physics experiments/detectors, but they will study the physics of heavy-ion collisions too. ALICE (A Large Ion Collider Experiment), on the other hand, is a dedicated heavy-ion collision experiment. Physicists from several Indian universities and institutions have contributed in a big way to the ALICE collaboration. They are responsible for, among other things, the designing, testing, installation and maintenance of the Photon Multiplicity Detector (PMD) in ALICE and future upgrades of it. PMD is a preshower detector with fine granularity, full azimuthal coverage and one unit of pseudo-rapidity coverage. It will be used to measure the multiplicity, spatial distribution and correlations of produced photons on an event-by-event basis. Since photons escape the QGP without interactions, such measurements can potentially provide a cleaner glimpse of the early QGP phase. The Indian community has also made significan contributions to the muon spectrometer of ALICE. The spectromemter will be useful in the investigations of the J/ψ and other quarkonia, discussed in subsection 3.4. These particles are detected via their dimuon decay channel. The muon tracks will be found with an accuracy of better than one-tenth of a millimetre, thanks to the state-of-the-art readout electronics, known as MANAS, which was developed indigenously. ALICE has decided to use a grid environment for their computing needs. India is a signatory to the worldwide LHC computing grid and some of the Department of Atomic Energy installations are designated as Tier-II centres for this purpose.

A workshop was organised in 2007 at CERN in order to collect all the existing predictions for heavy-ion collisions at the LHC. The proceedings [96] provide a broad overview of the field. Here we shall only present a few glimpses of what may be in store at the LHC.

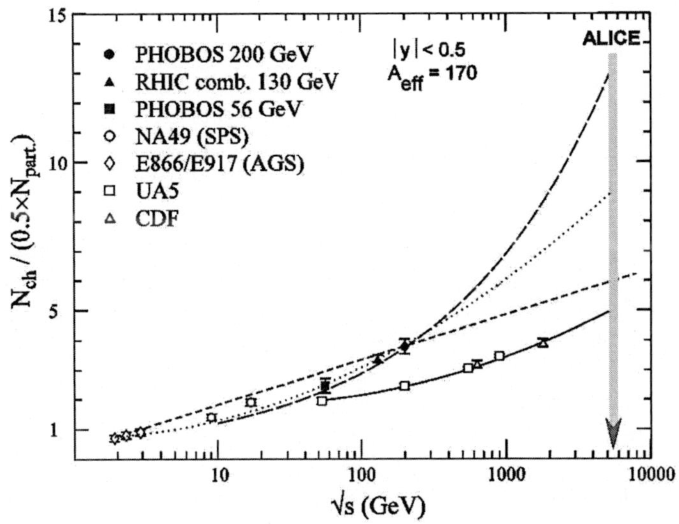

Figure 29. Charged-particle rapidity density per participant pair as a function of centre-of-mass energy for AA and pp collisions. Dashed line: a fit linear in $\ln(\sqrt{s})$, Dotted curve: a fit quadratic in $\ln(\sqrt{s})$, Long-dashed curve: based on the saturation model of [97]. From [66]

One of the first and easiest measurements at ALICE would be that of the charged-particle multiplicity in the mid-rapidity region. Particle production models and simple fits which are in agreement with the AGS, SPS, and RHIC data on this quantity differ substantially from each other when extrapolated to the LHC energy, as shown in Fig. 29. Thus this simple "first-day" measurement will test our understanding of the physics of

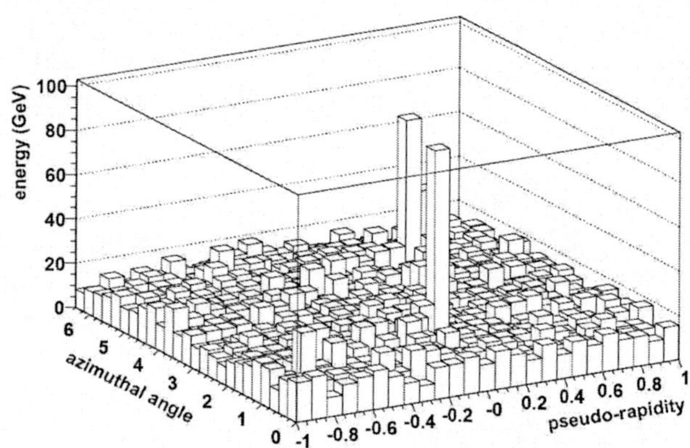

Figure 30. Pseudorapidity-azimuthal angle plot of Pb-Pb event at the LHC energy with two 100 GeV jets generated with HIJING and PYTHIA event generators. From [98]

multiparticle production. The charged-particle multiplicity provides a handle on the initial entropy production; the latter quantity is a necessary input in the hydrodynamic evolution of the produced matter.

Another relatively simple measurement at ALICE would be that of the elliptic flow v_2 which has played a crucial role at RHIC (Sec. **3.2**). The initial energy density (Eq. (4)) as well as the QGP lifetime are predicted to be higher at the LHC than those at RHIC. This is expected to raise the value of $v_2(p_T)$. On the other hand, the increased radial flow at the LHC is expected to lower it. (Recall the discussion on mass ordering in sec. **3.2**.) The net effect on $v_2(p_T)$ depends on the mass of the hadron: Minimum-bias $v_2(p_T)$ for pions (protons) is expected to be higher (lower) at the LHC than at RHIC, at low p_T; see Eskola et al. in [96]. Prediction by Kestin and Heinz is that $v_2(p_T)$ at a fixed impact parameter will be smaller at the LHC than at RHIC, for pions as well as protons [96]. However, p_T-integrated elliptic flow is expected to be higher for all hadrons due to the increased relative weight at large values of p_T.

In sec. **3.5** we have quoted the values of $T_{\rm ch}$ and μ_B for the SPS and RHIC energies. The latest predictions for the LHC are $T_{\rm ch} = 161 \pm 4$ MeV and $\mu_B = 0.8^{+1.2}_{-0.6}$ MeV [96].

Hard processes: Cross sections for the production of heavy flavours, $\sigma_{c\bar{c}}$ and $\sigma_{b\bar{b}}$, are expected to be about 10 and 100 times larger at the LHC than at RHIC. Cross sections for the production of jets with transverse energy in excess of 100 GeV are expected to be several orders of magnitude higher. Jet-photon events will also be abundant. Figure 30 displays the capability of ALICE to reconstruct the high-energy jets at the LHC in spite of the large soft-hadron background. Thus it would be possible to make detailed differential studies of heavy-quarkonium production, open-charm and open-beauty production, jet quenching, etc. at the LHC [96]. It will also be possible to study quark mass dependence and colour charge dependence of the energy loss of a parton as it traverses the medium.

Thus the LHC promises to be a valuable tool to test our models of ultrarelativistic heavy-ion collisions and deepen our understanding of QCD. For details, see [99].

REFERENCES

1. P. W. Higgs, Phys. Rev. Lett. **13** (1964) 508
2. C. Amsler et al., Phys. Lett. **B 667** (2008) 1
3. P. V. Landshoff, arXiv:0709.0395
4. U. M. Heller, MILC Collaboration, J. Phys. Conf. Ser. **9** (2005) 248
5. W.-M. Yao et al., J. Phys. **G 33** (2006) 1
6. C. Aubin, et al., Fermilab Lattice, MILC and HPQCD Collaborations, Phys. Rev. Lett. **95** (2005) 122002
7. M. Artuso et al., CLEO-c Collaboration, Phys. Rev. Lett. **95** (2005) 251801
8. B. Aubert et al., BABAR Collaboration, Phys. Rev. Lett. **98** (2007) 141801
9. Y. Aoki, G. Endrodi, Z. Fodor, S. D. Katz and K. K. Szabo, Nature **443** (2006) 675
10. Y. Aoki, Z. Fodor, S. D. Katz K. K. Szabo, Phys. Lett. **B 643** (2006) 46
11. M. Cheng et al., Phys. Rev. **D 74** (2006) 054507
12. M. Cheng et al., Phys. Rev. **D 77** (2008) 014511
13. R. V. Gavai, S. Gupta and S. Mukherjee, Pramana **71** (2008) 487
14. S. S. Gubser, I. R. Klebanov and A. A. Tseytlin, Nucl. Phys. **B 534** (1998) 202
15. A. Nakamura and S. Sakai, Phys. Rev. Lett. **94** (2005) 072305
16. P. Kovtun, D T. Son, A. O. Starinets, Phys. Rev. Lett. **94** (2005) 111601
17. H. B. Meyer, Phys. Rev. **D 76** (2007) 101701
18. R. V. Gavai and S. Gupta, Phys. Rev. **D 73** (2006) 014004
19. J. Cleymans, J. Phys. **G 28** (2002) 1575
20. V. Koch, A. Majumder and J. Randrup, Phys. Rev. Lett. **95** (2005) 182301
21. O. Kaczmarek and F. Zantow, Phys. Rev. **D 71** (2005) 114510
22. M. Asakawa and T. Hatsuda, Phys. Rev. Lett. **92** (2000) 120014; S. Datta, F. Karsch, P. Petreczky and I. Wetzorke, Phys. Rev. **D 69** (2004) 094507
23. T. Matsui and H. Satz, Phys. Lett. **B 178** (1986) 416
24. M. Buballa, Phys. Rept. **407** (2005) 205

25. K. Rajagopal and F. Wilczek, in "At the Frontier of Particle Physics / Handbook of QCD", Vol. 3, M. Shifman, ed., (World Scientific), p. 2061
26. L. McLerran and R. D. Pisarski, Nucl. Phys. **A 796** (2007) 83
27. R. V. Gavai, Phys. Rev. **D 32** (1985) 519
28. D. Banerjee, R. V. Gavai and S. Sharma, Phys. Rev. **D 78** 014506 (2008) and `e-Print: arXiv:0809.4535` [hep-lat]
29. Z. Fodor and S. Katz, J. H. E. P. **0203** (2002) 014
30. C. R. Allton et al. , Phys. Rev. **D 66** (2002) 074507; Ph. de Forcrand and O. Philipsen, Nucl. Phys. **B 642** (2002) 290; M.-P. Lombardo and M. d'Elia, Phys. Rev. **D 67** (2003) 014505. C. R. Allton et al. , Phys. Rev. **D 68** (2003) 014507; R. V. Gavai and S. Gupta, Phys. Rev. **D 68** (2003) 034506
31. R. V. Gavai and S. Gupta, Phys. Rev. **D 71** (2005) 114014
32. R. V. Gavai and S. Gupta, `e-Print: arXiv:0806.2233` [hep-lat]
33. J. D. Bjorken, in *Current Induced Reactions*, Lecture Notes in Physics vol. 56, New York: Springer, p. 93
34. J. D. Bjorken, Phys. Rev. **D 27** (1983) 140
35. T. Lappi and L. McLerran, Nucl. Phys. **A 772** (2006) 200
36. White paper by the STAR Collaboration, Nucl. Phys. **A 757** (2005) 102
37. M. Gyulassy, arXiv: nucl-th/0403032
38. N. Borghini, P. M. Dinh, J. Y. Ollitrault, Phys. Rev. **C 64** (2001) 054901
39. N. Borghini, P. M. Dinh, J. Y. Ollitrault, Phys. Rev. **C 66** (2002) 014905
40. R. S. Bhalerao, N. Borghini, J. Y. Ollitrault, Phys. Lett. **B 580** (2004) 157; Nucl. Phys. **A 727** (2003) 373; N. Borghini, R. S. Bhalerao, J. Y. Ollitrault, J. Phys. **G 30** (2004) S1213
41. S. A. Voloshin, A. M. Poskanzer, R. Snellings, arXiv:0809.2949
42. R. S. Bhalerao, J. P. Blaizot, N. Borghini, J. Y. Ollitrault, Phys. Lett. **B 627** (2005) 49; H.J. Drescher, A. Dumitru, C. Gombeaud, J.Y. Ollitrault, Phys. Rev. **C 76** (2007) 024905
43. M. D. Oldenberg (for the STAR Collaboration), J. Phys. **G 31** (2005) S437
44. B. Müller, Acta Phys. Pol. **B 38** (2007) 3705
45. See, e.g, R. D. Field, " Applications of Perturbative QCD", Addison-Wesley Publishing Company, The Advanced Book Program, 1989; Yu. Dokshitzer, V. Khoze, A. Mueller and S. Troyan, "Basics of perturbative QCD", Edition Frontiers 1991
46. C. A. Salgado and U. A. Wiedemann, Phys. Rev. **D 68** (2003) 014008
47. J. Adams, et al. (STAR Collaboration), Phys. Rev. Lett. **91** (2003) 072304
48. J. Adams, et al (STAR Collaboration), Phys. Rev. Lett. **97** (2006) 162301
49. R. Baier, D. Schiff and B. G. Zakharov, Ann. Rev. Nucl. Part. Sci. **50** (2000) 37
50. C. Loizides, Eur. Phys. J. **C 49** (2007) 339
51. A. Adare, et al. (PHENIX Collaboration), `e-Print: arXiv:0801.4020` [nucl-ex]
52. F. Karsch and R. Petronzio, Phys. Lett. **B 193** (1987) 105
53. R. V. Gavai and S. Gupta, Phys. Lett. **B 216** (1989) 239
54. R. V. Gavai, S. Gupta and K. Sridhar, Phys. Lett. **B 227** (1989) 161
55. C. Gerschel and J. Hufner, Ann. Rev. Nucl. Part. Sci. **49** (1999) 255; Phys. Lett. **B 207** (1988) 253
56. R. Gavai, et al.,Int. J. Mod. Phys. **A 10** (1995) 3043
57. B. Alessandro et al. (NA50 Collaboration), Eur. Phys. J. **C 39** (2005) 335
58. A. Capella and E. G. Ferreiro, Eur. Phys. J. **C 42** (2005) 419 and references therein; E. L. Bratkovskaya, A. P. Kostyuk, E. Cassing and H. Stöcker, Phys. Rev. **C 69** (2004) 054903
59. M. Gazdzicki and M. I. Gorenstein, Phys. Rev. Lett. **83** (1999) 4009
60. A. Adare, et al. (PHENIX Collaboration), Phys. Rev. Lett. **98** (2007) 232301
61. M. J. Tannenbaum, `e-Print: arXiv:nucl-ex/0702028`
62. H. Satz, J. Phys. **G 32** (2006) R25
63. P. Braun-Munziger anf J. Stachel, Phys. Lett. **B 490** (2000) 196; R. L. Thews, M. Schroedter and J. Rafelski, Phys. Rev. **C 63** (2001) 054905
64. J. Cleymans and H. Satz, Z. Phys. **C 57** (1993) 135; P. Braun-Munzinger, I. Heppe, J. Stachel, `e-Print: arXiv:nucl-th/9903010`
65. P. Braun-Munzinger, K. Redlich, J. Stachel, in *Quark-Gluon Plasma* 3, eds. R. C. Hwa and X. N. Wang, (World Scientific, Singapore, 2004)
66. F. Carminati et al. [ALICE Collaboration], J. Phys. **G 30** (2004) 1517
67. A. Andronic, P. Braun-Munzinger, J. Stachel, Nucl. Phys. **A 772** (2006) 167
68. STAR Collaboration, Phys. Rev. **C 72** (2005) 14904
69. T. Hirano et al., Phys. Lett. **B 636** (2006) 299; T. Hirano, Prog. Theor. Phys. Suppl. **168** (2007) 347; A. Adil et al. Phys. Rev. **C 74** (2006) 044905
70. O. Socolowski, F. Grassi, Y. Hama and T. Kodama, Phys. Rev. Lett. **93** (2004) 182301
71. M. Miller and R. Snellings, arXiv:nucl-ex/0312008; B. Alver et al. [PHOBOS Collaboration], Phy. Rev. Lett. **98** (2007) 242302; R. S. Bhalerao and J. Y. Ollitrault, Phys. Lett. **B 641** (2006) 260
72. R. J. Fries, J. Phys. **G 34** (2007) S851
73. P. Huovinen, Nucl. Phys. **A 761** (2005) 296
74. F. Karsch, D. Kharzeev, K. Tuchin, `e-Print: arXiv:0711.0914` [hep-ph]
75. L. P. Csernai, J. I. Kapusta, L. D. McLerran, Phys. Rev. Lett. **97** (2006) 152303

76. P. Danielewicz and M. Gyulassy, Phys. Rev. **D 31** (1985) 53
77. C. Eckart, Phys. Rev. **58** (1940) 919
78. L. D. Landau and E. M. Lifshitz, *Fluid Mechanics* (Pergamon, London, 1959)
79. I. Müller, Z. Phys. **198** (1967) 329; Living Rev. Relativity **2** (1999) 1
80. W. Israel and J. M. Stewart, Ann. Phys. (N.Y.) **118** (1979) 341
81. W. A. Hiscock and L. Lindblom, Ann. Phys. (N.Y.) **151** (1983) 466
82. A. Muronga, Phys. Rev. Lett. **88** (2002) 062302; Phys. Rev. **C 69** (2004) 034903
83. M. Natsuume, arXiv:hep-ph/0701201; D. T. Son and A. O. Starinets, arXiv: 0704.0240; D. Mateos, arXiv: 0709.1523
84. P.M. Morse and H. Feshbach, *Methods of Theoretical Physics* (McGraw-Hill, NY, 1953)
85. B. Carter, Proc. Roy. Soc. London, Ser. A, **433** (1991) 45; N. Andersson and G. L. Comer, Living Rev. Relativity **10** (2007) 1
86. M. Grmela and H. C. Öttinger, Phys. Rev. bf E 56 (1997) 6620
87. T. Koide, G. S. Denicol, P. Mota and T. Kodama, Phys. Rev. **C 75** (2007) 034909
88. P. Huovinen and D. Molnar, arXiv:0808.0953; M. Luzum and P. Romatschke, Phys. Rev. **C 78** (2008) 034915; H. Song and U. W. Heinz, Phys. Rev. **C 78** (2008) 024902; R. S. Bhalerao and S. Gupta, Phys. Rev. **C 77** (2008) 014902; A. K. Chaudhuri, arXiv:0801.3180
89. P. Romatschke and U. Romatschke, Phys. Rev. Lett. **99** (2007) 172301
90. M. Laine and Y. Schroder, Phys. Rev. **D 73** (2006) 085009
91. K. Dusling and D. Teaney, Phys. Rev. **C 77** (2008) 034905
92. P. Arnold, G. D. Moore and L.G. Yaffe (2000); JHEP **05** (2003) 051; P. Arnold, C. Dogan and G. D. Moore, Phys. Rev. **D 74** (2006) 085021
93. R. J. Fries, B. Müller and A. Schafer, arXiv:0807.4333
94. R. Baier et al., arXiv:0712.2451 (hep-th)
95. https://wiki.bnl.gov/TECHQM
96. N. Armesto et al. (Editors), J. Phys. **G 35** (2008) 054001
97. K.J. Eskola et al., Nucl. Phys. **B 570** (2000) 379
98. K. Šafařík, AIP Conf. Proc. **739** (2005) 346
99. ALICE: Physics Performance Report, vol. I, J. Phys. **G 30** (2004) 1517; vol. II, J. Phys. **G 32** (2006) 1295

Search and Study of Quark-gluon Plasma at the CERN-LHC

Tapan Nayak[a] *and Bikash Sinha*[b]

[a] *Variable Energy Cyclotron Centre, Kolkata-700064*

[b] *Variable Energy Cyclotron Centre, Kolkata-700064 and Saha Institute of Nuclear Physics, Kolkata-700064*

The major aim of nucleus-nucleus collisions at the LHC is to study the physics of strongly interacting matter and the Quark-gluon Plasma (QGP), formed in extreme conditions of temperature and energy density. We give a brief overview of the experimental programme and discuss the signatures and observables for a detailed study of QGP matter.

1. Introduction

The Large Hadron Collider (LHC) at CERN is designed to deliver colliding proton-proton (p-p) beams at centre-of-mass energies of 14 TeV and lead-lead (Pb-Pb) beams at 5.5 A.TeV. Collisions at these unprecedented energies offer outstanding opportunities for new physics in compliance with the standard model and beyond. The p-p collisions offer the tantalising possibility of discovering the Higgs boson, the missing link in the standard model. The collisions of two lead nuclei on the other hand, will create a speck of very high temperature and high energy density matter, called the Quark-gluon Plasma (QGP) where the properties of the system are governed by the quarks and gluons. According to conventional wisdom, the universe is believed to have been in this state only a few microseconds after the Big Bang. So the LHC is going to have a "peep" into the very early stages of the creation of the universe, and of course into the history of its evolution through space and time. A review of the recent status of QGP can be found in Ref. [1].

The LHC is going to answer several questions of fundamental interest. The ones which concern us in this report are the two most novel features of quantum chromodynamics (QCD), *viz.* the asymptotic freedom and quark confinement. Discovered in 1973, the asymptotic freedom tells that within the nucleons, quarks move mostly as free non-interacting particles. This has earned Gross, Wilczek and Politzer the Nobel Prize in physics in 2004. It is also known that colour-charged particles, such as quarks are confined within hadrons. Statistical QCD calculations which take into account these properties predict that strongly interacting systems at high temperature and/or energy density are composed of weakly interacting quarks and gluons. Such a phase consisting of (almost) free quarks and gluons is termed as the QGP. In 1974, Prof. T.D. Lee realised that it would be interesting to explore the phenomenon of QGP formation by distributing high energy and high density over a relatively large volume. Heavy-ion collisions at relativistic energies offer such a possibility of creating high energy and high density matter in the laboratory. Soon after, dedicated experimental programmes have started to search and study the QGP matter in great detail by colliding heavy-ions at dedicated accelerator facilities. The heavy-ion programme at CERN-LHC is the latest in this endeavour.

Heavy-ion physics is an integral part of the baseline programme of the LHC. The major aim of the experimental programme is to identify and access most of the QGP signatures for a detailed study of the QGP properties. The focus is to study how collective phenomena and microscopic properties, involving many degrees of freedom, emerge from microscopic laws of elementary particle physics. In addition with the LHC heavy-ions, one can access a novel range of Bjorken-x values where strong nuclear shadowing is expected. The initial density of the low-x gluons, accessible at the LHC energies is expected to be close to saturation. With these studies, the LHC will turn out to be a discovery machine for various types of new physics and will explore QCD phenomena in great detail.

2. QCD Phase Diagram

While embarking on the detailed experimental programme, it is essential to know the conditions required to create the QGP phase. Lattice QCD calculations have been performed by formulating QCD on the lattice and performing numerical Monte Carlo simulations [2,3]. Necessary conditions for the QGP phase transition can be obtained from these calculations. Results of such a calculation [4] is shown in Fig. 1 for energy density (ϵ) as a function of temperature (T). The energy density is seen to exhibit the typical behaviour of a system with phase transition, where an abrupt change is observed within a very narrow temperature range around the critical temperature, T_C. These calculations give a critical temperature, $T_C \sim 173 \pm 15$ MeV corresponding to the critical energy density of $\epsilon_C \sim 0.7$ GeV/fm^3.

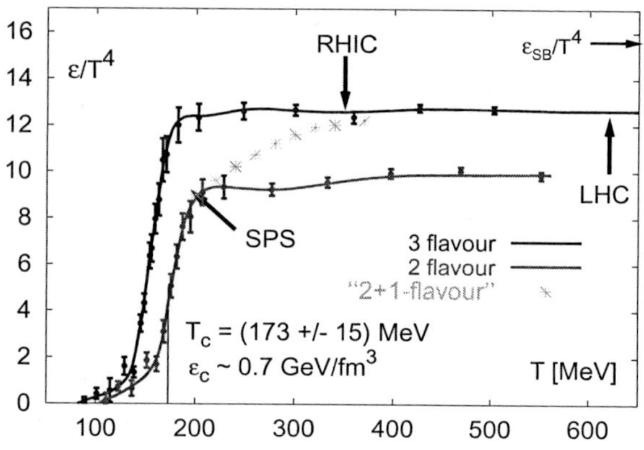

Figure 1. Lattice QCD calculations for energy density as a function of temperature

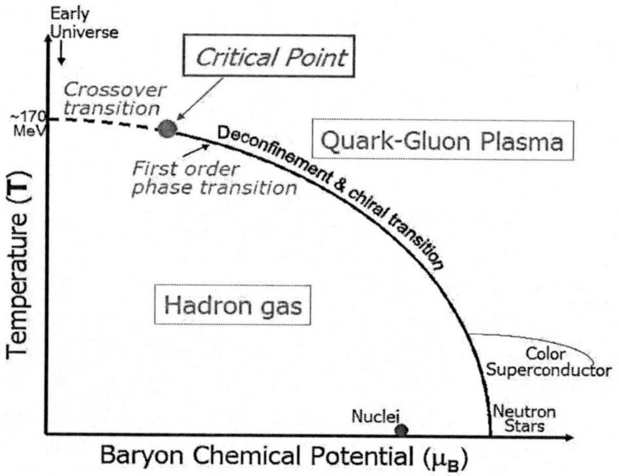

Figure 2. A schematic of the QCD phase diagram of nuclear matter

A schematic of the QCD phase diagram is shown in Fig. 2, as a function of the temperature (T) and the baryon chemical potential μ_B. This diagram signifies the separation of the QGP to hadronic phase of matter. Lattice QCD models predict a smooth crossover at high T and small μ_B while there are expectations for a first order transition at smaller T and larger μ_B. The existence of the critical point has also been predicted where a sharp transition between the QGP phase and the hadronic phase first appears. The exact location of the critical point is not known yet, but various calculations suggest that it might be within the reach of heavy-ion experiments.

3. Experimental Programme for QGP Search and the LHC

Dedicated programmes to create and study the QGP phase have started in early eighties with collisions of heavy-ions at relativistic energies. A tremendous amount of effort has been put for the development of four generations of experiments. Table 1 gives the list of some of the heavy-ion facilities. The quest for the search and study of QGP started first with the Au beam at 1 A.GeV at the Bevalac in Berkeley. The early success of the experiments in terms of bringing out the collective nature of the produced matter prompted the scientists at Brookhaven National Laboratory (BNL) and CERN to make concrete programmes for the future accelerator developments for heavy ions. The next milestone came with the acceleration of Au beam at 11.7 A.GeV at the BNL-AGS and Pb beam at 158 A.GeV at the CERN-SPS. First hints of the formation of a new state of matter has been obtained from the SPS data [5,6]. The Relativistic Heavy Ion Collider (RHIC) started becoming operational in the year 2000 with Au-Au collisions at $\sqrt{s_{\rm NN}} = 130$ GeV and soon after to top Au-Au energies of $\sqrt{s_{\rm NN}} = 200$ GeV. The experimental programme at RHIC included four experiments, two large and two small with the involvement of more than 1200 physicists. At present the RHIC experiments bring out highest quality data from p-p, d-Au, Cu-Cu and Au-Au at various energies. Strong evidence for the production of extreme hot and dense matter has been seen. The matter formed at RHIC has been termed as sQGP (strongly coupled QGP) [7]. The RHIC results, in combination with the ones from AGS and SPS, have enhanced our understandings of the QCD matter at different temperatures and densities.

In October 1990, in a workshop held at Aachen, Germany, the then CERN Director-General Carlo Rubbia, while discussing the case for the proposed the LHC, emphasised the idea of providing p-p collisions as well as heavy-ions [8]. In the CERN accelerator complex (Fig. 3), the ions are passed into the Low Energy Ion Ring (LEIR), then passed to the PS, the SPS and finally to the LHC.

The ALICE experiment at the LHC is designed specifically for heavy-ion physics [9]. The other two

Laboratory	Energy (A.GeV)	Δy
LBL-Bevelac	2.0 fixed target	1.81
Dubna	4.1 fixed target	2.36
BNL-AGS	11.7 (Au) fixed target	3.4
CERN-SPS	158 (Pb) fixed target	6.0
BNL-RHIC	200 (Au+Au) collider	11.7
CERN-LHC	5500 (Pb+Pb) collider	18.0

Table 1. Accelerator facilities for heavy-ions. Only top energies for given facilities are listed. The last column gives the accessible rapidity range.

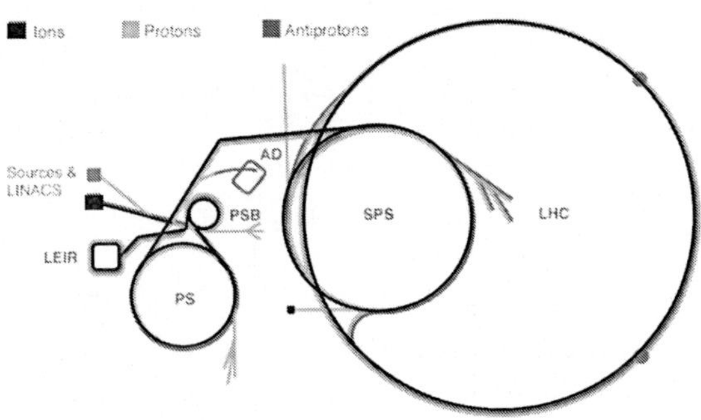

Figure 3. The CERN accelerator complex. The ions pass through the LIER, then to the PS, the SPS and finally to the LHC

major experiments ATLAS and CMS, also have heavy-ion programmes [10,11]. Both of these experiments will focus more on jets, photons and muon channels. The typical expected yearly running times of the LHC are of the order of 10^7 seconds for p-p collisions and 10^6 seconds for heavier systems [12]. Table 2 gives the centre-of-mass energy and expected luminosity at the LHC for some typical collision systems [12,13]. Due to the limited rate capability of the ALICE detector, the luminosity for p-p collisions will be limited to a maximum of $10^{31}\text{cm}^{-2}\text{s}^{-1}$ for the ALICE experiment.

A comparison of some of the basic parameters measured at the SPS and RHIC with expected values for the LHC is given in Table 3. The table lists the pseudorapidity density of charged particles ($dN_{ch}/d\eta$), formation time of QGP (τ_0), the energy density (ϵ) for $\tau_0 = 1$, the initial temperature in terms of the critical temperature (T_C), QGP life time (τ_{QGP}) and the freezeout time (τ_f) for the top centre-of-mass energies ($\sqrt{s_{NN}}$) of these accelerators. At the LHC, with the

System	$\sqrt{s_{NN}}$(TeV)	$L_0(\text{cm}^{-2}\text{s}^{-1})$	σ_{geom}(b)
p-p	14.0	10^{34}	~0.1
Pb-Pb	5.5	10^{27}	7.7
p-Pb	8.8	10^{29}	1.9
Ar-Ar	6.3	10^{29}	2.7

Table 2. Luminosities expected for different collision systems with different centre-of-mass energies [12].

increase of beam energy to about 28 times more compared to that of RHIC, the number of charged particles from the produced fireball increase by more than three times, the formation time of the QGP decreases significantly, the matter becomes much hotter, denser and long lived. All these conditions are conducive to opening up new physics domain at the LHC.

Condition	SPS	RHIC	LHC
$\sqrt{s_{NN}}$ (GeV)	17.3	200	5500
$dN_{ch}/d\eta$	450	600	1200–4000
τ_0 (fm/c)	~1	~0.2	~0.1
ϵ(GeV/fm^3)	2.5	4–5	10–40
T/T_C	1.1	1.9	3–4
τ_{QGP} (fm/c)	~1	2–4	>4
τ_f (fm/c)	~10	~20	~30

Table 3. Conditions created in central heavy-ion collisions at the top energies of SPS, RHIC and projections for the LHC.

4. The ALICE Experiment at the LHC

The experimental setup of the ALICE experiment [14, 15] is shown in Fig. 4. The detector setup can be broadly described by three groups of detectors: the central barrel, the forward detectors and the muon spectrometer [9]. Coverages of various detectors for charged particle measurements is shown in Fig. 5.

4.1. The Central Barrel

The central barrel consists of the Inner Tracking System (ITS), Time Projection Chamber (TPC), Transition Radiation Detector (TRD), the Time of Flight (TOF) detector and the newly proposed electromagnetic calorimeter (EMCAL). The design goal is to have low material budget and low magnetic field (B ≤ 0.5 T) in order to be sensitive to low-p_T particles.

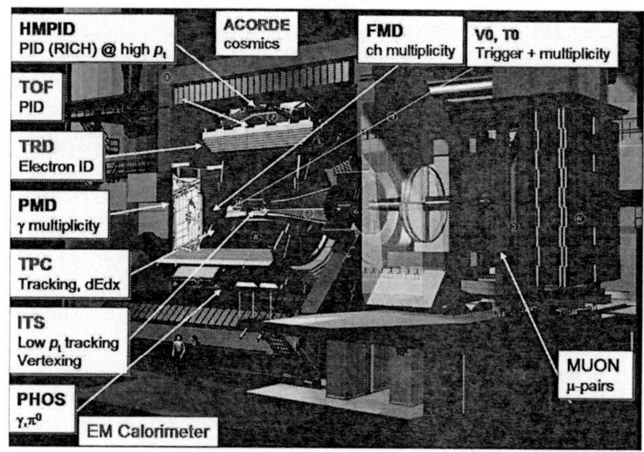

Figure 4. The experimental setup for the ALICE experiment [14,15] at the LHC

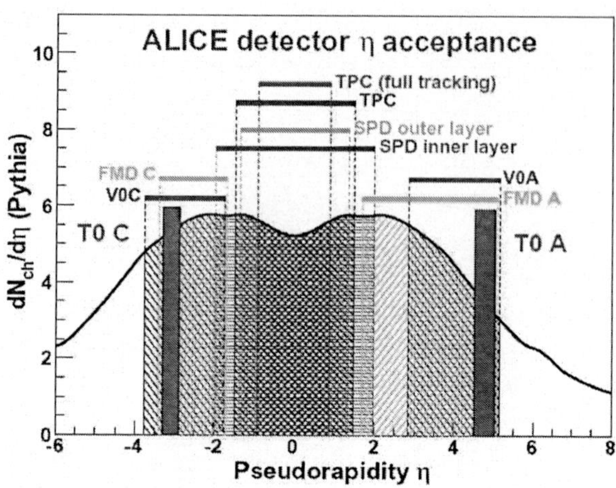

Figure 5. Coverages of various ALICE detectors for charged particle measurements. Superimposed is the Pythia event generator prediction for p-p collisions at 14 TeV

The ITS consists of six layers of silicon detectors. The two innermost layers consist of silicon pixel detectors (which covers up to the $|\eta| < 1.98$), followed by two layers of silicon drift detectors and the last two outer layers are of silicon strip detectors. The ITS is designed to handle high particle density expected in Pb-Pb collisions. The main uses of ITS will be to (a) determine the primary vertex of the collision, (b) localise the secondary vertices for decays of hyperons and D and B mesons, (c) track and identify particles with momentum below 200 MeV/c, (d) improve the tracking in the central barrel and (e) provide minimum bias and high multiplicity triggers.

The TPC is the main tracking detector of ALICE. It is cylindrical in shape. The active volume has an inner radius of about 85 cm, an outer radius of about 250 cm and an overall length along the beam direction of 500 cm. With this the coverage of TPC becomes $|\eta| < 0.9$. The TPC is made of large cylindrical field cage, filled with 90 m^3 of Ne/CO$_2$/N$_2$ (90/10/5), in which the primary electrons are transported over a distance of up to 2.5 m on either side of the central electrode to the end plates. Multi-wire proportional chambers with cathode pad readout are mounted into 18 trapezoidal sectors at each end plate. The TPC is the primary device for obtaining charged-particle momentum measurements and particle identification. It also provides an independent measure of primary vertex. The combination of ITS and a large TPC provides powerful tracking with excellent momentum resolutions (about 2–5%) from 100 MeV/c to 100 GeV/c.

The central barrel is equipped with a TRD for electron identification above 1 GeV/c, where the pion rejection capability from TPC is no longer sufficient. The TRD in combination with other central barrel detectors, will provide sufficient electron identification to measure the production of light and heavy vector meson resonances and the dilepton continuum produced in p-p and Pb-Pb collisions. It consists of 540 individual readout detector modules arranged in 18 super modules. Each module consists of a radiator of 4.8 cm thicknes, a multi-wire proportional readout chamber along with front-end elctronics. The gas mixture in the readout chamber is Xe/CO$_2$ (85%/15%).

The particle identification of hadrons in the intermediate momentum range (to about 4 GeV/c depending on the particle species) is improved significantly with the inclusion of a TOF system. The TOF system consists of Multi-gap Resistive-Plate Chambers (MRPC). The key aspect of these chambers is that the electric field is high and uniform over the whole sensitive gaseous volume of the detector. Inoziation produced by traversing charged particle starts a gas avalanche process which eventually generates the observed signals on the pick-up electrodes. The intrinsic time resolution of 40 ps has been achieved with an efficiency of close to 100%.

The central arm includes a High-Momentum Particle Identification Detector (HMPID) for the identification of hadrons at p_t >1 GeV/c. The detector is based on proximity-focusing Ring Imaging Cherenkov (RICH) counters and consists of seven modules of about 1.5×1.5 m^2 each, mounted in an independent support cradle. The coverage of HMPID is limited $-0.6 < \eta < 0.6$ with $1.2° < \phi < 58.8°$.

The measurements of low p_T direct photons and high-p_T π^0 are performed by a single arm high resolution photon (elctromagnetic) spectrometer (PHOS), consisting of lead-tungstate ($PbWO_4$) crystals. The major requirements of the PHOS include the ability to identify photons, discriminate direct photons from decay photons and perform momentum measurements over a wide dynamic range with high energy and spatial measurements. PHOS covers approximately a quarter of a unit of pseudorapidity, $-0.12 \leq \eta \leq 0.12$ and 100° in azimuthal angle. The total area is ~ 8 m^2.

A new EMCAL is being planned which will improve ALICE capability for measurement of high energy jets.

4.2. The Forward Detectors

The ALICE experiment is equipped with a set of forward detectors, such as the a Forward Multiplicity Detector (FMD), Photon Multiplicity Detector (PMD), Zero Degree Calorimeters (ZDCs) and detectors for trigger and timing (V0, T0).

The FMD consisting of several rings of silicon detectors, covers a very large range in pseudo-rapidity ($-3.4 < \eta < -1.7$ and $1.7 < \eta < 5.0$). The main function of FMD is to provide charged-particle multiplicity information.

The V0 detector consists of two arrays (each array with 32 segments) of scintillator counters called V0A and V0C, which are installed on both sides of the ALICE interaction point. With a time resolution of about 600 ps the detector provides minimum-bias triggers for in pp and A–A collisions. V0 detector helps in monitoring of the beam luminosity and reducing the beam-gas contributions.

The T0 detector consists of two arrays of Cherenkov counters (12 counters per array) and has an excellent time resolution (about 50 ps). The main purpose of T0 detector is to provide start time (T0) for the TOF detector. T0 detector also measures the vertex posision and performs several trigger related functions.

The PMD in ALICE has been conceived, designed and fabricated by an Indian collaboration comprising of Variable Energy Cyclotron Centre, Kolkata, Institute of Physics, Bhubaneswar, Indian Institute of Technology, Mumbai, Rajasthan University, Panjab University and Jammu Univerisity. The PMD is designed to measure the multiplicity and spatial distribution of photons in the forward rapidity region of $2.1 \leq \eta \leq 3.9$. The photon measurements along with those of charged particles from the FMD provide vital information in terms of the limiting fragmentation, the equation of state of matter from the determination of elliptic flow, information about phase transition and the formation of disoriented chiral condensates (DCC).

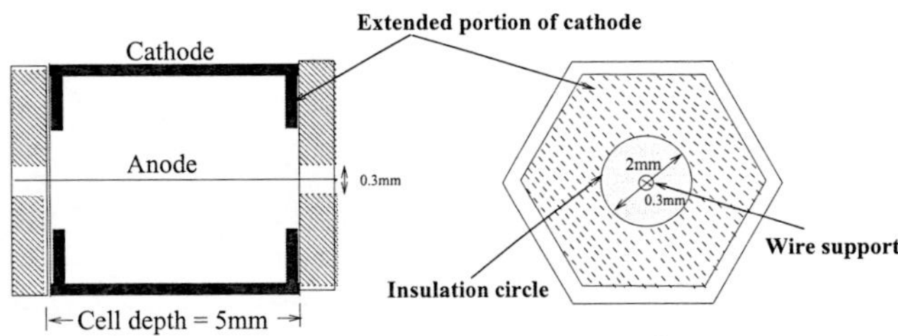

Figure 6. Schematic diagram of the cross section of a unit cell of the PMD in the ALICE experiment

Figure 7. Photograph of the partially installed PMD in the ALICE cavern

Because of the large particle density in the forward region, calorimetric techniques for photon measurements are not feasible. The PMD uses the preshower detector, where a three radiation length thick lead converter is sandwiched between two planes of highly granular gas proportional counters. The information from one of the gas detector planes, placed in front of the converter is used to veto the charged particles, whereas the preshower data from the other detector plane is used for photon identification. The granularity and the converter thickness of the PMD are optimised for high particle density so that the overlap of photon showers is minimal. The PMD consists of 221184 honeycomb shaped proportional counters, each of 0.25 cm^2

area. Each counter has a honeycomb shaped cathode extended towards a 20 μm thick gold-plated tungsten wire kept at a ground potential at the centre of each cell. The schematic diagram of the unit cell is shown in Fig. 6. The optimal operating voltage for the detector is -1400 volts which forms part of the plateau region of the proportional zone. The efficiency is about 96% for charged pions at this voltage.

The PMD is assembled in two equal halves. Each half has independent cooling, gas supply and electronics accessories. A partial installation with one-half of the PMD in the ALICE cavern is shown in Fig. 7. The front-end electronics consists of MANAS (Multiplexed ANAlog Signal Processor) chips (Fig. 8) for anode signal processing and the Cluster Read Out Concentrator Unit System (CROCUS) for data acquisition.

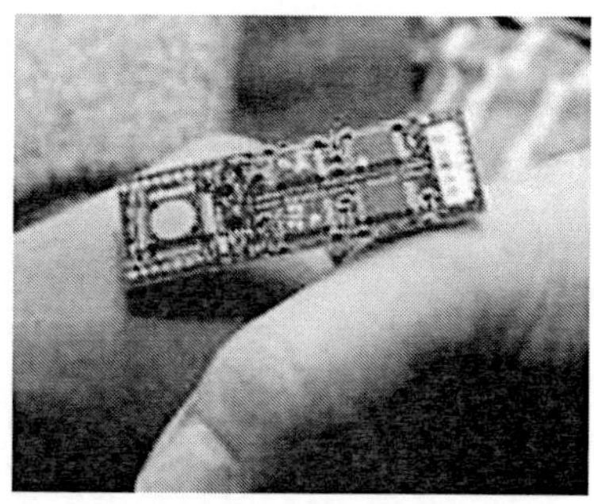

Figure 8. A closeup view of the electronic readout board of ALICE dimuon spectrometer showing four MANAS chips

4.3. The Dimuon Spectrometer

The ALICE forward muon spectrometer will study the complete spectrum of heavy quarkonia (J/ψ, ψ', Υ, Υ', Υ'') via their decay in the $\mu^+\mu^-$ channel. The spectrometer acceptance covers the pseudorapidity interval $2 \leq \eta \leq 4$ and the resonances can be detected down to zero transverse momentum. The invariant mass resolution is of the order of 70 MeV in the J/ψ region. The spectrometer consists of a front absorber ($\sim 10\lambda_{int}$) which absorbs hadrons and photons from the interaction vertex, a large dipole magnet, a highly granular tracking system and a trigger system placed behind a passive muon filter wall.

The dipole magnet is positioned at about 7 m from the interaction vertex. The magnetic field ($Bnom$ = 0.7 T, 3 Tm field integral) is what defined by the requirements of mass resolution.

Figure 9. Photograph of the second station of dimuon spectrometer

The muon tracking system is based on low thickness cathode pad chambers. The chambers are arranged in five stations (each of 2 chambers), two of the stations placed before, one inside and two after the dipole. To keep the occupancy at the 5% level, a high segmentaion of the readout pads is needed, leading to a total of about one million channels. The chambers have a position resolution of about 70 microns in the bending direction and 800 microns in the non-bending direction.

The frontend pulse processing of all the channels is done by MANAS chip which include preamplifier, a shaper and a multiplexer. A closeup view of the electronic readout board showing four MANAS chips is shown in Fig. 8. Saha Institute of Nuclear Physics, Kolkata is instrumental in the design and fabrication of the MANAS chip.

The station two muon tracking spectrometer has been designed and built in India by the the Saha Institute and Aligarh Muslim University. A photograph of the second muon station is shown in Fig. 9. The detector has been successfully installed in the ALICE cavern.

The trigger system is designed to select heavy quark resonance decays. The selection is made on the transverse momentum on individual muons. Four planes of resistive plate chambers (RPCs) arranged in two stations and positioned behind a passive muon filter provide the transverse momentum of each muon.

5. Signals and Observables

The QGP phase manifests itself in several different forms. The signals are not quite unique and one needs to make a comprehensive study of all available probes in

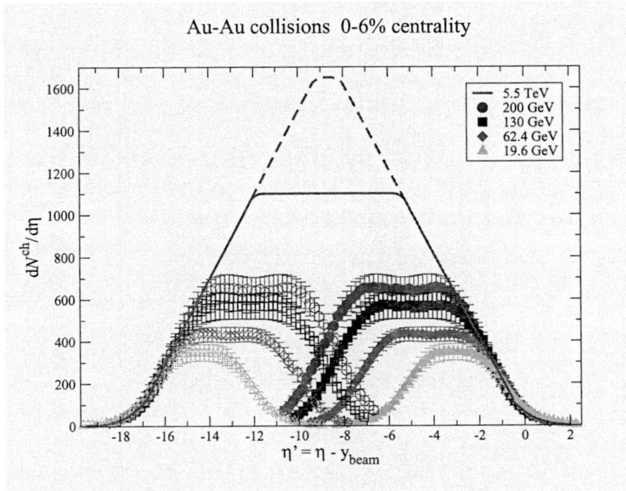

Figure 10. Charged particle multiplicity density scaled by their beam rapidities and extrapolation to the LHC energies [18]

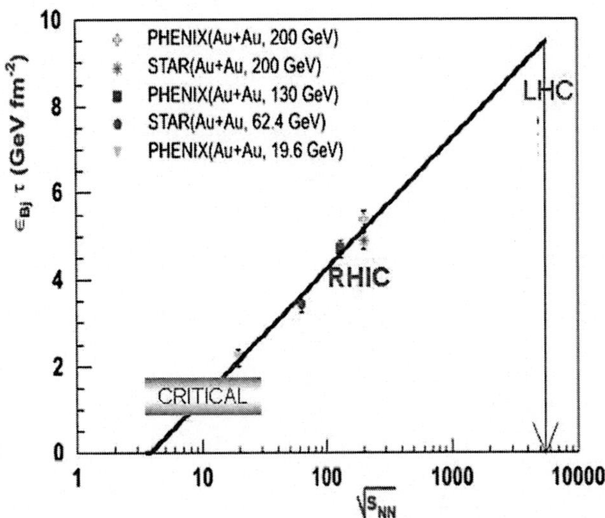

Figure 11. Energy density as a function of beam energy. The figure indicates the possible location of the critical point. A simple extrapolation is also made to the LHC energy [19], but the actual value is expected to be much higher

order to make any firm conclusion. The signals and observables can be categorised in terms of global observables, event-by-event fluctuations, DCC search, electromagnetic probes, heavy quarks and quarkonia, and to the physics at high p_T and jets. Finally possibilities at low Bjorken-x values will also tell us about the possibility of colour glass condensate picture [16]. A good volume on the predictions for Pb-Pb collisions at $\sqrt{s_{\rm NN}} = 5.5$ TeV may be found in [17]. Below we review some of the accessible QGP signatures for the LHC experiments.

5.1. Global Observables

The comprehensive study of global observables provides valuable information for thermal and chemical analysis of the freeze-out conditions. Some of these observables include the rapidity distributions of charged and identified particles, momentum spectra, particle ratios, flow and the size of the fireball.

Particle multiplicity measurements will constitute one of the first measurements in ALICE. This will be an eagerly awaited result during the first days of the LHC startup. There are large uncertainties in theoretical predictions [17] for rapidity density at central rapidity (the rapidity, y is defined in terms of energy and longitudinal momentum; alternatively one uses pseudorapidity, η, which is related to the angle of emission (θ) of the particle). Figure 10 shows the masured pseudorapidity distributions (shifted by the beam rapidity, $y_{\rm beam}$) of charged particles for several beam energies at RHIC. An extrapolation [18] to Pb-Pb collisions at the LHC energies may provide a good estimation. The extrapolated results, shown in Fig. 10, gives a value of ~ 1100 at mid rapidity. Other theoretical estimations come up with numbers between 1200 and 2500. The ALICE detector is optimised for the charged-particle density of 4000 and its performance is checked up to a value of 8000. Thus ALICE will be able to handle the large multiplicity data from the LHC.

Momentum spectra of charged particles and identified hadrons will constitute the next sets of measurements from ALICE. Fitting these spectra with specific models allow one to extract mean transverse momenta, temperature, radial flow and other observables [17].

The data on the charged particle multiplicity and mean transverse momenta allow one to get an estimation of the energy density. Figure 11 gives [19] the energy densities as a function of beam energy. The energy density where the critical point may occur is marked in the figure. The extrapolated value to the LHC shows the energy density to be about two times larger compared to the values at RHIC.

The measured net proton rapidity density distributions for AGS, SPS, RHIC energies [20] are shown in Fig. 12 with extrapolations to the LHC energies. At AGS energies the number of produced antiproton is very small and the net-proton distribution is similar to the proton distribution. At SPS and higher energies the net proton rapidity distribution shows double hump

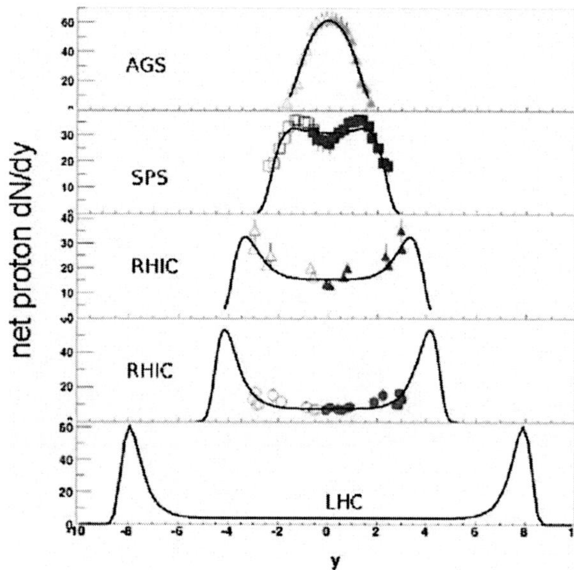

Figure 12. Net proton rapidity distributions at AGS, SPS and RHIC. Plots for two different energies at RHIC are shown. Extrapolation to the LHC energy shows a complete transparency for a large rapidity range [20]

with a dip around $y = 0$. This shows that beyond the SPS energy the reaction is beginning to be transparent in the sense that fewer of the original baryons are found at midrapidity after the collisions. A complete transparency can be expected at the LHC energies for a large rapidity range.

Estimation of freeze-out temperature and chemical potential is essential for mapping out the QGP phase space. Thermal model fits have been used [18] to estimate these values from the measured spectra and particle ratios. This is shown in Fig. 13. The chemical potential at top RHIC energies is between 20 and 40 MeV and at the LHC energies it is expected to quite low < 10 MeV.

An important measure of the collective dynamics of heavy-ion collisions is the elliptic flow (v_2). Figure 14 shows excitation function of v_2 for mid-central collisions. Because of the large values of v_2 at RHIC energies, in agreement with the value for an ideal fluid, the formation of a perfect liquid is ascertained at RHIC energies. A simple extrapolation of the v_2 value has been made for mid-central collisions at the LHC.

The dynamical evolution of the collision fireball and its spacetime structure has been traditionally studied using two-particle (HBT) correlations. The multiplicity and transverse momentum dependence of three-dimensional pion interferometric radii in Au-Au and Cu-Cu collisions at different RHIC energies [21] have

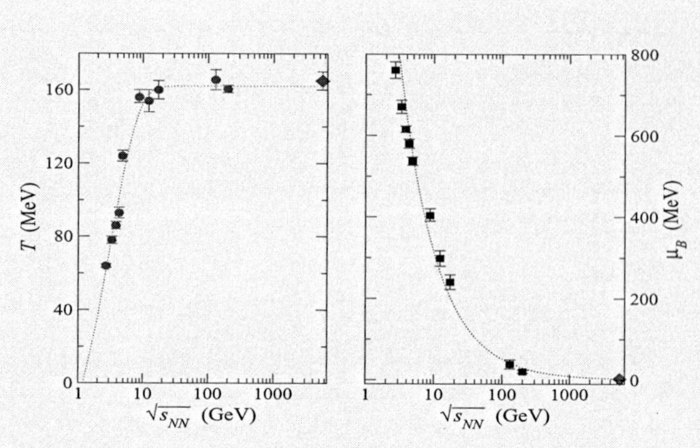

Figure 13. Estimation of freeze-out temperature and chemical potential from thermal model fits as a function of centre-of-mass energy of the collision [18]

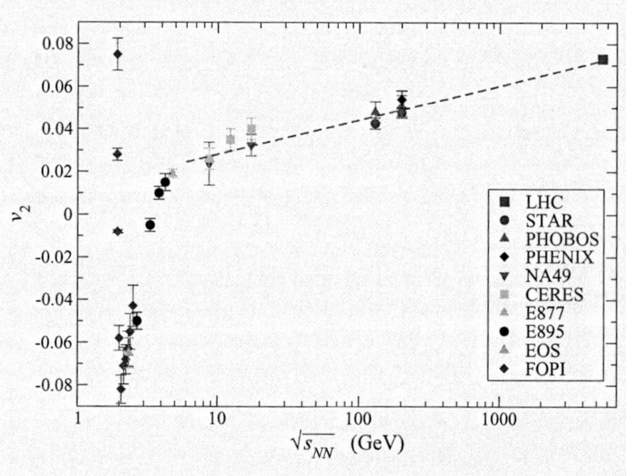

Figure 14. Elliptic flow, v_2 as a function of centre-of-mass energy of the collision [18] for existing data and extrapolated to the LHC

been shown in Fig. 15. The freeze-out volume estimates with charged pions measured from such studies, show linear dependence as a function of charge particle multiplicity indicating consistent behaviours with a universal mean-free-path at freeze-out. It will be interesting to find out what happens at the LHC as the fireball is supposed to be much larger and longer lived. At the LHC energies, because of the large number of produced particles, it may be possible to extract HBT radii on an event by event.

5.2. Event-by-event Physics

It is expected that the hot and dense system created in heavy-ion collisions at the LHC energies will show

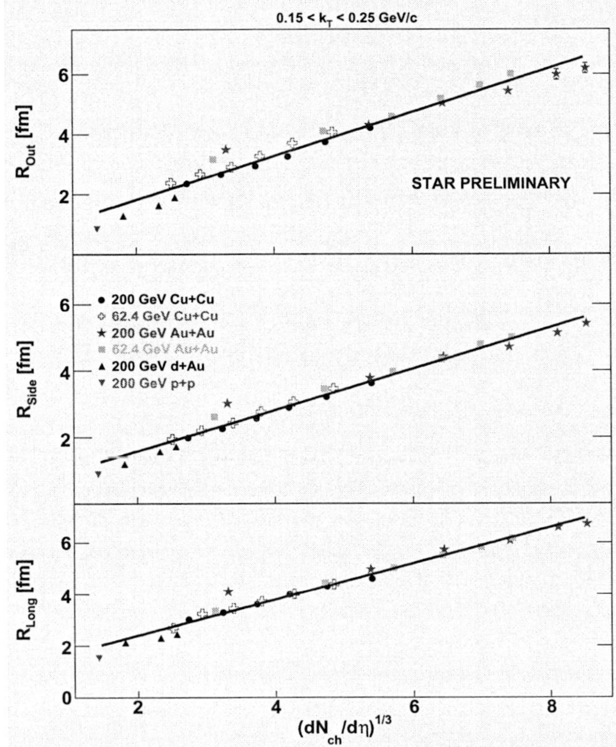

Figure 15. Pion source radii dependence on charged particle multiplicity. The lines are plotted to guide the eye and represent linear fits to the data [21]. Extrapolating to the LHC energies would mean a very large source size at the freeze-out

very characteristic behaviour of QGP phase transition, which may vary dramatically from one event to the other. As the number of particles produced at these energies is quite large, it is possible to study various observables in every single event and study their event-by-event fluctuations. The correlation and fluctuation measures provide possible ways to study variations in physical quantities from event to event.

Fluctuations of thermodynamic quantities such as temperature and entropy have been proposed to give evidence for the existence of QGP phase transition and also provide direct insight into the properties of the system created in heavy-ion collisions [15,22]. Active study of event-by-event fluctuations in heavy-ion collisions was initiated by experiments at the SPS and now very much applied to the data at RHIC energies. These studies include fluctuations in particle multiplicity, particle ratios, net charge, mean transverse momentum and formation of DCC domains. In order to be more sensitive to the origin of fluctuations, differential measures have been adopted where the analysis is performed at different scales (varying bins in η and ϕ).

For p-p collisions, the soft and semihard parts of the multi-particle production are successfully described in terms of colour strings stretched between the projectile and target. For nuclear collisions the number of strings grows with the growing energy and the number of participants in the collision. One has to take into account the interaction between strings in the form of their fusion and percolation [23]. Long range correlations were proposed as the main tool to study these phenomenon. The method of long-range correlation coefficients for different colliding systems and centralities will help to understand the critical fluctuation relevant to the string fusion and percolation phenomena.

The capabilities of the ALICE detector can be explored [15] in terms of measurements of temperature and p_T fluctuations, multiplicity and strangeness fluctuations, fluctuations of conserved quantities (such as net-charge and net-baryon), balance functions, fluctuation in azimuthal anisotropy, fluctuation in spacetime parameters and long range correlations. The effect of minijets and jets on the event-by-event studies will have to be clearly understood in order to make any conclusion. This is important in order to make any inference about the nature of event-by-event fluctuation as well as to understand the effect of jets passing through the high density medium created in heavy-ion collisions.

5.3. Disoriented Chiral Condensates

The QCD phase transition is predicted to be accompanied by chiral symmetry restoration at high temperature and densities. It leads to the formation of a chiral condensate in an extended domain such that the direction of the condensate is misaligned from that of the true vacuum. This DCC [24] results in the production of low momentum pions in a single isospin direction, leading to large fluctuations in the ratio (f) of neutral to charged pions. Normally a distribution of f would follow a binomial form with a mean of 1/3, whereas within a DCC domain it takes the form $1/2\sqrt{f}$. This feature of the f distribution can be used to characterise a DCC domain.

The formation of DCC has been hypothesised in the context of explaining observed abnormal events from cosmic ray experiments which had either excess of charged-particles (Centauro events) or excess of neutrals (anti-Centauro events) [25]. DCC search has been carried out by the miniMax experiment at Fermilab where no evidence was found [26]. A thorough search in Pb-Pb collisions at $\sqrt{s_{\rm NN}} = 17.3$ GeV was performed by the WA98 collaboration at CERN-SPS based on correlation study of photon and charged-particle multiplicities [27,28] using data from the preshower PMD and silicon pad multiplicity detector, respectively. A detailed analysis of centrality and acceptance dependence of multiplicity fluctuations had shown absence of any significance non-statistical fluctuations. Using the results from experimental data, mixed events and models which incorporate DCC domain, an upper limit on DCC production has been set. This is shown in Fig. 16.

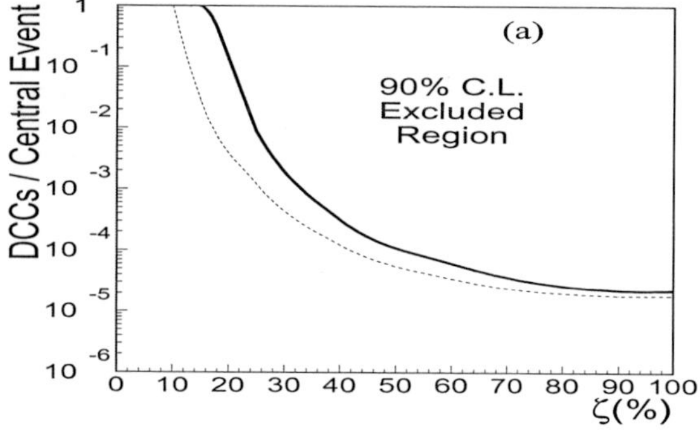

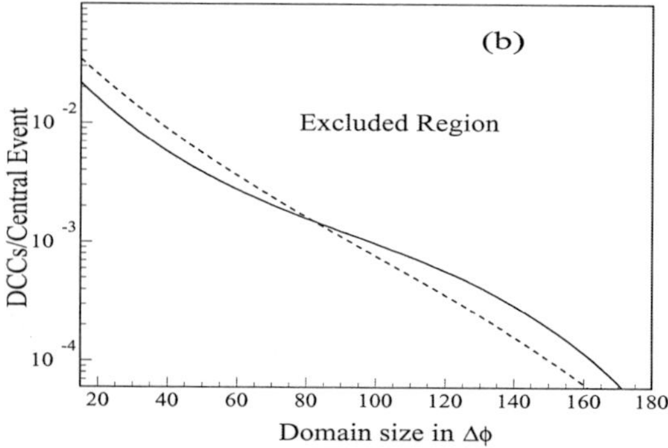

Figure 16. Upper limit (90% confidence level) of DCC production at SPS energies based on fluctuations of charged particles and photons. Panel (a) shows the limit as a function of the fraction of DCC pions [27], whereas panel (b) gives it as a function of DCC domain size in azimuthal angle [28]. The solid and dashed lines in lower panel correspond two centrality bins: top 5% and 5–10% of the minimum bias cross section

The search for DCC at the LHC energies will be possible by the ALICE experiment where the production of charged particles and photons will be studied in both the central and forward rapidity regions. In the central rapidity, a combination of the PHOS detector and TPC along with TOF will be used. In the forward rapidity region, PMD and FMD combination in a common coverage of $2.1 \leq \eta \leq 3.9$ will be used. The detection of DCC will mainly be done through the event-by-event fluctuation study [29]. Theoretical calculations also suggest that DCC formation might give rise to other signatures such as enhanced correlation of kaons and enhanced production of baryons, particularly, Ω and $\bar{\Omega}$. In addition DCC formation may be seen through the HBT correlations of identified pions and non-equilibrium photons in the sample of direct photons. Various analysis methods, adopted for DCC search in ALICE, may be found in [15].

5.4. Electromagnetic Probes

Interaction of the charged particles produced in the nucleus-nucleus collisions emit real and virtual photons (lepton pairs). Owing to the nature of the interaction they undergo minimal scatterings and are by far the best markers of the entire spacetime evolution of the collision [30–34].

The extraction of direct photons from experimental data is complicated by the huge amount of background from hadronic decays. The WA80 collaboration [35] provided the first interesting result with a p_T dependent upper limit on the direct photon production in S-Au collisions at lab energy of 200 A.GeV which was supported by theoretical calculations [36]. The high quality single photon data obtained from Pb-Pb collisions at CERN-SPS by the WA98 collaboration [37] have been the focus of considerable interest [38]. The direct photon spectra at low p_T from the PHENIX experiment [39] is seen to be consistent with rates calculated with thermal photon emission taken into account. The photon spectrum in ALICE will be measured with the PHOS spectrometer. The direct photons will be identified as an excess of photons when compared with the decay photons. The systematic error is expected to be about 8% [15].

A better insight into the nature of the evolving system can be provided by the HBT correlations of direct photons [40,41]. These photons are emitted during all stages of the collision and serve as a deep probe of the hot and dense matter. Although photon HBT has been suggested quite some time back, because of the difficulty in measurement (mainly because of low production cross section and large background) it took quite some years to finally have first results by the WA98 collaboration [42] and the final RHIC results [43] are expected soon. In ALICE direct photon HBT will be possible by the use of the PHOS detector.

The dilepton production has been studied since the start of the QGP programme. This is because the process of dilepton production is sensitive at low dilep-

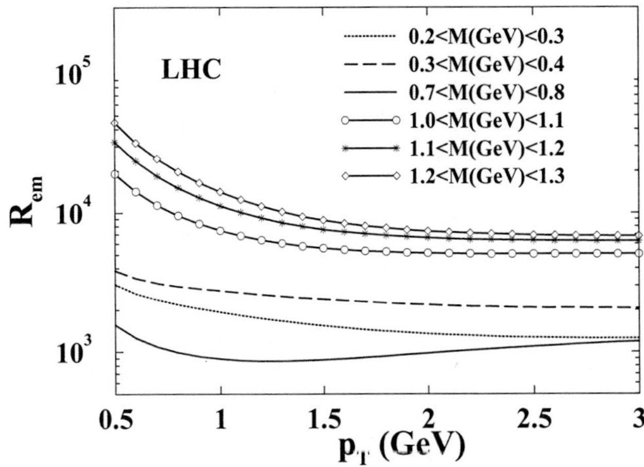

Figure 17. The thermal photon to dilepton ratio at the LHC energies as a function of transverse momentum, p_T, for various invariant mass windows [47]

ton masses to possible chiral restoration in dense matter, at intermediate masses to thermal dilepton production from the QGP and at higher masses it provides the best access to charmonium production and its predicted in-medium modification. At low masses $M < 1$ GeV, thermal dilepton production is mediated by broad vector meson, $\rho(770)$ in the hadronic phase. $\rho(770)$ is very short lived (lifetime of only 1.3 fm/c) and has a strong coupling to $\pi\pi$ channel. The 'in-medium' modifications of $\rho(770)$ mass and width close to the QCD phase boundary have long been considered as the prime signature for chiral symmetry restoration. Thermal dileptons have been studied at SPS [44] and RHIC [45] energies. The di-electron spectra, measured by the PHENIX experiment shows that the data are in good agreement with the cocktail of hadron decay sources over the full mass range. Di-electrons will be measured in ALICE with the central tracking system for masses above 0.2 GeV/c^2. Details of measurement and studies in ALICE may be found in [15].

It has been proposed since long back to look at the ratio of transverse momentum distribution of thermal photons to dileptons [46]. Recently it has been shown [47] that this ratio reaches a plateau beyond a certain value of transverse momentum as shown in Fig. 17. The ratio is sensitive to initial conditions and thus can be used to estimate the initial temperature of the system by properly selecting the transverse momentum and invariance mass windows.

5.5. Heavy Flavours and Quarkonia

The study of quarkonia production represents one of the most powerful methods to probe the medium of the fireball. Heavy quarkonia are sensitive to the collision dynamics at both short and long timescales and are expected to be sensitive to plasma formation. Suppressions of J/Ψ and Ψ' (bound states of charmonia) have long been considered to be the most direct signatures of QGP formation [48]. Indeed such suppressions have been observed in Pb-Pb collisions at the SPS energies [49]. Recent high accuracy measurement by the NA60 collaboration [50] for In-In collisions also shows such a suppression. There have been contrasting calculations which explain the suppressions even without invoking a scenario of QGP formation. Higher statistics data from RHIC experiments [51,52] are awaited for better distinctions between model predictions.

Several models predictions exist [17] at the LHC energies for production of heavy quarks and quarkonia in heavy-ion collisions. At the LHC, the much higher energy offers the possibility of measuring quarkonia production with significant statistics. ALICE will measure heavy quarkonia in both di-electron and dimuon channels. One of the main methods of physics analysis will be made by normalising quarkonia yields in heavy-ion collisions to reference processes, such as the production in pp collisions. The suppression of quarkonia, if any, will be clearly seen in these analyses.

5.6. High p_T and Jets

Jets are defined in QCD as cascades of consecutive emissions of partons in a narrow cone initiated by partons from an initial hard scattering. Because of QCD confinement, particles carrying a colour charge, such as quarks, cannot exist in free form. Therefore they fragment into hadrons before they can be directly detected, becoming jets. The measurement of jets help to determine the properties of the original quark and give indications of the properties of the hot and dense medium. High p_T partons produced in the initial stage of a nucleus-nucleus collision are expected to undergo multiple interactions inside the collision region prior to hadronisation. Thus the energy of the partons is reduced through collisional energy loss and medium induced gluon radiation. This so-called jet-quenching mechanism has been postulated as a tool to probe the properties of the QGP phase of matter. The measurement of γ-jet events is expected to provide an unique probe of parton energy loss [53,54].

At RHIC, the study of jets have been performed by introducing a nuclear modification factor and also by

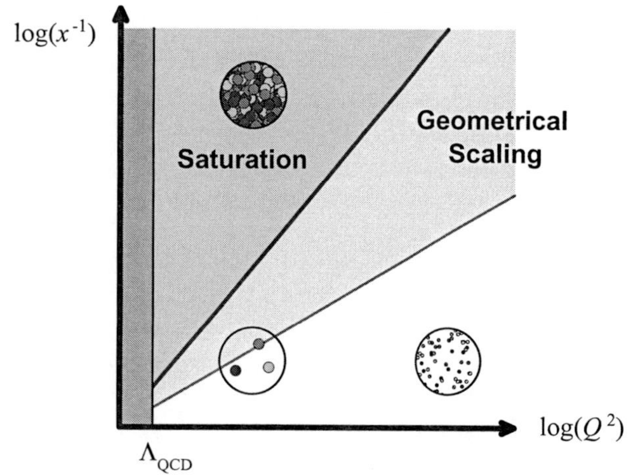

Figure 18. Saturation domain in the (Q^2, x) plane [58]

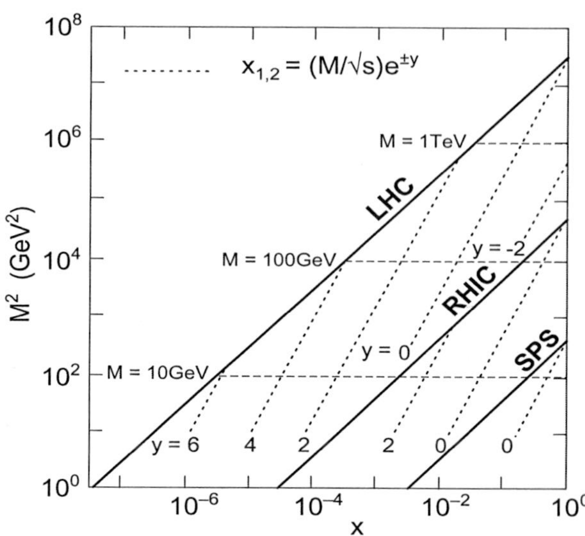

Figure 19. The range of Bjorken-x values accesible at heavy-ion collisions at the top SPS, RHIC and the LHC energies [14]

the method of back-to-back correlations. The nuclear modification factor R_{AA} has been defined as as the jet-yield within a given p_T bin for nucleus-nucleus collisions, normalised to corresponding yield for p-p collisions. The quantity, R_{AA} is defined in such a way that it becomes unity if the nucleus-nucleus collision were a mere superposition of number of nucleon-nucleon collisions. At RHIC energies, strong suppressions are observed in central Au-Au collisions corresponding to p-p [55,56]. The back-to-back correlation strength of high p_{T} hadrons is seen to be sensitive to the in-medium path length of the parton. The study by the STAR experiment over a broad range in transverse momenta [57] for d-Au and Au-Au collisions show that whereas there is no significant change in the near side peak with the increase of p_{T} of the associated particles, the away-side correlation strength decreases from d-Au to central Au-Au collisions. The strongest modifications of the correlated yields are seen at lower associated p_{T}.

At the LHC, the copious production of jets will enable us to understand the medium properties in a very detailed manner. Details of energy loss mechanism and predictions for the LHC energies may be found in [17]. Within the ALICE acceptance it is possible to map out the energy dependence of jet fragmentation over a wide kinematic range, up to a transverse energy of about 200 GeV [15]. Major problem in jet reconstruction in nucleus-nucleus collisions comes from the large background of the underlying event. Addition of the new EMCAL in ALICE is expected to improve the jet-energy resolution and thereby jet reconstruction capability.

5.7. QCD at Small x and Forward Physics

At large collision energy and relatively low momentum transfer (Q), one expects a new regime of QCD known as saturation [16,58]. This is described in a picture of colour glass condensate (CGC) where a saturation scale emerges naturally. This is pictorially depicted in Fig. 18 in terms of the saturation domain.

Figure 19 gives the accessible range of Bjorken-x values and M^2 relevant for particle production in nucleus-nucleus collisions at the top SPS, RHIC and the LHC energies [14,15]. The ALICE experiment will probe a continuous range of x as low as about 10^{-5}, accessing a novel regime where strong nuclear gluon shadowing is expected. The study of low x regime, especially at forward rapidities, will be most appropriate for getting to know the early stage of nuclear collision.

Two recent results from RHIC support the saturation phenomenon: (a) limiting fragmentation and (b) suppression of hadron spectra at froward rapidity. The limiting fragmentation has been observed [59,60] by combining data for rapidity distributions for several beam energies and colliding systems (Fig. 20). By shifting the rapidity axis by the beam rapidity, one can see that the distribution for produced particles at collisions of various energies and systems tend to an universal curve in the fragmentation region. This property naturally follows from the CGC framework. From the first measurement of rapidity distributions at the LHC, we will

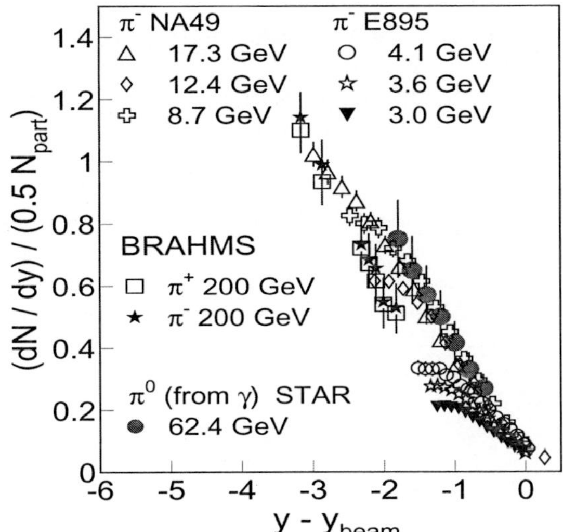

Figure 20. Limiting fragmentation as observed from the rapidity density distribution for different collision energies [59,60]

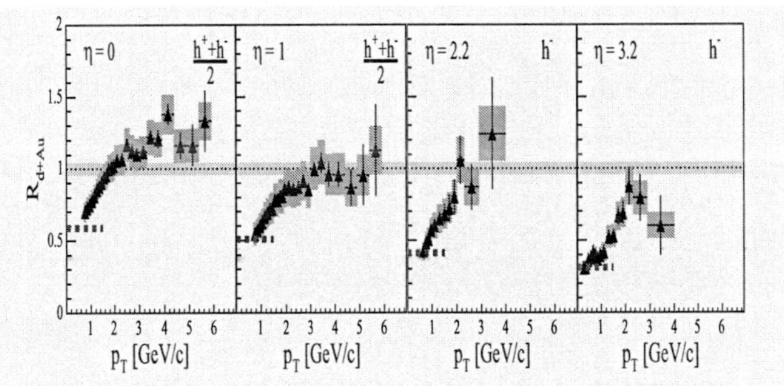

Figure 21. Forward suppression observed by BRAHMS for d+Au collisions at RHIC [61]

be able to confirm or dispute such a scenario. The suppression of hadron spectra at forward rapidity in d-Au collisions [61] from the BRAHMS experiment at RHIC is shown in Fig. 21. The suppression of the nuclear modification factor at forward rapidities is considered to be a consequence of the shadowing that builds up via the evolution in rapidity. A new proposal for a forward EMCAL is being considered which will enable ALICE to make detailed study of small x physics.

6. Summary

The field of QGP is set to enter a new era with the startup of the LHC. The experimental results from AGS, SPS and RHIC look very promising. The LHC is set to take this study to a completely new regime where the initial temperature is supposed to be much larger and energy density several times higher compared to what has been obtained so far. Three major experiments ALICE, ATLAS and CMS will take data during the heavy-ion collisions of the LHC. These three experiments are gearing up to study the new form of matter in great detail. The ALICE detector is designed to cope with the highest particle multiplicities above those anticipated for Pb-Pb collisions at $\sqrt{s_{\rm NN}}$ = 5.5 TeV. The detector has very large acceptance and excellent particle identification capabilities. The central barrel measures hadrons, electrons and photons. The forward muon spectrometer and PMD measure muons and photons, respectively, at forward rapidities. The ZDC, in addition to providing the beam luminosity for heavy-ion collisions, will also be used for event characterisation. Global observables will provide the bulk properties of the matter in terms of temperature, energy density, flow and size of the produced fireball. The ALICE experiment is well suited for precise event-by-event measurements of various observables. The fluctuation in thermodynamic quantities will provide the nature of the phase transition. Formation of DCC domains will signal the chiral symmetry restoration. The ALICE detector is sensitive to study the domain formation due to DCC. Transverse momentum distribution of thermal photons and dileptons and their ratios will be measured in order to gain insight to the spacetime evolution of the collision. The study of heavy-flavour and quarkonia production will allow for a comparison with QCD predictions and a study of the in-medium quenching of heavy quarks. The jet rates at the LHC will be quite large. Measurement of the longitudinal and transverse fragmentation functions of jets will provide details of the medium properties. Through the study of low-x physics one can also gain insight to the colour glass condensate initial conditions. The ATLAS and CMS detectors complement the ALICE programme by their excellent capabilities for jet and photon measurements at large transverse momenta. With the starting of the LHC's heavy-ion programme we expect a host of new fundamental physics within next decade.

REFERENCES

1. Proceedings of the Quark Matter 2008, Edited by: J. Alam, S. Chattopadhyay, T. Nayak B. Sinha and Y.P. Viyogi, J. Phys. G: Nucl. Part. Phys. **35** (2008)
2. O. Philipsen, arXiv:0808.0672v1 [hep-ph]
3. S. Gupta, J. Phys. G: Nucl. Part. Phys. **35** (2008) 104018

4. F. Karsch, E. Laermann and A. Peikert Nucl. Phys. **B605** (2001) 579
5. Uli Heinz and Maurice Jacob, arXiv:nucl-th/0002042v1
6. B. Mohanty, J. Alam, S. Sarkar, T.K. Nayak, B.K. Nandi, Phys. Rev. **C68** (2003) 021901
7. Berndt Muller, arXiv:0710.3366v3 [nucl-th]
8. CERN Courier, "Early days: Aachen: the case for LHC", October 2008
9. C. Fabjan *et al.*, ALICE Collaboration, J. Phys. G: Nucl. Part. Phys. **35** (2008) 104038
10. D. d'Enterria *et al.*, CMS Collaboration, J. Phys. G: Nucl. Part. Phys. **35** (2008) 104039
11. N. Grau *et al.*, ATLAS Collaboration, J. Phys. G: Nucl. Part. Phys. **35** (2008) 104040
12. F. Antinori (ALICE Collaboration), Nucl. Phys. (Proc. Suppl.) **B177** (2008) 156
13. R.M. Godbole *et al.*, Pramana **66** (2006) 657; e-Print: hep-ph/0604214
14. ALICE Collaboration, Physics Performance Report Vol. I J. Phys. **G30** (2004) 1517.
15. ALICE Collaboration, Physics Performance Report Vol. II J.Phys. **G32** (2006) 1295
16. Larry McLerran, arXiv:0804.1736v1 [hep-ph]
17. N. Armesto *et al.*, e-Print: arXiv:0711.0974
18. Nicolas Borghini and Urs Achim Wiedemann, *eprint: arXiv:0707.0564 [hep-ph]*
19. Raghunath Sahoo, Ph.D. thesis, Utkal University, eprint: arXiv:0804.1800 [nucl-ex] and arXiv:0805.4115 [nucl-ex]
20. S. Esumi, Jour. Phys. G: Nucl. Phys. **34** (2007) S575
21. Debasish Das *et al.* (STAR Collaboration) e-Print: nucl-ex/070204
22. T.K. Nayak, Jour. Phys. **G32** (2006) S187-S194
23. N. Armesto, L. McLerran, C. Pajares, Nucl. Phys. **A781** (2007) 201
24. B. Mohanty and J. Serreau, Phys. Rept. **414** (2005) 263
25. C.M.G. Lattes *et al.*, Phys. Rep. **65** (1980) 151
26. T.C. Brooks *et al.*, MiniMax Collaboration, Phys. Rev. **D55** (1997) 5667
27. M.M.Aggarwal *et al.*, WA98 Collaboration, Phys. Lett. **B420** (1998) 169
28. M.M.Aggarwal *et al.*, WA98 Collaboration, Phys. Rev. **C67** (2003) 044901
29. B. Mohanty, D.P. Mahapatra, T.K. Nayak, Phys. Rev. **C66** (2002) 044901
30. D.K. Srivastava, J. Phys. G: Nucl. Part. Phys. **35** (2008) 104026
31. L. D. McLerran and T. Toimela, Phys. Rev. D **31** (1985) 545
32. C. Gale and J.I. Kapusta, Nucl. Phys. B **357** (1991) 65
33. J. Alam, S. Raha and B. Sinha, Phys. Rep. **273** (1996) 243
34. J. Alam, S. Sarkar, P. Roy, T. Hatsuda and B. Sinha, Ann. Phys. **286** (200) 159
35. R. Albrecht *et al.* (WA80 Collaboration), Phys. Rev. Lett. **76** (1996) 3506
36. D.K. Srivastava, B.C. Sinha, e-Print: nucl-th/0103022
37. M.M. Aggarwal *et al.*, WA98 Collaboration, Phys. Rev. Lett. **85** (2000) 3595
38. J. Alam, S. Sarkar, T. Hatsuda, T. K. Nayak and B. Sinha, Phys. Rev. C **63** (2001) 021901 (R)
39. T. Dahms *et al.*, PHENIX Collaboration, J. Phys. G: Nucl. Part. Phys. **35** (2008) 104118
40. S.A. Bass, B. Muller, D.K. Srivastava, Phys. Rev. Lett. **93** (2004) 162301
41. J. Alam, B. Mohanty, P. Roy, S. Sarkar, B. Sinha, Phys. Rev. **C67** (2003) 054902
42. M.M.Aggarwal *et al.*, WA98 Collaboration, Phys. Rev. Lett. **93** (2004) 022301
43. D. Das, G. Lin, S. Chattopadhyay, A. Chikanian, E. Finch, T.K. Nayak, S.Y. Panitkin, J. Sandweiss, A. Suaide, H. Zhang (STAR Collaboration) e-Print: nucl-ex/0511055
44. S. Damjanovic *et al.*, NA60 Collaboration, J. Phys. G: Nucl. Part. Phys. **35** (2008) 104036
45. R. Averbeck *et al.*, PHENIX Collaboration, J. Phys. G: Nucl. Part. Phys. **34** (2007) S567
46. B. Sinha, Phys. Lett. B **128**, 91 (1983)
47. Jajati K. Nayak, J. Alam, S. Sarkar, B. Sinha, Phys. Rev. **C78** (2008) 034903
48. T. Matsui and H. Satz, Phys. Lett. **B178** (1986) 416
49. M.C. Abreu *et al.*, NA50 Collaboration, Phys. Lett. **B477** (2000) 28
50. G. Uasi *et al.*, NA60 Collaboration, J. Phys. G: Nucl. Part. Phys. **34** (2007) S233
51. S. Oda *et al.*, PHENIX Collaboration, J. Phys. G: Nucl. Part. Phys. **35** (2008) 104134
52. A. Tang *et al.*, STAR Collaboration, J. Phys. G: Nucl. Part. Phys. **35** (2008) 104135
53. R.J. Fries, B. Muller (Duke U.) , D.K. Srivastava, Phys. Rev. Lett. **90** (2003) 132301
54. T. Renk, Phys.Rev. **C74** (2006) 034906
55. S.S. Adler *et al.*, PHENIX Collaboration, Phys. Rev. **C69** (2004) 034910
56. J. Adams *et al.* (STAR Collaboration) Phys. Rev. Lett. **91** (2003) 172302
57. J. Adams *et al.* (STAR Collaboration) Phys. Rev. Lett. **97** (2006) 162301
58. F. Gelis, J. Phys. **G34** (2007) S421
59. J. Adams et al. (STAR Collaboration) Phys. Rev. Lett. **95** (2005) 062301
60. J. Adams et al. (STAR Collaboration) Phys. Rev. **C73** (2006) 034906
61. I.G. Bearden, Proceedings of Quark Matter 2006, J. Phys. **G34** (2007) S207

Strong Gauge Boson Scattering at the LHC

Saurabh D. Rindani

Theoretical Physics Division, Physical Research Laboratory, Navrangpura, Ahmedabad-380009, India

In the standard model with electroweak symmetry breaking through the Higgs mechanism, electroweak gauge boson scattering amplitudes are large if the Higgs boson is heavy and electroweak gauge interactions become strong. In theories with electroweak symmetry breaking through alternative mechanisms, there could be a strongly interacting gauge sector, possibly with resonances in an accessible energy region. In general, the scattering of longitudinally polarised massive gauge bosons can give information on the mechanism of spontaneous symmetry breaking. At energies below the symmetry breaking scale, the equivalence theorem relates the scattering amplitudes to those of the "would-be Goldstone" modes. In the absence of Higgs bosons, unitarity would be restored by some new physics which can be studied through WW scattering. Some representatives models are discussed. Isolating WW scattering at a hadron collider from other contributions involving W emission from parton lines needs a good understanding of the backgrounds. Resonances, if they exist below about a TeV, would be feasible of observation at the Large Hadron Collider (LHC).

1. Introduction

The Standard Model (SM) of strong, electromagnetic and weak interactions is based on local gauge invariance under the group $SU(2)_L \times U(1) \times SU(3)_C$, spontaneously broken to $U(1)_{\rm em} \times SU(3)_C$. The spontaneous breaking of gauge invariance gives masses to the gauge bosons $W^{\pm}$ and Z corresponding to three of the generators of the gauge group in a way which maintains the renormalisability of the theory. The gauge bosons corresponding to the $U(1)_{\rm em}$ gauge group of electromagnetism and the $SU(3)_C$ gauge group of colour remain massless. The spontaneous braking is also required to give masses to the fermions in the theory.

Spontaneous symmetry breaking (SSB) in the SM is achieved by means of the Higgs mechanism. An $SU(2)_L$ doublet of spin-0 fields is present in the theory, and the configuration which minimizes the action corresponds to a non-zero value for the neutral component of the doublet. Thus, the ground state (vacuum state) of the theory is not invariant under the $SU(2)_L$ group and the gauge symmetry is broken spontaneously. Since it is the neutral component of the scalar doublet which has a non-zero vacuum expectation value, $U(1)_{\rm em}$ remains an unbroken symmetry.

As a result of SSB, three of the four degrees of freedom of the complex scalar doublet, get absorbed resulting in the longitudinal degrees of freedom of the (massive) gauge bosons. These are the so-called "would-be Goldstone" modes, since the Goldstone mechanism for a spontaneously broken *global* invariance would have required these modes to propagate as massless particles. There remains one scalar field, the particle corresponding to which becomes massive. This massive spin-0 particle is known as the Higgs boson.

The introduction of spin-0 fields in the theory, while successful in breaking the electroweak gauge invariance to generate masses for gauge bosons and fermions, leads to certain problems. The main problem is that the mass of the Higgs boson cannot be made small in a natural way. As a result, alternative proposals for have been considered for the mechanism of SSB, some of which do not result in a light Higgs boson. The issue of the mechanism of spontaneous electroweak symmetry breaking is thus an important one. If the Higgs boson is discovered with properties predicted by SM, it would be a proof of Higgs mechanism. If the Higgs boson is not found, which would imply that it is either not present or very heavy, alternative mechanisms of SSB would have to be taken seriously.

The Large Hadron Collider (LHC) at CERN, Geneva, which in the very near future will collide protons on protons at a centre-of-mass energy of 14 TeV, is expected to unravel the mystery of the mechanism of SSB of the local gauge invariance responsible for the electroweak interactions. If the SM Higgs boson exists, and is light, it will be found. Alternatively, if the symmetry breaking is not due to a straightforward Higgs mechanism, or if the Higgs boson is not light, the LHC should show other effects, particularly in scattering of weak gauge bosons. Scattering of longitudinally polarised weak bosons play a dominant role and in the absence of a light Higgs boson, it is expected to become strong. Thus $W_L W_L$, $W_L Z_L$ and $Z_L Z_L$ scattering (where the subscript L denotes the longitudinal polarisation mode), will be an important tool to study the mechanism of SSB.

Here we discuss the importance of studying gauge boson scattering as a means of gaining knowledge about the electroweak symmetry breaking mechanisms, and also the feasibility of carrying out such a study at the LHC. There have been a number of studies on strong gauge boson scattering over the years. Since it is next to impossible to do an exhaustive review here, we will be forced to be selective.

We begin by discussing general features of scattering of massive spin-1 particles and specialise to the case when these particles are the gauge bosons of a spontaneously broken local gauge theory.

1.1. Longitudinal Gauge Boson Scattering

Amplitudes for scattering in theories with massive spin-1 particles can have bad high-energy behaviour. The propagator for a massive spin-1 field has a term which does not decrease with increasing momentum, leading to bad high-energy behaviour in amplitudes involving exchange of these vector particles. This can make the theory non-renormalisable. In theories with SSB, using the Higgs mechanism, renormalisability is maintained because Higgs exchange can cancel the bad high-energy behaviour arising from vector exchange. Amplitudes with massive vector particles in the initial and final states have bad high-energy behaviour because the polarisation vector has the form

$$\epsilon_L^\mu(k) \approx \frac{k^\mu}{m_V} \quad . \tag{1}$$

In the unitarity gauge, scattering of longitudinally polarised gauge bosons would have the largest amplitudes at high energy. In the high-energy limit, the amplitude for $W_L^+ W_L^-$ takes the form

$$\mathcal{M}^{\text{gauge}} = -\frac{g^2}{4\,m_W^2}\, u + \mathcal{O}((E/m_W)^0), \tag{2}$$

where u is the Mandelstam variable. Again, if the mass of the vector fields arises through a Higgs mechanism, the bad high-energy behaviour is cured. The contribution of $s-$ and $t-$channel Higgs exchanges is

$$\begin{aligned} \mathcal{M}^{\text{Higgs}} &= -\frac{g^2}{4\,m_W^2}\left[\frac{(s-m_W^2)^2}{s-m_H^2} + \frac{(t-m_W^2)^2}{t-m_H^2}\right] \\ &\approx \frac{g^2}{4\,m_W^2}\, u, \end{aligned} \tag{3}$$

in the limit $s >> m_H^2, m_W^2$, which cancels the bad high-energy behaviour of (2). However, if the Higgs mass is too high ($s << m_H^2$), the amplitude can be large, leading to a strongly-interacting gauge sector. The study of WW scattering can thus be important in studying the mechanism of symmetry breaking. (Here, as well in what follows, we use WW to denote WW, WZ as well as ZZ combinations.)

1.2. The Equivalence Theorem

Vector-boson scattering amplitudes at high energies may be calculated using the equivalence theorem [1]. The equivalence theorem relates the $W_L W_L$ scattering amplitude in the 't Hooft-Feynman gauge to the amplitude for scattering of the corresponding "would-be Goldstone" scalars at high energy ($s >> m_W^2$). Thus,

$$\begin{aligned} M(W_L W_L \to W_L W_L) &= M(ww \to ww) \\ M(Z_L Z_L \to Z_L Z_L) &= M(zz \to zz) \\ M(W_L Z_L \to W_L Z_L) &= M(wz \to wz) \\ M(W_L W_L \to Z_L Z_L) &= M(ww \to zz), \end{aligned} \tag{4}$$

where w and z are respectively the scalar modes which provide the longitudinal modes for W and Z respectively.

Goldstone boson interactions are governed by low-energy theorems for energy below the symmetry breaking scale ($s << m_{\text{SB}}^2$). Low-energy theorems are analogous to those obtained for $\pi\pi$ scattering in the chiral Lagrangian [2] and depend only on the symmetry of the theory. Thus, when there is no light Higgs (with $m_H < 1$ TeV or so), the low-energy theorems combined with equivalence theorem can predict $W_L W_L$ scattering amplitudes from the symmetries of the theory to leading order in s/m_{SB}^2. The specific theory for symmetry breaking then shows up at the next higher order in s/m_{SB}^2.

In the SM, we have the relations

$$m_H^2 = -2\mu^2 = 2\lambda v^2 = \lambda\sqrt{2}/G_F, \tag{5}$$

where λ is the quartic scalar coupling and μ is the mass parameter in the scalar potential. v is the vacuum expectation value of the neutral Higgs. Since v and G_F are fixed from experiment, large m_H means large λ. For $\lambda \gtrsim 4\pi$, i.e. $m_H \gtrsim (4\pi\sqrt{2}/G_F)^{\frac{1}{2}}$, perturbation theory is not valid. This corresponds to $m_H \approx 1.2$ TeV.

1.3. Limit from Unitarity

A limit may be obtained from unitarity, if the tree amplitudes are to be valid at high energies. In the absence of the Higgs, the amplitudes violate unitarity at high energies. The limit is of the order of the symmetry breaking scale. The Higgs cannot be very heavy, or else the unitarity limit would be crossed too soon. To see this, write the old-fashioned scattering amplitude f_{cm}, so that $|f_{cm}|^2 = \frac{d\sigma}{d\Omega}$. Then we have

$$f_{cm} = \frac{1}{8\pi\sqrt{s}}\mathcal{M} \; . \tag{6}$$

The scattering amplitude has the partial wave expansion

$$f(\theta) = \frac{1}{k}\sum_l (2l+1) P_l(\cos\theta) a_l \; , \tag{7}$$

where $a_l = e^{i\delta_l} \sin\delta_l$ is the partial wave amplitude written in terms of the phase shift δ_l. Unitarity is expressed by the optical theorem relation

$$\sigma = (4\pi/k)\,\mathrm{Im} f(0)\ . \tag{8}$$

For elastic scattering, the phase shift δ_l is real, but has a positive imaginary part if there is inelasticity. A convenient way to express elastic unitarity is

$$\mathrm{Im}[a_l^{-1}] = -1\ . \tag{9}$$

From this it follows that

$$|a_l| \leq 1;\ \ |\mathrm{Re}\, a_l| \leq \frac{1}{2}\ . \tag{10}$$

The $l = 0$ partial-wave unitarity for large s for WW scattering then gives, in the absence of the Higgs boson

$$\mathrm{Re}\ a_0 \equiv \mathrm{Re}\ \frac{G_F s}{16\pi\sqrt{2}} < 1/2\ . \tag{11}$$

Thus, for large enough s, unitarity is violated. The best bound comes from the isospin zero channel $\sqrt{1/6}\,(2W_L^+W_L^- + Z_L Z_L)$, viz.

$$s < \frac{4\pi\sqrt{2}}{G_F} \approx (1.2\ \mathrm{TeV})^2\ . \tag{12}$$

In the case of the standard Higgs mechanism, unitarity is restored by the exchange of a light Higgs boson. For unitarity to hold in the presence of the Higgs boson, the mass of the Higgs boson should satisfy

$$m_H^2 < \frac{4\sqrt{2}\pi}{G_F}\ . \tag{13}$$

This gives a limit of $m_H < 1.2$ TeV. Similar limits may be obtained by considering other partial-wave channels.

If there is no light Higgs, the perturbative unitarity limit is violated, and the gauge interactions are no longer weak. We then have a strong gauge sector. This could lead to the possibility of strong-interaction resonances, as seen in the case of the usual hadrons.

We will know discuss scenarios beyond the standard Higgs mechanism, where either there is no light Higgs, or the Higgs sector is more complicated than that in SM, so that the the WW scattering can become strong.

2. W_LW_L Scattering Beyond Higgs Mechanism

As we have seen, absence of a light Higgs boson leads to violation of unitarity. Violation of unitarity may be prevented or postponed to higher energy, in different ways, depending on the model. Models can have extra fermions and extra gauge interactions, which give additional contributions to W_LW_L scattering, e.g. resonances. An important consideration for many of the theories with a modified electroweak symmetry-breaking mechanism is conformity with precision tests. Specifically, they can have an unacceptably large S parameter of the oblique parameters S, T, U proposed by Peskin and Takeuchi [3], and a negative contribution to S is needed to restore agreement with experiment.

It is necessary to discriminate among mechanisms which give new effects in WW scattering. In particular, their treatment would be different depending on whether the gauge interaction remains perturbative or not. One possibility is that of a heavy Higgs, $m_H >> m_W^2$. In such a case, WW scattering would become large until the Higgs pole is crossed. If the Higgs mass is above the unitarity limit, the gauge boson interaction becomes non-perturbative. This needs special non-perturbative treatment of gauge boson interactions, and it is possible that the strong interaction produces resonances which would be seen. The second possibility is that there is no Higgs boson, and the symmetry is broken dynamically. In such a case there is a new sector with strong interactions, and this could produce its own resonances. A third possibility is that there is a light Higgs in an extension of the SM. Then again two cases arise: one where there are other Higgs bosons, and the dynamics still remains perturbative for WW centre-of-mass energy beyond the light Higgs resonance until the other Higgs resonances are crossed, or could become strong if the remaining Higgs bosons are heavy. These possibilities are discussed in the context of two-Higgs doublet models in [4,5]. The other case is when the light Higgs is light because it is a pseudo-Goldstone boson of a higher global symmetry, broken by the SM interactions [6]. In either case, WW scattering would be interesting to study in spite of a light Higgs boson.

We discuss a few of these models below. However, we begin with a model-independent approach.

2.1. Electroweak Chiral Lagrangian

In view of the variety of models, it would be preferable to have a unified, model-independent description of W_LW_L scattering. Effective Lagrangians may be used to provide such a description, valid up to a certain cut-off Λ. A no-resonance scenario is described in an electroweak chiral Lagrangian (EWCL) model, where one writes effective bosonic operators [7] in a series of increasing dimensions. The coefficients will have inverse powers of Λ. The effective Lagrangian is a low-energy expansion, and at the lowest non-trivial order, corresponding to operators of the lowest dimension, the scattering amplitudes for longitudinal gauge bosons, on using the equivalence theorem, would satisfy the low-energy theorems. However, coefficients of the higher-dimensional operators have to be fixed either on the basis of some underlying theory, or, in a purely phenomenological approach, from experiment.

Since the chiral Lagrangian is valid at low energies, an extrapolation to higher energies would lead to violation of unitarity. Unitarisation can be built in a somewhat *ad hoc* fashion by the use of Padé approximants, also known as the Inverse Amplitude Method (IAM), or the K-matrix method and this can generate a resonant behaviour [8]. The N/D method, developed for strong interactions [9], has also been applied for unitarisation of electroweak amplitudes [10].

Terms in the chiral Lagrangian must respect $SU(2)_L \times U(1)$ gauge symmetry, which is broken spontaneously. Experimental constraints require that the Higgs sector also approximately respect a larger $SU(2)_L \times SU(2)_C$ symmetry, where the second factor $SU(2)_C$ corresponds to the so-called custodial symmetry. The custodial symmetry is broken by the Yukawa couplings and the $U(1)$ gauge couplings. The global symmetry could in principle be even larger.

The chiral Lagrangian is thus constructed using the dimensionless unitary unimodular matrix field $U(x)$, which transforms under $SU(2)_L \times SU(2)_C$ as $(2,2)$. This corresponds to a non-linear realisation of the symmetry. $U(x)$ may be represented in terms of real (would-be Goldstone) fields π_a, $(a = 1, 2, 3)$, as follows:

$$U(x) = \exp \frac{i}{v} \sum_{a=1}^{3} \pi_a \tau_a \ , \tag{14}$$

where τ_a, $a = 1, 2, 3$ are the usual 2×2 Pauli matrices and v is the symmetry breaking scale[1].

The covariant derivative of $U(x)$ is

$$D_\mu U = \partial_\mu U + ig\frac{\vec{\tau}}{2} \cdot \vec{W}_\mu U - ig' U \frac{\tau_3}{2} B_\mu . \tag{15}$$

Further, the basic building blocks which are $SU(2)_L$ covariant and $U(1)_Y$ invariant are

$$T \equiv U \tau_3 U^\dagger \ , \quad V_\mu \equiv (D_\mu U) U^\dagger , \tag{16}$$

$$W_{\mu\nu} \equiv \partial_\mu W_\nu - \partial_\nu W_\mu + ig[W_\mu, W_\nu] \ , \tag{17}$$

where T, V_μ and $W_{\mu\nu}$ have dimensions zero, one and two respectively.

[1]These π_a should not be confused with the π-meson fields. The w, z scalar fields, corresponding to the longitudinal polarisation modes of the gauge bosons W, Z are linear combinations of the π_a.

Terms in the chiral Lagrangian in the $M_H \to \infty$ limit of the linear theory at tree level are

$$\begin{aligned} \mathcal{L}_0 &= \frac{1}{4} v^2 Tr[(D_\mu U)^\dagger (D^\mu U)] \\ &\quad - \frac{1}{4} B_{\mu\nu} B^{\mu\nu} - \frac{1}{2} Tr W_{\mu\nu} W^{\mu\nu} , \end{aligned} \tag{18}$$

where v is the symmetry breaking scale and $B_{\mu\nu} \equiv \partial_\mu B_\nu - \partial_\nu B_\mu$. The first term has dimension two, while the second two (kinetic energy) terms have dimension four. The gauge couplings to the quarks and leptons must also be added. The Yukawa couplings of the quarks and leptons should also be included in the symmetry breaking sector.

An additional dimension-two operator is allowed by the $SU(2)_L \times U(1)$ symmetry

$$\mathcal{L}_1^{'} = \frac{1}{4} \beta_1 g^2 v^2 [Tr(T V_\mu)]^2 \ . \tag{19}$$

This term does not emerge from the $M_H \to \infty$ limit of the renormalisable theory at tree level. It violates the $SU(2)_C$ custodial symmetry even in the absence of the gauge couplings. It is the low-energy description of custodial-symmetry breaking physics which has been integrated out, at energies above roughly $\Lambda \equiv 4\pi v \simeq$ 3 TeV. At tree level, $\mathcal{L}_1^{'}$ contributes to the deviation of the ρ parameter from unity.

At the dimension-four level, there are a variety of new operators that can be written down. Making use of the equations of motion, and restricting attention to CP-invariant operators, the list can be reduced to eleven independent terms [7]

$$\begin{aligned} &\mathcal{L}_1 = \tfrac{1}{2}\alpha_1 g g' B_{\mu\nu} Tr(T W^{\mu\nu}) \\ &\mathcal{L}_2 = \tfrac{1}{2} i \alpha_2 g' B_{\mu\nu} Tr(T[V^\mu, V^\nu]) \\ &\mathcal{L}_3 = i \alpha_3 g Tr(W_{\mu\nu}[V^\mu, V^\nu]) \\ &\mathcal{L}_4 = \alpha_4 [Tr(V_\mu V_\nu)]^2 \\ &\mathcal{L}_5 = \alpha_5 [Tr(V_\mu V^\mu)]^2 \\ &\mathcal{L}_6 = \alpha_6 Tr(V_\mu V_\nu) Tr(T V^\mu) Tr(T V^\nu) \\ &\mathcal{L}_7 = \alpha_7 Tr(V_\mu V^\mu) Tr(T V_\nu) Tr(T V^\nu) \\ &\mathcal{L}_8 = \tfrac{1}{4}\alpha_8 g^2 [Tr(T W_{\mu\nu})]^2 \\ &\mathcal{L}_9 = \tfrac{1}{2} i \alpha_9 g Tr(T W_{\mu\nu}) Tr(T[V^\mu, V^\nu]) \\ &\mathcal{L}_{10} = \tfrac{1}{2}\alpha_{10} [Tr(T V_\mu) Tr(T V_\nu)]^2 \\ &\mathcal{L}_{11} = \alpha_{11} g \epsilon^{\mu\nu\rho\lambda} Tr(T V_\mu) Tr(V_\nu W_{\rho\lambda}) \ . \end{aligned} \tag{20}$$

The EWCL incorporates the low-energy theorems, and therefore the piece $\mathcal{L}_0$ is completely fixed in terms of the symmetry-breaking scale v. However, the coefficients of the higher-dimensional operators have to be determined from experiment. Alternatively, they can be matched with a suitable underlying theory of one's choice.

The pieces of the EWCL contributing to gauge boson two-point functions have been tested in precision tests, and these contribute to the oblique parameters, S, T and U [3]. $\mathcal{L}_1^{'}$, $\mathcal{L}_1$ and $\mathcal{L}_8$ are directly related to the S, T and U parameters. By setting the Goldstone boson fields to zero in these operators, which is equivalent to going to the unitarity gauge, we get

$$S \equiv -16\pi \frac{d}{dq^2} \Pi_{3B}(q^2)|_{q^2=0} = -16\pi\alpha_1 \ , \tag{21}$$

$$\alpha T \equiv \frac{e^2}{c^2 s^2 m_Z^2} [\Pi_{11}(0) - \Pi_{33}(0)] = 2g^2 \beta_1 \ , \tag{22}$$

$$U \equiv 16\pi \frac{d}{dq^2} [\Pi_{11}(q^2) - \Pi_{33}(q^2)]|_{q^2=0} = -16\pi\alpha_8 \ . \tag{23}$$

The $\Delta\rho(\equiv \rho - 1)$ parameter is related to T by $\Delta\rho_{new} = \Delta\rho - \Delta\rho_{SM} = \alpha T$, where $\Delta\rho_{SM}$ is the contribution arising from SM corrections.

These relations have been used in [11] to study the possible values of the EWCL coefficients in the presence of a SM Higgs boson with a mass larger than the electroweak precision measurement limits. They find the following 68% CL allowed ranges for S and T, which could be converted to the ranges for the parameters α_1, β_1 using the Eqs. (21) and (22) above:

$$-0.37 < S < -0.17, \quad 0.34 < T < 0.58 \ . \tag{24}$$

The smallness of the parameter T can be understood in terms of the approximate $SU(2)_C$ custodial symmetry. This approximate symmetry implies that the coefficients $\alpha_{2,6,7,8,9,10,11}$ will be sub-dominant relative to the custodial-symmetry preserving ones. Gauge boson scattering can be considered to be dominated by the two coefficients α_4 and α_5.

There are constraints from unitarity, analyticity and causality of scattering amplitudes on the effective Lagrangian parameters [12,13]. The requirement of unitarity of the theory forces the cutoff of the EWCL to be about $\Lambda < 1.2$ TeV, but does not impose any constraints on the coefficients α_i, β_1. Causal and analytical structure of the amplitudes leads to bounds on the possible values of α_4 and α_5. These were first obtained in the context of chiral Lagrangians for strong interactions [14], and have been extended to the electroweak case. The second derivative with respect to the centre-of-mass energy of the forward elastic scattering amplitude of two Goldstone bosons is bounded from below by a positive integral of the total cross section for the transition $2\pi \to$ all [12]. However, it was shown by Vecchi [13] that stronger bounds are obtained from causality (we refer the reader to [13] for details):

$$\alpha_4(\mu) \geq \frac{1}{12\,(4\pi)^2} \ln \frac{\Lambda^2}{\mu^2} \ , \tag{25}$$

$$\alpha_4(\mu) + \alpha_5(\mu) \geq \frac{1}{8\,(4\pi)^2} \ln \frac{\Lambda^2}{\mu^2} \ . \tag{26}$$

Figure 1 taken from [15] shows the allowed regions in the plane of α_4 and α_5 after including constraints from causality and indirect bounds from precision experiments. The figure also includes a "black box", the region inaccessible at the LHC (see below).

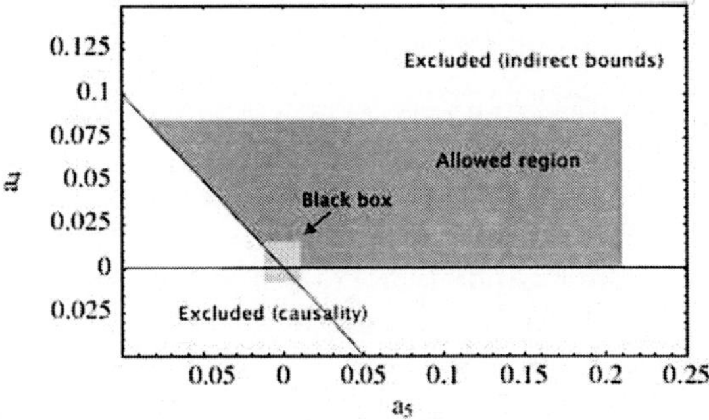

Figure 1. The region of allowed values in the $\alpha_4 - \alpha_5$ plane ($a_4 - a_5$ plane, in the notation of [15]), combining indirect bounds and causality constraints, is shown in gray. Also depicted is the region below which the LHC will not be able to resolve the coefficients ("black box"). From [15]

Further constraints on the parameters can be obtained from triple and quartic gauge boson couplings through direct experimental measurement. The quartic couplings, which correspond to gauge boson scattering, were studied in the LHC context in [16,17], where the second reference cited considered six-fermion final states at orders α_{em}^6 and $\alpha_{\text{em}}^4 \alpha_s^2$. It was found in [17] that with an integrated luminosity of 100 fb^{-1}, the LHC could place the following 99% confidence level bounds on the EWCL parameters α_4, α_5

$$-7.7 \times 10^{-3} < \alpha_4 < 15 \times 10^{-3} \ , \tag{27}$$

$$-12 \times 10^{-3} < \alpha_5 < 10 \times 10^{-3} \ . \tag{28}$$

These bounds correspond to assuming one parameter non-zero at a time.

In EWCL, one can build in unitarity at a given order by using a non-perturbative modification, which reduces to the original amplitude at the perturbative level [8]. One writes a low-energy expansion as $a(s) = a^{\rm LET}(s) + a^{(1)}(s)$, where $a^{\rm LET}$ represents the amplitude given by the low-energy theorems. At the lowest order the Padé approximant gives

$$a^{\rm Pade}(s) = \frac{a^{\rm LET}(s)}{1 - \frac{a^{(1)}(s)}{a^{\rm LET}(s)}} \,. \tag{29}$$

The K matrix method, which is simpler, consists in multiplying the amplitude by $-i$ and adding it to the denominator, so that the condition Im a^{-1} is satisfied. It gives

$$a^{K}(s) = \frac{a^{\rm LET}(s) + {\rm Re}a^{(1)}(s)}{1 - i(a^{\rm LET}(s) + {\rm Re}a^{(1)})} \,. \tag{30}$$

Both satisfy unitarity by construction to the relevant order.

A recent suggestion [18] has been to add resonances to the EWCL, in appropriate isospin channels so as to preserve custodial $SU(2)$ symmetry, parametrising them simply by their masses and widths. Ref. [18] uses a simple K matrix unitarisation. With an appropriate off-shell extension of the interactions, the authors have implemented the process of WW scattering in an event generator for six final-state fermions, it viz. WHIZARD [19].

2.2. Technicolour Models

Technicolour models are of the earliest proposals for dynamical breaking of the electroweak symmetry [20], without the introduction of elementary scalar fields. In analogy with QCD, local $SU(N)$ gauge invariance among technifermions in the fundamental representation leads to condensates which act as effective scalar fields and are responsible for breaking of the electroweak symmetry. The relevant coupling in technicolour theories is scaled up as compared to QCD to account for the electroweak scale. Thus, scattering of longitudinal gauge bosons can be strong.

The basic problem with technicolour, even when ordinary fermions acquire mass via an extended technicolour (ETC) sector [21], is that of large flavour-changing neutral currents and a large contribution to the S parameter. A proposal to address this problems has been walking technicolour (WT) [22] (in contrast to the usual theories which have renormalisation group "running"). Walking dynamics helps suppressing flavour-changing neutral currents without preventing ETC from yielding realistic fermion masses. Certain WT models are also in agreement with constraints imposed by precision electroweak data [23], because of the fact that walking dynamics itself can lower the contribution to the S parameter compared to a running theory. Various other modifications of WT theories have been considered recently to address different problems, and it seems that it is possible to construct working models of WT.

The scattering of Goldstone bosons representing the longitudinal gauge bosons was examined recently in [24,15]. In [24], the authors examine to what extent the violation of unitarity in WW scattering can be delayed in the case of a realistic WT model which satisfies existing phenomenological constraints. They consider three scenarios with values of the oblique parameter S 0.1, 0.2 and 0.3. Making use of Weinberg sum rules relating masses (M_V, M_A) and decay constants (F_V, F_A) of vector and axial-vector resonances, they find that unitarity violation can be delayed to as much as 3.7 TeV for $M_A = 1.5$ TeV. However, there is a critical value of the coupling $g \equiv 2M_V^2/F_V^2$ above which the theory violates unitarity at a much lower energy. Thus in the latter case, a spin-0 isospin-0 resonance would be required. On the other hand, for values of g below the critical value, the theory may very well be Higgsless. However it is entirely possible that the resonance being heavy may be difficult to observe.

2.3. Beyond SM Scenarios with a Light Higgs

There are two categories of theories which can have a light Higgs boson, with new physics in WW scattering above the light Higgs resonance. The first category corresponds to multi-Higgs models, where more than one doublet of scalar fields is introduced. In such a case WW scattering amplitude would grow after crossing the light Higgs resonance, but it could remain weak, if the heavier Higgs lies below the unitarity limit [4,5]. The other category corresponds to a scenario where there is one light Higgs boson, which is light because it is a pseudo-Goldstone boson corresponding to the breaking of a global symmetry. The Higgs could get mass only at the loop level. Above the light Higgs resonance, the WW scattering amplitude could be strong, and the resulting theory could have other resonances. The Little Higgs models, together with theories with holographic Higgs are studied in [6] under the category of "strongly-interacting light Higgs" theories.

In such theories, the new sector responsible for electroweak symmetry breaking, characterised by two parameters, a coupling g_ρ and a scale m_ρ describing the mass of heavy physical states. The coupling g_ρ, while larger than a typical SM coupling, is restricted to be below 4π to ensure that loop expansion remains dependable.

Experimental tests of such models would be observation of new states in addition to the light Higgs boson. However even in the absence of new states, $W_L W_L$ scattering amplitudes would be found to grow as E^2/f^2, where $f \equiv m_\rho/g_\rho$, since the Higgs boson moderates the high-energy behaviour only partially. Thus a study of gauge boson scattering would prove useful. It is also predicted that the there would be strong double-Higgs production through $W_L W_L \to HH$.

2.4. Higgsless Models

A number of Higgsless models have been proposed recently [25,26]. In these models, symmetry breaking is achieved by appropriate boundary conditions. The models differ in spatial dimensions, 5 in the original versions, 4 in the "deconstructed" versions. They also differ in embedding of SM fermions.

A feature common to such theories in five dimensions is that new weakly coupled particles, *viz.* the Kaluza-Klein excitations of the gauge bosons appear at TeV scale and cancel the bad high-energy behaviour of WW scattering amplitudes, and thus postpone unitarity violation. A version of the model with modified fermion sector can raise the scale of unitarity violation by at least a factor of 10 without running into conflict with precision electroweak constraints.

In the absence of Higgs, new massive vector boson (MVB) propagators contribute to WW scattering. The bad high energy behaviour of WZ scattering, for example, is cancelled by the contribution of the MVBs because of coupling constant sum rules [27]

$$g_{WWZZ} = g_{WWZ}^2 + \textstyle\sum_i (g_{WZV}^{(i)})^2 \,, \tag{31}$$

$$2(g_{WWZZ} - g_{WWZ}^2)(M_W^2 + M_Z^2) + g_{WWZ}^2 \tfrac{M_Z^4}{M_W^2} = \textstyle\sum_i (g_{WZV}^{(i)})^2 \left[3(M_i^{\pm})^2 - \tfrac{(M_Z^2 - M_W^2)^2}{(M_i^{\pm})^2} \right] \,. \tag{32}$$

Unitarity is violated at a scale

$$\Lambda \approx \frac{3\pi^4}{g^2} \frac{M_W^2}{M_1^{\pm}} \approx 5 - 10 \text{ TeV}. \tag{33}$$

The first MVB should appear below 1 TeV and thus accessible at the LHC. In the approximation that the first state V_1 saturates the sum rules, its partial width is given by

$$\Gamma(V_1^{\pm} \to W^{\pm} Z) \approx \frac{\alpha (M_1^{\pm})^3}{144 s_W^2 M_W^2} \,. \tag{34}$$

For $M_1^{\pm} = 700$ GeV, the width is about 15 GeV. In SM, there is no resonance in $W^{\pm} Z$ scattering

Even in Higgsless models, one has to distinguish between cases where $W_L W_L$ scattering is weak, and where it is strong. Thus, if the MVBs are light, with mass below 1 TeV, gauge boson scattering will be weak. However, if the MVBs are far above 1 TeV in mass, $W_L W_L$ scattering will be strong [28].

We now look at the practical aspects of actual observation of WW scattering at the LHC.

3. WW Scattering at the LHC

At a hadron collider like the LHC, WW scattering can occur with virtual W's emitted by the quarks in the hadrons. A W pair in the final state can be produced either through WW scattering diagrams, or through W emission from the partons of the initial hadrons. Figure 2 shows these two types of contributions. Figure 2(a) represents the genuine WW scattering diagrams, whereas Fig. 2(b) shows the "Bremsstrahlung" diagrams, which would be a background in the study of WW scattering.

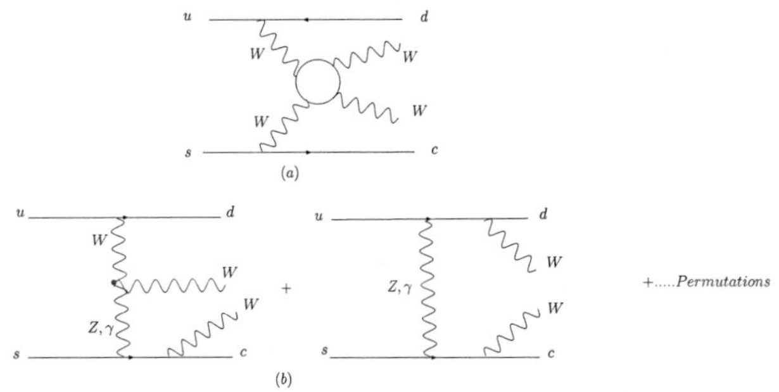

Figure 2. Main diagram topologies for the process $us \to cdW^+W^-$. From [29]

Thus, to study WW scattering at the LHC, one has to find ways of separating the genuine scattering contribution from the other "Bremsstrahlung" contributions, which is no mean task. In fact, as we shall see, the interference between these two classes of diagrams is large and huge cancellations can take place. Moreover, the gauge bosons taking part in the scattering are necessarily virtual (in fact, space-like). Thus the formalism has to take into account the virtuality of the gauge bosons.

A standard approach in the past to processes involving gauge bosons has been the equivalent vector-boson approximation, an extension of the equivalent-photon approximation used since early times [30].

3.1. Equivalent Vector-boson Approximation

The equivalent-photon approximation (Weizsäcker-Williams approximation) relates the cross section for a charged particle beam to interact with a target with a virtual photon exchange to the cross section for real photon beam to interact with the same target and produce the final state

$$\sigma = \int dx \sigma_\gamma(x) f_{e/\gamma}(x) . \quad (35)$$

Here the photon distribution with momentum fraction x in a charged-particle beam of energy E is given by

$$f_{e/\gamma}(x) = \frac{\alpha}{2\pi x} \ln\left(\frac{E}{m_e}\right) [x + (1-x)^2] . \quad (36)$$

This is generalised to what is known as an equivalent (or effective) vector-boson approximation (EVBA) for a process with weak bosons in place of photons [31,32]. The corresponding distributions of vector bosons V in a fermion f are given by [32]

$$\begin{aligned} f_{f/V_\pm}(x) &= \frac{\alpha}{2\pi x} \ln\left(\frac{E}{m_V}\right) \\ &\times \left[(v_f \mp a_f)^2 + (1-x)^2 (v_f \pm a_f)^2\right] \quad (37) \\ f_{f/V_L}(x) &= \frac{\alpha}{\pi x}(1-x)\left[v_f^2 + a_f^2\right] . \end{aligned}$$

Here the suffixes $\pm$ on V denote the helicities ± 1 of V and V_L is the state with helicity 0.

The use of EVBA entails (a) restricting to vector boson scattering diagrams, (b) neglecting diagrams of Bremsstrahlung type, (c) putting on-shell momenta of the vector bosons which take part in the scattering and (d) approximating the total cross section of the process $f_1 f_2 \to f_3 f_4 V_3 V_4$ by the convolution of the vector boson luminosities $\mathcal{L}^{V_1 V_2}_{s_1 s_2}(x)$ with the on-shell cross section

$$\begin{aligned} \sigma(f_1 f_2 \to f_3 f_4 V_3 V_4) &= \int dx \sum_{V_1, V_2} \sum_{s_1 s_2} \mathcal{L}^{V_1 V_2}_{s_1 s_2}(x) \\ &\times \sigma^{on}_{s_1 s_2}(V_1 V_2 \to V_3 V_4, x s_{qq}) . \end{aligned} \quad (38)$$

Here $x = M(V_1 V_2)^2 / s_{qq}$, while $M(V_1 V_2)$ is the vector boson pair invariant mass and s_{qq} is the square of the partonic c.m. energy, and s_1, s_2 are the spins of V_1, V_2. Note in the context of (c) above that the on-shell point $q^2_{1,2} = M^2_{V_{1,2}}$ is outside the physical region $q^2_{1,2} \leq 0$. Thus the extrapolation involved is more than that involved in the case of the equivalent-photon approximation.

Even if only the longitudinal polarisation, expected to be dominant, is kept, EVBA overestimates the true cross section. The transverse polarisation contribution is found to be comparable to the longitudinal one [32]. Improved EVBA [33], going beyond the leading approximation, still overestimates the cross section [34]. Further improvements in EVBA have been attempted [35].

3.2. Backgrounds

Backgrounds are of two types:

1. Bremsstrahlung processes – these are processes where the vector bosons are radiated by quark or anti-quark partons, and which do not contribute to VV scattering.
2. Processes which fake a VV final state.

It is important to understand the first inherent background, and device cuts which may enhance the signal. However, it may be possible to live with it – provided VV scattering signal is anyway enhanced because it is strong. In that case, one simply makes predictions for the combined process of $PP \to VV + X$. The second background is crucial to take care of, otherwise we do not know if we are seeing a VV pair in the final state or not.

Background processes are $q\bar{q} \to W^+W^-X$, $gg \to W^+W^-X$, $t\bar{t}$ + jet, with top decays giving W^+W^- pair. Electroweak-QCD process W^+ + jets can mimic the signal when the invariant mass of the two jets is around m_W. There is a potential background from QCD processes $q\bar{q}, gg \to t\bar{t}X, Wt\bar{b}$ and ($t\bar{t}$ +jets), in which a W can come from the decay of t or $\bar{t}$. W boson pairs produced from the intrinsic electroweak process $q\bar{q} \to q\bar{q}W^+W^-$ tend to be transversely polarised. Coupling to W^+ of incoming quark is purely left-handed. Helicity conservation implies that outgoing quark follows the direction of incoming quark for longitudinal W, and it goes opposite to direction of incoming quark for transverse (left-handed) W. Hence outgoing quark jet is less forward in background than in signal event, and tagging of the forward jet can help.

In addition, emission in the central region is favoured in the QCD background processes, whereas jet production in the central region is suppressed for WW scattering. Thus, a veto on additional jets in the central region would be a powerful discriminant between signal an background.

For a discussion of the WW scattering in the context of the LHC detectors [36].

3.3. Distinguishing the Signal

The feasibility of extracting WW scattering from experiment and comparison of EVBA with exact results was recently studied by Accomando and collaborators [29].

It is known that when W's are allowed to be off mass shell, the amplitude grows faster with energy, as compared to when they are on shell [37]. The problem of bad high-energy behaviour of WW scattering diagrams can be avoided by the use of the axial gauge [38].

Accomando *et al.* [29] have examined (a) the role of

choice of gauge in WW fusion, in particular, the axial gauge, (b) the reliability of EVBA, (c) the determination of regions of phase space, in suitable gauge, which are dominated by the signal (i.e. the WW scattering diagrams). Their results show that WW scattering diagrams do not constitute the dominant contribution in any gauge or phase space region. Thus, there is no substitute to the complete amplitude for studying WW fusion process at the LHC.

A similar discussion in the context of the EWCL including resonances, in contrast to the discussion of Accomando et al., is found in [18], where again the predictions of EVBA are compared with those obtained from the event generator WHIZARD.

We now describe some of the results of [29].

Table 1 shows the cross section in different gauges for the contribution of us partons to the full process $pp \to W^+W^-X$, to only the WW diagrams, and the ratio of these cross sections. The Higgs boson mass is assumed to be 200 GeV, and a cut on the WW invariant mass $M(WW)$ of 300 GeV is applied. It is clear that in the WW contribution is largely cancelled by the "Bremsstrahlung" type background contribution. Even in the axial gauge, in which the WW contribution is the least, it is still a factor of 2 larger than the actual cross section.

Gauge	$\sigma(pb)$ All diagrams	$\sigma(pb)$ WW diagrams	Ratio WW/all
Unitary	8.50×10^{-3}	6.5	765
Feynman	8.50×10^{-3}	0.221	26
Axial	8.50×10^{-3}	2.0×10^{-2}	2.3

Table 1. Contribution to the cross section from WW diagrams and all diagrams for Higgs mass of $M_H =$ 200 GeV and $M(WW) > 300$ GeV. Also shown is the ration of the WW contribution to the total contribution. From [29].

WW invariant-mass distributions are also obtained in [29]. Again the result including all diagrams does not give a true representation of the WW contribution alone. Accomando et al. have also made a comparison of the WW invariant mass distribution for the process $us \to dcW^+W^-$ in an improved EVBA with the exact complete result for the two cases of a very heavy Higgs and a Higgs of mass 250 GeV. EVBA exceeds the exact result except at the Higgs resonance. Ref. [29] also investigated the total cross section in EVBA and exact computation and their ratio for different cuts on the W scattering angle. The angular cuts serve to decrease the discrepancy between EVBA and exact computation.

The WW invariant mass distribution in the process $PP \to us \to cdW^+W^-$ is shown in Fig. 3.3, imposing the cuts E(quarks) > 20 GeV, P_T(quarks,W) $>$ 10 GeV, $2 < |\eta(\text{quark})| < 6.5$, $|\eta(\text{W})| < 3$. It is seen that for sufficiently large WW invariant mass it seems feasible to distinguish between light Higgs and heavy Higgs scenarios. It is reasonable to anticipate that this is the kinematic region where it may be possible to test non-standard scenarios of symmetry breaking.

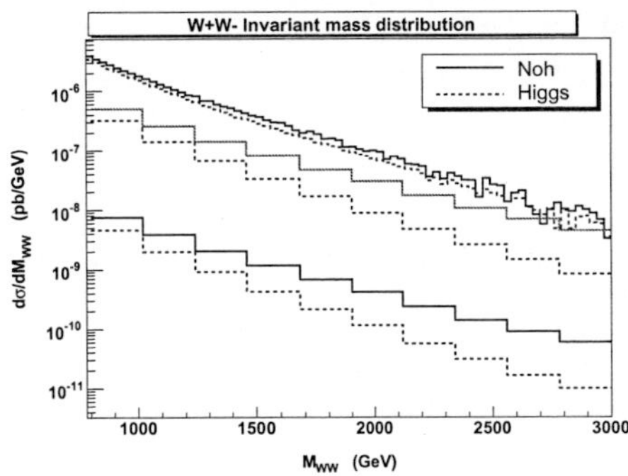

Figure 3. The WW invariant mass distribution in $PP \to us \to cdW^+W^-$ for no Higgs (solid curves) and for $M_h = 200$ GeV (dashed curve). The two intermediate (red) curves are obtained imposing cuts shown below. The two lowest (blue) curves refer to the process $PP \to us \to cd\mu^-\overline{\nu}_\mu e^+\nu_e$ with further acceptance cuts: $E_l > 20$ GeV, $p_T^\ell > 10$ GeV , $|\eta_l| < 3$

While the study of Accomando et al. is revealing, it may be argued that the alternative to SM with a light Higgs boson that they consider is a somewhat *ad hoc* and unphysical $m_H \to \infty$ limit of SM. In that context, the study carried out by Éboli, Gonzalez-Garcia and Mizukoshi [17] in the framework of EWCL may be considered more systematic. They impose jet acceptance cuts of $p_T^j > 20$ GeV, $|\eta_j| < 4.9$ and demand a rapidity gap between jets through $|\eta_{j1} - \eta_{j2}| > 3.8$, $\eta_{j1} \cdot \eta_{j2} < 0$. The lepton acceptance and isolation cuts used are $|\eta_\ell| <$ 2.5, $\eta^j_{\text{min}} < \eta_\ell < \eta^j_{\text{max}}$, with $\eta^j_{\text{min(max)}}$ as the minimum (maximum) rapidity of the tagging jets, $\Delta R_{\ell j} \geq 0.4$, $\Delta R_{\ell\ell} \geq 0.4$ and $p_T^\ell \geq p_T^{\text{min}}$, where $p_T^{\text{min}} = 100$ (30) GeV for opposite(equal) charge leptons. Since signal events contain neutrinos, they also require a missing transverse momentum $p_T^{\text{missing}} \geq 30$ GeV. In addition, to suppress the SM background as well as background from $t\bar{t}$ production, they impose additional cuts on the invariant masses of the pair of tagging jets, and of the W pair. We refer to the original paper for the details. However, here we limit ourselves to pointing out the sensitivity they obtain, *viz.* Eq. (27).

The results on the sensitivity obtained in Ref. [17] were made use of in [15] to discuss the feasibility of observation of two scenarios of SSB in the context of EWCL. One of these is the heavy Higgs scenario, where the Higgs mass is assumed to be between 2 and 2.5 TeV. While perturbative calculations are not reliable for such Higgs masses, it is presumed that some insight would be obtained into the strongly interacting behaviour. A calculation of the EWCL coefficients in this case yields values which, using Eq. (27) are too small to be observable at the LHC. A similar conclusion has been drawn in a generic technicolour type of model in the large N limit for an $SU(N)$ gauge theory for confinement of techni-fermions.

It seems that the prospects for the observation of non-trivial effects in WW scattering are limited to vector and scalar resonances, whatever the dynamics that produce the resonances. A concrete study in the context of EWCL with extra resonances is [18] and in the context of Higgsless models can be found in [39].

4. Discussion

In the absence of a light Higgs, WW interactions become strong at TeV scales leading to violation of perturbative unitarity. Study of WW scattering can give information of the electroweak symmetry breaking sector and discriminate between models.

There are a number of possible scenarios. In the SM with a light Higgs boson ($m_H \lesssim 1$ TeV), WW scattering is well-behaved at large energies. On the other hand, if the SM Higgs boson is heavier than about 1 TeV, WW interactions become strong. For extensions of SM, there are various possibilities. One possibility is that there are no elementary scalars and SSB is dynamical in origin. In that case, WW scattering would show new features which restore unitarity, possibly resonances. In theories with extra dimensions like Higgsless models, the violation of unitarity is delayed to higher energies because of the cancellation of the leading high-energy term by the exchange of Kaluza-Klein excitations of gauge bosons, which would be seen as resonances. A further possibility is that there is a light Higgs boson, which postpones unitarity violation, but there is new physics beyond. The new physics in its simplest form could be extra Higgs, and WW scattering could be weak or strong depending on the masses of these Higgs bosons. A more sophisticated possibility is that the light Higgs is a pseudo-Goldstone boson of some higher symmetry, and the WW scattering above the Higgs resonance would show interesting features. Thus WW scattering needs to be studied even if a light Higgs boson is found.

New physics could be modelled by means of an effective theory valid at low energies, *viz.* EWCL, whose lowest dimensional operators would be fixed by the low-energy theorems, but operators with higher dimensions would have coefficients fixed either from experiment, or on the basis of a detailed theory which describes the ultra-violet sector completely. Such a formalism has to be combined with a suitable method of unitarisation.

Two of the coefficients of higher dimensions are constrained by precision experiments. Two others would be constrained by the four-point couplings of the gauge bosons. Studies of popular scenarios of SSB beyond SM show that a determination of these latter couplings may pose a challenge to the LHC.

In general there are large cancellations between the scattering and Bremsstrahlung diagrams. Hence extraction of WW scattering contribution from the process $PP \to W^+W^-X$ needs considerable effort. EVBA overestimates the magnitude in most kinematic distributions. Appropriate cuts to reduce background are most essential. It is possible to extract information on WW scattering from hadronic experiments by concentrating on the large-invariant mass region.

Acknowledgements: I thank Andreas Nyffeler for comments and discussions. I also thank Namit Mahajan for discussions and pointing out Ref. [10].

REFERENCES

1. J. M. Cornwall, D. N. Levin and G. Tiktopoulos, *Phys. Rev. D* **10** (1974) 1145, *ibid.* **11** (1975) 972 (E); C. E. Vayonakis, *Lett. Nuovo Cim.* **17** (1976) 383; B. W. Lee, C. Quigg and H. B. Thacker, *Phys. Rev. D* **16** (1977) 1519; D. A. Dicus and V. S. Mathur, *Phys. Rev. D* **7** (1973) 3111; M. S. Chanowitz and M. K. Gaillard, *Nucl. Phys. B* **261** (1985) 379; H. G. J. Veltman, *Phys. Rev. D* **41** (1990) 2294. G. Gounaris, R. Kögerler, and H. Neufeld, *Phys. Rev. D* 34, 3257(1986); J. Bagger and C. Schmidt, *Phys. Rev. D* 41, 264 (1990); W. Kilgore, *Phys. Lett. B* 294, 257 (1992); H-J. He, Y-P. Kuang, X-Y. Li, *Phys. Rev. Lett.* 69, 2619 (1992); *Phys. Rev. D* 49, 4842 (1994); H-J. He and W. Kilgore, *Phys. Rev. D* 55, 1515 (1997) [arXiv:hep-ph/9609326]
2. S. Weinberg, *Phys. Rev.* **166** (1968) 1568
3. M. E. Peskin and T. Takeuchi, *Phys. Rev. D* **46**, 381

(1992); *Phys. Rev. Lett.* **65**, 964 (1990)
4. K. Cheung, C. W. Chiang and T. C. Yuan, arXiv:0803.2661 [hep-ph]
5. L. Randall, JHEP **0802** (2008) 084 [arXiv:0711.4360 [hep-ph]]
6. G. F. Giudice, C. Grojean, A. Pomarol and R. Rattazzi, *JHEP* **0706** (2007) 045 [arXiv:hep-ph/0703164]
7. T. Appelquist and C. W. Bernard, *Phys. Rev. D* **22** (1980) 200; A. C. Longhitano, *Nucl. Phys. B* **188** (1981) 118; *Phys. Rev. D* **22** (1980) 11; T. Appelquist and G. H. Wu, *Phys. Rev. D* **48** (1993) 3235 [arXiv:hep-ph/9304240]
8. A. Dobado, M. J. Herrero and J. Terron, Z. Phys. C **50** (1991) 205
9. G. F. Chew and S. Mandelstam, *Phys. Rev.* **119**, 467 (1960)
10. J. A. Oller, *Phys. Lett. B* **477**, 187 (2000) [arXiv:hep-ph/9908493]; K. i. Hikasa and K. Igi, *Phys. Rev. D* **48** (1993) 3055; *Phys. Lett. B* **261** (1991) 285 [Erratum-ibid. **270** (1991) 128]
11. J. A. Bagger, A. F. Falk and M. Swartz, *Phys. Rev. Lett.* **84** (2000) 1385 [arXiv:hep-ph/9908327]
12. J. Distler, B. Grinstein, R. A. Porto and I. Z. Rothstein, *Phys. Rev. Lett.* **98** (2007) 041601 [arXiv:hep-ph/0604255]
13. L. Vecchi, *JHEP* **0711** (2007) 054 [arXiv:0704.1900 [hep-ph]]
14. T. N. Pham and T. N. Truong, *Phys. Rev. D* **31**, 3027 (1985); B. Ananthanarayan, D. Toublan and G. Wanders, Phys. Rev. D **51**, 1093 (1995) [arXiv:hep-ph/9410302]; M. R. Pennington and J. Portoles, Phys. Lett. B **344**, 399 (1995) [arXiv:hep-ph/9409426]
15. M. Fabbrichesi and L. Vecchi, *Phys. Rev. D* **76**, 056002 (2007) [arXiv:hep-ph/0703236]
16. A. S. Belyaev, O. J. P. Éboli, M. C. Gonzalez-Garcia, J. K. Mizukoshi, S. F. Novaes and I. Zacharov, *Phys. Rev. D* **59** (1999) 015022 (1999) [arXiv:hep-ph/9805229]
17. O. J. P. Éboli, M. C. Gonzalez-Garcia and J. K. Mizukoshi, *Phys. Rev. D* **74** (2006) 073005 [arXiv:hep-ph/0606118]
18. A. Alboteanu, W. Kilian and J. Reuter, arXiv:0806.4145 [hep-ph]
19. W. Kilian, T. Ohl and J. Reuter, arXiv:0708.4233 [hep-ph]; T. Ohl, AIP Conf. Proc. **583**, 173 (2001) [arXiv:hep-ph/0011243]; M. Moretti, T. Ohl and J. Reuter, arXiv:hep-ph/0102195; J. Reuter, arXiv:hep-th/0212154
20. S. Weinberg, Phys. Rev. D **19**, 1277 (1979); L. Susskind, Phys. Rev. D **20**, 2619 (1979)
21. S. Dimopoulos and L. Susskind, *Nucl. Phys. B* **155**, 237 (1979); E. Eichten and K. D. Lane, *Phys. Lett. B* **90** (1980) 125
22. B. Holdom, *Phys. Rev. D* **24**, 1441 (1981); E. Eichten and K. D. Lane, *Phys. Lett. B* **90**, 125 (1980); K. D. Lane and E. Eichten, *Phys. Lett. B* **222**, 274 (1989)
23. R. Foadi, M. T. Frandsen and F. Sannino, *Phys. Rev. D* **77**, 097702 (2008) [arXiv:0712.1948 [hep-ph]]; R. Foadi, M. T. Frandsen, T. A. Ryttov and F. Sannino, *Phys. Rev. D* **76**, 055005 (2007) [arXiv:0706.1696 [hep-ph]]; D. D. Dietrich, F. Sannino and K. Tuominen, *Phys. Rev. D* **73**, 037701 (2006) [arXiv:hep-ph/0510217]
24. R. Foadi and F. Sannino, arXiv:0801.0663 [hep-ph]
25. C. Csaki, C. Grojean, H. Murayama, L. Pilo and J. Terning, *Phys. Rev. D* **69** (2004) 70 [hep-ph/0305237]; C. Csaki, C. Grojean, L. Pilo and J. Terning, *Phys. Rev. Lett.* **92** (2004) 101802 [hep-ph/0308038]
26. Y. Nomura, *JHEP* **11** (2003) 050 [arXiv:hep-ph/0309189]
27. A. Birkedal, K. Matchev and M. Perelstein, *Phys. Rev. Lett.* **94** (2005) 191803 [hep-ph/0412278]
28. M. S. Chanowitz, *Czech. J. Phys.* **55** (2005) B45 [arXiv:hep-ph/0412203]
29. E. Accomando, A. Ballestrero, A. Belhouari and E. Maina, *Phys. Rev. D* **74** (2006) 073010 [arXiv:hep-ph/0608019]
30. E. Fermi *Z. Phys.* **29** (1924) 315; C. F. von Weizsacker, *Z. Phys.* **88** (1934) 612; E. J. Williams, *Phys. Rev. D* **45** (1934) 729
31. S. Dawson, *Nucl. Phys. B* **249** (1985) 42; G. L. Kane, W. W. Repko and W. B. Rolnick, *Phys. Lett. B* **148** (1984) 367; J. Lindfors, Z. Phys. C **28** (1985) 427
32. R. M. Godbole and S. D. Rindani, *Phys. Lett. B* **190** (1987) 192; *Zeit. Phys. C* **36** (1987) 395
33. P. W. Johnson, F. I. Olness, and W.-K. Tung, *Phys. Rev. D* **36** (1987) 291
34. R. M. Godbole and F. I. Olness, *Int. J. Mod. Phys. A* **2** (1987) 1025
35. I. Kuss and H. Spiesberger, *Phys. Rev. D* **53** (1996) 6078 [arXiv:hep-ph/9507204]
36. ATLAS Collaboration, CERN/LHCC/99-15; CMS Collaboration, CERN/LHCC/94-38
37. R. Kleiss and W. J. Stirling, *Phys. Lett. B* **182** (1986) 75
38. Z. Kunszt and D. E. Soper, *Nucl. Phys. B* **296** (1988) 253
39. R. Malhotra, hep-ph/0611380

Signatures of Supersymmetry at the LHC

Amitava Datta[a], *Monoranjan Guchait*[b] *and Sourov Roy*[c]

[a]*Indian Institute of Science Education and Research, Salt Lake City, Kolkata-700106, India*

[b]*Department of High Energy Physics, Tata Institute of Fundamental Research, Mumbai-400005, India*

[c]*Department of Theoretical Physics and Centre for Theoretical Sciences, Indian Association for the Cultivation of Science, Kolkata-700032, India*

We briefly review the aspects of supersymmetry (SUSY) and SUSY breaking necessary for appreciating the prospect of discovering supersymmetric particles at the Large Hadron Collider (LHC), avoiding technicalities as far as practicable. After a brief survey of the generic signatures of SUSY at hadron colliders, we review the specific LHC signals of different models of SUSY breaking with emphasis on the *smoking gun* signals of each model, if any.

1. Introduction

Supersymmetry (SUSY) is a novel symmetry that relates bosons and fermions. Under a SUSY transformation a spin half electron, for example, changes into its superpartner: a spin-zero selectron. Historically SUSY was introduced purely for its theoretical elegance [1]. Quite remarkably subsequent researches revealed that it not only removes several shortcomings of the existing theories of particle physics but predicts a whole host of new phenomena which can be tested by experiments at high energy or by cosmological experiments [2]. The discovery of the superpartners as well as their properties is, therefore, a high priority programme for the physics at the Large Hadron Collider (LHC).

The Standard Model (SM) of particle physics (see below) has been very successful in explaining all the experimental data accumulated over the years. The list of such experiments is indeed impressive. It includes typical laboratory based low energy experiments like atomic parity violation [3] as well as accelerator-based experiments carried out at the highest energy attainable so far. The latter includes experiments at the Large Electron Positron (LEP) collider (CM energy approximately 200 GeV) at CERN which were completed a few years ago[1] and the ones at the $p\bar{p}$ collider Tevatron (CM energy approximately 2000 GeV or 2 TeV) which have been running at the Fermilab for the last 20 years or so [4].

One sector of the SM proposed by Glashow, Salam and Weinberg contains the unified theory of electroweak (EW) interactions. This sector known as the EW sector, contains three generations of spin-half quarks and leptons. The particles belonging to the first-generation are listed in Table 1. The symbols for the second-generation of particles are obtained by the replacements u (up) $\rightarrow$ c (charm), d (down) $\rightarrow$ s (strange) and e (electron) $\rightarrow$ μ (muon). The third-generation particles are denoted by u $\rightarrow$ t (top), d $\rightarrow$ b (bottom) and e $\rightarrow$ τ (tau-lepton).

Names	spin 0	spin 1/2
squarks, quarks	$(\tilde{u}_L\ \tilde{d}_L)$	$(u_L\ d_L)$
	$\tilde{u}_R$	u_R
	$\tilde{d}_R$	d_R
sleptons, leptons	$(\tilde{\nu}_e\ \tilde{e}_L)$	$(\nu_e\ e_L)$
	$\tilde{e}_R$	e_R

Table 1. Quarks and leptons belonging to the first-generation of the standard model and their superpartners.

There are also spin-one force carriers or gauge bosons. They are the Ws ($W^{\pm}, W^0$) and the B^0 (Table 2). The physical gauge bosons, which mediate EW interactions among the particles, are $W^{\pm}$, Z^0 and the photon (γ), the last two being orthogonal linear combinations of W^0 and B^0.

In addition there is the spin-zero Higgs boson[2] which generates the masses of all fermions and gauge bosons of the SM via spontaneous symmetry breaking. The masses of the heavier particles in the SM are $\sim$ 100 GeV. Qualitatively speaking, this mass or energy scale characterises the energy scale of the SM better known as the EW scale.

The remaining sector of the SM describes the strong

[1]For details see the article by A. Gurtu in this volume.

[2]For more about the Higgs bosons see the article by Djouadi and Godbole in this volume.

Names	spin 1/2	spin 1
gauginos, gauge bosons	$(\widetilde{W}^{\pm}\ \widetilde{W}^{0})$ $\widetilde{B}$ $\widetilde{g}$	$(W^{\pm}\ W^{0})$ B g

Table 2. Gauge bosons of the standard model and their superpartners.

interaction among the quarks only. This interaction is mediated by a set of eight gauge fields collectively called the gluons (g). The theory of quarks and gluons is known as quantum chromodynamics (QCD) [3]

The supersymmetric partners of the particles belonging to the SM – popularly known as the superparticles or simply the sparticles are also shown in Tables 1 and 2. The Higgs sector of any supersymmetric model is more complicated than simply adding the superpartner of the single neutral Higgs boson in the SM. Since the Higgs sector is not our main concern, we refer the reader to the article by Djouadi and Godbole in this volume for the details.

The triumph of SUSY began by alleviating a pathological feature of the Higgs sector of the SM now known as the mass hierarchy problem. As in any quantum field theory, the mass of the Higgs boson in the SM receives quantum corrections called self energy corrections. It is well known that such corrections, except for very exceptional cases, turn out to be infinite because the self energy is determined by certain integrals which diverge. Of course in a renormalisable field theory, like the SM, a finite result can be obtained by the renormalisation prescription. The infinite self energy of the Higgs boson by itself is, therefore not a problem technically.

The problem begins when we realise that the SM cannot be the last word on nature. Surely at least the gravitational interactions, which cannot be embedded in the SM or in any renormalisable field theory for that matter are always there. The energy at which the gravitational interactions become strong is the Planck scale ($M_P \sim 10^{18}$ GeV). Since the SM certainly is not the only relevant theory at this very high energy, the field theoretic techniques applicable at lower energies are not reliable any more. The sensitivity of the magnitude of the Higgs boson self energy to the new energy scale can be estimated by putting M_P as the upper limit of the above divergent integral. The result is a disaster: the correction turns out to be proportional to M_P^2!

On the other hand although the Higgs boson is yet to be discovered there are strong indications from the LEP experiments that its mass cannot be much more than a few hundred GeV. This run away behaviour of the Higgs boson mass in the presence of new physics at a energy scale much higher than the characteristic energy scale of the SM, is known as the hierarchy problem [5]. It should be emphasised that the self energies of the fermions or the gauge bosons in the SM are free from this problem because of appropriate symmetries of the theory. On the other hand there is no corresponding symmetry which protects the Higgs boson mass. A solution of this hierarchy problem in terms of a new symmetry is, therefore very welcome. The self energy of the Higgs boson (ϕ) can be diagrammatically represented, for example, by the Feynman diagram in Fig. 1a, where a fermion (f), say a quark, circulates in the loop. Suppose there is an additional contribution to the self energy from a diagram with a spin-0 boson ($\tilde{f}$) having exactly the same mass as f circulating in the loop (Fig. 1b). Moreover let the couplings of the boson and the fermion with ϕ denoted by λ_s and λ_f respectively, be related in a certain way. A little experience in QFT would then tell us that the two contributions would neatly cancel and the dangerous contribution proportional to M_P^2 disappears. This is exactly what SUSY provides for us: corresponding to every fermionic (bosonic) particle there is a bosonic (fermionic) superpartner of exactly the same mass! Moreover SUSY relates the coupling strengths of any particle and its superpartner in such a way that each and every dangerous contribution to the self energy of ϕ in the SM is canceled. The solution of the hierarchy problem [6] triggered a renewed interest in SUSY. Realistic models of SUSY like the minimal supersymmetric extension of the SM (MSSM) were constructed and testable predictions of the models were computed with great enthusiasm.

SUSY cannot be an exact symmetry. Had there been a spin zero selectron as light as the electron it would have been discovered long before. Thus SUSY must be a broken symmetry with the sparticle masses significantly larger than the particle masses. However even in this case the hierarchy problem provides an important guideline. If f and $\tilde{f}$ have different masses, as they should if the symmetry is broken, the dangerous $\sim M_P^2$ terms in the Higgs boson self energy still cancel. But the remaining dominant contribution to the Higgs boson self energy is $\sim m_{\tilde{f}}^2 ln(M_P/m_{\tilde{f}})$, where $m_{\tilde{f}}$ is assumed to be much larger than m_f. The logarithmic dependence on M_P obviously softens the hierarchy problem. But if $m_{\tilde{f}}$ is much larger than a few TeV,

[3]See the article by Mathews and Ravindran in this volume for further details and references.

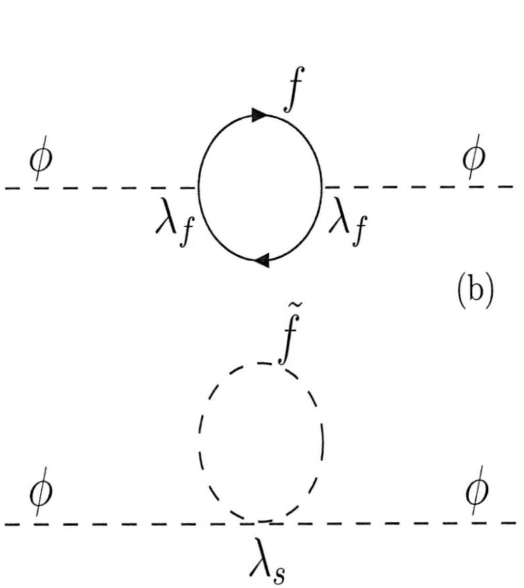

Figure 1. Cancellation of the dangerous contribution to the Higgs boson self energy in a supersymmetric theory (see the text for details)

one lands into a milder version of the hierarchy problem again and the Higgs boson mass tends to run away from the EW scale. This observation kindles the hope that if SUSY exists in nature, the sparticle masses cannot be much larger than the corresponding particle masses and hence the sparticles are likely to be accessible to the LHC with an unprecedented CM energy of 14 TeV.

The most general MSSM consistent with the symmetries of the SM turns out to be unacceptable since it contains baryon and lepton number violating interactions. Such interactions would spoil, for example, the stability of the proton. One can, however impose additional symmetries to get rid of these dangerous interactions. These symmetries introduce a new multiplicatively conserved quantum number called R-parity. It turns out that the SM particles have R-parity = 1 while the sparticles have R-parity = −1. A direct consequence of this new symmetry is that if a high energy collision of two particles (f_1 and f_2) produces sparticles, they must be produced in pairs, i.e.

$$f_1 + f_2 \to \tilde{f}_1 + \tilde{f}_2 \,, \tag{1}$$

where $\tilde{f}_1$ and $\tilde{f}_2$ are sparticles.

Another important consequence of the new symmetry is that any sparticle must decay into another sparticle along with one or more particles. This in turn implies that the lightest supersymmetric particle (LSP) must be stable. Now the LSP is very likely to be the superpartner of a neutral weakly interacting SM particle (Section 3.1). In that case this particle, if produced in colllider experiments, either directly or through the decays of heavier sparticles, will escape detection, resulting in apparent imbalance of energy and momentum. The presence of large missing energy in experiments involving the production and subsequent decay of sparticles is regarded as the hallmark of R-parity conserving SUSY. In this article we shall always assume R-parity conservation.

It is well-known that a large fraction of our universe is made of dark matter and dark energy. We know that the present universe is filled with left-overs from the evolution of the early universe following the Big Bang. The best known example of such relic particles is the cosmic microwave background. If the dark matter is also made of left-over particles of a certain type, they must be stable and weakly interacting (otherwise they would have been discovered long before). The LSP in the R-parity conserving SUSY appears to be tailor made as a dark matter candidate. Of course in order to account for the observed dark matter relic density quantitatively it must have mass and interaction strength in certain ranges. This pushes search for SUSY up by several rungs in the list of priorities for the LHC experiments. The discovery of SUSY at the LHC will be a great progress in our understanding of nature. The subsequent measurement of the properties of the sparticles - the LSP in particular - and quantitative verification of the conjecture of SUSY dark matter will be a cherry on the pudding[4].

The Achilles heel of any theory based on SUSY, however is the absence of a universally accepted guideline for SUSY breaking consistent with the presently available constraints on the sparticle masses. The implementation of spontaneous breakdown of SUSY in a realistic model without introducing additional unknown parameters would have been a great step forward. However, several attempts have failed to produce a consistent mass spectrum for the sparticles. For example, the SUSY breaking order parameter cannot belong to any of the MSSM supermultiplets, consisting of e.g. squark and quarks, because of certain mass sum rules. One such sum rule requires that the masses of the quarks and the squarks belonging to a generation satisfy the relation $m^2_{\tilde{u}_1} + m^2_{\tilde{u}_2} - 2m^2_u + m^2_{\tilde{d}_1} + m^2_{\tilde{d}_2} - 2m^2_d = 0$ (the

[4]For further details on dark matter and SUSY see the article by Baer and Tata in this volume.

notation will be further clarified in section 3). This is of course ruled out by experiments. However, this equation holds only at tree level and only for renormalisable theories. Then there exists a possibility that spontaneous SUSY breaking occurs in a sector which couples to the MSSM sector only via loops or via non-renormalisable interactions. These possibilities will be discussed in some details in the following sections.

At this stage the only option left is to add certain soft SUSY breaking terms to the theory without spoiling the symmetries of the SM and the renormalisability of the theory. Soft SUSY breaking masses - often called soft masses - which introduce the mass difference between particles and their superpartners are examples of such parameters. The number of new unknown parameters so added to the most general MSSM, however turns out to be more than hundred! The list is indeed too long for a theory claiming to be the fundamental description of nature.

Apart from the issue of aesthetics there are practical problems too. Arbitrary, random soft parameters in the MSSM Lagrangian potentially lead to flavour or CP violating processes with large probabilities. However, these processes are known to be severely suppressed even in comparison with the usual weak processes like μ decay and π decay. In order to understand this issue one should recall that once SUSY is allowed to be broken arbitrarily the masses of the sparticles also become arbitrary. This may lead to flavour-changing neutral current (FCNC) induced processes, such as $\mu \to e\gamma$, K^0–$\bar{K}^0$ mixing etc. The latter process, for example, can be suppressed if the masses of the $\tilde{u}$ and $\tilde{c}$ squarks happen to be degenerate to a very good approximation. Other FCNC processes can be similarly suppressed by requiring suitable pairs of other sparticles to be almost mass degenerate. Large CP violating effects can be avoided by assuming that the soft parameters do not introduce new complex phases.

From the physicist's point of view, however such degeneracies without an underlying symmetry is not satisfactory at all. One would like to find either a symmetry or a dynamical reason which would explain why apparently different parameters have nearly the same values. For example, these dangerous FCNC induced effects in the MSSM can be satisfactorily removed if one can show that the soft SUSY breaking parameters like the squark masses are universal because the dynamics at the SUSY breaking scale, which is much higher than the EW scale, is flavour blind (i.e. it does not distinguish between squarks of different types). It has also been noticed that if physics at some higher scale is indeed responsible for SUSY breaking, then a few other puzzles plaguing non-supersymmetic theories become non-issues in supersymmetric theories. Several models belonging to this class will be summarised in sections 2 and 4.

From the last paragraph it is clear that the sparticle mass spectrum and consequently, the production and decay of sparticles at the LHC are by and large model dependent. Thus the discovery of some generic SUSY-like signal, e.g. the observation of events having much larger missing energy than typical SM processes, though spectacular, will not automatically herald the nature of the underlying SUSY theory. Painstaking reconstruction of the sparticle masses, their spins and other properties will be required to identify the specific model of SUSY breaking. This makes the search for SUSY at the LHC even more exciting and challenging. Fortunately there are "smoking gun signals" of some of these models. The observation of these would strongly hint to the specific model without waiting for the full reconstruction of the model parameters. In section 3, we shall first summarise the generic SUSY signals. Then we shall take up the issue of signatures of specific models including the smoking gun signals, if any, in sections 3 and 4.

2. SUSY Breaking at a High Scale

In quantum field theory parameters like mass and charge (or any other coupling constants) are energy dependent. This prediction has indeed been verified by experiments. Low energy experiments, e.g. have measured the fine structure constant (α) very accurately and its value is well known ($\approx 1/137$). However the value of α as measured by the LEP experiments at a CM energy of approximately 100 GeV is significantly larger ($\approx 1/128$). Another well known example is the variation of the coupling strength of quark-gluon interactions in QCD. This coupling, however decreases with increasing energy and is rather small for quark-gluon interactions at very high energies (footnote 3).

The variation of the masses and coupling constants with energy in a particular theory can be theoretically studied by a set of coupled differential equations known as the renormalisation group (RG) equations [7]. The predictions of the RG equations of various theories - QCD in particular - have been verified by a large number of experiments operating at different energies.

That SUSY may be more attractive in presence of new physics at much higher energies is indicated by the supersymmetric generalisation of any grand unified theory (GUT) [8] of the simplest type [2].[5] The cou-

[5]These are the so called grand desert type theories in which there

pling constants of strong, electromagnetic and weak interactions measured at currently attainable energies of course have widely different magnitudes. However the weaker (stronger) couplings increase (decrease) with energy (see the examples discussed above). This suggests the interesting possibility that these interactions have a common strength at a much higher energy called the GUT scale ($M_G \sim 10^{16}$ GeV). This coupling constant unification, however fails in non-supersymmetric GUTs of the simplest kind. It is well-known that unification can be achieved in supersymmetric GUTs of the simplest variety provided the sparticle masses are ~ 1 TeV. The essential reason is that the sparticles contribute to the RG evolution of the couplings and these contributions are important to make them unify at the right scale (M_G).

The idea that the impact of physics at very high energies, much higher than the currently attainable energies, on physics at currently available energies can be studied via the RG equations, have led to several interesting theories of SUSY breaking. The main point is that certain interactions at very high energies may introduce a rather simplified pattern of SUSY breaking involving only a few free parameters. The prediction for the low energy spectrum may then be obtained by the RG equations. Alternatively one may determine the masses of the sparticles from experiments at a lower energy. One can then evolve these masses to a high scale via the RG equations and check whether they exhibit any special feature. This information may eventually reveal the underlying SUSY breaking model.

The models of high scale SUSY breaking have a generic feature. Each consists of a 'hidden sector' which does not interact with the 'observable sector' consisting of the particles belonging to the MSSM. The hidden sector particles are assumed to be too heavy to be observed in accelerator based experiments in the near future. SUSY is broken in the hidden sector spontaneously. The central question of SUSY has now been changed : "How does the *observable sector* know about SUSY breaking?" The mechanism of mediating SUSY breaking from the hidden sector to the observable sector via some 'messenger fields' which couples to both the hidden and the observable sectors are different in different models. This leads to characteristic MSSM soft terms and, consequently, sparticle masses and collider signals in different models.

The most popular example of mediating SUSY breaking from the hidden to the visible sector is via gravitational interactions which play the role of the messenger. Here the new physics enters near the Planck Scale (M_P). If SUSY is broken in the hidden sector by a vacuum expectation value (VEV) $\langle F \rangle$, where F is a hidden sector field, then the soft terms in the visible sector should be $m_{\text{soft}} \sim \langle F \rangle / M_{\text{P}}$, by dimensional analysis. Hence, if we demand m_{soft} of the order of a TeV, then the scale associated with the origin of SUSY breaking in the hidden sector should be $\sqrt{\langle F \rangle} \sim 10^{10}$ or 10^{11} GeV.

One can argue, somewhat naively, that since gravitational interactions does not distinguish between different scalar particles in the MSSM, all such particles (the squarks, sleptons and the Higgs bosons) acquire the same mass (m_0) at a high scale via this mechanism. This is already encouraging because a common mass of the sfermions belonging to the first two generations strongly hints towards a natural mechanism for the suppression of the dangerous FCNC processes discussed in the introduction. A similar argument would indicate that the soft breaking masses for all the spin-1/2 gauge fermions or the gauginos would be the same ($m_{1/2}$). Admittedly this argument is oversimplified. In order to realise universal masses for the scalars and the gauginos one needs additional simplifying assumptions for hidden sector interactions. This model is popularly known as the minimal supergravity model or the mSUGRA model [9].

In the mSUGRA model, the free parameters at high energy, commonly chosen as the GUT scale are

$$m_0, m_{1/2}, A_0, \tan\beta, sign(\mu)\,,$$

where m_0 and $m_{1/2}$ are the common mass of the scalars and gauginos respectively. The parameter A_0 is another soft SUSY breaking parameter known as the trilinear coupling. The ratio of the VEVs of the two neutral Higgs bosons (footnote 2) is denoted by $\tan\beta$. The parameter μ respects SUSY and is commonly referred to as the Higgsino mass parameter as it contributes to the masses of the superpartners of the Higgs boson. The masses of sparticles and couplings at the energy scale of experimental interest can be obtained via the RG evolutions of these parameters [10]. Some example of masses at the EW scale are

$$m_{\tilde{g}} \approx 2.7 m_{1/2}\,, \tag{2}$$

where $M_{\tilde{g}}$ is the mass of the gluino, the superpartner of the gluon. The squark masses of the first two generations are

$$m^2_{\tilde{u}_L} \approx m_0^2 + 4.9 m^2_{1/2}, \quad m^2_{\tilde{u}_R} \approx m_0^2 + 4.6 m^2_{1/2}\,. \tag{3}$$

It turns out that the up-type squarks belonging to the second generation ($\tilde{c}_L$ and $\tilde{c}_R$) have exactly the same

is no new physics between the EW scale and the GUT scale M_G.

mass. The masses of the down type squarks ($\tilde{d}_L$, $\tilde{d}_R$, $\tilde{s}_L$ and $\tilde{s}_R$) have the same mass as the corresponding up type squark except for relatively small EW symmetry breaking corrections. The mass degeneracy between u and c type squarks and that between d and s type squarks are sufficient to suppress the FCNC processes. This is one of the attractive features of the mSUGRA model.

The numerical coefficients in Eqs. (2) and (3) are determined by the gauge interactions of the sparticle concerned and the universal strong interaction determines their magnitudes to a large extent although there are sub-dominant EW contributions. This in fact is the main reason for the near degeneracy of all squarks belonging to the first two generations.

In principle the Yukawa interactions of the fermions in the SM also contribute to the RG evolutions of the soft masses of their superpartners. But these interactions proportional to the fermion masses are practically negligible for the light quarks of the first two generations. However the situation is very different for the third-generation. For example, large m_t (the top quark mass) ensures that the contribution of the top quark Yukawa coupling to the evolution of the masses, e.g. of the top squarks ($\tilde{t}_L$ and $\tilde{t}_R$) may make these sparticles significantly lighter than the squarks belonging to the first two generations. The experimental constraints on the FCNC processes mediated by the $\tilde{t}$ squarks are not very stringent and relatively light $\tilde{t}_{L,R}$ are consistent with the data.

We recall that in the SM a negative mass squared parameter of the Higgs boson is introduced rather arbitrarily. This leads to the spontaneous breakdown of EW symmetry which yields positive masses to all particles of the SM including the Higgs boson. In mSUGRA the Higgs mechanism occurs naturally. The mass squared of the Higgs boson (m_0^2) is positive at the GUT scale. However the subsequent RG evolution makes it negative at the EW scale. This triggers spontaneous break down of EW symmetry. This radiative EW symmetry breaking mechanism [11] adds one more feather to the cap of the mSUGRA model.

In the next section we shall review the expected signatures of the mSUGRA model at the LHC and in section 4 we shall do the same for others models of SUSY breaking.

3. SUSY Searches at the LHC

Two collaborations ATLAS and CMS[6] will independently look for signals of SUSY and other new physics candidates, if any, with all purpose detectors. At the LHC, two protons each of energy 7 TeV, will collide head on at a centre of mass energy $\sqrt{S}$ = 14 TeV and they will break up into constituents of the proton, namely, gluons, quarks and antiquarks, collectively known as partons. It is mainly the strong interaction of the partons via QCD, which leads to observable processes with large cross sections. However as we shall see below, the EW interactions of quarks and antiquarks also produce important signals al beit with smaller cross sections.

At the LHC the particles of the SM like top quarks, gauge bosons (W,Z), etc. will be produced in very large numbers. The sparticles, if they exist will also be produced, but with somewhat smaller cross sections since they are likely to be significantly heavier than the corresponding particles. Separating the signatures of sparticle production from the huge SM background is the main challenge confronting the SUSY searchers at the LHC.

In order to fully appreciate the mass reach of the LHC, let us briefly recollect the sparticle mass limits from LEP and Tevatron Run II – the best limits obtained to date. The highest CM energy attained by LEP was 209 GeV and the mass limts on most of the sparticles accessible to LEP were extended almost to the kinematic limit (i.e. ≈ 100 GeV) in the clean environment of an $e^+ - e^-$ collider[7] [12,13].

The CDF and D0 collaborations[8] have been looking for the sparticles since the dawn of the Tevatron experiments nearly 20 years ago [13]. As noted earlier these limits are usually model dependent. Assuming that there are five flavours of squarks of L and R type and each has approximately the same mass as the gluino ($m_{\tilde{q}} \approx m_{\tilde{g}} = \tilde{m}$) the CDF collaboration obtained the limit $\tilde{m} > 392$ GeV. For heavier squarks $m_{\tilde{q}} = 600$ GeV, the gluino mass limit is $m_{\tilde{g}} > 280$ GeV [14]. The D0 collaboration obtained similar limits [15].

3.1. Sparticle Production at the LHC

If the mass of the strongly interacting sparticles (the gluinos ($\tilde{g}$) and the squarks ($\tilde{q}$), the superpartners of gluons and quarks respectively) are within the kinematic reach of the LHC, which roughly corresponds to $m_{\tilde{g}}, m_{\tilde{q}} \leq 2.5$ TeV, they will produce observable signals. As already noted in the introduction these sparticles

[6]A large number of Indian physicists from different research institutes and universities are members of the CMS collaboration.

[7]The limits on the LSP and the sneutrino masses are more model dependent and may be well below the kinematic limit in some scenarios.

[8]A large number of Indian physicists from different universities and institutes have been participating in the D0 experiment.

will be produced in pairs according to R-parity conserving SUSY models. At the partonic level some typical reactions are

$$q\bar{q},\ gg \to \tilde{g}\tilde{g}, \tilde{q}\tilde{q}^*, \tag{4}$$

$$qq \to \tilde{q}\tilde{q}, \tag{5}$$

$$qg \to \tilde{q}\tilde{g}, \tag{6}$$

$$q\bar{q},\ gg \to \tilde{t}_1\tilde{t}_1^*. \tag{7}$$

Since the dynamics of QCD at high energies is fairly well known the cross sections of the above processes are essentially controlled by the masses of the final state sparticles [16]. These masses, in turn are determined by the SUSY breaking mechanism. Signals of sparticle production is, therefore somewhat model dependent. In this section we shall, however concentrate on some generic features of sparticle production which are fairly model independent.

The inclusion of the process in Eq. (7), which is nothing but a special example of $\tilde{q}\tilde{q}^*$ production, deserves further clarifications. In the MSSM, the third-generation squarks ($\tilde{q}_L, \tilde{q}_R$; Q = t,b) require special attention. After soft SUSY breaking, many of the sparticles in Tables 1 and 2 do not represent the physical states. In fact suitable mixtures of the states listed in the above tables correspond to the observable states expected to be produced at the LHC or other future colliders. Consider the L and R type sfermions $\tilde{f}_L$ and $\tilde{f}_R$, which are the superpartners of the chiral fermions f_L and f_R, where f represents any quark or lepton flavour in the SM. These sfermions can mix via the mass squared matrix

$$M^2_{\tilde{f}} = \begin{bmatrix} m^2_{LL} & m_f\tilde{A}_f \\ m_f\tilde{A}_f & m^2_{RR} \end{bmatrix}, \tag{8}$$

where the matrix elements are given by

$$\begin{aligned} m^2_{LL} &= m^2_f + m^2_{\tilde{f}_L} + (I^3_f - q_f\sin^2\theta_W)\cos 2\beta\, M^2_Z \\ m^2_{RR} &= m^2_f + m^2_{\tilde{f}_R} + q_f\sin^2\theta_W\ \cos 2\beta\, M^2_Z \\ \tilde{A}_f &= A_f - \mu(\tan\beta)^{-2I^3_f}. \end{aligned} \tag{9}$$

Here $m_{\tilde{f}_L,\tilde{f}_R}$ are the soft masses for the L and R type sfermions, m_f is the mass of the corresponding fermion (e.g. a quark or a lepton) and A_f is another SUSY breaking parameter. The parameter μ and tan β have already been defined. The other SM parameters appearing in Eqs. (8) and (9) are not particularly relevant for our present discussion. The effects of mixing depend on m_f in the off diagonal terms. These terms are therefore, much smaller than $m_{\tilde{f}_L,\tilde{f}_R}$ for the first two generations. Only for the mixing between the superpartners of the top quark the off diagonal terms in the mass matrix could be important due to large m_t (the soft breaking term A_t may also be large). This may lead to sizable splitting between two physical states $\tilde{t}_1$ and $\tilde{t}_2$.[9] In mSUGRA type models the top squarks may also be light due RG evolution of their masses (Section 2). For a large region of the parameter space the lighter physical state ($\tilde{t}_1$) can be significantly lighter than the other squarks due to these two effects. Therefore light $\tilde{t}_1$ pairs can be produced with substantial rates at the LHC.

In Fig. 2 the lowest order cross sections for gluino and squark pair production in various combinations are shown along with that for $\tilde{t}_1 - \tilde{t}_1{}^*$ pair production [17]. In this figure $\tilde{q}$ collectively represent degenerate L and R type squarks of all flavours except $\tilde{t}$. It is also assumed that $m_{\tilde{g}} = m_{\tilde{q}}$. Notice that the cross sections fall rapidly with the masses. For $m_{\tilde{g}}, m_{\tilde{q}}$ = 1 TeV, e.g. the total squark-gluino pair production cross section is $\sim$ 4–5 pb. Thus for an integrated luminosity of $\mathcal{L}$ = 10 fb^{-1} about 40,000–50,000 squark-gluino pairs will be produced.[10] For the same $m_{\tilde{t}_1}$, about 100 stop squark pairs are expected.

The existence of a light top squark may lead to several other important consequences. For example in some models $\tilde{t}_1$ may turn out to be the next-to-lightest supersymmetric particle (NLSP). This leads to the interesting possibility that the $\tilde{t}_1$ - LSP annihilation might have been the dominant dark matter producing process in the early universe (see the article by Baer and Tata in this volume for further details). Other important consequences of the light $\tilde{t}_1$ scenario will be reviewed later.

In addition to the dominant production processes, Eqs. (4, 5, 6 and 7), the EW gauginos - two charginos ($\tilde{\chi}^{\mp}_i$; i = 1–2) and four neutralinos ($\tilde{\chi}^0_j$; j = 1–4) may also be produced with smaller but observable rates. The indices i and j increase in the order of increasing mass of the sparticle. The charginos are linear

[9]The mixing between the superpartners of the bottom quark and the τ lepton may also be large especially if tan β is large.

[10]Luminosity is an important parameter of any collider experiment. The product of the cross section of a process and luminosity gives the number of events produced per second. Integrated luminosity is the luminosity accumulated over some interval of time. At the LHC $\mathcal{L}$ = 10 fb^{-1} will be accumulated in about two years (1 fb^{-1} = 1000 pb^{-1}) inspite of the low luminosity phase during the first two-three years. This should be compared with the rather optimistic goal of collecting 9 fb^{-1} at the Tevatron Run II by the end of 2010 after approximately nine years of running. In the high luminosity phase the LHC is expected to collect 100 fb^{-1} per year.

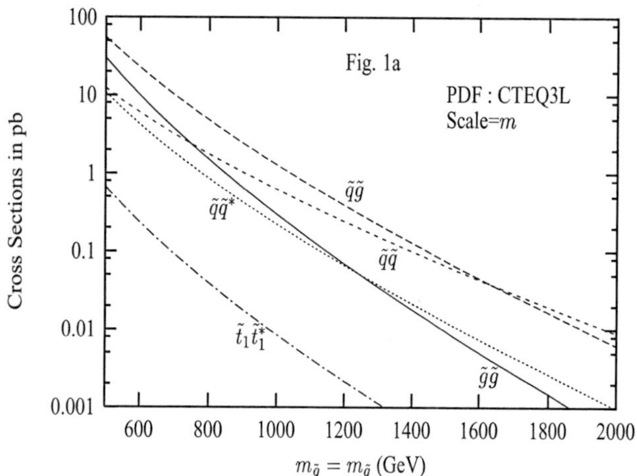

Figure 2. Gluino and squark pair production cross sections at the LHC, from [17]

combinations of $\tilde{W}^{\pm}$ (the superpartners of the charged W bosons) and $\tilde{H}^{\pm}$ (the superpartners of the charged Higgs bosons). On the other hand the neutralinos are linear combinations of $\tilde{W}_3$ (the superpartners of the neutral W boson), $\tilde{B}$ (the superpartners of the neutral U(1) gauge boson B) and $\tilde{H}_{1,2}$ (the superpartners of the two neutral Higgs bosons). In most models the lightest neutralino ($\tilde{\chi}_1^0$) is assumed to be the LSP.

They are produced by quark and anti-quark annihilations, mediated by photons, W/Z gauge bosons and squarks. The initial quark and, in particular, the anti-quark flux in the proton are rather low at the LHC. Moreover the underlying interactions are EW in nature. As a result the production rates are comparatively low at the LHC. Although, in principle, all combinations of charginos and neutralions can be produced, the dominant ones are

$$q\bar{q}' \to \tilde{\chi}_1^{\mp}\tilde{\chi}_1^{\pm}, \quad \tilde{\chi}_1^{\mp}\tilde{\chi}_2^0, \quad \tilde{\chi}_1^0\tilde{\chi}_2^0 . \tag{10}$$

It should be noted that the cross sections of the above processes are not controlled by the masses of the final state particles alone. The compositions of the charginos and neutralinos also determine the magnitude of cross sections. These compositions in turn are controlled by the parameters

$$M_1, M_2, \mu, \tan\beta , \tag{11}$$

where M_2, M_1 - the soft masses for SU(2) and U(1) gauginos respectively, μ and $\tan\beta$ are already defined. Obviously the cross sections for these processes are more parameter space dependent than that of squark-gluino production. As an example we note that, the cross section for the $\tilde{\chi}_1^{\mp}\tilde{\chi}_2^0$ chargino-neutralino pair production is at the sub pb level for $m_{\tilde{\chi}_1^{\mp}}, m_{\tilde{\chi}_2^0}$ ~200 GeV.

The associated production of $\tilde{g}$ and $\tilde{q}$ with charginos and neutralinos are also possible. For example,

$$gq \to \tilde{q}\tilde{\chi} \tag{12}$$

is an allowed process, where $\tilde{\chi}$ stands for either charginos ($\tilde{\chi}_i^{\mp}$; i = 1–2) or neutralinos ($\tilde{\chi}_j^0$; j = 1–4). The cross sections for the associated production are also dependent on the masses of produced sparticles and as well as the SUSY parameters in Eq. (11) due to the presence of one EW vertex. Typical cross sections of these processes are at a few pico barn level for the range of masses ~200–300 GeV.

Finally it would be incomplete unless we mention the pair production of sleptons via the Drell Yan like mechanisms

$$q\bar{q} \to \tilde{\ell}\tilde{\ell}^*, \quad \tilde{\nu}_\ell\tilde{\nu}_\ell^* (\ell = \mathrm{e}, \mu, \tau) \tag{13}$$

$$q\bar{q}' \to \tilde{\nu}_\ell\tilde{\ell}^* + \tilde{\nu}_\ell^*\tilde{\ell} \tag{14}$$

mediated by Z/γ/W through s-channel. Because of the low initial quark flux and energy suppression as these interactions are s-channel processes, the production cross sections are rather tiny ($\sim \mathcal{O}$(fb)) for $m_{\tilde{\ell}}$, $m_{\tilde{\nu}} \sim$ 200 GeV.

One important point to be noted here is that the leading order (LO) cross sections as shown in Fig. 1 have been obtained in the lowest order in perturbation theory. They involve significant theoretical uncertainties due to the choice of the QCD scale and the parameterisation of parton distributions inside the proton. One can reduce some of these uncertainties by including more terms in the perturbative series for the cross section.

Since the dominant sparticle productions take place via strong interaction these higher order corrections are expected to be substantial. Generally, the relative importance of these corrections is given by the K-factor defined by the ratio K = $\frac{\sigma_{NLO}}{\sigma_{LO}}$, where σ_{NLO} is the cross section including the next-to-leading order terms. The K factors can be substantially large ranging from ~ 1.1–1.6 [18]. In predicting the sparticle production rates one should take care of this K-factor appropriately.

3.2. SUSY Signals at the LHC

Once SUSY particles, dominantly gluinos and squarks are produced, they eventually decay to almost massless leptons and quarks along with the lightest neutralino ($\tilde{\chi}_1^0$) - the LSP - via a long decay chain involving the charginos and the neutralinos in the intermediate states. The long cascade decay processes, of course, depend on the relative magnitudes of $m_{\tilde{g}}$ and $m_{\tilde{q}}$. Accordingly, there are two possibilities: (a) $m_{\tilde{g}} \gtrsim m_{\tilde{q}}$ (b) $m_{\tilde{g}} \lesssim m_{\tilde{q}}$. The decay chain for case (a) is

$$\tilde{g} \to q\tilde{q}; \quad \tilde{q} \to q\tilde{\chi}_i^{\mp}, q\tilde{\chi}_j^0 \tag{15}$$

and in case (b) we have

$$\tilde{q} \to q\tilde{g}; \quad \tilde{g} \to q\bar{q}'\tilde{\chi}_i^{\mp}, q\bar{q}\tilde{\chi}_j^0. \tag{16}$$

Finally, the chargino and neutralino states decay to ordinary light fermions and the lightest neutralino

$$\tilde{\chi}_i^{\pm} \to f_1\bar{f}_2\tilde{\chi}_1^0, \tag{17}$$

$$\tilde{\chi}_j^0 \to f_3\bar{f}_3\tilde{\chi}_1^0. \tag{18}$$

Here f_i's stand for appropriate quarks and leptons. In Eqs. (15, 16) the $\tilde{g}/\tilde{q}$ decay branching ratios (BRs) depend on their masses and on the SUSY parameters space Eq. (11). Typically the BRs of the gluino are $\tilde{g} \to q\bar{q}'\tilde{\chi}_1^{\pm}$ (50–60%), $q\bar{q}\tilde{\chi}_2^0$ (35–30%), $q\bar{q}\tilde{\chi}_1^0$ (15–10%) for a wide region of the parameter space. Squark decays to charginos and neutralinos have almost the same relative rates. The lighter chargino ($\tilde{\chi}_1^{\pm}$) decays, Eq. (17), follow roughly the same branching ratio as W decays but the BRs of $\tilde{\chi}_2^0$ decay depend sensitively on the SUSY parameter space. For example, if the sleptons are light, i.e if $m_{\tilde{l}}$ is small and $\tilde{\chi}_2^0$ is gaugino dominated then, this neutralino will have relatively large BR to leptonic decay channels. The heavier chargino and neutralino states also decay to lighter particles accompanied by Higgs scalars [17]

$$\begin{aligned} \tilde{\chi}_i^0 &\to \phi\tilde{\chi}_k^0, (i > k)(\phi = h, H, A,), \\ \tilde{\chi}_2^{\pm} &\to H^{\pm}\tilde{\chi}_1^0 \\ &\to W^{\pm}\tilde{\chi}_1^{\mp}, \end{aligned}$$

where h, H and A are the three neutral Higgs bosons and $H^{\pm}$ is the charged Higgs boson in the MSSM. The sleptons which are produced via Eq. (14) also end up with light fermions and $\tilde{\chi}_1^0$, $\tilde{\ell} \to \ell\tilde{\chi}_1^0$.

As has already been noted the quarks and leptons in the final state will indicate an apparent imbalance of energy-momentum. This happens because the LSP escapes detection. In hadron colliders the momentum imbalance in the longitudinal direction, i.e. along the beam direction, cannot be tested as the fragments of colliding hadrons follow that direction and remain undetected. However in the transverse direction, i.e. perpendicular to the beam direction, the missing transverse energy ($\not{E}_T$), which is an experimental observable related to the missing transverse momentum, can be measured. This momentum imbalance is a characteristic signature of SUSY.

The quarks in the final stage of SUSY cascade decays cannot be observed directly. In fact each of them goes through non-perturbative processes like fragmentation and hadronisation and end up into a narrow stream of charged and neutral hadrons. These particles enter into a part of the detector called the hadron calorimeter and deposit energy in a cluster like forms. These clusters are called jets. These jets are the important observables in collider experiments. Photons and electrons (muons) are detected in the electromagnetic calorimeter (muon chambers).

Generically, any sparticle pair production followed by the cascade decays (Eqs. (15–18)) of the members of the pair at hadron colliders lead to the signal

$$\text{n} - \text{leptons} + \text{m} - \text{jets} + \not{E}_T; n, m = 0, 1, 2, 3..... \tag{19}$$

The long sequence of steps involved in production and detection of the SUSY signals are studied by Monte Carlo simulation which is implemented by packages called event generators [19]. Some of the commonly used and publicly available generators which can simulate a large number of SM or SUSY processes are HERWIG [20], ISAJET [21], PYTHIA [22] etc. In addition there are generators dedicated to specific processes. The prospects of detecting SUSY of various shades and stripes at the LHC experiments have been assessed by simulations using these generators. It should, however be noted that the non-perturbative processes mentioned above cannot be computed by using the standard techniques of field theory. Various models have been developed to study them. The event generators employ these models. Although many of the generators have so far been successful in explaining data up to Tevatron Run II, the reliability of these generators at the LHC has to be tested against the data on well understood SM processes collected during the early stages of the experiment. Only after a generator has been validated in the above way it can be used to study signals of new physics with confidence.

There are experimental challenges as well. For example, measuring $\not{E}_T$ in an experiment with the desired accuracy is a non-trivial task. Moreover, there are various sources of fake $\not{E}_T$. For example, detector effects like instrumental noise, hot or dead channels or cracks in between different parts of the detectors give some amount of fake $\not{E}_T$. Therefore, in order to establish SUSY signals based on $\not{E}_T$ convincingly, it is crucial to understand very precisely the fake $\not{E}_T$ and to develop a method for eliminating it.

In the LHC experiment a huge number of events will

be produced. Only a small fraction of them will be of interest and will be stored. These are known as physics events. The ratio of the events stored to the total number of events produced is $\sim 10^{-10}$–10^{-9}. Even the physics events will mostly consist of the known particles like the W, Z, top quark, bottom quark, etc. If we are lucky and nature is kind to us, then some new particles may be produced with comparatively low rates. It is indeed a very challenging task to pick up the events signaling new physics out of the debris, called background or noise, mainly due to the SM processes.

The standard technique is to apply some selection criteria based on the kinematic properties of the signals from the new particles. As for example, a lepton, electron or muon, as well as a jet is selected off-line provided it transverse momentum (P_T) or energy (E_T) is above a certain minimum value which is quite common in the SM processes. This value is often referred to as a kinematical cut or simply a cut. As already discussed a strong cut on $\not{E}_T$ is a powerful weapon in the arsenal of a SUSY searcher. Of course, the magnitude of these cuts depend on the nature of the signal and and the SM background events. In addition there are many other kinematic observables, like angular correlation among different final state particles, sum of their energies, etc. Suitable cuts are applied on these observable with a goal to optimise the signal to background ratio. One should also worry about the instrumental backgrounds and design suitable selection cuts to eliminate them. The dominant SM backgrounds which can give the same type of event topology as the SUSY signals are

$$pp \to W/Z + jets\,, \tag{20}$$

$$pp \to t\bar{t} + jets\,, \tag{21}$$

$$pp \to QCD\,(jets) \tag{22}$$

with leptonic or hadronic decay of W/Z and top quark decay, $t \to bW \to bf\bar{f}'$(f=ℓ, q). Moreover, there are other sub-dominant SM process e.g. WW/WZ/ZZ + jets, etc. which can mimic the SUSY signals. For the clean detection of signal it is often needed to suppress the SM backgrounds by factor of $\sim 10^4$–10^5. We will see in a later section that using some experimental techniques and some kinematic selection cuts, the level of SM backgrounds can indeed be brought down to a negligible level.

The discovery potential of SUSY is usually studied in the literature on the basis of a variety of models predicting different type of final states (Sections 2 and 4). However for illustration we will discuss in this section SUSY signals at the LHC within the framework of the mSUGRA model. The SUSY signals in other scenarios will be taken up in a later section.

In the mSUGRA model, as discussed in section 2, there are only five free parameters specified at some unification scale and the relevant parameters at the energy of experimental interest can be obtained via the RG equations [10]. A set of these parameters, therefore completely determines the SUSY signals corresponding to a point in the mSUGRA parameter space. Several representative points, better known as the benchmark points are chosen such that the members of the set reflects different characteristics of SUSY signals. Both the CMS and the ATLAS collaborations have performed Monte Carlo studies on the feasibility of discovering SUSY with their detectors in this way. In the following we shall present results mainly based on the analysis by the CMS group [23].

We present in Fig. 3 the simulated $\not{E}_T$ distribution for SUSY events (the upper black curve) with only jets in the final state corresponding to one such bench mark point (LM1). This plot is made by full simulation[11] [23]. Along with the signal distribution, the distributions of the same observable for several SM backgrounds, Eqs. (20–22) are also shown. The signal is found to be well above the total background.

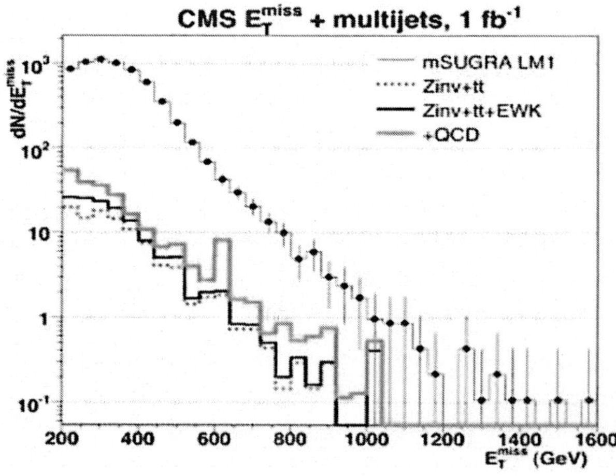

Figure 3. Missing E_T distribution for SUSY events and SM backgrounds with purely hadronic final states [23]

[11] In full simulation all possible detector effects are taken into account.

Recall that due to the long cascade decay chains in the signal events comparatively large multiplicities of jets and leptons appear. They are also harder as they originate from comparatively heavy sparticles like $\tilde{g}$, $\tilde{q}$. Therefore, a variable, called the effective mass, defined by the scalar sum of $\not{E}_T$ and the E_T of the four leading jets

$$M_{eff} = \sum_{j=1}^{4} |E_T^j| + |\not{E}_T| \tag{23}$$

will show a striking difference in its distribution in SUSY signal and SM background events. In Fig. 4 we present the distribution of M_{eff} obtained by the ATLAS collaboration [24] for illustration. In this figure

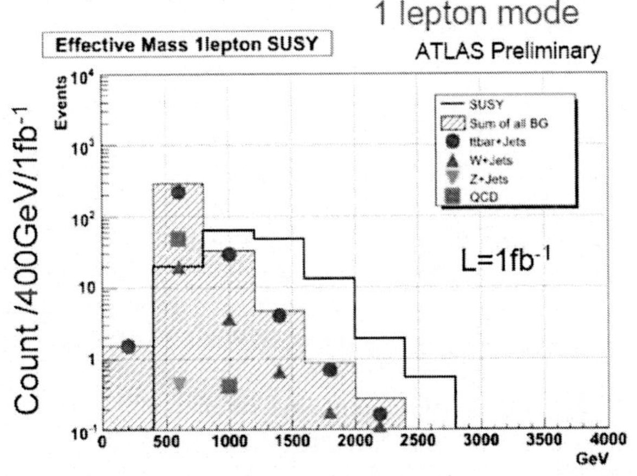

Figure 4. Effective mass M_{eff} distribution for SUSY events and SM backgrounds with 1 lepton + jets in the final states [24]

both the signal and the total background contributions are shown. Clearly the $\not{E}_T$ and as well as M_{eff} distributions will indicate unambiguously the existence of SUSY provided we have a complete understanding of the sources of fake $\not{E}_T$.

We present in Fig. 5, the CMS discovery reach of SUSY in the m_0–$m_{1/2}$ plane for different final state topologies corresponding to an integrated luminosity of 1 fb^{-1} [23] which will be accumulated during the first few months of data taking (footnote 10). The search channels with promising discovery potentials are now described briefly.

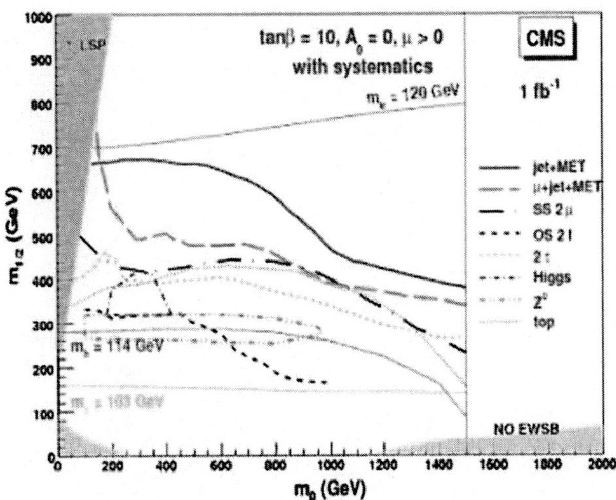

Figure 5. Discovery potential of the CMS experiment for 1 fb^{-1} luminosity [23]

• Jets + missing transverse energy ($\not{E}_T$).
This inclusive signal corresponding to the best discovery reach (see the solid line in Fig. 5) is due to gluino and squark production followed by their cascade decays to purely hadronic final states (Eqs. (15–18)). The dominant SM background is due to the Z + jets events with $Z \to \nu\nu$ giving missing energy like signal events. Other sources of SM backgrounds are $t\bar{t}$, WW, single top and pure QCD events. These backgrounds are suppressed by judiciously chosen kinematic cuts. From the squark and gluino mass formulae in terms of m_0 and $m_{1/2}$ (Section 2), it follows that for $m_0 \lesssim 1$ TeV, gluinos with mass $m_{\tilde{g}} \sim 1300$–1400 GeV can be observed even at this low luminosity. However for $m_0 \gtrsim 1$ TeV, the squark mass becomes large yielding low rates leading a drop in the discovery reach.

• Single μ + Jets + $\not{E}_T$
In this signal final states containing at least one muon are considered. This muon may arise in the cascade decay chain due to the leptonic decay of a chargino Eq. (17). or a neutralino Eq. (18), where muons may also come from W or Z decays appearing in the decay cascade. Because of the relatively low BR of leptonic decays of $\tilde{\chi}_1^{\mp}$, which is about 11% for each species of letpons, the signal rates are depleted to some extent. However due to the presence of a high p_T muon the SM backgrounds, especially the background from pure QCD jets are small and consequently, the signal to background ratio large. In Fig. 5 the discovery reach

for this channel is shown by the long dashed line. At low values of m_0 squarks are light making $\tilde{\chi}_1^{\mp}\tilde{\chi}_1^{\pm}$ and $\tilde{\chi}_1^{\mp}\tilde{\chi}_2^0$ production rates large. They contribute dominantly to this final state along with $\tilde{g}$ and $\tilde{q}$ pair production making this channel the main discovery channel. On the whole this channel gives the best discovery reach after jet + $\not{E}_T$ channel.

In summary squark and gluino masses of about 1.5 TeV are within the striking range of the LHC for a nominal integrated luminosity of 1 fb^{-1}. This is already far above the mass reach of the Tevatron, the most powerful collider before the advent of the LHC. The above reach can be further extended to 2 TeV for an integrated luminosity of about 10 fb^{-1} [23]. With accumulating integrated luminosity larger regions of the SUSY parameter space will be accessible to the LHC experiments. For 300 fb^{-1} squark gluino masses upto 2.5 TeV can be probed but the precise mass reach is rather model dependent.

It is also seen from the Fig. 5 that several other search channels have reasonable reaches although they do not look like discovery channels. In fact large regions of the m_0–$m_{1/2}$ parameter space are covered by more than one search channel. The observation of the signal in multiple channels may provide additional information about the underlying theory. For example, the existence of a Z or the lightest Higgs boson in the final state would strongly suggest the presence of a $\tilde{\chi}_2^0$ in the decay cascade.

So far we have focused on generic SUSY signals which are events of the type n-lepton + m-jet + $\not{E}_T$. It is encouraging to note that the relative sizes of the signals corresponding to different m and n can distinguish between different regions of the mSUGRA parameter space. For example, the region corresponding to low m_0 allowed by dark matter data has three distinct regions. These regions can be distinguished by the above characteristics [25]. Since the ratio of the number of events in two different channels originating from the same production processes are practically independent of major theoretical uncertainties like the QCD scale dependence, this approach can indeed be helpful in obtaining additional information.

The particles belonging to the third-generation lead to special collider signatures. For example, the jets from the decay of B-hadrons (hadrons containing a b-quark) do not point to the collision vertex since the B, with a relatively large life time, travels a measurable distance away from this point before it decays. The jets from the hadronic decays of a τ contains very few hadrons compared to other jets. The jets coming from B or τ decays can be identified by the LHC experiments with large efficiency. Using the b-jet tagging or τ tagging facilities at the LHC, important additional information can be extracted from SUSY signals. For example $\tilde{\chi}_2^0$ can decay into a τ^+–τ^- pair with very large BR particularly for large tan β. In this case the τ slepton turns out to be much lighter than the other sleptons due to the RG effects and mixing as discussed in sections 2 and 3. Thus $\tilde{\chi}_2^0$ decays mediated by this sparticle dominantly goes to di-tau pairs. The members of the τ-pair are often polarised. This polarisation can be further exploited to improve the prospect of SUSY search at the LHC [26].[12]

CMS has also looked for SUSY signals in the di-tau channel and have found a reasonable reach. However even for moderately large tan β some of the low m_0 regions of the parameter space allowed by the dark matter data discussed above can lead to final states with many tagged τ and b-jets [25,27]. The importance of flavour tagging in identifying SUSY signals has been noted by several groups [28].

Another interesting region of the parameter space in the mSUGRA model is the 'focus point' region [29]. This is one of the few regions in the mSUGRA model consistent with the dark matter data (footnote 4). In this region characterised by large m_0, the squarks and the sleptons are too heavy to be detected at the LHC. However the gluino may well be within the reach of the LHC. Even in this region the top squark can be relatively light due to RG evolution (Section 2)[13]. As a result the gluino decays, though mediated by all squarks in principle, will be dominantly mediated by the lighter top squark. Thus they primarily decay into channels involving large number of t and b quarks. Some of these b quarks come directly from gluino decays and others from t decays. Consequently b-jet tagging can again be employed to distinguish the focus point signals from the ones from other mSUGRA regions as well as from the SM background [30].

3.3. Measurements of Masses at the LHC

So far we have discussed the prospects of discovering SUSY via different signals at the LHC. However the mere existence of these signals is not a conclusive evidence for SUSY. It is necessary to measure the mass, charge and other quantum numbers of each superparticle to identify the SUSY model chosen by nature. In this section we describe briefly the first few steps of achieving this goal. In our example, the second lightest

[12]See the article by Guchait and Roy in this volume for further details.

[13]Mixing effects (Section 3) are unlikely to be very significant here since $m_0 >> m_t$.

neutralino in a squark decay cascade decays to dileptons

$$\tilde{\chi}_2^0 \to \ell^+\ell^-\tilde{\chi}_1^0$$

which is a generic decay in many SUSY models (Fig. 6). This decay may occur via a two body decay mode if $m_{\tilde{\chi}_2^0} > m_{\tilde{\ell}}$ $(m_Z + m_{\tilde{\chi}_1^0})$, $\tilde{\chi}_2^0 \to \tilde{\ell}\ell \to \ell\ell\tilde{\chi}_1^0$ ($\tilde{\chi}_2^0 \to Z\tilde{\chi}_1^0 \to \ell\ell\tilde{\chi}_1^0$), otherwise it goes into three body modes mediated by an off-shell slepton or Z boson. In the

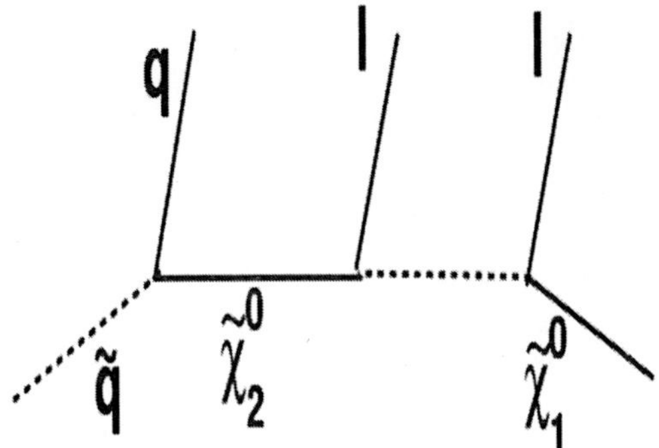

Figure 6. A simple decay chain containing a $\tilde{\chi}_2^0$ decay into the dilepton channel

case of three body decay, the distribution of dilepton invariant mass $m_{\ell\ell}$ shows a sharp edge and the position of the end point is exactly equal to $m_{\tilde{\chi}_2^0} - m_{\tilde{\chi}_1^0}$. At the end points the dilepton system and $\tilde{\chi}_1^0$ are at rest in the $\tilde{\chi}_2^0$ rest frame, so from the measurements of dilepton four vectors, it is possible to determine the four vector of $\tilde{\chi}_1^0$ and hence the four vectors of $\tilde{\chi}_2^0$. The mass of $\tilde{\chi}_1^0$ can also be determined from the kinematics. And eventually adding the four momentum of the correct jet in the event with that of $\tilde{\chi}_2^0$ it is possible to reconstruct the mass of the origin of the $\tilde{\chi}_2^0$ (either a $\tilde{g}$ or a $\tilde{q}$).

In case of two body decay of $\tilde{\chi}_2^0$, the edge of dilepton invariant mass distribution can be expressed in terms of $m_{\tilde{\ell}}$ and the neutralino masses

$$m_{\ell\ell}^{max} = m_{\tilde{\chi}_2^0}\sqrt{1 - \frac{m_{\tilde{\ell}}^2}{m_{\tilde{\chi}_2^0}^2}}\sqrt{1 - \frac{m_{\tilde{\chi}_1^0}^2}{m_{\tilde{\ell}}^2}}\,.$$

In case of $\tilde{q} \to q\tilde{\chi}_2^0$ decay, there are also similar end points for the $q\ell$ and qll. These end pints are expected to be measured within a good accuracy $\sim$ 10% [23]. Once these measurements are performed, it is possible to determine the masses of $m_{\tilde{\chi}_1^0}, m_{\tilde{\chi}_2^0}, m_{\tilde{\ell}}, m_{\tilde{q}}$. Recently, quite a few new ideas have been proposed to reconstruct the sparticle masses from different observables [31].

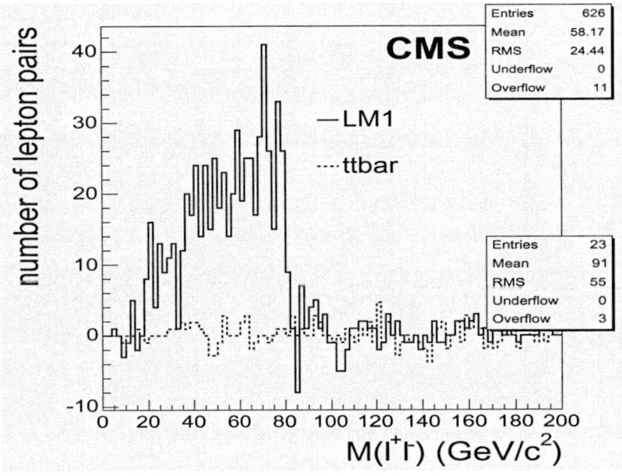

Figure 7. Invariant mass distribution of same flavour distributions [23]

Figure 7 presents the distribution of dilepton invariant mass of same flavour leptons after subtracting the background due to SUSY and SM. It is shown after a full simulation that end points can be measured within an uncertainty of $\sim$ 1 GeV [23].

4. Other Models of Mediation of SUSY Breaking

We have seen in the previous sections that the phenomenology of minimal supergravity has been very well studied with the lightest neutralino as the LSP over most of the parameter space. In this section we review some other mechanisms of SUSY breaking where the modes of communicating this breaking from the hidden sector to the observable sector MSSM are different. In particular, we discuss the LHC signatures of gauge-mediated SUSY breaking (GMSB) and anomaly-mediated SUSY breaking (AMSB) as well as signatures of models which are generalisation of mSUGRA, like models with non-universal scalar and gaugino masses and split-SUSY.

4.1. Gauge Mediated Supersymmetry Breaking

In gauge mediated SUSY breaking [32] the basic idea is to introduce new supermultiplets, not included in the

MSSM, as the messenger fields (Section 2).[14] They couple to the hidden sector which is the source of SUSY breaking. These messengers carry SM quantum numbers and, consequently couple directly to the SM gauge fields and the corresponding gauginos. Although the messenger fields do not couple to the sfermions in the lowest order, these couplings arise in higher orders of perturbation theory. The SUSY breaking in the hidden sector is thus communicated to the observable sector and soft masses and other SUSY breaking terms are generated.

One of the attractive features of gauge mediation is that the soft masses of squarks and sleptons depend only on their gauge quantum numbers. Thus for example, squarks of the same type (say, $\tilde{u}$ and $\tilde{c}$) are mass degenerate. In this way mass degeneracies required for suppressing FCNC effects in all sectors are guarantied.

For the sake of completeness we now review very briefly the theoretical ingredients of GMSB. The readers mostly interested in the LHC phenomenology may skip the next two paragraphs and directly move to the sparticle spectrum and collider signatures. In the simplest case, let us assume that the messenger fields are a set of left-handed chiral supermultiplets $\hat{\psi}$, $\hat{\bar{\psi}}$ and they have specific transformation properties under the SM gauge group. They are taken to be vectorlike with respect to the SM gauge interaction (i.e. $\hat{\bar{\psi}}\hat{\psi}$ are SM singlet). We assume that they couple to a gauge-singlet superfield S through a superpotential: $W_{\text{mess}} = yS\hat{\psi}\hat{\bar{\psi}}$. Both scalar and auxiliary components of the superfield S acquires VEVs, denoted by $\langle S\rangle$ and $\langle F_S\rangle$ respectively. This way the fermionic and scalar component of the messengers get different masses: $m^2_{\text{fermions}} = |y\langle S\rangle|^2$, $m^2_{\text{scalars}} = |y\langle S\rangle|^2 \pm |y\langle F_S\rangle|$. In the messenger spectrum $\langle F_S\rangle \neq 0$ leads to SUSY violation and since the messenger fields are charged under the SM gauge groups, the gauginos of the MSSM can receive masses at one-loop. The scalars of the MSSM get leading contribution to their masses at the two-loop level.

If we assume that the messengers come in complete multiplets of the $SU(5)$ global symmetry, and are very close in mass, then approximate unification of gauge couplings will occur at the GUT scale M_U. In such a case if we consider the messengers to consist of N_5 copies of the $\mathbf{5} + \bar{\mathbf{5}}$ of $SU(5)$, then we have for the gaugino masses
$M_a = \frac{\alpha_a}{4\pi}\Lambda N_5 \;\; (a = 1, 2, 3)$.
Here we have introduced one more mass scale $\Lambda \equiv \langle F_S\rangle/\langle S\rangle$. Here $\alpha_a = g_a^2/4\pi$ and the g_a are the three gauge couplings of the SM. Theses gaugino masses are the running gaugino masses at an RG scale Q_0 corresponding to the average mass of the heavy messenger particles: $Q_0 \sim M_{\text{mess}} \sim y\langle S\rangle$. For the MSSM scalars we have
$m^2_{\phi_i}(Q_0) = 2\Lambda^2 N_5 \sum_{a=1}^{3}\left(\frac{\alpha_a}{4\pi}\right)^2 C_a(i)$,
where $C_a(i)$ are the quadratic Casimir invariants of the representation of ϕ_i under the SM group. One should note here that the gaugino and scalar masses are in the same order in α. The requirement of gauge coupling unification demands that for messenger masses of order 10^6 GeV or less, one needs $N_5 \leq 4$. The trilinear soft terms are much smaller and one can assume them to be zero at the messenger scale Q_0. The LEP constraints on the sparticle masses and the requirement that the gluino mass is $\lesssim$ 1 TeV restrict Λ in the range 30 TeV $\lesssim \Lambda \lesssim$ 120 TeV.

In the GMSB spectrum the strongly interacting sparticles, the squarks and the gluino, are the heaviest since their soft masses are generated by the strong gauge interactions with the messenger sector. The soft masses of the sleptons and the EW gauginos are generated by weaker interactions. These sparticles are therefore, somewhat lighter. However the most distinctive feature of GMSB is that the gravitino ($\tilde{G}$) is the LSP. This can have very important consequences for collider physics [33] (see below). The mass of the gravitino is roughly in the range 1 eV $\lesssim m_{3/2} \lesssim$ 1 GeV. Since gravity is too weak, the direct production of gravitino in experiments is highly suppressed. However since this is the LSP, all the other SUSY particles will eventually decay into final states containing the gravitino leading to missing energy.

The decay width of a sparticle $\tilde{X}$ into its SM partner X and a gravitino (more precisely the longitudinal goldstino component) is

$$\Gamma(\tilde{X} \to X\tilde{G}) = \frac{m^5_{\tilde{X}}}{16\pi F^2}\left(1 - m^2_X/m^2_{\tilde{X}}\right)^4. \qquad (24)$$

Here $\sqrt{F}$ is the fundamental scale of SUSY breaking and is typically $\sim$ 100 TeV. The decay width (the decay length) of $\tilde{X}$ is larger (smaller) for smaller F.

The collider signals in GMSB depends sensitively on the nature of the next-to-lightest SUSY particle

[14] A supermultiplet consists of a particle and its superpartner (e.g. the u_L quark and the squark $\tilde{u}_L$ form a supermultiplet of MSSM). More formally a spin-1/2 particle (a quark or a lepton) and the corresponding sparticle can be looked upon as the components of a 'matter' superfield or a chiral superfield of L or R type. However a superfield also consists of unphysical particles called auxiliary components. Similarly a SM gauge boson, the corresponding gaugino and unphysical particles are components of a gauge superfield or a vector superfield. Usually the MSSM Lagrangian is written interms of these superfields, since a superfield as a whole has simple properties under SUSY transformations.

(NLSP). As we have discussed in an earlier paragraph, the answer to this question is model dependent. Nevertheless, there are only a few distinct possibilities. We have noted above that the gaugino masses scale like N_5, while the scalar masses like $\sqrt{N_5}$. As a result the sleptons will tend to be lighter than the EW gauginos for larger values of N_5. Thus for lower values of N_5, say, $N_5 = 1$, the lightest neutralino, which is usually bino like is the LSP over most of the parameter space. The NLSPs can also be the right (R)-type sleptons ($\tilde{e}_R$, $\tilde{\mu}_R$ and $\tilde{\tau}_R$). This possibility is favoured if N_5 is larger. For large value of $\tan\beta$ the stau ($\tilde{\tau}$) turns out to be lighter than the other two R-sleptons due to mixing and RG effects and becomes the sole NLSP in most cases. There is another possibility called the neutralino-stau co-NLSP scenario where signatures of neutralino NLSP and stau NLSP are both present.

The decay length of the (Eq. (24)) NLSP depends on the fundamental SUSY breaking scale and so a measurement of the decay length gives a measurement of the fundamental SUSY breaking scale $\sqrt{F}$. Depending on the decay length, the NLSP can decay inside the detector or outside. In the former case it may either decay promptly with a small life time or its life time may be large enough to produce an observable secondary vertex which can be reconstructed.

If the NLSP is the lightest neutralino then the dominant decay mode is $\tilde{\chi}_1^0 \to \gamma + \tilde{G}$. For a neutralino NLSP, its prompt decay into a photon and a gravitino (which escapes detection) can produce the inclusive $\gamma\gamma + \not{E}_T$ events at the LHC [34] and at the Tevatron [35]. For relatively light squarks and gluinos the production of these strongly interacting sparticles dominantly contribute to the signal. If on the other hand the squarks and gluinos are very heavy, the dominant contributions come from chargino and neutralino production. It has been shown that with an integrated luminosity of 10 fb^{-1}, the LHC experiments can probe values of $\Lambda \lesssim 400$ TeV, corresponding to $m_{\tilde{g}} \le 2.8$ TeV. Here $m_{\tilde{g}}$ is computed from its theoretical mass formula. This mass reach is certainly comparable to what we have in mSUGRA.

If the neutralino NLSP decays inside the detector with a long enough decay length then displaced secondary vertices appear. As a result the highly energetic photon from the neutralino NLSP decay is in general non-pointing (i.e. it may not be pointing to the interaction point). In this case the signal is again $\gamma\gamma + \not{E}_T$ events along with measurable decay lengths. This could be considered as the 'smoking gun' signature of GMSB.

The masses of the slepton ($\tilde{\ell}$) and the neutralino can be determined at the LHC from events with a lepton and a non-pointing photon arising from the cascade decay $\tilde{\ell} \to \ell\tilde{\chi}_1^0 \to \ell\gamma\tilde{G}$ [36]. The slepton $\tilde{\ell}$ may be copiously produced at the LHC from the gluino or squark cascades involving $\tilde{\chi}_2^0$ and $\tilde{\chi}_1^{\pm}$. A Monte Carlo simulation with the ATLAS detector shows that the masses could be measured with an error of 3% for $\mathcal{O}(100)$ $\ell\gamma$ pairs. The fitted masses are shown in Fig. 8. The fit results are $m_{\tilde{\ell}} = 162.1$ GeV and $m_{\tilde{\chi}_1^0} = 117.0$ GeV, while the input values are 161.7 GeV and 117.0 GeV, respectively. In Ref. [37] it has been shown that the

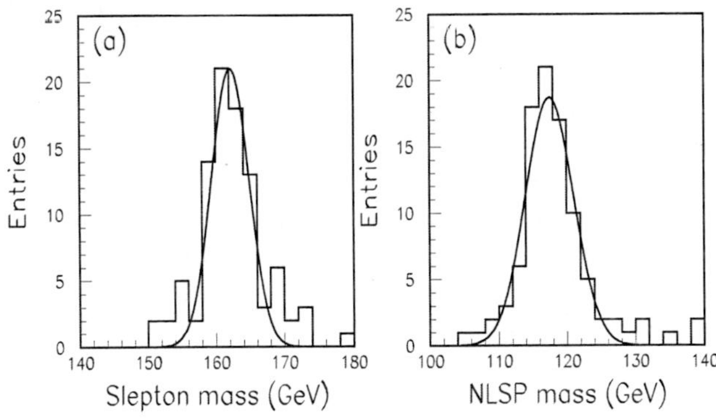

Figure 8. Distributions of the fit results of (a) the slepton mass $m_{\tilde{\ell}}$ and (b) the neutralino mass $m_{\tilde{\chi}_1^0}$. Results of Gaussian fitting are also shown. Taken from Ref. [36]

well studied $\gamma\gamma$ inclusive Higgs signal can be used at the LHC to test GMSB models in which a heavy neutral Higgs boson decays into two light neutralinos, the latter yielding two photons and missing energy.

If on the other hand a right handed slepton happens to be the NLSP it decays by $\tilde{\ell}_R \to \ell + \tilde{G}$. At the LHC pair production of NLSP sleptons with prompt decay to gravitino can give rise to final states $\ell^+\ell^- + \not{E}_T$. However, this signal suffers from large irreducible backgrounds. On the other hand, production of heavier states cascading to $\tilde{\ell}_R$ can give clean signatures with multiple leptons. For example, $\tilde{\ell}_L\tilde{\ell}_L$ pair production followed by the cascade decays $\tilde{\ell}_L^{\pm} \to \tilde{\ell}_R^{+}\ell^-\ell^{\pm}$ can produce final states $6\ell + \not{E}_T$ which do not suffer from SM backgrounds [38].

If the charged slepton decay takes place inside the detector with a measurable decay length, then one can observe very clear heavily ionising charged tracks with a kink. The "kinks" are sudden turns in the track of a charged particle. In this case the kinks appear where the charged lepton and the gravitino are being emmitted. As in the case of neutralino NLSP, a measurement of the decay length distribution can give a direct measure of the fundamental SUSY breaking scale. If the decay $\tilde{\ell} \to \ell + \tilde{G}$ takes place outside the detector, then the signature of GMSB would be two heavily ionizing charged tracks without missing energy. This is a very non-standard signature of SUSY and is easily detectable [39].

4.2. Anomaly-mediated Supersymmetry Breaking Models

Anomaly-mediated supersymmetry breaking (AMSB) involves a higher dimensional supergravity theory where the hidden sector and visible sector superfields are localised on two distinct parallel three-branes separated by a distance $\sim r_c$ (r_c is the compactification radius) in the extra dimension [40]. Assuming one single extra dimension, the flavour-violating terms are suppressed by factors like $e^{-r_c M_5}$, where M_5 is the 5-dimensional Planck scale. So as long as $r_c M_5 \gg 1$, the dangerous FCNC terms are absent.

The AMSB scenario can be described in terms of a 4-dimensional effective theory below the scale μ_c ($\sim r_c^{-1}$) where only 4-dimensional supergravity fields are assumed to propagate in the bulk. The whole process can be studied with the help of a "compensator" superfield Φ whose scalar and auxiliary components get VEVs according to $\langle \Phi \rangle = 1 + \langle F_\Phi \rangle \theta^2$. The auxiliary field acquires VEV ($\langle F_\Phi \rangle$) through its coupling to SUSY breaking sector. It also couples to the visible sector fields due to an anomalous violation of super-conformal invariance at the quantum level (super-Weyl anomaly). This causes SUSY breaking to show up in the visible sector and hence the name anomaly-mediated SUSY breaking. Denoting the VEV of the auxiliary component as F_Φ from now on, the resulting soft terms are

$$M_a = \frac{\beta_{g_a}}{g_a} F_\Phi; \quad a_y = \frac{\beta_y}{y} F_\Phi;$$
$$m_{\tilde{f}}^2 = -\frac{1}{4}\left(\frac{\partial \gamma}{\partial g}\beta_g + \frac{\partial \gamma}{\partial y}\beta_y\right)|F_\Phi|^2, \quad (25)$$

where appropriate beta functions ($\beta_{g,y}$) and anomalous dimensions (γ) are to be considered. It should also be noted that $F_\Phi \sim m_{3/2}$, the gravitino mass.

The analytic expressions for the scalar and gaugino masses are renormalisation group invariant, and thus, can be computed at the low-energy scale. However at low energies, it predicts the existence of tachyonic sleptons. Several solutions to this problem exist but here we will consider the minimal AMSB model wherein a constant term m_0^2 is added to all the scalar squared masses thereby making the slepton squared masses sufficiently positive. As a consequence, the RG invariance is lost and one needs to consider the expressions in Eq. (25) as the boundary conditions at the unification scale (M_U).

The minimal AMSB (mAMSB) model is described by just four parameters: $m_{3/2}$, m_0, $\tan\beta$ (the ratio of two Higgs VEVs) and sign(μ). The magnitude of μ is determined by requiring correct EW symmetry breaking. A particularly interesting feature of the mAMSB model is that the ratios of the respective $U(1)$, $SU(2)$ and $SU(3)$ gaugino mass parameters M_1, M_2 and M_3, at low energies turn out to be $|M_1| : |M_2| : |M_3| :: 2.8 : 1 : 7.1$. An immediate consequence is that the lighter chargino and the lightest neutralino are both almost exclusively a Wino and, hence, nearly degenerate in mass (each with mass approximately equal to M_2). A very small mass splitting ($\Delta M_{\tilde{\chi}_1} \sim 200$ MeV) is generated at the tree level as well as from the one-loop corrections. The gravitino mass is much larger than the MSSM soft terms. Left and right charged sleptons of first two-generations are nearly mass degenerate. The staus are somewhat split in mass with almost maximal mixing. In this model, the squarks are always much heavier than the sleptons due to larger contributions from the strong coupling constant.

The lighter chargino is long-lived with the dominant decay mode $\tilde{\chi}_1^\pm \to \tilde{\chi}_1^0 + \pi^\pm$. This would typically result in a short heavily ionising charged track and/or a characteristic soft pion in the detector [41]. This is the 'smoking gun' signature of anomaly mediation with a Wino LSP, and is applicable beyond the minimal model. These charginos can be triggered on when produced in association with jets. It has been shown that for Tevatron Run II with 2 fb^{-1} luminosity, one could discover chargino masses up to 300 GeV [41]. In order to distinguish this scenario from GMSB with long-lived sleptons, correlations between particle masses and cross sections may be used. A similar analysis has been performed also at the LHC for the high luminosity run [42].

It has been shown that with an integrated luminosity of 10 fb^{-1}, the dilepton plus jets plus $\not{E}_T$ channel offers the reach at the LHC to values of the gluino mass ($m_{\tilde{g}}$) $\sim$ 2.5 TeV for low values of m_0. For large m_0, the reach is $m_{\tilde{g}} \sim 1.3$ TeV where the best signature is typically 0 or 1 isolated lepton plus jets plus $\not{E}_T$ [43]. The presence of terminating tracks due to the long-lived chargino in the signal events could help to distinguish the mAMSB model from other SUSY breaking mediation model. The discovery potential of the LHC has also been investigated for the mAMSB model, using the ATLAS fast detector simulator, including track reconstruction and particle identification. It has been found that with 100 fb^{-1} of integrated luminosity the search will reach upto 2.8 TeV in the squark mass and 2.1 TeV in the gluino mass for a 5σ discovery [44]. In Fig. 9, the number of $\tilde{\chi}_1^+$s which would be produced at the LHC and decay within a fiducial volume in the active material of the ATLAS tracker is plotted as a function of m_0 and $m_{3/2}$.

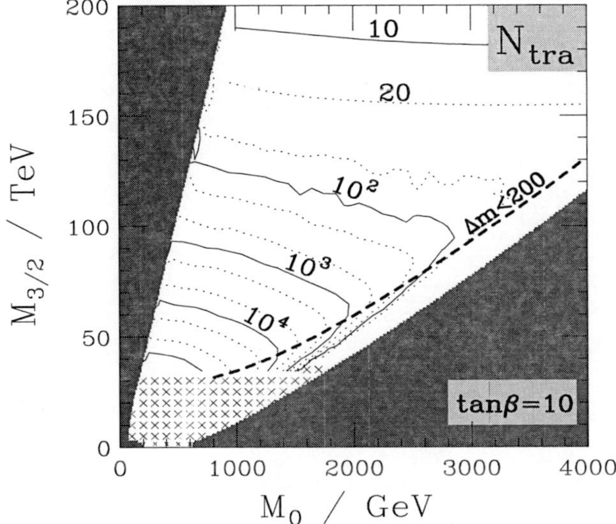

Figure 9. The number of $\tilde{\chi}_1^\pm \to \tilde{\chi}_1^0$ decays expected within the central region of the ATLAS detector ($|\eta| < 2$) with transverse decay vertices between 100 mm and 800 mm from the interaction point for integrated luminosity of 100 fb^{-1}. The initial track from the chargino is required have $p_T > 10$ GeV. $\Delta M_{\tilde{\chi}_1}$ is < 200 MeV. Figure is taken from Ref. [44]

4.3. Non-universal Gaugino and Scalar Masses

Most studies on the signatures of SUSY at the LHC, are based on the assumption that the gaugino masses are universal at the unification scale as in the mSUGRA model. There is, however no compelling theoretical reason for such a choice. In a supersymmetric grand unified theory (SUSYGUT) like the SU(5) grand unified model, gaugino masses are generated by the VEV of the auxiliary scalar component F_Φ of a superfield Φ in the gauge kinetic function $f(\Phi)$ (see below). When the scalar field is chosen to be a singlet under the GUT group the gaugino masses turn out to be universal usualy parametrised by a single parameter $m_{1/2}$. If on the other hand a GUT non-singlet scalar is chosen non-universal gaugino masses emerge but these masses are related by calculable group theoretic factors (Table 3).

For the more theoretically oriented readers we summarise in the next few lines the basic mechanism of generating non-universal gaugino masses. The readers only interested in the LHC signatures can directly go to Table 3 and the discussions following Eq. (28).

The function $f(\Phi)$ is an analytic function of the chiral superfields Φ in the theory [45]. It should be noted that the chiral superfields Φ consist of a set of gauge singlet superfields Φ^s and gauge non-singlet superfields Φ^n, respectively, under the grand unified group. If the auxiliary part F_Φ of a chiral superfield Φ in the $f(\Phi)$ gets a VEV, then gaugino masses arise from the coupling of $f(\Phi)$ with the field strength superfield W^a. The Lagrangian for the coupling of gauge kinetic function with the gauge field strength is written as

$$\mathcal{L}_{gk} = \int d^2\theta f_{ab}(\Phi) W^a W^b + H.c., \tag{26}$$

where a and b are gauge group indices [for example, $a, b = 1, 2, ..., 24$ for SU(5)] and repeated indices are summed over. The gauge kinetic function $f_{ab}(\Phi)$ is

$$f_{ab}(\Phi) = f_0(\Phi^s)\delta_{ab} + \sum_n f_n(\Phi^s)\frac{\Phi^n_{ab}}{M_P} + \cdots, \tag{27}$$

where as described above the Φ^s and Φ^n are the singlet and non-singlet chiral superfields, respectively. Here $f_0(\Phi^s)$ and $f_n(\Phi^s)$ are functions of gauge singlet superfields Φ^s and M_P is some large scale. When F_Φ gets a VEV $\langle F_\Phi \rangle$, the interaction (26) gives rise to gaugino masses

$$\mathcal{L}_{gk} \supset \frac{\langle F_\Phi \rangle_{ab}}{M_P} \lambda^a \lambda^b + H.c., \tag{28}$$

where $\lambda^{a,b}$ are gaugino fields.

We now illustrate the generation of non-universal gaugino masses in the framework of the group SU(5). Since the gauginos belong to the **24**-dimensional adjoint representation of the gauge group, F_Φ can belong to any of the following representations appearing in the symmetric product of the two **24**-dimensional representations of SU(5):

$$(\mathbf{24} \otimes \mathbf{24})_{Symm} = \mathbf{1} \oplus \mathbf{24} \oplus \mathbf{75} \oplus \mathbf{200}. \tag{29}$$

In the minimal case, F_Φ is assumed to be in the singlet representation of SU(5). This corresponds to equal gaugino masses at the GUT scale. However, Φ can belong to any of the non-singlet representations **24**, **75**, and **200** of SU(5). In that case, the gaugino masses are unequal but related to one another via the representation invariants [46]. It should be kept in mind that an

F_Φ	M_1^G	M_2^G	M_3^G	$M_1^{\rm EW}$	$M_2^{\rm EW}$	$M_3^{\rm EW}$
1	1	1	1	0.14	0.29	1
24	–0.5	–1.5	1	–0.07	–0.43	1
75	–5	3	1	–0.72	0.87	1
200	10	2	1	1.44	0.58	1

Table 3. *Ratios of gaugino mass parameters at the GUT scale in the normalisation $M_3^{\rm G}$ = 1 and at the electroweak scale in the normalisation $M_3^{\rm EW}$ = 1 at the one-loop level.*

arbitrary combination of these different representations is also allowed.

In Table 3, we display the ratios of the resulting gaugino masses at tree level for F_Φ belonging to different representations of SU(5). Clearly, the non-singlet representations have characteristic mass relationships for the gauginos at the GUT scale. The resulting relations at the EW scale, using the renormalisation group (RG) evolution (Section 2) at the one-loop level are also displayed.

The phenomenology of supersymmetric models with non-universal gaugino masses has been considered e.g. in [47–52] and very recently in [53–56] in the context of the LHC. The phenomenology of supersymmetric models depends crucially on the compositions of neutralinos and charginos. Hence, it is extremely important to investigate the changes in the experimental signatures with the changes in the composition of neutralinos and charginos which arise because of the non-universal gaugino masses at the GUT scale.

In Ref. [56] a multichannel analysis of SUSY signals has been carried out at the LHC for a number of non-universal representations breaking the SU(5) and SO(10) GUT groups. These channels include *jets* + $\not\!\!E_T$, *same-sign-dileptons* (SSD), *opposite-sign-dileptons* (OSD), *trileptons* + *jets* + $\not\!\!E_T$ and *single lepton* + *jets* + $\not\!\!E_T$. The results have been compared with those coming from the universal gaugino masses. It has been found that the most useful way to discriminate among the various cases is to look for the ratios of events rates in various channels. Hadronically quiet trilepton channels were studied in [54].

If the squarks and gluinos are light enough, their production cross sections are large at the LHC. The light neutralinos $\tilde\chi_{1,2}$ are typical decay products of $\tilde{\rm g}$ and $\tilde{\rm q}$. Following [17] it was already noted in section 3.2 that the neutral Higgs bosons can be copiously produced in the decay of $\tilde\chi_2$, if the mass difference between $\tilde\chi_2$ and $\tilde\chi_1$ is large enough. At the LHC heavy higgs (H/A) search may be difficult for low and medium $\tan\beta$ values [17,57]. The cross section for H/A production decreases rapidly for masses around $\sim$ 280 GeV/c^2. Thus the discovery region is not expected to cover these high Higgs mass values even with high luminosity unless the cross section is boosted by high $\tan\beta$. As the squark, gluino production rates are largely independent of the value of $\tan\beta$, Higgs production via $\tilde\chi_2$ decays have been found to be particularly interesting. It is, therefore, worthwhile to revisit the prospect of Higgs search in these channels in models with non-universal gaugino masses.

Production of Higgs bosons via $\tilde\chi_2 \to \tilde\chi_1$ h/H/A in models where gaugino mass patterns are as in the singlet and **24** representation (Table 3) was studied in [47]. It was found for sample parameters that only the lightest Higgs could be produced in the model with singlet, while also the heavy Higgses, H and A, could be produced from the cascade in the model with **24** representation. Recently the role of gaugino mass non-universality (in a model independent framework) in the context of Higgs productions has been discussed in [55].

In [53] the discovery potential of the neutral Higgs bosons in the representation 24 was studied and the results were compared with those obtained with universal gaugino masses. The event selection is based on the requirement of four energetic jets, large missing $\rm E_T$, separation of the jets into two hemispheres and the reconstruction of the Higgs boson mass from two jets tagged as b jets. The event generation and simulation were performed and the results were made public in the CMS framework [58]. The discovery potential in the (m_A–$\tan\beta$) plane is shown in Fig. 10, where m_A is the mass of the CP-odd Higgs boson. The discovery region extends to H/A masses of $\sim$210 GeV/c^2 around $\tan\beta = 10$ and to $\sim$190 GeV/c^2 around $\tan\beta = 30$. For $\sim$ 190 GeV/c^2 the low $\tan\beta$ region was covered with $> 5\sigma$ significance down to $\tan\beta = 2$.

There is also scope for non-universality in scalar masses which can have interesting implications for the signatures at the LHC. It is known for quite some time that non-universal soft SUSY breaking terms for the scalars at a very high scale, may arise naturally within the supergravity framework [59–61]. For example, non-universal soft SUSY breaking terms may arise at the GUT scale due to renormalisation group evolutions from the Planck scale to the GUT scale [60]. On the other hand, if the rank of a GUT group is reduced by spontaneous symmetry breaking, one may obtain D-term contributions to scalar masses [61]. The size of these new contributions to scalar masses can be comparable to the universal mass m_0 in mSUGRA. These

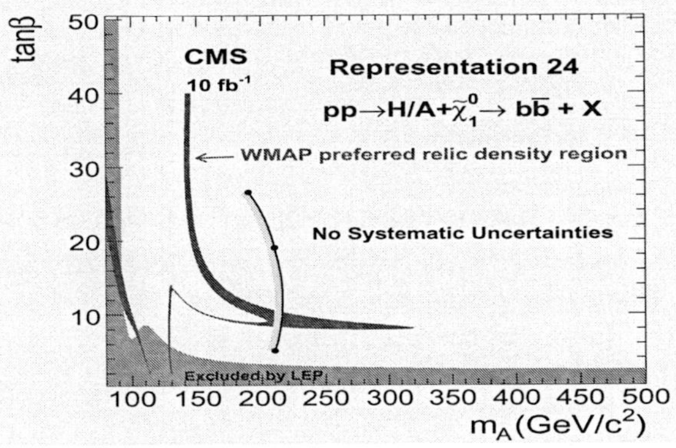

Figure 10. CMS discovery potential at 5σ level for $\tilde{q}, \tilde{g} \rightarrow \tilde{\chi}_1 H/A \rightarrow \tilde{\chi}_1 b\bar{b}$ in the representation **24** with an integrated luminosity of 10 fb^{-1}. No systematic uncertainties are included

contributions will in general have different values for different members of the same GUT multiplet which can lead to non-universal squark and slepton masses at the GUT scale. These non-universal terms are generation independent and do not lead to additional problems related to FCNCs. The low energy spectrum in these cases evolves from these high scale inputs. Because of these non-universalities of the scalar masses at a very high scale, the low energy sparticle spectrum can be very different from the universal scalar mass (m_0) case and the signatures at the LHC can also become significantly distinct. Non-universalities in the scalar masses may affect the SUSY signals through the gluino branching ratios as well as through the total squark-gluino production cross section [62]. Recently, the signatures at the LHC for the cases with high-scale scalar non-universality have been studied in various channels such as like and opposite sign dileptons, inclusive and hadronically quiet trileptons, inclusive jets, etc. [54,63]. It has been observed that these non-universalities affect the ratios of various final states produced through the decay cascades of sparticles. A more detailed discussion of these issues is beyond the scope of this article.

4.4. Split Supersymmetry

Let us discuss a more recent and very interesting scenario, namely split SUSY which assumes a very large splitting between the soft scalar and fermion masses in the MSSM. In order to understand this scenario better, let us first recall that the naturalness criterion has been one of the guiding principles in the formulation of the MSSM. This criterion demands that the masses of the superpartners should be somewhere around 1 TeV or so. However, the naturalness criterion is incompatible with the tiny cosmological constant. An important question, therefore arises. Is it possible to abandon the principle of naturalness and maintain the nice phenomenological aspects of the MSSM at the same time?

It has been argued [64] that the successful unification of gauge couplings of the MSSM can be retained even when all the scalars of the theory, except one fine-tuned light Higgs boson lie far above the EW scale. In this scenario, despite the loss of the original motivation to cure the hierarchy problem, one can still find a supersymmetric theory which is free of many of the undesirable features of the MSSM such as the flavour and CP problem, fast proton decay via dimension five operators, a tightly constrained mass of the lightest Higgs, etc. The gauginos and Higgsinos of this theory are chosen to lie near the TeV scale to ensure gauge coupling unification at $M_U \sim 10^{16}$ GeV as well as a stable LSP in the desirable mass range. These features describe the scenario of split SUSY.

Split SUSY is an effective theory in which the heavy scalars are integrated out and assumed to have degenerate mass $\tilde{m}$. The coupling constants at the scale $\tilde{m}$ are obtained by matching the Lagrangian describing the effective theory with the interaction terms of the SUSY Higgs doublets H_u and H_d. The Higgs doublet $h = -cos\beta\epsilon H_d^* + sin\beta H_u$ is fine-tuned to have a small mass term. One can identify a minimal split SUSY model described by four parameters: (1) the common mass $\tilde{m}$, (2) $\tan\beta$, (3) the Higgsino mass parameters $\mu(M_U)$ at the GUT scale and (4) the gluino mass $m_{\tilde{g}}$. It has been shown that certain special constraints [65] are imposed on the parameter space of the minimal split SUSY model by the infrared fixed point of the top Yukawa coupling. In order to study the RG evolution of various masses and couplings from the unification scale down to the EW scale, one should remember that between the unified scale M_U and the scale of heavy scalars $\tilde{m}$, the theory is described by the MSSM fields and we should use the MSSM RG equations in that region. Below the scale $\tilde{m}$, one should use the spectrum of split SUSY with gauginos and higgsinos included in the two-loop evolution.

Phenomenologically the most interesting feature of this scenario is that the gluinos are long-lived due to large squark masses which mediate their decays. Negative searches for anomalously heavy isotopes suggest that $\tilde{m} \lesssim 10^{13}$ GeV for 1 TeV gluino. In colliders these long-lived gluinos can produced displaced vertices. These gluinos can also hadronise into colour singlet states (called R-hadrons). If these states are neutral then they lose energy through hadronic interactions and if they are charged then they can deposit energy in the form of ionisation [66]. These types of signatures of the long-lived gluions are the 'smoking gun' signatures of split-SUSY. Collider phenomenology of split-SUSY models were also considered in [67]. It was found that the long-lived gluino can be discovered at the LHC even if their masses are above 2 TeV.

5. Conclusion

In this work we have reviewed how various SUSY breaking mechanisms lead to different soft masses of the sparticles at very high energies (Sections 2, 4.1–4.4). These masses in turn lead to model dependant sparticle spectrum at the energy scale of experimental interest (the EW scale). The nature of the lightest supersymmetric particle which carries the missing energy as well as the decay cascades of the heavier sparticles are also different in various SUSY breaking scenarios of contemporay interest. This has very interesting implications for cosmology and collider physics.

The expected SUSY signatures at the LHC have been studied by Monte carlo simulations. The generic $\mathrm{n-leptons + m-jets} + \not{E}_T$; $n, m = 0, 1, 2, 3..$ signatures are primarily due to squark-gluino production with large cross sections followed by cascade decays of these sparticle. Even if the squarks and gluinos have masses as large as 2.5 TeV, they are likely to be within the striking range of the LHC in most models and in most cases the jets + $\not{E}_T$ signal appears to be the most potent search channel. However, a characteristic signature of GMSB (Section 4.1), which distinguishes it from other SUSY breaking mechanisms, is the additional abundance of photons in the final states - a smoking gun signal.

The other search channels with different values of m and n may provide complementary information. The flavour tagging facilities at the LHC like b-jet tagging and τ-jet tagging may be helpful in identifying specific regions of the parameter space (Section 3.2 and references there in).

In order to establish SUSY beyond doubt it is essential to reconstruct the masses and other quantum numbers of the sparticles - a daunting task indeed. Mass reconstruction is possible in some favourable scenarios (Section 3.3 and 4.1 and references there in).

Several mechanisms of SUSY breaking predicts long lived sparticles providing 'smoking gun' signals of the underlying models. Examples are the next-to-lightest-supersymmetric particle of the GMSB model (either the lightest neutralino or a τ-slepton), the lighter chargino of the AMSB model and the gluino in split SUSY. The charged tracks or displaced vertices associated with these long lived spartcles will be tantalising to observe.

Physics at the LHC promises to be stimulating for many years to come!

Acknowledgement: The work of AD was supported by the Research Project SR/S2/HEP-18/2003 granted by the Department of Science and Technology, Government of India. He thanks Professor R. Bhattacharya for a careful reading of the manuscript.

REFERENCES

1. P. Ramond, Phys. Rev. D3 (1971) 2415; A. Neveu and J.H. Schwarz, Nucl. Phys. B31 (1971) 86; J.L. Gervais and B. Sakita, Nucl. Phys. B34 (1971) 632; Yu. A. Gol'fand and E.P. Likhtman, JETP Lett.13 (1971) 323; D.V. Volkov and V.P. Akulov, Phys. Lett. B46 (1973)109; J. Wess and B. Zumino, Nucl. Phys. B70 (1974) 39
2. For textbooks see, e.g., *Perspectives in Supersymmetry,* ed. G.L.Kane, World Scientific(1998); *Theory and Phenomenology of sparticles*, M. Drees, R.M. Godbole and P.Roy; World Scientific(2004); *Weak scale Supersymmetry: From superfields to scattering events,*, H. Baer and X. Tata, Cambridge UK, Univ. Press(2006). For reviews, see, e.g., H. P. Nilles, Phys. Rept. **110** (1984) 1; H. E. Haber and G. Kane, Phys. Rept. **117** (1985) 75; S.P. Martin, hep-ph/9709356 (published in the first reference, ed. G. L. Kane)
3. For recent reviews see, e.g., A. Derevianko, S. G. Porsev, Eur.Phys.J.A32 (2007); J. Guena, M. Lintz and M. Bouchiat, Mod.Phys.Lett.A20 (2005) 375
4. for details see Tevatron Electroweak Working Group http:tevewwg.fnal.gov/; see also Precision electroweak physics at the Tevatron, by CDF and D0 Collaborations (Eric B. James for the collaboration), hep-ex/0701003
5. S. Weinberg, Phys. Rev. D13 (1976) 974; Phys. Rev. D19 (1979) 1277; E. Gildener, Pjhys. Rev. D14 (1976) 1667; L. Susskind, Phys. Rev. D20 (1979) 2619; G. 't Hooft, in *Recent developments in gauge theories*, Proceedings of the NATO Advanced Summer Institute, Cargese 1979, (Plenum, 1980)
6. S. Dimopoulos and S. Raby, Nucl. Phys. B192 (1981) 353; E. Witten, Nucl. Phys. B188 (1981) 513; M. Dine, W. Fischler, and M. Srednicki, Nucl. Phys. B189 (1981) 575; S. Dimopoulos and H. Georgi, Nucl. Phys. B193 (1981) 150; N. Sakai, Z. Phys. C11 (1981) 153; R. K. Kaul, Phys.Lett.B109 (1982) 19; R.K. Kaul and P. Majumdar, Nucl. Phys. B199 (1982) 36
7. See any modern text book on quntum field theory like *An Introduction to Quantum Field Theory*, M. E. Peskin and D. V. Schroeder, Westview Press(1995). First Indian edition Levant Books, Kolkata (2005)
8. Some standard text books are *Grand Unified Theo-*

ries, G. G. Ross, Benjamin-Cummins (1984); *Unification and Supersymmetry*, R. N. Mohapatra, Springer Verlag (1986)
9. A.H. Chamseddine, R. Arnowitt, and P. Nath, Phys. Rev. Lett. 49 (1982) 970; R. Barbieri, S. Ferrara, and C.A. Savoy, Phys. Lett. B119 (1982) 343; L. J. Hall, J. Lykken and S. Weinberg, Phys. Rev. **D27** (1982) 2359; N. Ohta, Prog. Theor. Phys. **70**, 542 (1983).
10. K. Inoue, A. Kakuto, H. Komatsu and S. Takeshita, Prog. Theor. Phys. **67** (1982) 1889; ibid. 68 (1982) 927; Erratum-ibid. **70** (1983) 330
11. L. E. Ibanez and G. G. Ross, Phys. Lett. B **110**, (1982) 215; K. Inoue, A. Kakuto, H. Komatsu and S. Takeshita, Prog. Theor. Phys. **68**, (1982) 927; Erratum-ibid. **70**, (1983) 330; J. R. Ellis, J. S. Hagelin, D. V. Nanopoulos and K. Tamvakis, Supergravity," Phys. Lett. B **125**, (1983) 275; L. Alvarez-Gaume, J. Polchinski and M. B. Wise, Nucl. Phys. **B221**, (1983) 495
12. For the latest limits on the sparticle masses from LEP experiments see: http://lepsusy.web.cern.ch/lepsusy/
13. For a recent review with almost upto date experimental limits on sparticle masses see, L. Pape and D. Treille, Rep. Prog. Phys. **69** 2843
14. For the latest results from CDF see http://www-cdf.fnal.gov/physics/exotic/exotic.html
15. For D0 results on new phenomena see http://www-d0.fnal.gov/Run2Physics/WWW/results/np.html
16. G.L. Kane and J.P. Leveille, Phys. Lett. B112 (1982) 227; P.R. Harrison and C.H.Llewellyn-Smith, Nucl. Phys. B213 (1983) 223 (Erratum: Nucl. Phys. B223 (1983) 542); S. Dawson, E. Eichten and C. Quigg, Phys. Rev. D31, (1985) 1581; E. Reya and D.P. Roy, Phys. Lett. **B141** (1984)442; Phys. Rev. D32 (1985) 645
17. Aseshkrishna Datta, A. Djouadi, M. Guchait, F. Moortgat, Nucl.Phys. B681 (2004) 31
18. W. Beenaker, R. Hopker, M. Spira and P. M. Zerwas, Nucl.Phys. B492 (1917) 51
19. For a review see, e.g., T. Sjostrand, hep-ph/0611247
20. G. Corcella *et al*, J. High. Energy Phys. **JHEP01** (2001) 010
21. F. E. Paige, S. D. Protopopescu, H. Baer and X. Tata, hep-ph/0312045
22. T. Sostrand *et al*, Comput. Phys. Commoun. , **135** (2001) 238
23. The CMS collaboration, CERN-LHCC-2006-021(2006); J. Phys G: Nucl. Part. Phys. 34 (2007) 995
24. ATLAS Collaboration, Expected Performance of the ATLAS Experiment, Detector, Trigger and Physics, CERN-OPEN-2008-020, Geneva, 2008, to appear
25. N. Bhattacharyya, Amitava Datta and S. Poddar, arXiv:0807.0278 (to appear in Phys. Rev. D)
26. M.Guchait and D.P.Roy, Phys. Lett. B541(2002)356
27. U. Chattopadhyay, D. Das, Amitava Datta, S. Poddar, Phys.Rev.D76 (2007) 055008
28. R.H.K. Kadala, P.G. Mercadante, J.K. Mizukoshi, Xerxes Tata, Eur.Phys.J. **C56** (2008) 511; Daniel Feldman, Zuowei Liu, Pran Nath, Phys.Rev.D78 (2008) 083523
29. K.L.Chan, U. Chattopadhyay and P.Nath, Phys. Rev. D58 (1998) 096004; J. Feng, K. Matchev and T. Moroi, Phys. Rev. Lett. 84 (2000) 2322, Phys. Rev D61 (2000) 075005
30. U. Chattopadhyay, Amitava Datta, Aseshkrishna Datta, Anindya Datta and D.P.Roy, Phys. Lett. B493 (2000) 127; P.G.Mercadante, J.K.Mizukoshi and X. Tata, Phys. Rev D72 (2005) 035009; H. Baer, V. Barger, G. Shaughnessy, H. Summy and L.-t. Wang, Phys.Rev. **D75** (2007) 095010; S.P.Das, Amitava Datta, M.Guchait, M. Maity, S. Mukherjee, Eur. Phys. J C54 (2008) 645
31. B.C.Allanach, C.G.Lester, M. A. Parker and B.R.Webber, JHEP 0009 (2000) 004; M. M. Nojiri,Y. Shimizu, S. Okada and K. Kawagoe, JHEP 806 (2003) 035. W.S. Cho, K. Choi, Y.G.Kim, C. B. Park, JHEP 0802 (2008) 035; G.G.Ross, M. Serna, Phys. Lette B665 (2008) 212; D. J. Miller, P. Osland and A.R.Raklev, JHEP 03 (2006) 034; M. M. Nojiri, G. Polesello and D.R.Tovey, JHEP 0805 (2008) 14
32. M. Dine and W. Fischler, Phys. Lett. B**110**, (1982) 227; C.R. Nappi and B.A. Ovrut, Phys. Lett. B**113**, (1982) 175; L. Alvarez-Gaumé, M. Claudson and M.B. Wise, Nucl. Phys. B **207**, (1982) 96; For reviews, see, e.g. C. Kolda, ArXiv: hep-ph/9707450; G.F. Giudice and R. Rattazzi, Phys. Rept. **322**, (1999) 419
33. See, e.g., S. Dimopoulos, S. Thomas, and J.D. Wells, Phys. Rev. D **54**, (1996) 3283; Nucl. Phys. B**488**, (1997) 39; B. Mukhopadhyaya and S. Roy, Phys. Rev. D **57**, (1998) 6793; J.A. Bagger *et al.*, Phys. Rev. D**55**, (1997) 3188
34. H. Baer *et al.*, Phys. Lett. B **435**, (1998) 109
35. S. Ambrosanio *et al.*, Phys. Rev. D **54**, (1996) 5395
36. K. Kawagoe *et al.*, Phys. Rev. D**69**, (2004) 035003
37. J.L. Diaz-Cruz, D.K. Ghosh, and S. Moretti, Phys. Rev. D**68**, (2003) 014019
38. S. Dimopoulos, M. Dine, S. Raby, S. Thomas, J.D. Wells, Nucl. Phys. B (Proc. Suppl.) **52A**, (1997) 38
39. See the second reference in [34]
40. L. Randall and R. Sundrum, Nucl. Phys. B**557**, (1999) 79; G.F. Giudice *et al.*, J. High Energy Phys. **12**, (1998) 027
41. J.L. Feng *et al.*, Phys. Rev. Lett. **83**, (1999) 1731; S. Roy, Mod. Phys. Lett. **A19**, (2004) 83 and references therein. T. Gherghetta, G.F. Giudice, and J.D. Wells, Nucl. Phys. B **559**, (1999) 27
42. M. Ibe, T. Moroi, and T.T. Yanagida, Phys. Lett. B**644**, (2007) 355; S. Asai *et al.*, Phys. Lett. B**653**, (2007) 81
43. H. Baer, J.K. Mizukoshi, and X. Tata, Phys. Lett. B**488**, (2000) 367
44. A.J. Barr, B.C. Allanach, C.G. Lester, M.A. Parker, and P. Richardson, J. High Energy Phys. **03**, (2003) 045
45. E. Cremmer, S. Ferrara, L. Girardello, and A. Van Proeyen, Phys. Lett. B**116**, (1982) 231

46. J. Ellis, K. Enqvist, D.V. Nanopoulos, and K. Tamvakis, Phys. Lett. B**155**, (1985) 381; M. Drees, Phys. Lett. B**158**, (1985) 409; G. Anderson *et al* hep-ph/9609457
47. K. Huitu, J. Laamanen, P.N. Pandita, and S. Roy, Phys. Rev. D**72**, (2005) 055013
48. G. Anderson, H. Baer, C.H. Chen, and X. Tata, Phys. Rev. D**61**, (2000) 095005
49. K. Huitu, Y. Kawamura, T. Kobayashi, and K. Puolamaki, Phys. Rev. D**61**, (2000) 035001
50. G. Belanger, F. Boudjema, A. Cottrant, A. Pukhov, and A. Semenov, Nucl. Phys. **B706**, (2005) 411
51. A. Djouadi, Y. Mambrini, and M. Muhlleitner, Eur. Phys. J. **C20**, (2001) 563
52. S.I. Bityukov and N.V. Krasnikov, Phys. Atom. Nucl. **65**, (2002) 1341; Yad. Fiz. **65**, (2002) 1374
53. K. Huitu, R. Kinnunen, J. Laamanen, S. Lehti, S. Roy, and T. Salminen, arXiv:0808.3094 [hep-ph]
54. S. Bhattacharya, Aseshkrishna Datta, and B. Mukhopadhyaya, arXiv:0809.2012 [hep-ph]
55. P. Bandyopadhyay, Aseshkrishna Datta, and B. Mukhopadhyaya, arXiv:0806.2367 [hep-ph]
56. S. Bhattacharya, Asehkrishna Datta, and B. Mukhopadhyaya, J. High Energy Phys. **10**, (2007) 080
57. H. Baer, M. Bisset, X. Tata, and J. Woodside, Phys. Rev. D**46**, (1992) 303
58. CMS PRS, http://cmsdoc.cern.ch/cms/PRS/results/susybsm/rep24/rep24.html
59. See, e.g., M. Olechowski and S. Pokorski, Phys. Lett. B**344**, (1995) 201; T. Kobayashi, D. Suematsu, K. Yamada, and Y. Yamagishi, Phys. Lett. B**348**, (1995) 402
60. P. Moxhay and K. Yamamoto, Nucl. Phys. **B256**,(1985) 130; B. Gato, Nucl. Phys. **B278**, (1986) 189; N. Polonsky and A. Pomarol, Phys. Rev. D**51**, (1995) 6532
61. M. Drees, Phys. Lett. B**181**, (1986) 279; J.S. Hagelin and S. Kelly, Nucl. Phys. **B342**, (1990) 95; Y. Kawamura, H. Murayama, and M. Yamaguchi, Phys. Rev. D**51**, (1995) 1337
62. Amitava Datta, Aseshkrishna Datta, and M.K. Parida, Phys. Lett. B**431**, (1998) 347; Amitava Datta, Aseshkrishna Datta, M. Drees, and D.P. Roy, Phys. Rev. D**61**, (2000) 055003
63. S. Bhattacharya, Aseshkrishna Datta, and B. Mukhopadhyaya, Phys. Rev. D**78**, (2008) 035011
64. N. Arkani-Hamed and S. Dimopoulos, J. High Energy Phys. **06**, (2005) 073; G.F. Giudice and A. Romanino, Nucl. Phys. B**699**, (2004) 65; *Erratum-ibid.* B **706**, (2005) 487
65. K. Huitu, J. Laamanen, P. Roy, and S. Roy, Phys. Rev. D**72**, (2005) 055002
66. J.L. Hewett *et al.*, J. High Energy Phys. **09**, (2004) 070
67. W. Kilian, T. Plehn, P. Richardson, and E. Schmidt, Eur. Phys. J. **C39**, (2005) 229

Dark Matter and the LHC

Howard Baer[a] *and Xerxes Tata*[b]

[a]*Department of Physics, Florida State University, Tallahassee, FL, 32306 USA*

[b]*Department of Physics and Astronomy, University of Hawaii, Honolulu, HI, 96822 USA*

An abundance of astrophysical evidence indicates that the bulk of matter in the universe is made up of massive, electrically neutral particles that form the dark matter (DM). While the density of DM has been precisely measured, the identity of the DM particle (or particles) is a complete mystery. In fact, within the laws of physics as we know them (the Standard Model, or SM), none of the particles have the right properties to make up DM. Remarkably, many new physics extensions of the SM – designed to address theoretical issues with the electroweak symmetry breaking sector – require the introduction of new particles, some of which are excellent DM candidates. As the LHC era begins, there are high hopes that DM particles, along with their associated new matter states, will be produced in *pp* collisions. We discuss how the LHC experiments, along with other DM searches, may serve to determine the identity of DM particles and elucidate the associated physics. Most of our discussion centres around theories with weak-scale supersymmetry, and allows for several different DM candidate particles.

1. Introduction

The LHC programme has been described as the greatest experiment ever to be mounted in physics. Certainly this seems to be true on many different levels: the largest, costliest, most massive detectors; the most collaborators per experiment; the highest energy reach of any accelerator experiment. The intellectual stakes of the LHC programme are enormous: on the theory side, the extreme sensitivity of the scalar sector of the Standard Model (SM) to very high scale physics beckons for new physics at the weak scale ($\sim 100-1000$ GeV), possibly ushering in a new paradigm for the laws of physics.

We discuss how the LHC experiments may serve to validate the extended Copernican principle. In previous times, we have learned that the earth is *not* the centre of the solar system, that our galaxy is *not* the entire universe, and that we do *not* live in any special place or time. Now, due to an impressive accumulation of astrophysical data, we learn that our star, our planet and ourselves are *not even made up of the dominant form of matter in the universe.* It now appears that most of the matter in the universe – the so-called *dark matter* (DM) – must consist of massive, electrically and (likely) colour neutral particles that were produced with non-relativistic velocities (cold DM or CDM) in the early universe. None of the particles of the SM have the right properties to make up CDM. Thus, CDM constitutes decisive evidence for physics beyond the SM [1]!

Compelling arguments suggest the CDM particle is linked to the weak nuclear interactions, and further that it has a mass of order the weak scale: $\sim 100-1000$ GeV. This is often referred to as the *WIMP miracle*, and the DM particles referred to as WIMPs (weakly interacting mass particles). Many attractive theoretical scenarios designed to ameliorate the extreme sensitivity of the scalar sector of the SM to radiative corrections, naturally include candidates for CDM particles with weak scale masses that interact with ordinary matter with cross sections comparable to those for weak nuclear interactions. Regardless of its origin, if CDM is composed of WIMPs, then it may be possible to produce and study the DM particle(s) directly at the LHC. In fact, the LHC may well turn out to be a DM factory, where the nature of DM particles and their properties might be studied in a controlled environment. In any collider experiment, WIMPs would be like neutrinos in that they would escape the detector without depositing any energy in the experimental apparatus, resulting in an *apparent imbalance of energy and momentum* in collider events. While WIMPs would manifest themselves only as *missing (transverse) energy* at collider experiments, it should nevertheless be possible to study the *visible* particles produced in WIMP-related production and decay processes to study the new physics associated with the WIMP-sector.

Indeed, there exists a real possibility that much of the mystery surrounding DM and its properties can be cleared up in the next decade by a *variety* of experiments already operating or soon-to-be deployed. In this effort, experiments at the LHC will play a crucial role. There are – in tandem with the LHC – a variety of other DM search experiments already in operation, or in a deployment or planning phase. *Direct Detection* (DD) experiments seek to directly measure relic DM particles left over from early stages of the Big Bang. These DD experiments range from terrestrial microwave cavities that search for axions via their conversion to photons, to crystaline or noble liquid targets located deep underground that search for WIMP-nucleon collisions.

DM can also be searched for in *indirect detection* (ID) experiments. In ID experiments, one searches for WIMP-WIMP annihilation into various SM particles including neutrinos, gamma rays and anti-matter. Clearly, this technique applies only if the DM is self-conjugate, or if DM particles and anti-particles are roughly equally abundant. One ID search method involves the use of neutrino telescopes mounted deep under water or in polar ice. The idea is that if relic WIMPs are the DM in our galactic halo, the sun (or earth) will sweep them up as they traverse their galactic orbits, and gravitationally trap these in the central core where they can accumulate, essentially at rest, to densities much higher than in the Milky Way halo. These accumulated WIMPs can then annihilate one with another into SM particles with energies $E \lesssim m_{\rm WIMP}$. Most SM particles would be immediately absorbed by the solar material. However neutrinos can easily escape the sun. Thus, WIMP annihilation in the sun results in an isotropic flux of *high energy* neutrinos from the solar core – these energies are impossible to produce via conventional nuclear reactions in the sun – some of which would make it to earth. These neutrinos ocassionally interact with nuclei in ocean water or ice and convert to a high energy muon, which could then be detected via Cerenkov radiation by photomultiplier tubes that are parts of neutrino telescopes located within the medium.

Another possibility for ID is to search for the by-products of WIMP annihilation in various regions of our galactic halo. Even though the halo number density of WIMPs would be quite low, the volume of the galaxy is large. Ocassionally one expects relic WIMP-WIMP annihilation to SM particles. The trick is then to look for rare anti-matter production or high energy gamma ray production from these WIMP halo annihilations. A variety of land-based, high altitude and space-based anti-matter and gamma ray detectors have been or are being deployed. The space-based Pamela experiment is searching for positrons and anti-protons. The land-based HESS telescope will soon be joined by the GLAST satellite in the search for high energy gamma rays. While high energy anti-particles would provide a striking signal, these lose energy upon deflection when traversing the complicated galactic magnetic field and so can only be detected over limited distances. Gamma rays, on the other hand, are undeflected by magnetic fields and so have an enormous range. Moreover, these would point back to their point of origin. Thus, the galactic centre, where DM is expected to accumulate at a high density, might be a good source of GeV-scale gamma rays resulting from WIMP-WIMP annihilation to vector boson ($V = W, Z$) pairs or to quark jets, followed by $(V \to)q \to \pi^0 \to \gamma\gamma$ after hadronisation and decay.

If WIMPs and their associated particles are discovered at the LHC and/or at DD or ID search experiments, it will be a revolutionary discovery. But it will only be the beginning of the story as it will usher in a new era of *DM astronomy*! The next logical step would be the construction of an e^+e^- collider of sufficient energy so that WIMP (and related particles) can be produced and studied with high precision in a clean, well-controlled experimental environment. The precise determination of particle physics quantities associated with WIMP physics will allow us to *deduce* the relic density of these WIMPs within the standard Big Bang cosmology. If this turns out to be in agreement with the measured relic density, we would have direct evidence that DM consists of a single component. If the predicted relic density is too small, it could make the case for multiple components in the DM sector. If the predicted density is too large, we would be forced to abandon the simplest picture and seek more complicated (non-thermal) mechanisms to account for the measurement, or deduce that this detected WIMP itself is unstable. The determination of the properties of the DM sector will also serve as a tool for a detailed measurement of astrophysical quantities such as the galactic and local WIMP density and local velocity profiles, which could shed light on the formation of galaxies and on the evolution of the universe.

2. Evidence for Dark Matter

DM in the universe was first proposed in the 1930s by astronomer Fritz Zwicky [2]. In the 1970s and on evidence for DM accrued at an accelerating pace. Here we discuss the major classes of evidence for DM in the universe.

- *Galactic clusters*: In the 1930s, Zwicky studied nearby clusters of galaxies, bound to each other by gravity in spite of the expansion of the universe. Using arguments based on the virial theorem from classical mechanics, Zwicky concluded there was not enough visible mass within the galactic clusters to successfully bind them; he thus concluded that there must be large amounts of non-luminous or DM, existing within the clusters.

- *Rotation curves*: In the 1970s, V. Rubin and W.K. Ford [3] began an intensive study of the rotation curves of galaxies. They were able to measure stellar velocity as a function of distance from the galactic centre. With most of the visible matter concentrated in or around the galactic centre, one expects the stellar rotational velocities to fall off with distance from the galactic centre in accord with Newtonian gravitation. Instead, the stellar velocities tended to flatness out to the furthest distances which could be probed. This is in accord with a diffuse halo of dark particles surrounding the galaxy out to the furthest distances.

- *Lensing*: In General Relativity, the path of light through spacetime is bent, or "lensed" as it passes by a large mass distribution. Lensing effects are observed when light from distant galaxies or clusters passes by large mass distributions. Numerous studies of both strong and weak (statistical) lensing show the presence of large quantities of DM in the universe.

- *Hot gas in clusters*: Hot gas bound to clusters of galaxies can be mapped out by the emitted x-rays. The visible mass in these galaxies would not have enough gravity to bind the hot gas, which requires additional binding from putative DM.

- *Cosmic microwave background (CMB)*: Detailed studies of anisotropies in the CMB has resulted a very precisely measured CMB power spectrum. The peaks and valleys in this spectrum are extremely sensitive to the composition of the universe, and independently show that the universe is comprised of about 70% dark energy (DE), 25% DM and 4% baryons, along with tiny fractions of neutrinos and photons. Thus the "known stuff" makes up just about 5% of the content of our universe.

- *Large scale structure*: Measurements of large scale structure, when compared to simulations of the evolution of structure in the universe, match very well with a universe composed of both CDM (possibly with some warm DM) and DE.

- *Big Bang nucleosynthesis*: One of the triumphs of Big Bang cosmology is that given an initially hot, dense universe, one can calculate the abundances of the light elements produced via nucleosynthesis during the first few hundred seconds. The measured abundances agree with observation if the baryon-to-photon ratio $\eta_B \equiv n_B/n_\gamma \sim 6 \times 10^{-10}$. The photon number density is known from thermodynamics, so this implies a baryonic mass density of the universe of about $\sim$ 4%, consistent with the value independently obtained from CMB data discussed above.

- *Distant supernovae probes*: Probes of distant supernovae [4] have allowed an extension of the Hubble diagram out to redshifts of $z \sim 1$. A best fit match to the Hubble diagram indicates the presence of both DE and DM in the universe.

- *Colliding galactic clusters*: Observation of colliding clusters of galaxies – a recent example comes from the so-called bullet cluster – shows an actual separation of DM (deduced from lensing) from the gaseous halo made of baryonic matter. This is exactly what is expected if a vast halo of non-interacting DM accompanies the luminous matter and gas in galactic clusters.

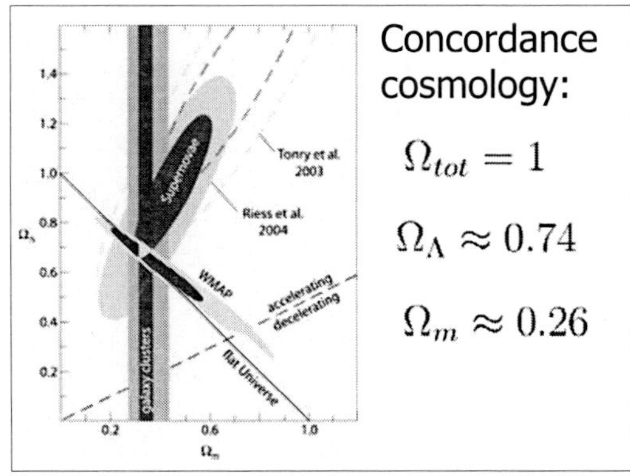

Figure 1. Measurements from CMB, large scale structure and supernovae plotted in the $\Omega_\Lambda\, vs.\, \Omega_{matter}$ plane. Adapted from *http://www.astro.washington.edu/astro323/WebLectures/*

The ΛCDM universe: Collating all the data together, especially that from CMB, red shifts of high-z supernovae, and large scale structure, allows one to fit to the composition of the universe. We see from Fig. 1 that these very diverse data find consistency amongst themselves, leading to the so-called "concordance" model for the universe, the ΛCDM model. (Here, Λ stands for Einstein's cosmological constant, which may be the source of the DE.) In the ΛCDM model, the universe is composed of about 70% DE, 25% DM, 4% baryons with a tiny fraction of neutrinos and radiation. The measured abundance of CDM in our universe [5],

$$\Omega_{CDM}h^2 = 0.111^{+0.011}_{-0.015} \ (2\sigma), \tag{1}$$

where $\Omega_{CDM} = \rho_{CDM}/\rho_c$, with ρ_{CDM} the CDM mass density, ρ_c the critical closure density and h is the scaled Hubble parameter, serves as a severe constraint on all particle physics theories that include a DM candidate. Since DM may well consist of more than one component, strictly speaking the relic density serves as an upper bound $\Omega_X h^2 \leq 0.122$ on the density of any single component X. We now turn to a discussion of some of the particle physics candidates for the DM particle X.

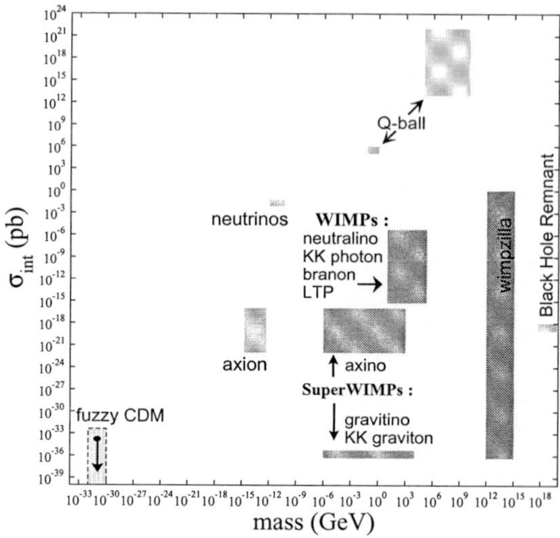

Figure 2. Dark matter candidates in the mass vs. interaction strength plane, taken from Ref. [7]

3. DM Candidates

While the evidence for the existence of DM in the universe is now very convincing, and while the density of DM in the universe is becoming precisely known, the identity of the DM particle(s) is a complete mystery. None of the particles in the SM have the right properties to make up CDM. Many candidates, however, have been proposed in the theoretical literature. To appreciate the variety of candidate particles proposed, we list a number of possibilities. The range of masses and interaction strengths of many of these candidates is shown in Fig. 2.

- *Neutrinos*: Massive neutrinos are weakly interacting neutral massive particles and so are natural candidates for the DM in the universe [6]. It is now known that the usual active neutrinos are so light that they could not give rise to the observed structure in the universe because these would move faster than the typical galactic escape velocity, and so cannot cause the clumping that large scale structure simulations require. They are usually referred to as *hot* DM or HDM, and are likely to be a subdominant component of the DM in the universe. There are, however proposals for much heavier, CDM gauge singlet neutrinos that are not part of the SM [8].

- *Planck mass black hole remnants*: It is possible many tiny black holes (BHs) were produced in the early universe. Ordinarily, these BHs would decay via Hawking radiation. However, it has been suggested that once they reach the Planck mass, quantum gravity effects forbid further radiation, making them stable and hence good CDM candidates [9].

- *Q-balls*: These objects are topological solitons that occur in quantum field theory [10,11].

- *Wimpzillas*: These very massive beasts were proposed to show that viable DM candidates could have masses far beyond the weak scale [12].

- *Axions*: The symmetries of the QCD Lagrangian allow the term $-\mathcal{L} \ni \frac{\theta_{QCD}}{32\pi^2}F^{\mu\nu}\tilde{F}_{\mu\nu}$ – which gives rise to CP violation in the strong interactions. However measurements of the neutron electric dipole moment (EDM) require $\theta_{QCD} \lesssim 10^{-10}$. Why this parameter is so much smaller than its natural value of ~ 1 is referred to as the strong CP problem. The most compelling solution to the strong CP problem – the Peccei-Quinn-Weinberg-Wilczek solution [13] – effectively replaces the parameter θ_{QCD} by a quantum field, and the potential energy allows the field to relax to near zero strength. However, a remnant of this procedure is that a physical pseudoscalar boson – the axion a – remains in the spectrum. The axion is an excellent candidate for CDM in the universe [14]. Its favoured mass range is $m_a \sim 10^{-5}$–10^{-3} eV, where the lower bound gives too high a relic density, and the upper bound comes from limits on stellar cooling. Axions have a very weak but possibly observable coupling to two photons. They are at present being searched for in terrestrial microwave cavity experiments such as ADMX [15]. Since they have little direct impact on the LHC physics, we will not dwell on them in as much detail as some other possible candidates.

- *WIMPs and the WIMP miracle*: Weakly interacting neutral, massive particles occur in many particle physics models where the SM is extended to address the physics associated with electroweak symmetry breaking (EWSB). If the associated new particles sector has a conserved "parity-like" quantum number that distinguishes it from the SM sector, the lightest particle in this new sector is stable and (if electrically and colour neutral) frequently makes an excellent DM candidate. Examples of WIMP particles come from 1. lightest neutralino state in SUSY theories with conserved R-parity [16], 2. lightest Kaluza-Klein excitations from extra-dimensional theories with conserved KK-parity [17,18] and 3. lightest T-odd particles in Little Higgs theories with conserved T-parity [19–22].[1]

  It is possible to calculate the *thermal* WIMP abundance from the Big Bang using very general principles. The initial condition is that at early universe temperatures $T > m_{WIMP}$, the WIMPs would have been in thermal equilibrium with the cosmic soup. In this case, their abundance follows straightforwardly from equilibrium statistical mechanics. As the universe expands and cools, ultimately the WIMPs fall out of thermal equilibrium at a temperature where the expansion rate of the universe equals the WIMP annihilation rate, because then the WIMPs are unable to find one another to annihilate fast enough: this is known as the freeze-out temperature T_F. As a result, the WIMP density does not drop exponentially as the universe continues to cool, but reduces only as R^{-3} due to the expansion of the universe. The WIMP abundance after freeze-out can be found by solving the Boltzmann equation in a Friedman-Robertson-Walker universe for the WIMP number density. The WIMP mass density today, $\rho(T_0)$, is then given by

$$\rho(T_0) = \left(\frac{T_0}{T_\gamma}\right)^3 T_\gamma^3 \sqrt{\frac{4\pi^3 g_* G_N}{45}} \left[\int_0^{x_F} \langle \sigma v_{\rm rel}\rangle dx\right]^{-1}$$

  where $T_\gamma = 2.72$ K is the current temperature of the CMB, T_0 is the corresponding neutralino temperature, $g_* \sim 100$ is the number of relativistic degrees of freedom at WIMP freeze-out, $\langle\sigma v\rangle$ is the thermally averaged WIMP annihilation cross section times relative velocity and $x_F = T_F/m_{\rm WIMP} \simeq 1/20$ is the scaled freeze-out temperature. But for the fact that photons are reheated as various species decouple, the temperatures of the WIMPs and photons would have been the same. Since the reheating process is assumed to be isentropic, the ratio $\left(\frac{T_\gamma}{T_0}\right)^3$ is simply given by the ratio of the number of effective degrees of freedom at freeze-out to that today and is about 20. Dividing by the closure density $\rho_c = 8.1 \times 10^{-47} h^2$ GeV^4 then gives us $\Omega_{WIMP}h^2$, where h is the Hubble parameter in units of 100 km/s/Mpc. For s-wave annihilation, $\langle\sigma v\rangle$ is independent of x; then, for $\Omega h^2 \sim 0.1$, we find it is about 10 pb – about the size of an electroweak cross section for annihilation of non-relativistic particles with a mass of about 50 GeV, not far from the weak scale! This provides independent astrophysical evidence that new physics – the DM particle – may well be lurking at the weak scale! The co-incidence of the scale of DM with the scale of EWSB is sometimes referred to as *the WIMP miracle*, and suggests that the new physics that governs EWSB may coincide with the DM sector, and inspires many to believe that WIMPs are the prime candidate to constitute the CDM of the universe.[2]

- *SuperWIMPs*: SuperWIMPs are electrically and colour neutral stable DM candidates that interact with much smaller strength (perhaps only gravitationally) than WIMPs. Such particles often occur in particle physics theories that include WIMPs. Examples include 1. the lightest $n = 1$ level KK *graviton* $G^1_{\mu\nu}$ in extra-dimensional theories, 2. the *gravitino* $\tilde{G}$ (the superpartner of the graviton) in SUSY theories and 3. the *axino* $\tilde{a}$ (the fermionic member of the axion supermultiplet). Since superWIMP interactions with

[1]We point out that it has recently been argued [23] that T-parity is generically not conserved because of anomalies in the quantum theory. It has, however, been pointed out that whether T-parity is or is not conserved can only be definitively addressed only in the context of a UV-completion of the model [24].

[2]See, however, Ref. [25].

ordinary matter have strengths far below conventional weak interaction strengths, they are not expected to yield observable signals in DD or ID search experiments. However, they can lead to intriguing new phenomena at collider experiments such as the LHC and ILC. If every WIMP decays to a superWIMP, then superWIMPs inherit the thermally produced number density of WIMPs, and their contribution to $\Omega_{CDM}h^2$ is reduced from the corresponding would-be WIMP contribution by the ratio of the superWIMP to WIMP masses. The superWIMPs produced from WIMP decay may be either warm or CDM depending on the WIMP lifetime and WIMP-superWIMP mass gap [26]. SuperWIMPs may also be produced during the re-heating of the universe after inflation; this component of their relic abundance is cold, and its magnitude depends on the reheating temperature T_R.

Of the possibilities mentioned above, supersymmetry stands out for several reasons. Weak scale supersymmetry provides an elegant mechanism to stabilise the weak scale against runaway quantum corrections to the Higgs scalar mass that arise when the SM is embedded into a larger theory that includes particles with masses hierarchically larger than the weak scale, *e.g.* grand unified theories (GUTs). Unless the Higgs boson mass parameter is tuned with uncanny precision, these corrections drive the weak scale as well as the physical Higgs boson mass to the GUT scale. The supersymmetric extension of the SM, with weak scale superpartners requires no such a fine tuning, and (unlike many examples discussed above) provides a framework that is perturbatively valid all the way up to the GUT or Planck scale.

SUSY theories thus naturally meld with GUTs, preserving many of their successes, and providing successful predictions where non-SUSY GUTS appear to fail. The latter include the celebrated unification of gauge couplings and the value of the ratio m_b/m_τ. In many SUSY models with unified values of scalar mass parameters renormalised at an ultra-high energy scale, radiative corrections drive the weak scale squared Higgs boson mass parameter to negative values triggering EWSB if the top quark mass is in the range 150-200 GeV. This radiative EWSB mechanism was discovered in the mid-1980s, well before the top mass was determined to be $\sim$ 172 GeV by experiments at the Fermilab Tevatron. In addition, fits to precision electroweak measurements – plotted on the m_t *vs.* M_W plane – now indicate a slight preference for SUSY (with light sparticles) over the SM [27].

Although weak scale SUSY theories have the very attractive features noted above, the presence of many new scalar fields also gives rise to potential new problems not present in the SM. If supersymmetry is broken in an *ad hoc* manner, flavour-changing processes (that do not also change electric charge) occur at unacceptably large rates, as do some CP-violating processes. This is probably a clue about the (presently unknown) mechanism by which the superpartners acquire SUSY-breaking masses. But the most severe problem caused by the appearance of scalars is that we can write *renormalisable* interactions that violate baryon and/or lepton number conservation. These interactions would cause the proton to decay within a fraction of a second, in sharp contrast to a lower limit on its life-time in excess of 10^{29} years (independent of the mode of decay)! To forbid these potentially disastrous interactions, we need to posit an additional conservation law, which is often taken to be the conservation of a parity-like quantum number (referred to as R-parity) taken to be $+1$ for ordinary particles and -1 for their SUSY partners. As a result, the lightest SUSY particle must be stable (since all lighter particles have $R = +1$).

Unlike the SM, SUSY theories with a conserved R-parity naturally include several candidates for DM. All that is needed is that the lightest superpartner be electrically and colour neutral. These include, but are not limited to: 1. the lightest neutralino $\widetilde{Z}_1$, a true WIMP candidate, 2. the gravitino $\tilde{G}$, a gravitationally interacting spin-$\frac{3}{2}$ superWIMP candidate, 3. the spin-$\frac{1}{2}$ axino $\tilde{a}$, which is the superpartner of the axion and 4. the superpartner of a sterile neutrino. The superpartner of ordinary neutrinos is excluded as galactic DM because it would already have been detected by direct searches for DM. The axino interaction strength is between that of a true WIMP and a gravitino superWIMP.

Finally we remark here that the SM does not include a viable mechanism for *baryogenesis* in the early universe, primarily because the CP violation is too small. In SUSY theories, with their added richness, several mechanisms appear to be possible: electroweak baryogenesis, leptogenesis (which is connected to GUT theories and neutrino mass), so-called Affleck-Dine baryogenesis involving decay of flat directions of the SUSY scalar potential and finally, the possibility of inflaton decay to heavy neutrino states.

Despite the lack of direct evidence for SUSY, its many attractive features lead many theorists to expect weak scale supersymmetry to manifest itself as the next paradigm for the laws of physics. While SUSY could have fortuitously revealed itself in experiments at LEP or the Tevatron, the LHC is the first facility designed to directly probe the weak scale energy regime where superpartners are naturally expected. We will, for the most part, discuss supersymmetric theories in the remainder of this article and show that data from the LHC as well as from other DD and ID experiments will incisively test the weak scale SUSY idea. We will briefly return to other ideas with non-SUSY WIMPs in Sec. 7.

4. Supersymmetric Theories

The representations of the SM make a clear distinction between the "matter" and "force" sectors of the theory. The spin-half matter particles have different gauge quantum numbers from the spin-one gauge bosons (which necessarily must be in the adjoint representation of the gauge group) that mediate the strong and electroweak interactions. Spin-zero fields, which are essential for spontaneous EWSB (and which mediate a non-gauge force between particles), belong to yet another representation. In supersymmetric theories, where bosons and fermions belong to the same super-multiplet, bosons and fermions transform the same way, providing a level of synthesis never previously attained.

The *superfield* formalism, where bosonic and fermionic fields are combined into a single superfield, provides a convenient way for constructing supersymmetric models of particle physics. This is analogous to the familiar isospin formalism where particles of different charge are combined to form an isomultiplet. Chiral scalar superfields include one chiral-component of a spin-half fermion, together with a complex scalar field, the superpartner of this chiral fermion. A massive Dirac fermion necessarily has two chiral components, and so needs two chiral superfields to describe it. For example, the Dirac electron therefore has *two* complex scalar superpartners (denoted by $\tilde{e}_L$ and $\tilde{e}_R$), one corresponding to each chirality of the electron/positron. Notice that the number of polarisation states for fermions (four, because there are two polarisations each for the electron and positron) is exactly the same as the number of bosonic polarisation states (each complex spin-zero field corresponds to two polarisation states, one for the spin-zero particle, and one for the spin-zero antiparticle). This equality of bosonic and fermionic degrees of freedom is a general feature of SUSY models. Moreover, the gauge quantum numbers for the spin-zero partners of the chiral fermion fields must be the same as for the corresponding fermions, so that *the usual minimal coupling prescription completely fixes the gauge interactions of these particles.*

Gauge superfields include spin-1 gauge bosons along with spin-$\frac{1}{2}$ self-conjugate (or Majorana) gauginos, both tranforming under the adjoint representation. Finally, there are gravitational supermultiplets containing massless spin-2 graviton fields and spin-$\frac{3}{2}$ gravitinos. These are *all* representations of $N = 1$ supersymmetry, where there is just one super-charge. We will focus here only on $N = 1$ SUSY since it leads most directly to phenomenologically viable models with chiral fermions.

The superfield formalism [28–30] facilitates the construction of a supersymmetric version of the SM, known as the Minimal Supersymmetric Standard Model, or MSSM. As explained above, for each quark and lepton of the SM, the MSSM necessarily includes spin-0 superpartners $\tilde{q}_L$ and $\tilde{q}_R$ along with $\tilde{\ell}_L$ and $\tilde{\ell}_R$, whose gauge quantum numbers are fixed to be the known gauge quantum numbers of the corresponding fermions. Thus, for example, the right-handed up quark scalar (usually denoted by $\tilde{u}_R$) is a colour-triplet, weak isosinglet with the same weak hypercharge 4/3 as the right-handed up-quark. The MSSM thus includes a plethora of new scalar states: $\tilde{e}_L$, $\tilde{e}_R$, $\tilde{\nu}_{eL}$, $\tilde{u}_L$, $\tilde{u}_R$, $\tilde{d}_L$, $\tilde{d}_R$ in the first-generation, together with analogus states for the other two generations. Spin-zero *squark* partners of quarks with large Yukawa couplings undergo left-right mixing: thus, the $\tilde{t}_L$ and $\tilde{t}_R$ states mix to form mass eigenstates – $\tilde{t}_1$ and $\tilde{t}_2$ – ordered from lowest to highest mass.

The spin-0 Higgs bosons are embedded in Higgs superfields, so that the MSSM also includes spin-$\frac{1}{2}$ higgsinos. Unlike in the SM, the same Higgs doublet cannot give a mass to both up- and down-type fermions without catastrophically breaking the underlying supersymmetry. Thus the MSSM includes two Higgs doublets instead of one as in the SM. This gives rise to a richer spectrum of physical Higgs particles, including neutral light h and heavy H scalars, a pseudoscalar A and a pair of charged Higgs bosons $H^{\pm}$.

The gauge sector of the MSSM contains gauge bosons along with spin-half gauginos in the adjoint representation of the gauge group: thus, along with eight coloured gluons, the MSSM contains eight coloured spin-$\frac{1}{2}$ gluinos. Upon EWSB the four gauginos of $SU(2)_L \times U(1)_Y$ mix (just as the $SU(2)_L$ and $U(1)_Y$ gauge bosons mix) amongst themselves *and* the higgsinos, to form charginos – $\widetilde{W}_1^{\pm}$ and $\widetilde{W}_2^{\pm}$ – and neutralinos – $\widetilde{Z}_1$, $\widetilde{Z}_2$, $\widetilde{Z}_3$ and $\widetilde{Z}_4$. The $\widetilde{Z}_1$ state, the lightest neutralino, is often the lightest supersymmetric particle (LSP), and turns out to be an excellent WIMP candidate for CDM in the universe.

If nature is perfectly supersymmetric, then *spin-0 superpartners would have exactly the same mass* as the corresponding fermions. Charged spin-0 partners of the electron with a mass of 0.51 MeV could not have evaded experimental detection. Their non-observation leads us to conclude that SUSY must be a broken symmetry. In the MSSM, SUSY is broken explicitly by including so-called soft SUSY breaking (SSB) terms in the Lagrangian. The SSB terms preserve the desirable features of SUSY, such as the stabilisation of the scalar sector in the presence of radiative corrections, while lifting the superpartner masses in accord with what is necessary from experiment. It is important to note that the equality of dimensionless couplings between particles and their superpartners is still preserved (modulo small effects of radiative corrections): in particular, phenomenologically important gauge interactions of superpartners and the corresponding interactions of gauginos remain (largely) unaffected by the SSB terms.

The addition of the SSB Lagrangian terms may seem *ad-hoc* and ugly. It would be elegant if instead supersymmetry could be spontaneously broken. But it was recognised in the early to mid-1980s that models where global SUSY is spontaneously broken at the weak scale ran into serious difficulties. The situation is very different if we elevate SUSY from a global symmetry to a *local* one. In local SUSY, we are *forced to include the graviton/gravitino super-multiplet* into the theory, in much the same way that we have to include spin-1 gauge fields to maintain local gauge invariance of Yang-Mills theories. Theories with local SUSY are known as *supergravity* (SUGRA) theories because they are supersymmetric and necessarily include gravity. Moreover, the gravitational sector of the theory reduces to general relativity in the classical limit. Within the framework of SUGRA it is possible to add an additional sector whose dynamics spontaneously breaks SUSY but which interacts with SM particles and their superpartners only via gravity (the so-called hidden sector). The spontaneous breakdown of supersymmetry results in a mass for the gravitino in the same way that in local gauge theories gauge bosons acquire mass by the Higgs mechanism. This is, therefore, referred to as the *super-Higgs mechanism.* The remarkable thing is that because of the gravitational couplng between the hidden and the MSSM sectors, the effects of spontaneous supersymmetry breaking in the hidden sector are conveyed to the MSSM sector and (provided the SUSY-breaking scale in the hidden sector is appropriately chosen) weak scale SSB terms that lift the undesirable degeneracies between the masses of SM particles and their superpartners are automatically induced. Indeed, in the limit where $M_{\rm Pl} \to \infty$ (keeping the gravitino mass fixed), we recover a global SUSY theory along with the desired SSB terms! The gravitino typically has a weak scale mass and decouples from particle physics experiments because of its tiny gravitational couplings. For reasons that we cannot discuss here, these locally supersymmetric models are free [28–30] of the above mentioned difficulties that plague globally supersymmetric models.

Motivated by the successful unification of gauge couplings at a scale $M_{\rm GUT} \sim 2\times10^{16}$ GeV in the MSSM, we are led to construct a GUT based on local supersymmetry. In this case, the theory renormalised at $Q = M_{\rm GUT}$ contains just one gaugino mass parameter $m_{1/2}$. Renormalisation effects then split the physical gaugino masses in the same way the measured values of the gauge couplings arise from a single unified GUT scale gauge coupling. In general, supergravity models give rise to complicated mass matrices for the scalar superpartners of quarks and leptons, with concomitant flavour violation beyond acceptable levels. However, in models with *universal* soft SUSY breaking terms, a super-GIM mechanism suppresses flavour violating processes [31]. In what has come to be known as the minimal supergravity (mSUGRA) model, a universal scalar mass m_0 and also a universal SSB scalar coupling A_0 are assumed to exist at a high scale $Q = M_{\rm GUT} - M_{\rm Pl}$. The physical masses of squarks and sleptons are split after renormalisation, and can be calculated using renormalisation group techniques. Typically, in the mSUGRA model, we have $m_{\tilde{q}} \gtrsim m_{\tilde{\ell}_L} \gtrsim m_{\tilde{\ell}_R}$. Although the Higgs scalar mass parameters also start off at the common value m_0 at the high scale, the large value of the top quark Yukawa coupling drives the corresponding squared mass parameter to negative values and EWSB is radiatively broken as we have already discussed. Within this framework, the masses and couplings required for phenomenology are fixed by just a handful of parameters which are usually taken to be

$$m_0,\ m_{1/2},\ A_0,\ \tan\beta \text{ and } sign(\mu). \tag{2}$$

Here $\tan\beta$ is the ratio of the vacuum expectation values of the Higgs fields that give masses to up and down type fermions, and μ is the supersymmetric higgsino mass parameter whose magnitude is fixed to reproduce the measured value of M_Z. If all parameters are real, then potentially large CP-violating effects are suppressed as well. Computers codes such as Isajet, SuSpect, SoftSUSY and Spheno that calculate the full spectrum of sparticle and Higgs boson masses are publicly available [32].

The mSUGRA model (sometimes referred to as the constrained MSSM or CMSSM) serves as a paradigm for many SUSY phenomenological analyses. However, it is important to remember that it is based on many assumptions that can be tested in future collider experiments but which may prove to be incorrect. For instance, in many GUT theories, it is common to get *non-universal* SSB parameters. In addition, there are other messenger mechanisms besides gravity. In gauge-mediated SUSY breaking (GMSB) models [33], a special messenger sector is included, so gravitinos may be much lighter than all other sparticles, with implications for both collider physics and cosmology. In anomaly-mediated SUSY breaking (AMSB) models [34], gravitational anomalies induce SSB terms, and the gravitino can be much heavier than the weak scale. There are yet other models [35] where SSB parameters get comparable contributions from gravity-mediated as well as from anomaly-mediated sources and very recently, also from gauge-mediation [36]. The pattern of superpartner masses is sensitive to the mediation-mechanism, so that we can expect collider experiments to reveal which of the various mechanisms that have been proposed are actually realised in nature. We also mention that in both the GMSB and AMSB models, it is somewhat less natural (but still possible!) to obtain the required amount of SUSY DM in the universe. Although these are all viable scenarios, they have not been as well scrutinised as the mSUGRA model.

5. Supersymmetric Dark Matter

5.1. Neutralino Relic Density

Once a SUSY model is specified, then given a set of input parameters, it is possible to all compute superpartner masses and couplings necessary for phenomenology. We can then use these to calculate scattering cross sections and sparticle decay patterns to evaluate SUSY signals (and corresponding SM backgrounds) in collider experiments. We can also check whether the model is allowed or excluded by experimental constraints, either from direct SUSY searches, *e.g.* at LEP2 which requires that $m_{\widetilde{W}_1} > 103.5$ GeV, $m_{\tilde{e}} \gtrsim 100$ GeV and $m_h > 114.4$ GeV (for a SM-like light SUSY Higgs boson h) or from indirect searches through loop effects from SUSY particles in low energy measurements such as $B(b \to s\gamma)$ or $(g-2)_\mu$. We can also calculate the expected thermal LSP relic density. To begin our discussion, we will first assume that the lightest neutralino $\widetilde{Z}_1$ is the candidate DM particle.

As mentioned above, the relic density calculation involves solving the Boltzmann equation, where the neutralino density changes due to both the expansion of the universe and because of neutralino annihilation into SM particles, determined by the thermally averaged $\widetilde{Z}_1\widetilde{Z}_1$ annihilation cross section. An added complication occurs if neutralino *co-annihilation* is possible. Co-annihilation occurs if there is another SUSY particle close in mass to the $\widetilde{Z}_1$, whose thermal relic density (usually suppressed by the Boltzmann factor $exp\frac{-\Delta M}{T}$) is also significant. In the mSUGRA model, co-annihilation may occur from a stau, $\tilde{\tau}_1$, a stop $\tilde{t}_1$ or the lighter chargino $\widetilde{W}_1$. For instance, in some mSUGRA parameter-space regions the $\tilde{\tau}_1$ and $\widetilde{Z}_1$ are almost degenerate, so that they both have a significant density in the early universe, and reactions such as $\widetilde{Z}_1\tilde{\tau}_1 \to \tau\gamma$ occur. Since the electrically charged $\tilde{\tau}_1$ can also annihilate efficiently via electromagnetic interactions, this process also alters the equilibrium density of neutralinos. All in all, there are well over a thousand neutralino annihilation and co-annihilation reactions that need to be computed, involving of order 7000 Feynman diagrams. There exist several publicly avalable computer codes that compute the neutralino relic density: these include DarkSUSY [37], MicroMegas [38] and IsaReD [39] (a part of the Isatools package of Isajet [40]).

As an example, we show in Fig. 3 the m_0 *vs.* $m_{1/2}$ plane from the mSUGRA model, where we take $A_0 = 0$, $\mu > 0$, $m_t = 171.4$ GeV and $\tan\beta = 10$. The red-shaded regions are not allowed because either the $\tilde{\tau}_1$ becomes the lightest SUSY particle, in contradiction to negative searches for long lived, charged relics (left edge), or EWSB is not correctly obtained (lower-right region). The blue-shaded region is excluded by LEP2 searches for chargino pair production ($m_{\widetilde{W}_1} <$ 103.5 GeV). We show contours of squark (solid) and gluino (dashed) mass (which are nearly invariant under change of A_0 and $\tan\beta$). Below the magenta contour near $m_{1/2} \sim 200$ GeV, $m_h < 110$ GeV, which is roughly the LEP2 lower limit on m_h in the model. The thin green regions at the edge of the unshaded white region has $\Omega_{\widetilde{Z}_1}h^2$: 0.094–0.129 where the neutralino saturates the observed relic density. In the adjoining yellow regions, $\Omega_{\widetilde{Z}_1}h^2 < 0.094$, so these regions require multiple DM components. The white regions all have $\Omega_{\widetilde{Z}_1}h^2 > 0.129$ and so give too much thermal DM: they are excluded in the standard Big Bang cosmology.

The DM-allowed regions are classified as follows:

- At very low m_0 and low $m_{1/2}$ values is the so-called *bulk* annihilation region [41]. Here, sleptons are quite light, so $\widetilde{Z}_1\widetilde{Z}_1 \to \ell\bar{\ell}$ via t-channel slepton exchange. In years past (when $\Omega_{\rm CDM}h^2 \sim 0.3$ was quite consistent with data), this was regarded as the favoured region. But today LEP2 sparticle search limits have increased the LEP2-forbidden

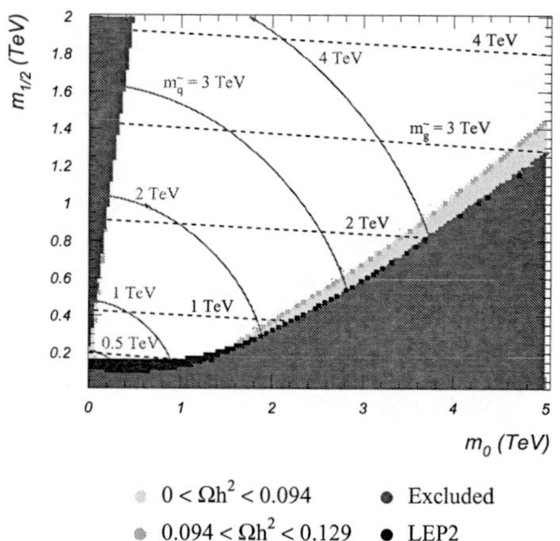

Figure 3. DM-allowed regions in the $m_0 - m_{1/2}$ plane of the mSUGRA model for $\tan\beta = 10$ with $A_0 = 0$ and $\mu > 0$

region from below, while the stringent bound $\Omega_{CDM}h^2 \leq 0.13$ has pushed the DM-allowed region down. Now hardly any bulk region survives in the mSUGRA model.

- At low m_0 and moderate $m_{1/2}$, there is a thin strip of (barely discernable) allowed region adjacent to the stau-LSP region where the neutralino and the lighter stau were in thermal equilibrium in the early universe. Here co-annihilation with the light stau serves to bring the neutralino relic density down to its observed value [42].
- At large m_0, adjacent to the EWSB excluded region on the right, is the hyperbolic branch/focus point (HB/FP) region, where the superpotential μ parameter becomes small and the higgsino-content of $\widetilde{Z}_1$ increases significantly. Then $\widetilde{Z}_1$ can annihilate efficiently via gauge coupling to its higgsino component and becomes mixed higgsino-bino DM. If $m_{\widetilde{Z}_1} > M_W,\ M_Z$, then $\widetilde{Z}_1\widetilde{Z}_1 \to WW,\ ZZ,\ Zh$ is enhanced, and one finds the correct measured relic density [43].

We show the corresponding situation for $\tan\beta = 52$ in Fig. 4. While the stau co-annihilation and the HB/FP regions are clearly visible, we see that now a large DM consistent region now appears.

- In this region, the value of m_A is small enough so that $\widetilde{Z}_1\widetilde{Z}_1$ can annihilate into $b\bar{b}$ pairs through s-channel A (and also H) resonance. This region has been dubbed the A-funnel [44]. It can be quite broad at large $\tan\beta$ because the width Γ_A can be quite wide due to the very large b- and τ-Yukawa couplings. If $\tan\beta$ is increased further, then $\widetilde{Z}_1\widetilde{Z}_1$ annihilation through the (virtual) A^* is large all over parameter space, and most of the theoretically-allowed parameter space becomes DM-consisten. For even higher $\tan\beta$ values, the parameter space collapses due to a lack of appropriate EWSB.

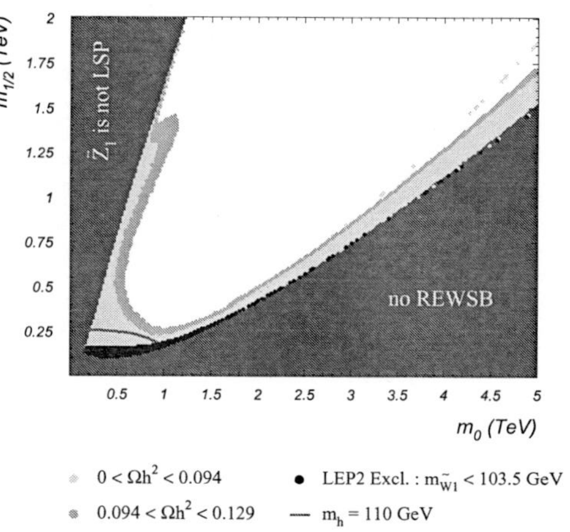

Figure 4. DM-allowed regions in the $m_0 - m_{1/2}$ plane of the mSUGRA model for $\tan\beta = 52$ with $A_0 = 0$ and $\mu > 0$. The various colours of shading is as in Fig. 3

It is also possible at low $m_{1/2}$ values that a light Higgs h resonance annihilation region can occur just above the LEP2 excluded region [45]. Finally, if A_0 is large and negative, then the $\tilde{t}_1$ can become light and $m_{\tilde{t}_1} \sim m_{\widetilde{Z}_1}$, so that stop-neutralino co-annihilation [46] can occur.

Up to now, we have confined our discussion to the mSUGRA framework in which compatibility with (1) is obtained only over selected portions of the $m_0 - m_{1/2}$ plane. The reader may well wonder what happens if we relax the untested universality assumptions that underlie mSUGRA. Without going into details, we only mention here that in many simple one-parameter extensions of mSUGRA where the universality of mass parameters is relaxed in any one of the matter scalar, the Higgs scalar, or the gaugino sectors, *all points in the* $m_0 - m_{1/2}$ *plane become compatible with the relic density constraint* due to a variety of mechanisms: these are catalogued in Ref. [48]. Implications of the relic density measurement for collider searches must thus be drawn with care.

5.2. Neutralino Direct Detection

Fits to galactic rotation curves imply a local relic density of $\rho_{CDM} \sim 0.3$ GeV/cm^3. For a 100 GeV WIMP, this translates to about one WIMP per coffee mug volume at our location in the galaxy. The goal of DD experiments is to detect the very rare WIMP-nucleus collisions that should be occuring as the earth, together with the WIMP detector, moves through the DM halo.

DD experiments are usually located deep underground to shield the experimental apparatus from background due to cosmic rays and ambient radiation from the environment or from radioactivity induced by cosmic ray exposure. One technique is to use cryogenic crystals cooled to near absolute zero, and look for phonon and ionisation signals from nuclei recoiling from a WIMP collision. In the case of the CDMS experiment [49] at the Soudan iron mine, target materials include germanium and silicon. Another technique uses noble gases cooled to a liquid state as the target. Here, the signal is scintillation light picked up by photomultiplier tubes and ionisation. Target materials include xenon [50], argon and perhaps neon. These noble liquid detectors can be scaled up to large volumes at relatively low cost. They have the advantage of fiducialisation, wherein the outer layers of the detector act as an active veto against cosmic rays or neutrons coming from phototubes or detector walls: only single scatters from the inner fiducial volume qualify as signal events. A third technique, typified by the COUPP experiment [51], involves use of superheated liquids such as CF^3I located in a transparent vessel. The nuclear recoil from a WIMP-nucleon collision then serves as a nucleation site, so that a bubble forms. The vessel is monitored visually by cameras. Background events are typically located close to the vessel wall, while neutron interactions are likely to cause several bubbles to form, instead of just one, as in a WIMP collision. This technique allows for the use of various target liquids, including those containing elements such as fluorine, which is sensitive to *spin-dependent* interactions.

The cross section for WIMP-nucleon collisions can be calculated, and in the low velocity limit separates into a coherent spin-independent component (from scattering mediated by scalar quarks and scalar Higgs bosons) which scales as nuclear mass squared, and a spin-dependent component from scattering mediated by the Z boson or by squarks, which depends on the WIMP and nuclear spins[29]. The scattering cross section per nucleon versus m_{WIMP} serves as a figure of merit and facilitates the comparison of the sensitivity of various experiments using different target materials.

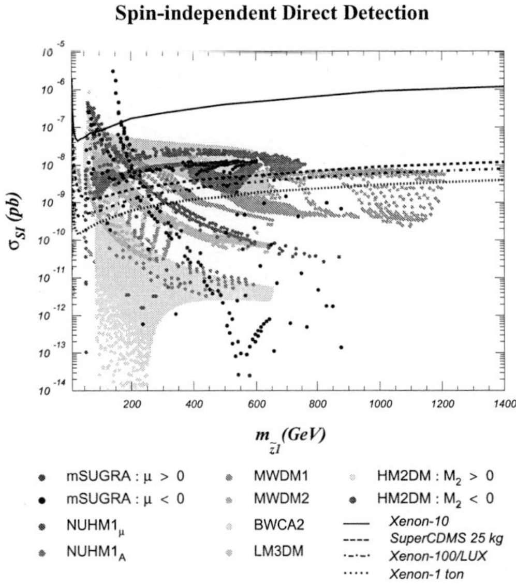

Figure 5. The spin-independent neutralino-proton scattering cross-section vs $m_{\widetilde{Z}_1}$ in a variety of SUSY models, compatible with collider constraints where thermally produced Big Bang neutralinos saturate the observed dark matter density

In Fig. 5, we show the spin-independent $\widetilde{Z}_1 p$ cross section versus $m_{\widetilde{Z}_1}$ for a large number of SUSY models (including mSUGRA). Every colour represents a different model. For each model, parameters are chosen so that current collider constraints on sparticle masses are satisfied, and further, that the lightest neutralino (assumed to be the LSP) saturates the observed relic abundance of CDM. Also shown is the sensitivity of current experiments together with projected sensitivity of proposed searches at superCDMS, Xenon-100, LUX, WARP and at a ton-sized noble liquid detector. The details of the various models are unimportant for our present purpose. The key thing to note is that while the various models have a branch where $\sigma_{SI}(p\widetilde{Z}_1)$ falls off with $m_{\widetilde{Z}_1}$, there is another branch where this cross-section asymptotes to just under 10^{-8} pb [47,48,52,53]. Points in this branch (which includes the HB/FP region of mSUGRA) are consistent with (1) because $\widetilde{Z}_1$ has a significant higgsino component. Neutralinos with an enhanced higgsino content can annihilate efficiently in the early universe via gauge interactions. Moreover, since the spin-independent DD amplitude is mostly determined by the Higgs boson-higgsino-gaugino coupling, it is large in models with MHDM which has both gaugino and higgsino components. Thus the enhanced higgsino component of MHDM increases both the neutralino annihilation in the early universe as well as the spin-independent DD rate. The exciting thing is that the experiments currently being deployed – such as Xenon-100, LUX and WARP – will have the sensitivity to probe this class of models. To go further will require ton-size or greater target material.

We note here that if $m_{\rm WIMP} \lesssim 150$ GeV, then it may be possible to extract the WIMP mass by measuring the energy spectrum of the recoiling nuclear targets [54]. Typically, of order 100 or more events are needed for such a determination to 10–20%. For higher WIMP masses, the recoil energy spectrum varies little, and WIMP mass extraction is much more difficult. Since the energy tranfer from the WIMP to a nucleus is maximised when the two have the same mass, DD experiments with several target nuclei ranging over a wide range of masses would facilitate the distinction between somewhat light and relatively heavy WIMPs, and so, potentially serve to establish the existence of multiple WIMP components in our halo.

5.3. Indirect Detection of Neutralinos

As explained in Sec. 1, there are also a number of indirect WIMP search techniques that attempt to detect the decay products from WIMP annihilation at either the centre of the sun, at the galactic centre or within the galactic halo.

5.3.1. Neutrino Telescopes

Neutrino telescopes such as ANTARES or IceCube can search for high energy neutrinos produced from WIMP-WIMP annihilation into SM particles in the core of the sun (or possibly the earth). The technique involves detection of multi-tens of GeV muons produced by ν_μ interactions with polar ice (IceCube) or ocean water (ANTARES). The muons travel at a speeds greater than the speed of light in the medium, thus leaving a tell-tale signal of Cerenkov light which is picked up by arrays of phototubes. The IceCube experiment, currently being deployed at the south pole, will monitor a cubic kilometre of ice in search of $\nu_\mu \to \mu$ conversions. It should be fully deployed by 2011. The experiment is mainly sensitive to muons with $E_\mu > 50$ GeV.

In the case of neutralinos of SUSY, mixed higgsino dark matter (MHDM) has a large (spin-dependent) cross-section to scatter from hydrogen nuclei via Z-exchange and so is readily captured. Thus, in the HB/FP region of mSUGRA, or in other SUSY models with MHDM, we expect observable levels of signal exceeding 40 events/km^2/yr with $E_\mu > 50$ GeV. For the mSUGRA model, the IceCube signal region is shown beneath the magenta contour labelled μ in Fig. 6 [55]. These results were obtained using the Isajet-DarkSUSY interface [37]. Notice that DD signals are also observable in much the same region (below the contour labelled DD) where the neutralino is MHDM.

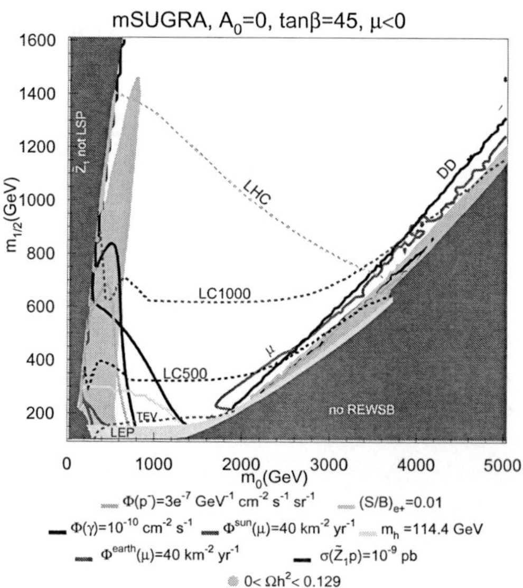

Figure 6. The projected reach of various colliders, direct and indirect DM search experiments in the mSUGRA model. For the indirect search results we have adopted the conservative default DarkSUSY isotropic DM halo density distribution. Plot is from Ref. [55]

5.3.2. Anti-matter from WIMP Halo Annihilations

WIMP annihilation in the galactic halo offers a different possibility for indirect DM searches. Halo WIMPs annihilate equally to matter and anti-matter, so the rare presence of high energy anti-matter in cosmic ray events – positrons e^+, anti-protons $\bar{p}$ or even anti-deuterons $\bar{D}$ – offer possible signatures. Positrons produced in WIMP annihilations must originate relatively close by, or else they will find cosmic electrons to annihilate against, or lose energy via bremsstrahlung. Anti-protons and anti-deuterons could originate further from us because, being heavier, they are deflected less and so lose much less energy. The expected signal rate depends on the WIMP annihilation rate into anti-matter, the model for the propogation of the anti-matter from its point of origin to the earth, and finally on the assumed profile of the DM in the galactic halo. Several possible halo density profiles are shown in Fig. 7. We see that while the *local* WIMP density is inferred to a factor of $\sim$ 2–3 (we are at about 8 kpc from the galactic centre), the DM density at the galactic centre is highly model-dependent close to the core. Since the ID signal should scale as the square of the WIMP density at the source, positron signals will be uncertain by a factor of a few with somewhat larger uncertainty for $\bar{p}$ and $\overline{D}$ signals that originate further away. Anti-particle propagation through the not-so-well-known magnetic field leads to an additional uncertainty in these predictions. The recently launched Pamela space-based anti-matter telescope can look for e^+ or $\bar{p}$ events while the balloon-borne GAPS experiment will be designed to search for anti-deuterons. Anti-matter signals tend to be largest in the case of SUSY models with MHDM or when neutralinos annihilate through the A-resonance [56].

5.3.3. Gamma Rays from WIMP Halo Annihilations

As mentioned in the introduction, high energy gamma rays from WIMP annihilation offer some advantages over the signal from charged antiparticles. Gamma rays would point to the source and would degrade much less in energy during their journey to us. This offers the possibility of the line signal from $\widetilde{Z}_1\widetilde{Z}_1 \to \gamma\gamma$ processes that occur via box an triangle diagrams. While this reaction is loop-suppressed, it yields monoenergetic photons with $E_\gamma \simeq m_{\rm WIMP}$, and so can provide a measure of the WIMP mass. Another possibility is to look for continuum gamma rays from WIMP annihilation to hadrons where, for instance, the gamma is the the result of π^0 decays. Since the halo WIMPs are essentially at rest, we expect a diffuse spectrum of gamma

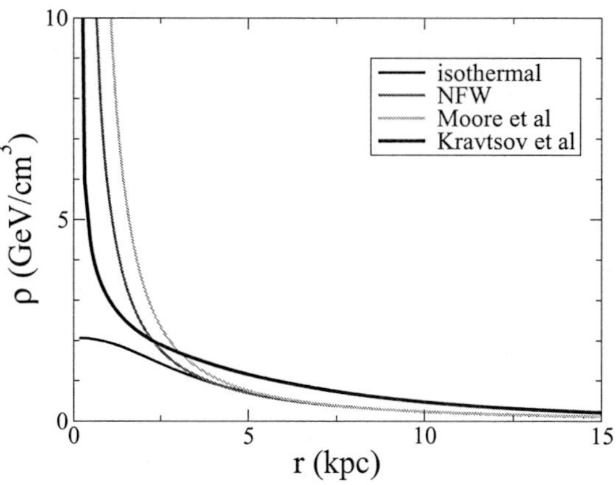

Figure 7. Various predictions for the DM halo in the Milky Way as a function of distance from the galactic centre. The earth is located at $r \sim 8$ kpc

rays, but with $E_\gamma < m_{\rm WIMP}$. Because gamma rays can traverse large distances, a good place to look at is the galactic centre, where the WIMP density (Fig. 7) is expected to be very high. Unfortunately, the density at the core is also very uncertain, making predictions for the gamma ray flux uncertain by as much as four orders of magnitude. Indeed, detection of WIMP halo signals may serve to provide information about the DM distribution in our galaxy.

Anomalies have been reported in the cosmic gamma ray spectrum. In one example, the Egret experiment [57] sees an excess of gamma rays with $E_\gamma > 1$ GeV. Explanations for the Egret GeV anomaly range from $\widetilde{Z}_1\widetilde{Z}_1 \to b\bar{b} \to \gamma$ with $m_{\widetilde{Z}_1} \sim 60$ GeV [58], to mis-calibration of the Egret experiment [59]. The GLAST gamma ray observatory is scheduled for lift-off in 2008 and should help resolve this issue, as will the upcoming LHC experiment [60].

5.4. Gravitino Dark Matter

In gravity-mediated SUSY breaking models, gravitinos typically have weak scale masses and because they only have tiny gravitational couplings, are usually assumed to be irrelevant for particle physics phenomenology. Cosmological considerations, however lead to the *gravitino problem*, wherein overproduction of gravitinos, followed by their late decays into SM particles, can disrupt the successful predictions of Big Bang nucleosynthesis. The gravitino problem can be overcome by choosing an appropriate range for $m_{\tilde{G}}$ and a low enough re-heat temperature for the universe after inflation [61] as illustrated in Fig. 8, or by hypothesising that the $\tilde{G}$ is in fact the stable LSP, and thus constitutes the DM [62].

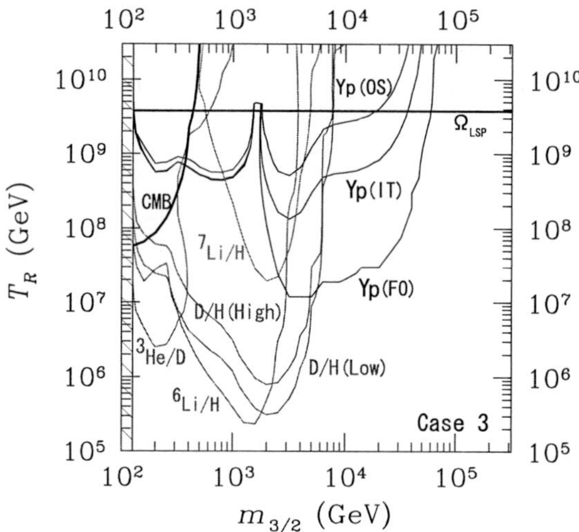

Figure 8. An illustration of constraints from Big Bang nucleosynthesis which require T_R to be below the various curves, for the HB/FP region of the mSUGRA model with $m_0 = 2397$ GeV, $m_{1/2} = 300$ GeV, $A_0 = 0$ and $\tan\beta = 30$, from Kohri *et al.* [61] to which we refer the reader for more details

Here, we consider the consequences of a gravitino LSP in SUGRA models. If gravitinos are produced in the pre-inflation epoch, then their number density will be diluted away during inflation. After the universe inflates, it enters a re-heating period wherein all particles can be thermally produced. However, the couplings of the gravitino are so weak that though gravitinos can be produced by the particles that do partake of thermal equilibrium, gravitinos themselves never attain thermal equilibrium: indeed their density is so low that gravitino annihilation processes can be neglected in the calculation of their relic density. The thermal production (TP) of gravitinos in the early universe has been calculated, and including EW contributions, is given by the approximate expression (valid for $m_{\tilde{G}} \ll M_i$ [63]):

$$\Omega_{\tilde{G}}^{TP} h^2 \simeq 0.32 \left(\frac{10\ GeV}{m_{\tilde{G}}}\right) \left(\frac{m_{1/2}}{1\ \text{TeV}}\right)^2 \left(\frac{T_R}{10^8\ \text{GeV}}\right), \quad (3)$$

where T_R is the re-heat temperature.

Gravitinos can also be produced by decay of the next-to-lightest SUSY particle, the NLSP. In the case of a long-lived neutralino NLSP, the neutralinos will be produced as usual with a thermal relic abundance in the early universe. Later, they will each decay as $\widetilde{Z}_1 \to \gamma\tilde{G}$, $Z\tilde{G}$ or $h\tilde{G}$. The total relic abundance is then

$$\Omega_{\tilde{G}} h^2 = \Omega_{\tilde{G}}^{TP} h^2 + \frac{m_{\tilde{G}}}{m_{\widetilde{Z}_1}} \Omega_{\widetilde{Z}_1} h^2. \quad (4)$$

The $\tilde{G}$ from NLSP decay may constitute warm/hot DM depending in the $\widetilde{Z}_1 - \tilde{G}$ mass gap, while the thermally produced $\tilde{G}$ will be CDM [26].

The lifetime for neutralino decay to the photon and a gravitino is given by [64]

$$\tau(\widetilde{Z}_1 \to \gamma\tilde{G}) \simeq \frac{48\pi M_P^2}{m_{\widetilde{Z}_1}^3} A^2 \frac{r^2}{(1-r^2)^3(1+3r^2)}$$
$$\sim 5.8 \times 10^8\ \text{s} \left(\frac{100\ \text{GeV}}{m_{\widetilde{Z}_1}}\right)^3 \frac{1}{A^2} \frac{r^2}{(1-r^2)^3(1+3r^2)}, \quad (5)$$

where $A = (v_4^{(1)} \cos\theta_W + v_3^{(1)} \sin\theta_W)^{-1}$, with $v_{3,4}^{(1)}$ being the wino and bino components of the $\widetilde{Z}_1$[28], M_P is the reduced Planck mass and $r = m_{\tilde{G}}/m_{\widetilde{Z}_1}$. Similar formulae (with different mixing angle and r-dependence) hold for decays to the gravitino plus a Z or h boson. We see that – except when the gravitino is very much lighter than the neutralino as may be the case in GMSB models with a low SUSY breaking scale – the NLSP decays well after Big Bang nucleosynthesis. Such decays would inject high energy gammas and/or hadrons into the cosmic soup post-nucleosynthesis, which could break up the nuclei, thus conflicting with the successful BBN predictions of Big Bang cosmology. For this reason, gravitino LSP scenarios usually favour a stau NLSP, since the BBN constraints in this case are much weaker.

Finally, we remark here upon the interesting interplay of baryogenesis via leptogenesis with the nature of the LSP and NLSP. For successful thermal leptogenesis to take place, it is found that the re-heat temperature of the universe must exceed $\sim 10^{10}$ GeV [65]. If this is so, then gravitinos would be produced thermally with a huge abundance, and then decay late, destroying BBN predictions. For this reason, some adherents of leptogenesis tend to favour scenarios with a gravitino LSP, but with a stau NLSP [66].

5.5. Axino Dark Matter

If we adopt the MSSM as the effective theory below $M_{\rm GUT}$, and then seek to solve the strong CP problem via the Peccei-Quinn solution [13], we must introduce not only an axion but also a spin-$\frac{1}{2}$ *axino* $\tilde{a}$ into the theory. The axino mass is found to be in the range of keV-GeV [67], but its coupling is suppressed by the Peccei-Quinn breaking scale f_a, which is usually taken to be of order $10^9 - 10^{12}$ GeV: thus, the axino interacts more weakly than a WIMP, but not as weakly as a gravitino. The axino can be an compelling choice for DM in the universe [68].

Like the gravitino, the axino will likely not be in thermal equilibrium in the early universe, but can still be produced thermally via particle scattering. The thermal production abundance is given by [68,69]

$$\Omega_{\tilde{a}}^{TP} h^2 \simeq 5.5 g_s^6 \log\left(\frac{1.108}{g_s}\right)\left(\frac{10^{11}\ \text{GeV}}{f_a/N}\right)^2 \times \left(\frac{m_{\tilde{a}}}{100\ \text{MeV}}\right)\left(\frac{T_R}{10^4\ \text{GeV}}\right), \quad (6)$$

where f_a is the PQ scale, N is a model-dependent colour anomaly factor that enters only as f_a/N and g_s is the strong coupling at the reheating scale.

Also like the gravitino, the axino can be produced non-thermally by NLSP decays, where the NLSP abundance is given by the standard relic density calculation. Thus

$$\Omega_{\tilde{a}} h^2 = \Omega_{\tilde{a}}^{TP} h^2 + \frac{m_{\tilde{a}}}{m_{NLSP}} \Omega_{\text{NLSP}} h^2. \quad (7)$$

In this case, the thermally produced axinos will be CDM for $m_{\tilde{a}} \gtrsim 0.1$ MeV [68], while the axinos produced in NLSP decay will constitute hot/warm DM [26]. Since the PQ scale is considerably lower than the Planck scale, the lifetime for decays such as $\widetilde{Z}_1 \to \gamma\tilde{a}$ are of order ~ 0.03 sec – well before BBN. Thus, the axino DM scenario is much less constrained than gravitino DM.

Note also that if axinos are the CDM of the universe, then models with very large $\Omega_{\widetilde{Z}_1} h^2 \sim 100 - 1000$ can be readily accommodated, since there is a huge reduction in relic density upon $\widetilde{Z}_1$ decay to the axino. This possibility occurs in models with multi-TeV scalars (and hence a multi-TeV gravitino) and a bino-like $\widetilde{Z}_1$. In this case with very large $m_{\tilde{G}}$ there is no gravitino problem as long as the re-heat temperature $T_R \sim 10^6 - 10^8$ GeV. This range of T_R is also what is needed to obtain successful *non-thermal* leptongenesis (involving heavy neutrino N production via inflaton decay) [70] along with the correct abundance of axino DM [71]. A scenario along these lines has been proposed [72] to reconcile Yukawa-unified SUSY models, which usually predict a vast over-abundance of neutralino DM, with the measured relic density.

6. SUSY DM at the LHC

6.1. Sparticle Production at the LHC

Direct production of neutralino DM at the LHC ($pp \to \widetilde{Z}_1\widetilde{Z}_1 X$, where X stands for assorted hadronic debris) is of little interest since the high p_T final state particles all escape the detector, and there is little if anything to trigger an event record. Detectable events come from the production of the heavier superpartners, which in turn decay via a multi-step cascade which ends in the stable LSP.

In many models, the strongly interacting squarks and/or gluinos are among the heaviest states. Unless these are extremely heavy, these will have large production cross sections at the LHC. Strong interaction production mechanisms for their production include, 1. gluino pair production $\tilde{g}\tilde{g}$, 2. squark pair production $\tilde{q}\tilde{q}$ and 3. squark-gluino associated production $\tilde{q}\tilde{g}$. Note here that the reactions involving squarks include a huge number of subprocess reactions to cover the many flavours, types (left and right) and also the anti-squarks. The various possibilities each have different angular dependence in the production cross sections [73], and the different flavours/types of squarks each have different decay modes [74]. These all have to be kept track of in order to obtain a reliable picture of the implications of SUSY in the LHC detector environment. Squarks and gluinos can also be produced in association with charginos and neutralinos [75]. Associated gluino production occurs via squark exchange in the t or u channels and is suppressed if squarks are very heavy.

If coloured sparticles are very heavy, then electroweak production of charginos and neutralinos may be the dominant sparticle production mechanism at the LHC. The most important processes are pair production of charginos, $\widetilde{W}_i^{\pm}\widetilde{W}_j^{\mp}$ where $i,j = 1,2$ and chargino-neutralino production, $\widetilde{W}_i^{\pm}\widetilde{Z}_j$, with $i = 1,2$ and $j = 1-4$. In models with unified GUT scale gaugino masses and large $|\mu|$, $Z\widetilde{W}_1\widetilde{W}_1$ and $W\widetilde{Z}_2\widetilde{W}_1$ couplings are large so that $\widetilde{W}_1\widetilde{W}_1$ and $\widetilde{W}_1\widetilde{Z}_2$ production occurs at significant rates. The latter process can lead to the gold-plated trilepton signature at the LHC [76]. Neutralino pair production ($pp \to \widetilde{Z}_i\widetilde{Z}_j X$ where $i,j = 1-4$) is also possible. This reaction occurs at low rates at the LHC unless $|\mu| \simeq M_{1,2}$ (as in the case of MHDM). Finally, we mention slepton pair production: $\tilde{\ell}^+\tilde{\ell}^-$, $\tilde{\nu}_\ell\tilde{\ell}$ and $\tilde{\nu}_\ell\bar{\tilde{\nu}}_\ell$, which can give detectable dilepton signals if $m_{\tilde{\ell}} \lesssim 300$ GeV [77].

In Fig. 9 we show various sparticle production cross sections at the LHC as a function of $m_{\tilde{g}}$. Strong interaction production mechanisms dominate at low mass, while electroweak processes dominate at high mass. The associated production mechanisms are never dominant. The expected LHC integrated luminosity in the first year of running is expected to be around 0.1 fb^{-1}, while several tens of fb^{-1} of data is expected to be recorded in the first several years of operation. The ultimate goal is to accumulate around 500–1000 fb^{-1}, correponding to $10^5 - 10^6$ SUSY events for $m_{\tilde{g}} \sim 1$ TeV.

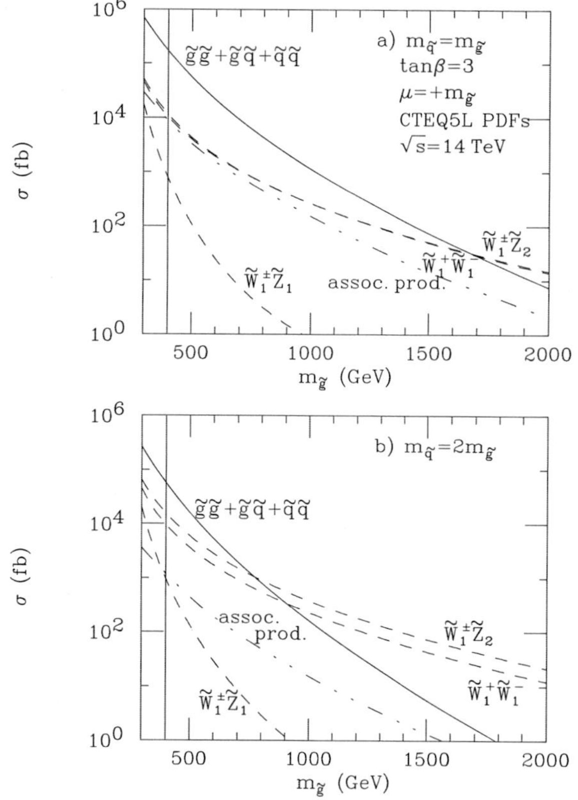

Figure 9. Cross sections for production of various sparticles at the LHC. Gaugino mass unification is assumed

6.2. Sparticle Cascade Decays

In R-parity conserving models, sparticles decay to lighter sparticles until the decay terminates in the LSP [74]. Frequently, the direct decay to the LSP is either forbidden or occurs with only a small branching fraction. Since gravitational interactions are negligible, gluinos can only decay via $\tilde{g} \to q\tilde{q}$, where the q and $\tilde{q}$ can be of any flavour or type. If two-body decay modes are closed, the squark will be virtual, and the gluino will decay via three-body modes $\tilde{g} \to q\bar{q}\widetilde{Z}_i,\ q\bar{q}'\widetilde{W}_j$. If squarks are degenerate, and Yukawa coupling effects negligible, three-body decays to the wino-like chargino and neutralino usually have larger branching fractions on account of the larger gauge coupling. If $|\mu| < M_2$, gluinos and squarks may thus decay most of the time to the heavier charginos and neutralinos, resulting in lengthy cascade decay chains at the LHC.

Squarks decay always to two-body modes: $\tilde{q} \to q\tilde{g}$ if it is kinematically allowed or $\tilde{q}_L \to q'\widetilde{W}_i,\ q\widetilde{Z}_j$, while $\tilde{q}_R \to q\widetilde{Z}_j$ only, since right-squarks do not couple to charginos. Sleptons do not have strong interactions so cannot decay to gluinos. Their electroweak decays are similar to corresponding decays of squarks $\tilde{\ell}_L \to \ell'\widetilde{W}_i,\ \ell\widetilde{Z}_j$ while $\tilde{\ell}_R \to \ell\widetilde{Z}_j$ only.

Charginos may decay via two-body modes: $\widetilde{W}_i \to W\widetilde{Z}_j,\ \tilde{\ell}\nu_\ell,\ \ell\tilde{\nu}_\ell,\ Z\widetilde{W}_j$ or even to $\phi\widetilde{W}_j$ or $H^-\widetilde{Z}_j$, where $\phi = h, H, A$. If two-body modes are inaccessible, then three-body decays dominate: $\widetilde{W}_i \to \widetilde{Z}_j f\bar{f}'$, where f and f' are SM fermions which couple to the W. Frequently, the decay amplitude is dominated by the virtual W so that the three-body decays of $\widetilde{W}_1$ have the same branching fractions as those of the W. Neutralinos decay via $\widetilde{Z}_i \to W\widetilde{W}_j,\ H^+\widetilde{W}_j,\ Z\widetilde{Z}_j,\ \phi\widetilde{Z}_j$ or $f\tilde{f}$. If two body neutralino decays are closed, then $\widetilde{Z}_i \to \widetilde{Z}_j f\bar{f}$, where f are the SM fermions. In some models, the branching fraction for radiative decays $\widetilde{Z}_i \to \widetilde{Z}_j\gamma$ (that only occurs at the one-loop level) may be significant [78]. The cascade decay modes of neutralinos depend sensitively on model parameters [79].

If $\tan\beta$ is large, then b and τ Yukawa coupling effects become important, enhancing three body decays of $\tilde{g}$, $\widetilde{W}_i$ and $\widetilde{Z}_j$ to third-generation fermions [80]. For very large values of $\tan\beta$ these decays can even dominate, resulting in large rates for b-jet and τ-jet production in SUSY events [81]. Finally, the various Higgs bosons can be produced both directly and via sparticle cascades at the LHC [82]. Indeed, it may be possible that h is first discovered in SUSY events because in a sample of events enriched for SUSY, it is possible to identify h via its dominant $h \to b\bar{b}$ decays rather than via its subdominant decay modes, as required for conventional searches [82]. The heavier Higgs bosons decay to a variety of SM modes, but also to SUSY particles if these latter decays are kinematically allowed, leading to novel signatures such as $H,\ A \to \widetilde{Z}_2\widetilde{Z}_2 \to 4\ell + E_T^{\mathrm{miss}}$ [83].

The cascade decays terminate in the LSP. In the case of a $\widetilde{Z}_1$ LSP, the $\widetilde{Z}_1$ is a DM candidate, and leaves its imprint via E_T^{miss}. In the case of a weak scale $\widetilde{G}$ or $\tilde{a}$ LSP, then $\widetilde{Z}_1$ will decay as discussed above. In these cases, the $\widetilde{Z}_1$ lifetime is long enough that it decays outside the detector, so one still expects large $E_T^{\rm miss}$ in the collider events. An exception arises for the case of super-light gravitinos (with masses in the eV to keV range) that are possible in GMSB models: see (5). Then, the decay may take place inside inside the detector, possibly with a large vertex separation. It is also possible that the NLSP is charged and quasi-stable, in which case collider events may include highly ionising tracks instead of, or in addition to $E_T^{\rm miss}$.

The decay branching fractions depend on the entire spectrum of SUSY particle masses and their mixings. They are pre-programmed in several codes: Isajet [40], SDECAY [84] and Spheno [85].

6.3. Event Generation for the LHC

Once sparticle production cross sections and decay branching fractions have been computed, it is useful to embed these into event generator programmes to simulate what SUSY collider events will look like at the LHC. There are several steps involved:

- Calculate all sparticle pair production cross sections. Once all initial and final states are accounted for, this involves over a thousand individual subprocess reactions. In event generation, a particular reaction is selected on a probabilistic basis, with a weight proportional to its differential cross-section.
- Sparticle decays are selected probabilistically into all the allowed modes in proportion to the corresponding branching fractions.
- Initial and final state quark and gluon radiation are usually dealt with using the parton shower (PS) algorithm, which allows for probabilistic parton emission based on approximate collinear QCD emission matrix elements, but exact kinematics. The PS is also applied at each step of the cascade decays, which may lead to additional jet production in SUSY collider events.
- A hadronisation algorithm provides a model for turning various quarks and gluons into mesons and baryons. Unstable hadrons must be further decayed.
- The beam remnants – proton constituents not taking part in the hard scattering – must be showered and hadronised, usually with an independent algorithm, so that energy deposition in the forward detector region may be reliably calculated.

At this stage, the output of an event generator programme is a listing of particle types and their associated four-vectors. The resulting event can then be interfaced with detector simulation programmes to model what the actual events containing DM will look like in the environment of a collider detector.

Several programmes are available, including Isajet [40], Pythia [86] and Herwig [87]. Other programmes such as Madevent [88], CompHEP/CalcHEP [89] and Whizard [90] can generate various $2 \to n$ processes including SUSY particles. The output of these programmes may then be used as input to Pythia or Herwig for showering and hadronisation. Likewise, parton level Isajet SUSY production followed by cascade decays can be input to Pythia and Herwig via the Les Houches Event format [91].

6.4. Signatures for Sparticle Production

Unless coloured sparticles are very heavy, the SUSY events at the LHC mainly result in gluino and squark production, followed by their possibly lengthy cascade decays. These events, therefore, typically contain very hard jets (from the primary decay of the squark and/or gluino) together with other jets and isolated electrons, muons and taus (identified as narrow one- and three-prong jets), and sometimes also photons, from the decays of secondary charginos and neutralinos, along with $E_T^{\rm miss}$ that arises from the escaping DM particles (as well as from neutrinos). In models with a superlight gravitino, there may also be additional isolated photons, leptons or jets from the decay of the NLSP. The relative rates for various n-jet + m-lepton + k-photon $+E_T^{\rm miss}$ event topologies is sensitive to the model as well as to the parameter values, and so provide a useful handle for phenomenological analyses.

Within the SM, the physics background to the classic $jets + E_T^{\rm miss}$ signal comes from neutrinos escaping the detector. Thus, the dominant SM backgrounds come from $W + jets$ and $Z + jets$ production, $t\bar{t}$ production, QCD multijet production (including $b\bar{b}$ and $c\bar{c}$ production), WW, WZ, ZZ production plus a variety of $2 \to n$ processes which are not usually included in event generators. These latter would include processes such as $t\bar{t}t\bar{t}$, $t\bar{t}b\bar{b}$, $t\bar{t}W$, WWW, WWZ production, etc. Decays of electroweak gauge bosons and the t-quark are the main source of isolated leptons in the SM. Various additional effects – uninstrumented regions, energy mis-measurement, cosmic rays, beam-gas events – can also lead to $E_T^{\rm miss}$ events.

In contrast to the SM, SUSY events naturally tend to have large jet multiplicities and frequently an observable rate for high multiplicity lepton events with large

$E_T^{\rm miss}$. Thus, if one plots signal and background versus multiplicity of any of these quantities, as one steps out to large multiplicity, the expected SUSY events should increase in importance, and even dominate the high multiplicity channels in some cases. This is especially true of isolated multi-lepton signatures, and in fact it is convenient to classify SUSY signal according to lepton multiplicity [92]:

- zero lepton $+jets + E_T^{\rm miss}$ events
- one lepton $+jets + E_T^{\rm miss}$ events
- two opposite sign leptons $+jets + E_T^{\rm miss}$ events (OS)
  - same flavour (OSSF)
  - different flavour (OSDF)
- two same sign leptons $+jets + E_T^{\rm miss}$ events (SS)
- three leptons $+jets + E_T^{\rm miss}$ events (3ℓ)
- four (or more) leptons $+jets + E_T^{\rm miss}$ events (4ℓ).

6.5. The LHC Reach for SUSY

Event generators, together with detector simulation programmes can be used to project the SUSY discovery reach of the LHC. Given a specific model, one may first generate a grid of points that samples the parameter (sub)space where signals rates are expected to vary significantly. A large number of SUSY collider events can then be generated at every point on the grid along with the various SM backgrounds to the SUSY signal mentioned above. Next, these signal and background events are passed through a detector simulation programme and a jet-finding algorithm is implemented to determine the number of jets per event above some $E_T(jet)$ threshold (usually taken to be $E_T(jet) > 50 - 100$ GeV for the LHC). Finally, *analysis cuts* are imposed which are designed to reject mainly SM BG while retaining the signal. These cuts may include both topological and kinematic selection criteria. For observability with an assumed integrated luminosity, we require that the signal exceed the chance 5 standard deviation upward fluctuation of the background, together with a minimum value of ($\sim 25\%$) the signal to background ratio, to allow for the fact that the background is not perfectly known. For lower sparticle masses, softer kinematic cuts are used, but for high sparticle masses, the lower cross sections but higher energy release demand hard cuts to optimise signal over background.

In Fig. 10, we illustrate the SUSY reach of the LHC within the mSUGRA model assuming an integrated luminosity of 100 fb^{-1}. We show the result in the $m_0 - m_{1/2}$ plane, taking $A_0 = 0$, $\tan\beta = 10$ and $\mu > 0$. The signal is observable over background in the corresponding topology below the corresponding curve. We note the following.

1. Unless sparticles are very heavy, there is an observable signal in several different event topologies. This will help add confidence that one is actually seeing new physics, and may help to sort out the production and decay mechanisms.
2. The reach at low m_0 extends to $m_{1/2} \sim$ 1400 GeV. This corresponds to a reach for $m_{\tilde{q}} \sim m_{\tilde{g}} \sim 3.1$ TeV.
3. At large m_0, squarks and sleptons are in the $4-5$ TeV range, and are too heavy to be produced at significant rates at the LHC. Here, the reach comes mainly from just gluino pair production. In this range, the LHC reach is up to $m_{1/2} \sim$ 700 GeV, corresponding to a reach in $m_{\tilde{g}}$ of about 1.8 TeV, and may be extended by $\sim$ 15–20% by b-jet tagging [93].

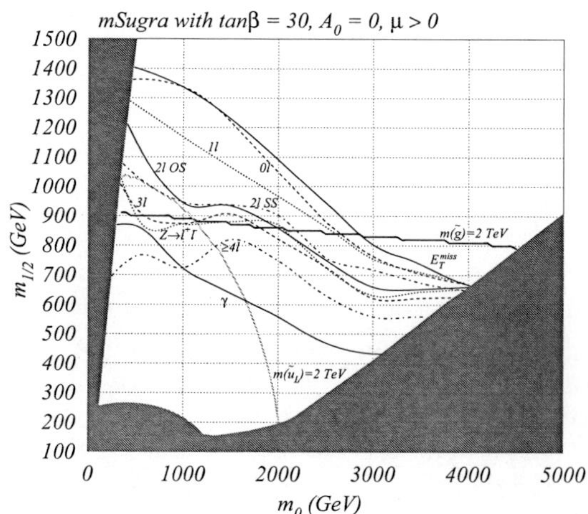

Figure 10. The 100 fb^{-1} fb reach of the LHC for SUSY in the mSUGRA model. For each event topology, the signal is observable below the corresponding contour

In Fig. 6 we can see a comparison of the LHC reach (notice that it is insensitive to $\tan\beta$ and sign(μ)) with that of the Tevatron (for clean 3ℓ events with 10 fb^{-1}), and the proposed e^+e^- International Linear Collider (ILC), with $\sqrt{s} = 0.5$ or 1 TeV along with various dark matter DD and ID search experiments. We remark that:

- While the LHC can cover most of the relic density allowed region, the HB/FP region emerges far beyond the LHC reach
- As already noted, the DD and ID experiments have the greatest sensitivity in the HB/FP region where the neutralino is MHDM. In this sense, DD and ID experiments *complement* the LHC searches for SUSY
- The ILC reach is everywhere lower than the LHC, except in the HB/FP region. In this region, while gluinos and squarks can be extremely heavy, the μ parameter is small, leading to a relatively light spectrum of charginos and neutralinos. These are not detectable at the LHC because the visible decay products are too soft. However, since chargino pair production is detectable at ILC even if the energy release in chargino decays is small, the ILC reach extends beyond the LHC in this region [94].

Finally, we note here that while the results presented above are for the LHC reach in the mSUGRA model, the LHC reach (measured in terms of $m_{\tilde{g}}$ and $m_{\tilde{q}}$) tends to be relatively insensitive to the details of the model chosen, as long as gluino and squark production followed by cascade decays to the DM particle occur.

6.6. Early Discovery of SUSY at the LHC Without $E_T^{\rm miss}$

Recently, it has been pointed out that a SUSY search using the traditional $jets + E_T^{\rm miss}$ signature may not be possible for a while after start-up due to various detector calibration issues. In this case, it is possible to abandon using the $E_T^{\rm miss}$ cut, and instead require a *high multiplicity* of isolated leptons: SS, OSSF, OSDF, 3ℓ. The high lepton multiplicity requirement severely reduces SM background while maintaining large enough signal rates. In Ref. [95], it is claimed an LHC reach of $m_{\tilde{g}} \sim 750$ GeV is possible with just 0.1 fb^{-1} of integrated luminosity, *without* using an $E_T^{\rm miss}$ cut.

6.7. Determination of Sparticle Properties

Once a putative signal for new physics emerges at the LHC, the next step is to establish its origin. This will entail detailed measurements of cross sections and distributions in various event topologies to gain insight into the identity of the new particles being produced, their masses, decay patterns, spins, couplings (gauge quantum numbers) and ultimately mixing angles. These measurements are not straightforward in the LHC environment because of numerous possible SUSY production reactions occurring simultaneously, a plethora of sparticle cascade decay possibilities, hadronic debris from initial state radiation and lack of invariant mass reconstruction due to the presence of $E_T^{\rm miss}$. All these lead to ambiguities and combinatoric problems in reconstructing exactly what sort of signal reactions are taking place. In contrast, at the ILC, the inital state is simple, the beam energy is tunable and beam polarisation can be used to select out specific processes.

While it seems clear that the ILC is better suited for a systematic programme of precision sparticle measurements, studies have shown (albeit in special cases) that interesting measurements are also possible at the LHC. We go into just a subset of all details here in order to give the reader an idea of some of the possibilities suggested in the literature.

One suggested starting point is the distribution of effective mass $M_{\rm eff} = E_T^{\rm miss} + E_T(j1) + E_T(j2) + E_T(j3) + E_T(j4)$ in the inclusive SUSY sample, which sets the approximate mass scale $M_{\rm SUSY} \equiv \min(m_{\tilde{g}}, m_{\tilde{q}})$ for the strongly interacting sparticles are being produced [96] and provides a measure of $M_{\rm SUSY}$ to 10–15%.

More detailed information on sparticle masses may be accessed by studying specific event topologies. For instance, the mass of dileptons from $\widetilde{Z}_2 \to \ell^+\ell^-\widetilde{Z}_1$ decays is bounded by $m_{\widetilde{Z}_2} - m_{\widetilde{Z}_1}$ (this bound is even more restrictive if $\widetilde{Z}_2$ decays via an on-shell slepton) [97]. We therefore expect an OSSF invariant mass distribution to exhibit an edge at $m_{\widetilde{Z}_2} - m_{\widetilde{Z}_1}$ (or below) in any sample of SUSY events so long as the "spoiler" decay modes $\widetilde{Z}_2 \to \widetilde{Z}_1 Z$ or $\widetilde{Z}_1 h$ are closed. Contamination from chargino production can be statistically removed by subtracting out the distribution of OSDF dileptons. In MHDM models, there may be more than one visible mass edge because the $\widetilde{Z}_3$ may also be accessible in cascade decays.

In the happy circumstance where production of gluinos or a single type of squark is dominant, followed by a string of two-body decays, then further invariant mass edges are possible. One example comes from $\tilde{g} \to b\bar{\tilde{b}}_1 \to b\bar{b}\widetilde{Z}_2 \to b\bar{b}\ell\bar{\ell}\widetilde{Z}_1$; then one can try to combine a b-jet with the dilepton pair to reconstruct the squark-neutralino mass edge: $m(b\ell\bar{\ell}) < m_{\tilde{b}_1} - m_{\widetilde{Z}_1}$. Next, combining with another b-jet can yield a gluino-neutralino edge: $m(b\bar{b}\ell\bar{\ell}) < m_{\tilde{g}} - m_{\widetilde{Z}_1}$. The reconstruction of such a decay chain may be possible as shown in Ref. [96], where other sequences of two-body decays are also examined. In practice, such fortuitous circumstances may not exist, and there are many combinatoric issues to overcome as well. A different study [98] shows that end-point measurements at the LHC will make it possible to access the mass difference between the LSP and the stau in a mSUGRA scenario where the stau co-annihilation mechanism is operative.

These end-point measurements generally give mass differences, not masses. However, by an analysis of the decay chain $\tilde{q}_L \to q\widetilde{Z}_2 \to q\tilde{\ell}^{\pm}\ell^{\mp} \to q\ell^{\pm}\ell^{\mp}\widetilde{Z}_1$, it has been argued [99] that reconstruction of masses may be possible under fortuitous circumstances. More recently, it has been suggested that it may be possible to directly access the gluino and/or squark masses (not mass differences) via the introduction of the so-called m_{T2} variable. We will refer the reader to the literature for details [100].

Mass measurements allow us to check consistency of specific SUSY models with a handful of parameters, and together with other measurements can readily exclude such models. But these are not the only interesting measurements at the LHC. It has been shown that if the NLSP of GMSB models decays into a superlight gravitino, it may be possible to determine its lifetime, and hence the gravitino mass at the LHC [101]. This will then allow one to infer the underlying SUSY breaking scale, a scale at least as important as the weak scale! A recent study [102] suggests that this is possible even when the the decay length of the NLSP is too short to be measured. While linear collider experiments will ultimately allow the precision measurements that will directly determine the new physics to be softly broken supersymmetry [103], it will be exciting to analyse the real LHC data that will soon be available to unravel many of the specific details about how (or if) SUSY is actually implemented in nature.

6.8. Measuring DM Properties at the LHC and ILC

SUSY discovery will undoubtedly be followed by a program (as outlined in Sec. 6.7) to reconstruct sparticle properties. What will we be able to say about DM in light of these measurements? Such a study was made by Baltz *et al.* [104] where four mSUGRA case study points (one each in the bulk region, the HB/FP region, the stau coanihilation region and the A-funnel region) were examined for the precision with which measurements of sparticle properties that could be made at the LHC, and also at a $\sqrt{s} = 0.5$ and 1 TeV e^+e^- collider. They then adopted a 24-parameter version of the MSSM and fit its parameters to these projected measurements. The model was then used to predict several quantities relevant to astrophysics and cosmology: the DM relic density $\Omega_{\widetilde{Z}_1}h^2$, the spin-independent neutralino-nucleon scattering cross section $\sigma_{SI}(\widetilde{Z}_1 p)$, and the neutralino annihilation cross section times relative velocity, in the limit that $v \to 0$: $\langle\sigma v\rangle|_{v\to 0}$. The last quantity is the crucial particle physics input for estimating signal strength from neutralino annihilation to anti-matter or gammas in the galactic halo. What this yields then is a *collider measurement* of these key DM quantities.

As an illustration, we show in Fig. 11 (taken from Ref. [104]) the precision with which the neutralino relic density is constrained by collider measurements for the LCC2 point which is in the HB/FP region of the mSUGRA model. Measurements at the LHC cannot fix the LSP composition, and so unable to resolve the degeneracy between a wino-LSP solution (which gives a tiny relic density) and the true solution with MHDM. Determinations of chargino production cross sections at the ILC can easily resolve the difference. It is nonetheless striking that up to this degeneracy ambiguity, experiments at the LHC can pin down the relic density to within $\sim 50\%$ (a remarkable result, given that there are sensible models where the predicted relic density may differ by orders of magnitude!). This improves to 10–20% if we can combine the LHC and ILC measurements.

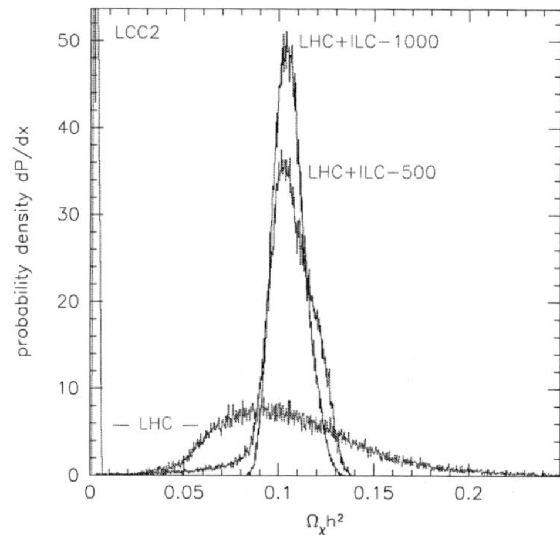

Figure 11. Determination of neutralino relic abundance via measurements at the LHC and ILC, taken from Ref. [104]

This collider determination of the relic density is very important. If it agrees with the cosmological measurement it would establish that the DM is dominantly thermal neutralinos from the Big Bang. If the neutralino relic density from colliders falls significantly below (1), it would provide direct evidence for multi-component DM – perhaps neutralinos plus axions or other exotica. Alternatively, if the collider determination gives a much larger value of $\Omega_{\widetilde{Z}_1} h^2$, it could point to a long-lived but unstable neutralino and/or non-thermal DM.

The collider determination of model parameters would also pin down the neutralino-nucleon scattering cross section. Then if a WIMP signal is actually observed in DD experiments, one might be able to determine the local DM density of neutralinos and aspects of their velocity distribution based on the DD signal rate. This density should agree with that obtained from astrophysics if the DM in our galaxy is comprised only of neutralinos.

Finally, a collider determination of $\langle \sigma v \rangle |_{v \to 0}$ would eliminate uncertainty on the particle physics side of projections for any ID signal from annihilation of neutralinos in the galactic halo. Thus the observation of a gamma ray and/or anti-matter signal from neutralino halo annihilations would facilitate the determination of the galactic halo DM density distribution.

7. Some Non-SUSY WIMPs at the LHC

7.1. B^1_μ State from Universal Extra Dimensions

Models with Universal Extra Dimensions, or UED are interesting constructs which provide a foil for SUSY search analyses [18]. In the 5-D UED theory, one posits that the fields of the SM actually live in a 5-D brane world. The extra dimension is "universal" since *all* the SM particles propagate in the 5-D bulk. The single extra dimension is assumed to be compactified on a S_1/Z_2 orbifold (line segment). After compactification, the 4-D effective theory includes the usual SM particles, together with an infinite tower of Kaluza-Klein (KK) excitations. The masses of the excitations depend on the radius of the compactified dimension, and the first ($n = 1$) KK excitations can be taken to be of order the weak scale. In these theories, KK-parity $(-1)^n$ can be a conserved quantum number. If this so-called KK-parity is exact, then the lightest odd KK parity state will be stable and can be a DM candidate. At tree-level, all the KK excitations in a given level are essentially degenerate. Radiative corrections break the degeneracy, leaving coloured excitations as the heaviest excited states and the $n = 1$ KK excitation of the SM $U(1)_Y$ gauge boson B^1_μ as the lightest [105] KK odd state: in the UED case, therefore, the DM particle has spin-1. The splitting caused by the radiative corrections is also essential to assess how the KK excitations decay, and hence are crucial for collider phenomenology [106].

The relic density of B^1_μ particles has been computed, and found to be compatible with observation for certain mass ranges of B^1_μ [107]. Also, in UED, the coloured excitations can be produced with large cross sections at the LHC, and decay via a cascade to the B^1_μ final state. Thus, the collider signatures are somewhat reminiscent of SUSY, and it is interesting to ask whether it is possible to distinguish a $jets + leptons + E_T^{\rm miss}$ signal in UED from that in SUSY. Several studies [108] answer affirmatively, and in fact provide strong motivation for the measurement of the spins of the produced new particles [109]. UED DM generally leads to a large rate in IceCube, and may also give an observable signal in anti-protons and possibly also in photons and positrons [18,110]. DD is also possible but the SI cross section is typically smaller than 10^{-9} pb.

7.2. Little Higgs Models

Little Higgs models [19,22] provide an alternative method compared to SUSY to evade the quadratic sensitivity of the scalar Higgs sector to ultra-violet (UV) physics. In this framework, the Higgs boson is a pseudo-Goldstone boson of a spontaneously broken global symmetry that is not completely broken by any one coupling, but is broken when *all* couplings are included. This then implies that there is quadratic sensitivity to UV physics, but only at the multi-loop level. Specific models where the quadratic sensitivity enters at the two-loop level should, therefore, be regarded as low energy effective theories valid up to a scale $\Lambda \sim 10$ TeV, at which a currently unknown, and perhaps strongly-coupled UV completion of the theory is assumed to exist. Models that realise this idea require new TeV-scale degrees of freedom that can be searched for at the LHC: new gauge bosons, a heavy top-like quark, and new spin-zero particles, all with couplings to the SM. These models, however, run into phenomenological difficulties with precision EW constraints, unless a discrete symmetry – dubbed T-parity [20] – is included. SM particles are then T-even, while the new particles are T-odd.

We will set aside the issue (mentioned earlier) of whether T-parity conservation is violated by anomalies [23] and assume that a conserved T-parity can be introduced [24]. In this case, the lightest T-odd particle A_H – the Little Higgs partner of the hypercharge gauge boson with a small admixture of the neutral W_{3H} boson – is stable and yields the observed amount of DM for a reasonable range of model parameters [110]. In this case, the DM particle has spin-1, though other cases with either a spin-$\frac{1}{2}$ or spin-0 heavy particle may also be possible. A_H can either annihilate with itself into vector boson pairs or $t\bar{t}$ pairs via s-channel Higgs exchange, or into top pairs via exchange of the heavy T-odd quark in the t-channel. Co-annihilation may also be possible if the heavy quark and A_H are sufficiently close in mass. Signals at the LHC [111] mainly come from pair production of heavy quarks, and from single production of the heavy quark in association with A_H. These lead to low jet multiplicity events plus $E_T^{\rm miss}$. The $E_T^{\rm miss}$ comes from the escaping A_H particle, which must be the endpoint of all T-odd particle decays.[3] If A_H is the dominant component of galactic DM, we will generally expect small DD and ID rates for much the same reasons that the signals from the bino LSP tend to be small [110]: see, however, Ref. [112] for a different model with large direct detection rate.

8. Outlook

The union of particle physics, astrophysics and cosmology has reached an unprecedented stage. Today we are certain that the bulk of the matter in the universe is non-luminous, not made of any of the known particles, but instead made of one or more *new physics* particles that do not appear in the SM. And though we know just how much of this unknown DM there is, we have no idea *what* it is. Today, many theoretical speculations which seek to answer one of the most pressing particle physics puzzles, "What is the origin of EWSB and how can we embed this into a unified theory of particle interactions?" automatically also point to a resolution of this 75-year-old puzzle as to what the dominant matter component of our universe might be. Particle physicists have made many provocative suggestions for the origin of DM, including supersymmetry and extra spatial dimensions, ideas that will completely change the scientific paradigm if they prove to be right.

The exciting thing is that many of these speculations will be *directly tested* by a variety of particle physics experiments along with astrophysical and cosmological searches. The LHC scheduled to commence operation in 2008, will directly study particle interactions at a scale of 1 TeV where new matter states are anticated to exist for sound theoretical reasons. These new states may well be connected the DM sector, and so in this way the LHC can make crucial contributions to not only particle physics, but also to cosmology.

Any discovery at the LHC of new particles at the TeV scale will make a compelling case for the construction of a lepton collider to study the properties of these particles in detail and to elucidate the underlying physics. Complementary to the LHC, there are a variety of searches for signals from relic DM particles either locally or dispersed throughout the galactic halo. The truly unprecedented thing about this programme is that if our ideas connecting DM and the question of EWSB are correct, measurements of the properties of new particles produced at the LHC (possibly complemented by measurements at an electron-positron linear collider) may allow us to independently infer just how much DM there is in the universe, and quantitatively predict what other searches for DM should find.[4]

Particle physics, cosmology and astrophysics are rapidly obliterating their boundaries and merging into a single discipline. The ΛCDM model that has emerged posits that 70% of the energy budget of the universe is contained in so-called DE weird stuff with negative pressure that is completely different from anything that we have ever encountered! Thus, not only are the particles we are made of a small fraction of the total matter content of the universe, most of the energy of the universe appears to be in non-material DE extending even further the Copernican principle.[5] This ΛCDM framework is being incisively probed by observation, and may possibly need modification. The nature of DE is a completely open question. Experiments over the next decade or two will, we expect, reveal the identity of DM and, we hope, will provide clues as to the origin of DE. This unprecendented synthesis of the physics of both the largest and smallest scales observable in nature should make the next twenty years very exciting!

Acknowledgement: This research was supported in part by the United States Department of Energy.

REFERENCES

1. For reviews, see *e.g.* C. Jungman, M. Kamionkowski and K. Griest,*Phys. Rept.* **267** (1996) 195; A. Lahanas, N. Mavromatos and D. Nanopoulos, *Int. J. Mod.*

[3]We note here that it is also possible to construct so-called twin-Higgs models [113] where the Higgs sector is stabilised via new particles that couple to the SM Higgs doublet, but are singlets under the SM gauge group. In this case, there would be no obvious new physics signals at the LHC.

[4]These studies have only just begun, and have only been carried out in the context of supersymetry, which unlike extra-dimensional or Little Higgs models, is a complete theory, valid up to very high energy.

[5]Our colleagues who subscribe to the multiverse view carry this yet further, suggesting that our universe is just one of many. Unlike for the ideas discussed here, we are not aware of possible tests for this view.

Phys. **D 12** (2003) 1529; M. Drees, hep-ph/0410113; K. Olive, "Tasi Lectures on Astroparticle Physics", astro-ph/0503065; G. Bertone, D. Hooper and J. Silk, *Phys. Rept.* **405** (2005) 279
2. F. Zwicky, *Helvetica Physica Acta* **6** (1933) 110; see also *Astrophys. J.* **86** (1937) 217
3. V. Rubin and W. K. Ford, *Astrophys. J.* **159** (1970) 359; V. Rubin, N. Thonnard and W. K. Ford, *Astrophys. J.* **238** (1980) 471
4. A. Conley *et al. Astrophys. J.* **644** (2006) 1
5. D. N. Spergel *et al.* (WMAP Collaboration), *Astrophys. J. Supp.*, **170** (2007) 377
6. B. Lee and S. Weinberg, *Phys. Rev. Lett.* **39** (1977) 165
7. Report of the DMSAG panel on http://www.science.doe.gov/hep/hepap_reports.shtm. See also, L. Roszkowski,*Pramama*, **62** (2004) 389
8. A. Kusenko, *Phys. Rev. Lett.* **97** (2006) 241301
9. P. Chen, *Mod. Phys. Lett.***A 19** (2004) 1047
10. S. Coleman, *Nucl. Phys.* **B 262** (1985) 263, [Erratum-ibid. **B269** (1986) 744
11. A. Kusenko and M. Shaposhnikov, *Phys. Lett.* **B 418** (1998) 46; A. Kusenko, V. Kuzmin and M. Shaposhnikov, *Phys. Rev. Lett.* **80** (1998) 3185
12. E. Kolb, D. Chung and A. Riotto, *Phys. Rev.* **D 59** (1999) 023501 and hep-ph/9810361
13. R. Peccei and H. Quinn, *Phys. Rev. Lett.* **38** (1977) 1440 and *Phys. Rev.* **D 16** (1977) 1791; S. Weinberg, *Phys. Rev. Lett.* **40** (1978) 223; F. Wilczek, *Phys. Rev. Lett.* **40** (1978) 279
14. For a recent review, see *e.g.* S. J. Asztalos, L. Rosenberg, K. van Bibber, P. Sikivie and K. Zioutas, *Ann. Rev. Nucl. Part. Sci.***56** (2006) 293
15. For recent results, see L. Duffy *et al.*, *Phys. Rev. Lett.* **95** (2005) 091304
16. H. Goldberg, *Phys. Rev. Lett.* **50** (1983) 1419; J. Ellis, J. Hagelin, D. Nanopoulos and M. Srednicki, *Phys. Lett.* **B 127** (1983) 233; J. Ellis, J. Hagelin, D. Nanopoulos, K. Olive and M. Srednicki, *Nucl. Phys.* **B 238** (1984) 453
17. G. Servant and T. Tait, *New Jou. Phys.***4** (2002) 99
18. For a review, see D. Hooper and S. Profumo, *Phys. Rept.* **453** (2007) 29
19. N. Arkani-Hamed, A. Cohen and H. Georgi,*Phys. Lett.* **B 513** (2001) 232; N. Arkani-Hamed, A. Cohen, E. Katz and A. Nelson, *J. High Energy Phys.* **07** (2002) 034
20. H. C. Cheng and I. Low, *J. High Energy Phys.* **0309** (2003) 051
21. J. Hubisz and P. Meade, *Phys. Rev.* **D 71** (2005) 035016
22. For reviews, see M. Schmaltz, *Ann. Rev. Nucl. Part. Sci.***55** (2005) 229 and M. Perelstein, *Prog. Part. Nucl. Phys.***58** (2007) 247
23. C. Hill and R. Hill *Phys. Rev.* **D 75** (2007) 115009
24. H.-C. Cheng, arXiv:0710.3407 [hep-ph]
25. J. Feng and J. Kumar, arXiv:0803.4196 [hep-ph]
26. K. Jedamzik, M. LeMoine and G. Moultaka, JCAP**0607** (2006) 010
27. S. Heinemeyer, W. Holik, D. Stockinger, A. Weber and G. Weiglein,*J. High Energy Phys.* **0608** (2006) 052
28. H. Baer and X. Tata, *Weak Scale Supersymmetry: From Superfields to Scattering Events*, (Cambridge University Press, 2006)
29. M. Drees, R. Godbole and P. Roy, *Sparticles*, (World Scientific, 2004)
30. P. Binétruy, *Supersymmetry* (Oxford, 2006)
31. S. Dimopoulos and H. Georgi, *Nucl. Phys.* **B 193** (1981) 150
32. B.C. Allanach, S. Kraml and W. Porod, *J. High Energy Phys.* **03** (2003) 016; G. Belanger, S. Kraml and A. Pukhov, *Phys. Rev.* **D 72** (2005) 015003; S. Kraml and S. Sekmen in: *Physics at TeV Colliders 2007, BSM working group report*, in prep.; see `http://cern.ch/kraml/comparison`
33. M. Dine, A. Nelson, Y. Nir and Y. Shirman, *Phys. Rev.* **D 53** (1996) 2658; for a review, see G. Giudice and R. Rattazzi, *Phys. Rept.* **322** (1999) 419
34. L. Randall and R. Sundrum, *Nucl. Phys.* **B 557** (1999) 79; G. Giudice, M. Luty, H. Murayama and R. Rattazzi, *J. High Energy Phys.* **9812** (1998) 027
35. S. Kachru, R. Kallosh, A. Linde and S. P. Trivedi, *Phys. Rev.* **D 68** (2003) 046005; K. Choi, A. Falkowski, H. P. Nilles, M. Olechowski and S. Pokorski, *J. High Energy Phys.* **0411** (2004) 076; K. Choi, A. Falkowski, H. P. Nilles and M. Olechowski, *Nucl. Phys.* **B 718** (2005) 113; K. Choi, K-S. Jeong and K. Okumura, *J. High Energy Phys.* **0509** (2005) 039; H. Baer, E. Park, X. Tata and T. Wang, *J. High Energy Phys.* **0706** (2007) 033, and references therein
36. L. Everett, I.-W. Kim and K. Zurek, arXiv:0804:0592
37. P. Gondolo, J. Edsjo, P. Ullio, L. Bergstrom, M. Schelke and E. A. Baltz, JCAP **0407** (2004) 008
38. G. Belanger, F. Boudjema, A. Pukhov and A. Semenov, *Comput. Phys. Commun.***174** (2006) 577; *Comput. Phys. Commun.***176** (2007) 367
39. IsaRED, by H. Baer, C. Balazs and A. Belyaev, *J. High Energy Phys.* **0203** (2002) 042
40. ISAJET, by H. Baer, F. Paige, S. Protopopescu and X. Tata, hep-ph/0312045; see also H. Baer, J. Ferrandis, S. Kraml and W. Porod, *Phys. Rev.* **D 73** (2006) 015010
41. H. Baer and M. Brhlik, *Phys. Rev.* **D 53** (1996) 597; V. Barger and C. Kao, *Phys. Rev.* **D 57** (1998) 3131
42. J. Ellis, T. Falk and K. Olive, *Phys. Lett.* **B 444** (1998) 367; J. Ellis, T. Falk, K. Olive and M. Srednicki, *Astropart. Phys.* **13** (2000) 181; M.E. Gómez, G. Lazarides and C. Pallis, *Phys. Rev.* **D 61** (2000) 123512 and *Phys. Lett.* **B 487** (2000) 313; A. Lahanas, D. V. Nanopoulos and V. Spanos, *Phys. Rev.* **D 62** (2000) 023515; R. Arnowitt, B. Dutta and Y. Santoso, *Nucl. Phys.* **B 606** (2001) 59; see also Ref. [39]
43. K. L. Chan, U. Chattopadhyay and P. Nath, *Phys. Rev.* **D 58** (1998) 096004; J. Feng, K. Matchev and T. Moroi, *Phys. Rev. Lett.* **84** (2000) 2322 and *Phys. Rev.* **D 61** (2000) 075005; see also H. Baer, C. H. Chen,

F. Paige and X. Tata, *Phys. Rev.* **D 52** (1995) 2746 and *Phys. Rev.* **D 53** (1996) 6241; H. Baer, C. H. Chen, M. Drees, F. Paige and X. Tata, *Phys. Rev.* **D 59** (1999) 055014; for a model-independent approach, see H. Baer, T. Krupovnickas, S. Profumo and P. Ullio, *J. High Energy Phys.* **0510** (2005) 020

44. M. Drees and M. Nojiri, *Phys. Rev.* **D 47** (1993) 376; H. Baer and M. Brhlik, *Phys. Rev.* **D 57** (1998) 567; H. Baer, M. Brhlik, M. Diaz, J. Ferrandis, P. Mercadante, P. Quintana and X. Tata, *Phys. Rev.* **D 63** (2001) 015007; J. Ellis, T. Falk, G. Ganis, K. Olive and M. Srednicki, *Phys. Lett.* **B 510** (2001) 236; L. Roszkowski, R. Ruiz de Austri and T. Nihei, *J. High Energy Phys.* **0108** (2001) 024; A. Djouadi, M. Drees and J. L. Kneur, *J. High Energy Phys.* **0108** (2001) 055; A. Lahanas and V. Spanos, *Eur. Phys. J.* **C 23** (2002) 185
45. R. Arnowitt and P. Nath, *Phys. Rev. Lett.* **70** (1993) 3696; H. Baer and M. Brhlik, Ref. [41]; A. Djouadi, M. Drees and J. Kneur, *Phys. Lett.* **B 624** (2005) 60
46. C. Böhm, A. Djouadi and M. Drees, *Phys. Rev.* **D 30** (2000) 035012; J. R. Ellis, K. A. Olive and Y. Santoso, *Astropart. Phys.* **18** (2003) 395; J. Edsjö, *et al.*, JCAP **0304** (2003) 001
47. H. Baer, A. Mustafayev, E. Park and X. Tata, JCAP**0701**, 017 (2007)
48. H. Baer, A. Mustafayev, E. Park and X. Tata, arXiv:0802:3384
49. Z. Ahmed *et al.*, arXiv:0802:3530 [astro-ph]
50. J. Angle *et al.*, *Phys. Rev. Lett.* **100** (2008) 021303
51. E. Behnk *et al.*, arXiv:0804:2886 [astro-ph]
52. D. Feldman, Z. Liu and P. Nath, *Phys. Lett.* **B 662** (2008) 190
53. H. Baer, C. Balazs, A. Belyaev and J. O'Farrill, JCAP **0309**, (2003) 007
54. R. Schnee, (CDMS Collaboration); A. M. Green, JCAP **0708** (2007) 022; C-L. Shan and M. Drees, arXiv:0710.4296 [hep-ph]
55. H. Baer, A. Belyaev, T. Krupovnickas and J. O'Farrill, JCAP **0408** (2004) 005
56. H. Baer and J. O'Farrill, JCAP**0404**, 005 (2004)
57. P. Sreekumar *et al.* [EGRET Collaboration], Astrophys. J. **494**, 523 (1998) [arXiv:astro-ph/9709257]
58. W. de Boer, M. Herold, C. Sander, V. Zhukov, A. V. Gladyshev and D. I. Kazakov, arXiv:astro-ph/0408272
59. F. W. Stecker, S. D. Hunter and D. A. Kniffen, arXiv:0705.4311 [astro-ph]
60. H. Baer, A. Belyaev and H. Summy, arXiv:0802.3127 [hep-ph]
61. S. Weinberg, *Phys. Rev. Lett.* **48** (1982) 1303; R. H. Cyburt, J. Ellis, B. D. Fields and K. A. Olive, *Phys. Rev.* **D 67** (2003) 103521; K. Jedamzik, *Phys. Rev.* **D 70** (2004) 063524; M. Kawasaki, K. Kohri and T. Moroi, *Phys. Lett.* **B 625** (2005) 7 and *Phys. Rev.* **D 71** (2005) 083502. K. Kohri, T. Moroi and A. Yotsuyanagi, *Phys. Rev.* **D 73** (2006) 123511
62. H. Pagels and J. Primack, *Phys. Rev. Lett.* **48** (1982) 223; J. Feng, A. Rajaraman and F. Takayama, *Phys. Rev. Lett.* **91** (2003) 011302 and *Phys. Rev.* **D 68** (2003) 085018
63. M. Bolz, A. Brandenburg and W. Buchmuller, *Nucl. Phys.* **B 606** (2001) 518; J. Pradler and F. Steffen, hep-ph/0608344
64. J. Feng, S. Su and F. Takayama, *Phys. Rev.* **D 70** (2004) 075019
65. W. Buchmuller, P. Di Bari and M. Plumacher, Annal. Phys. **315** (2005) 305
66. W. Buchmuller, L. Covi, J. Kersten, K. Schmidt-Hoberg, JCAP**0611** (2006) 007; W. Buchmuller, L. Covi, K. Hamaguchi, A. Ibarra and T. Yanagida, *J. High Energy Phys.* **0703** (2007) 037
67. J. E. Kim and H. P. Nilles
68. L. Covi, J. E. Kim and L. Roszkowski, *Phys. Rev. Lett.* **82** (1999) 4180; L. Covi, H. B. Kim, J. E. Kim and L.Roszkowski, *J. High Energy Phys.* **0105** (2001) 033
69. A. Brandenburg and F. Steffen, JCAP**0408** (2004) 008
70. G. Lazarides and Q. Shafi, *Phys. Lett.* **B 258** (1991) 305; K. Kumekawa, T. Moroi and T. Yanagida, *Prog. Theor. Phys.* **92** (1994) 437; T. Asaka, K. Hamaguchi, M. Kawasaki and T. Yanagida, *Phys. Lett.* **B 464** (1999) 12
71. H. Baer and H. Summy,arXiv:0803.0510 (2008)
72. H. Baer, S. Kraml, S. Sekmen and H. Summy, JHEP**0803**, 056 (2008)
73. H. Baer and X. Tata, *Phys. Lett.* **B 160** (1985) 159
74. H. Baer, J. Ellis, G. Gelmini, D. V. Nanopoulos and X. Tata, *Phys. Lett.* **B 161** (1985) 175; G. Gamberini, *Z. Physik* **C 30** (1986) 605; H. Baer, V. Barger, D. Karatas and X. Tata, *Phys. Rev.* **D 36** (1987) 96; H. Baer, X. Tata and J. Woodside, *Phys. Rev.* **D 45** (1992) 142
75. H. Baer, D. Dzialo-Karatas and X. Tata, *Phys. Rev.* **D 42** (1990) 2259
76. H. Baer, C. H. Chen, F. Paige and X. Tata, *Phys. Rev.* **D 50** (1994) 4508
77. H. Baer, C. H. Chen, F. Paige and X. Tata, *Phys. Rev.* **D 49** (1994) 3283
78. H. E. Haber and D. Wyler, *Nucl. Phys.* **B 323** (1989) 267; S. Ambrosanio and B. Mele, *Phys. Rev.* **D 53** (1996) 2541 and *Phys. Rev.* **D 55** (1997) 1399 [Erratum-ibid. D**56**, 3157 (1997)]; H. Baer and T. Krupovnickas, *J. High Energy Phys.* **0209** (2002) 038
79. H. Baer and X. Tata, *Phys. Rev.* **D 47** (1993) 2739
80. H. Baer, C. Chen, M. Drees, F. Paige and X. Tata, *Phys. Rev. Lett.* **79** (1997) 986
81. H. Baer, C. Chen, M. Drees, F. Paige and X. Tata, *Phys. Rev.* **D 59** (1999) 015010
82. H. Baer, M. Bisset, X. Tata and J. Woodside, *Phys. Rev.* **D 46** (1992) 303
83. H. Baer, M. Bisset, D. Dicus, C. Kao and X. Tata, *Phys. Rev.* **D 47** (1993) 1062; H. Baer, M. Bisset, C. Kao and X. Tata, *Phys. Rev.* **D 50** (1994) 316

84. M. Muhlleitner, A. Djouadi and Y. Mambrini, *Comput. Phys. Commun.***168** (2005) 46
85. W. Porod, *Comput. Phys. Commun.***153** (2003) 275
86. T. Sjostrand, S. Mrenna and P. Skands, *J. High Energy Phys.* **0605** (2006) 026
87. G. Corcella *et al.*, *J. High Energy Phys.* **0101** (2001) 010
88. F. Maltoni and T. Stelzer, *J. High Energy Phys.* **0302** (2003) 027; J. Alwall *et al.*, *J. High Energy Phys.* **0709** (2007) 028
89. A. Pukhov *et al.*, hep-ph/9908288
90. W. Kilian, T. Ohl and J.Reuter, arXiv:0708.4233
91. J. Alwall *et al.*, *Comput. Phys. Commun.***176** (2007) 300
92. H. Baer, X. Tata and J. Woodside, Ref.[74]
93. P. Mercadante, J. K. Mizukoshi and X. Tata, *Phys. Rev.* **D 72** (2005) 035009; S. P. Das *et al.* arXiv:0708:2048 [hep-ph]; R. Kadala, P. Mercadante, J. K. Mizukoshi and X. Tata, arXiv:0803:0001 [hep-ph]
94. H. Baer, A. Belyaev, T. Krupovnickas and X. Tata, *J. High Energy Phys.* **0402** (2004) 007; H. Baer, T. Krupovnickas and X. Tata, *J. High Energy Phys.* **0406** (2004) 061
95. H. Baer, H. Prosper and H. Summy, *Phys. Rev.* **D 77** (2008) 055017
96. I. Hinchliffe *et al.*, *Phys. Rev.* **D 55** (1997) 5520 and *Phys. Rev.* **D 60** (1999) 095002
97. H. Baer, K. Hagiwara and X. Tata, *Phys. Rev.* **D 35** (1987) 1598; H. Baer, D. Dzialo-Karatas and X. Tata, *Phys. Rev.* **D 42** (1990) 2259; H. Baer, C. Kao and X. Tata, *Phys. Rev.* **D 48** (1993) 5175; H. Baer, C. H. Chen, F. Paige and X. Tata, *Phys. Rev.* **D 50** (1994) 4508
98. R. Arnowitt *et al. Phys. Lett.* **B 639** (2006) 46 and *Phys. Lett.* **B 649** (2007) 73
99. H. Bachacou, I. Hinchliffe and F. Paige, *Phys. Rev.* **D 62** (2000) 015009; Atlas Collaboration, LHCC 99-14/15
100. C. Lester and D Summers, *Phys. Lett.* **B 463** (1999) 99; A. Barr, C. Lester and P. Stephens, J. Phys. bf G29 (2003) 2343; C. Lester and A. Barr, *J. High Energy Phys.* **0712** (2007) 102; W. Cho, K. Choi, Y. Kim and C. Park, arXiv:0711.4526
101. K. Kawagoe *et al. Phys. Rev.* **D 69** (2004) 035003; S. Ambrosanio *et al. J. High Energy Phys.* **0101** (2001) 014
102. H. Hamaguchi, S. Shirai and T. Yanagida, arXiv:0712.2462
103. J. Feng *et al. Phys. Rev.* **D 52** (1995) 1418; M. Nojiri, K. Fujii and T. Tsukamoto, *Phys. Rev.* **D 54** (1996) 6756
104. E. Baltz, M. Battaglia, M. Peskin and T. Wizansky, *Phys. Rev.* **D 74** (2006) 103521. See also R. Arnowitt *et al.*, arXiv:0802.2968 [hep-ph] for a similar study in the stau co-annihilation region
105. H. C. Cheng, K. Matchev and M. Schmaltz, *Phys. Rev.* **D 66** (2002) 036005
106. T. Rizzo, *Phys. Rev.* **D 64** (2001) 095010; C. Macescanu, C. McMullen and S. Nandi, *Phys. Rev.* **D 66** (2002) 015009
107. See Ref. [17]; K. Kong and K. Matchev, *J. High Energy Phys.* **0601** (2006) 038
108. H. C. Cheng, K. Matchev and M. Schmaltz, *Phys. Rev.* **D 66** (2002) 056006; A. Datta, K. Kong and K. Matchev, *Phys. Rev.* **D 72** (2005) 096006
109. A. Alves, O. Eboli and T. Plehn, *Phys. Rev.* **D 74** (2006) 095010
110. A. Birkedal, A Noble, M. Perelstein and A. Spray, *Phys. Rev.* **D 74** (2006) 035002; D. Hooper and G. Zaharijas, *Phys. Rev.* **D 75** (2007) 035010
111. C.-S. Chen, K. Cheung and T. C. Yuan, *Phys. Lett.* **B 664** (2007) 158; T. Han, H. Logan and L.-T. Wang, *J. High Energy Phys.* **0601** (2006) 099
112. Y. Bai, arXiv:0801:1662
113. Z. Chacko, H-S. Goh and R. Harnik, *Phys. Rev. Lett.* **96** (2006) 231802; see E. Dolle and S. Su, *Phys. Rev.* **D 77** (2008) 075013 for an analysis of DM in a twin-Higgs scenario

Using Tau Polarisation for Charged Higgs Boson and SUSY Searches at the LHC

Monoranjan Guchait[a] *and D. P. Roy*[b]

[a]*Department of High Energy Physics, Tata Institute of Fundamental Research, Homi Bhabha Road, Mumbai-400005, India*

[b]*Homi Bhabha's Centre for Science Foudantion, Tata Institute of Fundamental Research, V.N. Purav Marg, Mumbai-400088, India*

The τ polarisation can be easily measured at the LHC in the 1-prong hadronic τ decay channel by measuring what fraction of the τ-jet momentum is carried by the charged track. A simple cut requiring this fraction to be > 0.8 retains most of the $P_\tau = +1$ τ-jet signal while suppressing the $P_\tau = -1$ τ-jet background and practically eliminating the fake τ background. This can be utilised to extract the charged Higgs signal. It can be also utilised to extract the SUSY signal in the stau NLSP region, and in particular the stau co-annihilaton region.

1. Introduction

It is easy to measure τ polarisation P_τ as it is reflected in the kinematic distribution of its decay products. Moreover the best channel for measuring τ polarisation is also the best channel for τ identification, i.e. the 1-prong hadronic τ decay channel. In particular a simple kinematic cut, requiring the single charged prong to carry $> 80\%$ of the hadronic τ-jet momentum retains most of the $P_\tau = +1$ τ-jet events, while suppressing the $P_\tau = -1$ τ-jet background and practically eliminating the fake τ background from standard hadronic jets. Interestingly the most important channel for charged Higgs boson search at the LHC is its τ decay channel, $H^- \to \tau_R^- \bar{\nu}_R$, giving $P_\tau = +1$. Similarly a very important part of the parameter space of the minimal supergravity (mSUGRA) model has $\tilde{B}$ as the lightest superparticle, while the next to the lightest one is a stau ($\tilde{\tau}_1$) with a dominant $\tilde{\tau}_R$ component. In this case one expects the supersymmetric (SUSY) signal at the LHC to contain a $P_\tau = +1$ τ from the cascade decay of squarks and gluinos via $\tilde{\tau}_1 \to \tau_R^- \tilde{B}$. In both cases one can use the above kinematic cut to enhance the $P_\tau = +1$ signal over the $P_\tau = -1$ background as well as the fake τ background.

The paper is organised as follows. In section 2 we summarise the formalism of τ polarisation in the 1-prong hadronic decay channel and discuss how the above mentioned kinematic cut retains most of the detectable $P_\tau = +1$ τ-jet signal while supressing the $P_\tau = -1$ τ-jet as well as the fake τ-jet backgrounds. Section 3 briefly introduces the SUSY search programme at the LHC via SUSY as well as SUSY Higgs (and in particular $H^\pm$) signals. In section 4, we describe the most important $H^\pm$ signal in both $m_{H^\pm} < m_t$ and $m_{H^\pm} > m_t$ regions, which contains a hard τ with $P_\tau = +1$ from the above mentioned $H^\pm$ decay. In section 5 we show Monte Carlo simulations using the above kinematic cut for extraction of the $H^\pm$ signal at the LHC for both the $m_{H^\pm} < m_t$ and $m_{H^\pm} > m_t$ regions. In the latter case we also briefly discuss a corresponding kinematic cut for extracting the $m_{H^\pm}$ signal in the 3-prong hadronic decay channel of τ. In section 6 we briefly describe the SUSY signal coming from the above mentioned cascade decay propcess. We also emphasize a very important part of the SUSY parameter space, called the stau co-annihilation region, where the signal contains a soft τ with $P_\tau = +1$. In section 7 we show the use of the kinematic cut for extracting the SUSY signal at the LHC in the 1-prong hadronic τ-decay channel, with particular emaphsis on the stau co-annihilation region.

2. τ-Polarisation

The best channel for τ-polarisation is its 1-prong hadronic decay channel, accounting for 50% of its decay width. Over 90% this comes from

$$\tau \to \pi^\pm \nu(12.5\%), \rho^\pm \nu(26\%), a_1^\pm \nu(7.5\%), \quad (1)$$

where the branching fraction for π and ρ include the small K and K* contributions, which have identical polarisation effects [1]. The CM angular distributions of τ decay into π and vector meson v (=ρ, a_1) is simply given in terms of its polarisation as

$$\frac{1}{\Gamma_\pi}\frac{d\Gamma_\pi}{d\cos\theta} = \frac{1}{2}(1 + P_\tau \cos\theta)$$

$$\frac{1}{\Gamma_v}\frac{d\Gamma_{vL,T}}{d\cos\theta} = \frac{\frac{1}{2}m_\tau^2, m_v^2}{m_\tau^2 + 2m_v^2}(1 \pm P_\tau \cos\theta)\,, \tag{2}$$

where L,T denote the longitudinal and transverse polarisation states of the vector meson. The fraction x of the τ laboratory momentum carried by its decay meson, i.e the (visible) τ-jet is related to the angle θ via

$$x = \frac{1}{2}(1+\cos\theta) + \frac{m^2_{\pi,v}}{2m_\tau^2}(1-\cos\theta) \tag{3}$$

in the collinear approximation ($p_\tau \gg m_\tau$). It is clear from Eqs. 2 and 3 that the relatively hard part of the signal (P_τ=+1) τ-jet comes from the π, ρ_L and a_{1L} contributions, while for the background ($P_\tau = -1$) τ-jet it comes from the ρ_T and a_{1T} contributions [2]. Note that this is the important part that would pass the p_T threshold for detecting τ-jets.

One can simply understand the above feature from angular momentum conservation. For $\tau_{R(L)} \to \nu_L\pi^-, v^-_{\lambda=0}$ it favours forward (backward) emission of π or longitudinal vector meson, while it is the other way around for transverse vector meson emission, $\tau^-_{R(L)} \to \nu_L v^-_{\lambda=-1}$. After boosting back to the laboratory frame the forward emitted meson becomes the leading particle, giving a hard τ-jet.

Now the ρ_T and a_{1T} decays favour equal sharing of the momentum among the decay pions, while the ρ_L and a_{1L} decays favour unequal sharing, where the charged pion carries either very little or most of the τ-jet momentum. Thus plotted as a function of the momentum fraction carried by the charged pion

$$R = \frac{p_{\pi^\pm}}{p_{\tau-jet}} \tag{4}$$

the longitudinal ρ and a_1 contributions peak at very low of high R ($\lesssim 0.2$ or $\gtrsim 0.8$), while the transverse contributions peak in the middle [2,3]. This is shown in Fig. 1 [3]. Note that the $\tau^\pm \to \pi^\pm\nu$ contribution would appear as a delta function at $R = 1$ in this figure. The low R peaks of the longitudinal ρ and a_1 contributions are not detectable because of the minimum p_T requirement on the charged track for τ-identification ($R \gtrsim 0.2$). Now moving the R cut from 0.2–0.8 cuts out the transverse ρ and a_1 peaks, while retaining the detectable longitudinal peak along with the single $\pi^\pm$ contribution. Thanks to the complimentarity of these two sets of contributions, one can effectively suppress the former while retaining most of the latter by a simple cut on the ratio

$$R > 0.8. \tag{5}$$

Thus one can suppress the hard part of the τ jet backround ($P_\tau = -1$) while retaining most of it for

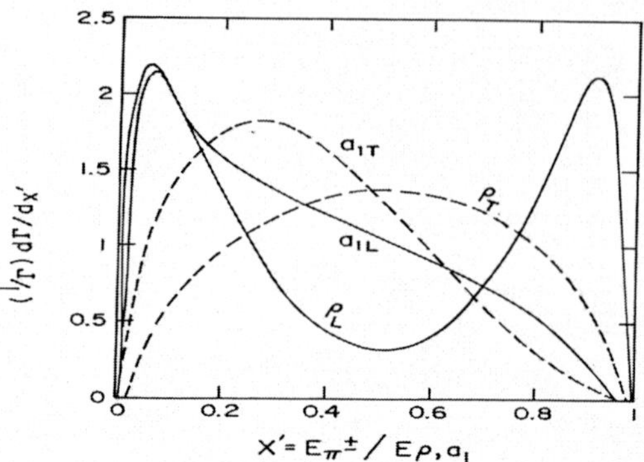

Figure 1. Distributions of $\tau \to \rho\nu, a_1\nu$ events in the fractional hadron(τ-jet) energy-momentum carried by the charged prong X′ ≡ R [3,4]. Note that $\tau^\pm \to \pi^\pm\nu$ contribution corresponds to a delta function at X′ = 1

the detectable signal ($P_\tau = +1$), even without separating the different meson contributions from one another [3]. This is a simple but very powerful result particularly for hadron colliders, where one cannot isolate the different meson contributions to the τ-jet in (1). The result holds equally well for a more exact simulation of the τ-jet including the nonresonant contributions. It should be noted here that the simple polarisation cut (5) suppresses not only the $p_\tau = -1$ τ-jet background, but also the fake τ-jet background from common hadronic jets. This is particularly important for τ-jets with low p_T threshold of 15–20 GeV, as we shall need for SUSY search in the stau co-annihilation region in section 7. Imposing this cut reduces the faking efficiency of hadronic jets from 5–10% level to about 0.2%. The reason is that a common hadronic jet can fake an 1-prong τ-jet by a rare fluctuation, when all but one of the constituent particles (mostly pions) are neutral. Then requiring the single charged particle to carry more than 80% of the total jet energy requires a second fluctuation which is even rarer.

3. SUSY and SUSY Higgs Searches at the LHC

The minimal suupersymmetric standard model (MSSM) has been the most popular extension of the Standard Model (SM) for four reasons. It provides (a) a natural solution to the hierarchy problem of the electroweak symmetry breaking (EWSB) scale of the SM, (b) a natural (radiative) mechanism for EWSB, (c) a natural candidate for the dark matter of the universe in terms of the lightest superparticle (LSP) and (d) unification of the gauge couplings at the grand unification (GUT) scale. Therefore, there is a great deal of current interest in probing this model at the LHC. This is based on a two-prong search strategy. On the one hand we are looking for the signal of supersymmetric (SUSY) particle production at the LHC. On the other hand we are also looking for the signal of the extended Higgs bosn sector of the MSSM and in particular the charged Higgs boson($H^{\pm}$). We shall see below that the τ channel plays a very important role for both SUSY and the $H^{\pm}$ signals and one can use the above mentioned τ polarisation effect in extracting both these signals at the LHC.

4. $H^{\pm}$ Signal

As mentioned above the MSSM contains two Higgs doublets H_u and H_d, the ratio of whose vevs is denoted by $\tan\beta$. The two complex doublets correspond to 8 degrees of freedom, 3 of which are absorbed as Goldstone bosons to give masses and longitudinal components to the $W^{\pm}$ and Z bosons. This leaves 5 physical states: two neutral scalars h and H, a pseudo scalar A and a pair of charged Higgs bosons

$$H^{\pm} = H_u^{\pm}\cos\beta + H_d^{\pm}\sin\beta. \tag{6}$$

While it may be hard to distinguish any of these neutral Higgs bosons from that of the SM, the $H^{\pm}$ pair carry the distinctive hallmark of the MSSM. Hence the $H^{\pm}$ search plays a very important role in probing the SUSY Higgs sector [4]. All the tree level masses and couplings of the MSSM Higgs bosons are given in terms of $\tan\beta$ and any one of their masses, usually taken to be m_A. It is simply related to $m_{H^{\pm}}$ via

$$m^2_{H^{\pm}} = m^2_A + m^2_W. \tag{7}$$

The most important $H^{\pm}$ couplings are

$$H^{\pm}tb(cs): \frac{g}{\sqrt{2}M_W}(m_{t(c)}\cot\beta + m_{b(s)}\tan\beta),$$

$$H^{\pm}\tau\nu: \frac{g}{\sqrt{2}M_W}m_\tau\tan\beta. \tag{8}$$

Assuming the $H^{\pm}tb$ coupling to remain perturbative up to the GUT scale implies $1 < \tan\beta < m_t/m_b$.

For $m_{H^{\pm}} < m_t$, Eq. (8) imply large branching fractions for

$$t \to bH^{\pm} \tag{9}$$

decay at the two ends of the above range, $\tan\beta \sim 1$ and $\tan\beta \sim m_t/m_b \simeq 50$, driven by the m_t and m_b terms respectively. But there is a huge dip in the intermediate region around

$$\tan\beta \sim \sqrt{m_t/m_b} \sim 7 \tag{10}$$

which is overwhelmed by the SM decay $t \to bW$. Eq. (8) also implies that the dominant decay mode for this $H^{\pm}$ over the theoretically favoured region of $\tan\beta > 1$ is

$$H^- \to \tau^-\bar{\nu}_R; \quad P_\tau = +1, \tag{11}$$

where the polarisation follows simply from angular momentum conservation, requiring the τ^- to be right handed. It implies the opposite polarisation for the SM process

$$W^- \to \tau^-\bar{\nu}_R; \quad P_\tau = -1 \tag{12}$$

since the τ^- is now required to be left-handed. One can use the opposite polarisations to distinguish the $H^{\pm}$ signal from the SM background [2,3]. In particular one can use the kinematic cut, mentioned in the introduction, to enhance the signal/background ratio and extend the $H^{\pm}$ search at the LHC over the intermediate $\tan\beta$ range (10), which would not be possible otherwise [3].

For $m_{H^{\pm}} > m_t$ the dominant production process at the LHC is the LO process

$$gb \to tH^- + h.c. \tag{13}$$

The dominant decay channel is $H^- \to \bar{t}b$, which has unfortunately a very large QCD background. By far the most viable signal comes from the second largest decay channel (11), which has a branching fraction of $\gtrsim$10% in the moderate to large $\tan\beta$ ($\gtrsim$10) region. The largest background comes from $t\bar{t}$ production, followed by the decay of one of the top quarks into the SM channel (12). One can again exploit the opposite τ polarisations to enhance the signal/background ratio and extend the $H^{\pm}$ search to several hundreds of GeV for $\tan\beta \gtrsim 10$ [5–7]. This will be discussed in detail in the next section.

5. τ Polarisation in the $H^{\pm}$ Search

A parton level Mone Carlo simulation of the $H^{\pm}$ signal in the $m_{H^{\pm}} < m_t$ region [3] showed that using the polarisation cut (5) enhances the signal/background ration substantially and makes it possible to extend the $H^{\pm}$ search at the LHC over most of the intermediate $\tan\beta$ region (10). This has been confirmed now by more exact simulations with particle level event generators. Figure 2 shows the $H^{\pm}$ discovery contours at the LHC using this polarisation cut [7]. The vertical contour on left shows $H^{\pm}$ discovery contour via $t \to bH$ decay. The mild dip in the middle shows the remaining gap in this intermediate $\tan\beta$ region.

For $m_{H^{\pm}} > m_t$ the signal comes from (11) and (13) while the background comes from $t\bar{t}$ production, followed by the decay of one top into (12). To start

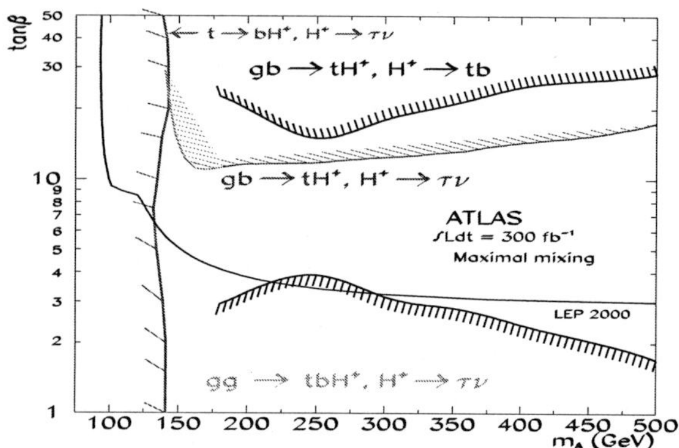

Figure 2. The 5σ $H^{\pm}$ boson discovery contours of the ATLAS experiment at the LHC from $t \to bH^{\pm}, H^{\pm} \to \tau^{\pm}\nu$ (vertical); $gb \to tH^{-}; H^{-} \to \tau^{-}\nu$(middle horizontal) and $gb \to tH^{-}; H^{-} \to \bar{t}b$ channel [7]

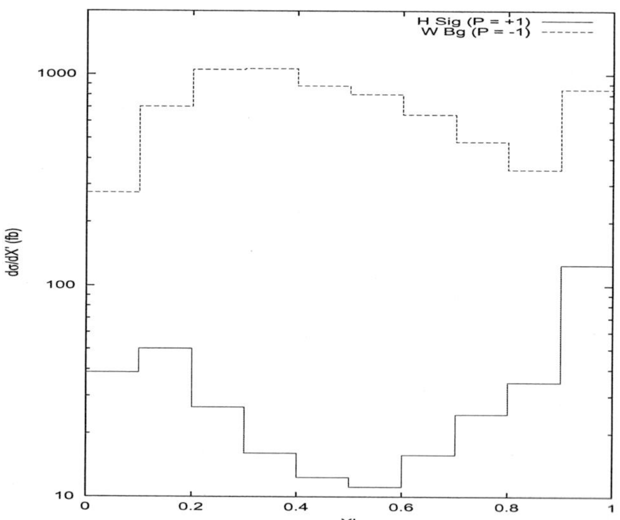

Figure 3. The LHC cross section for a 300 GeV $H^{\pm}$ signal at $\tan\beta = 40$ shown along with the $t\bar{t}$ background in the 1-prong τ-jet channel, as function of the τ-jet momentum fraction X$'$(R) carried by the charged pion [4]

with the background is over two orders of magnitude larger than the signal; but the signal has a harder τ-jet. Thus a $p_T^{\tau-jet} > 100$ GeV cut improves the signal/background ratio. Figure 3 shows the R(X$'$) distribution of the resulting signal and background. One can see that increasing the R cut from 0.2–0.8 suppresses the background substantially while retaining most of the detectable ($R > 0.2$) signal events. The remaining signal and background can be separated by looking at their distributions in the transverse mass of the τ-jet with the missing p_T, coming from the accompanying ν. Figure 4 shows these distributions from a recent simulation [6] using PYTHIA Monte Carlo event generation [8], interfaced with TAUOLA [9] for handling τ decay. One can clearly separate the $H^{\pm}$ signal from the W background and also measure the $H^{\pm}$ mass using this plot.

Finally, one can also use the polarisation effect in the 3-prong hadronic τ-decay channel

$$\tau^{\pm} \to \pi^{\pm}\pi^{\pm}\pi^{\mp}\nu \tag{14}$$

with no neutrals. This has a branching fraction of 10%, which acounts for 2/3rd of inclusive 3-prong τ-decay (including neutrals). Excluding neutrals effectively eliminates the fake τ-jet background from common hadronic jets. About 3/4 of the branching fraction for Eq. (14) comes from a_1. The momentum fraction R of $\pi^{\pm}\pi^{0}\pi^{0}$ channel is equivalent to the momentum fraction carried by the unlike sign pion in $a_1 \to \pi^{\pm}\pi^{\pm}\pi^{\mp}$ channel. Thus one sees from Fig. 1 that one can retain the $a_{1\mathrm{L}}$ peak while suppressing a_{1T} by restricting this momentum fraction to <0.2, which is accessible in this case. This will suppress the hard τ-jet background events from $P_\tau = -1$ while retaining them for the $P_\tau = +1$ signal. This simple result holds even after the inclusion of the non-resonant contribution.

Figure 5 shows the $H^{\pm}$ discovery contours of the LHC using 1-prong and (1 + 3)-prong channels [6]. One sees a modest improvement of the discovery reach by including the 3-prong channel. Note also that the 1-prong $H^{\pm}$ discovery contour for $100 fb^{-1}$ luminosity is consistent with that of Fig. 2 for the ultimate $300 fb^{-1}$ luminosity of the LHC.

6. SUSY Signal

We shall concentrate in the mSUGRA model as a simple and well-motivated parametrisation of the MSSM. This is described by four and half parameters [10]

$$m_0, m_{1/2}, A_0, \tan\beta \text{ and sgn}(\mu) \tag{15}$$

the first three representing the common gaugino and scalar masses and trilinear coupling at the GUT scale. The $\tan\beta$ stands for the ratio of the two Higgs vacuum expectation values. The last one denotes the sign of the higgsino mass parameter whose magnitude is fixed by the radiative EWSB condition described below. All the weak scale superparticle masses are given in terms of these parameters by the renormalisation group evolution (RGE). In particular the gaugino masses evolve

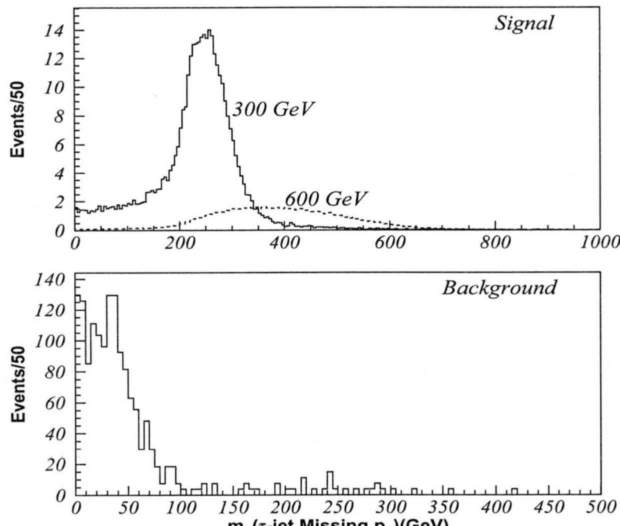

Figure 4. The number of events are shown against transverse mass m_T for signal and background for 1-prong decay channel of τ-jets. These are subject to $p_T^{\tau-jet} > 100$ GeV, $R > 0.8$ and $\not{E}_T > 100$ GeV. The masses of charged Higgs are 300 GeV and 600 GeV and $\tan\beta = 40$ [6]

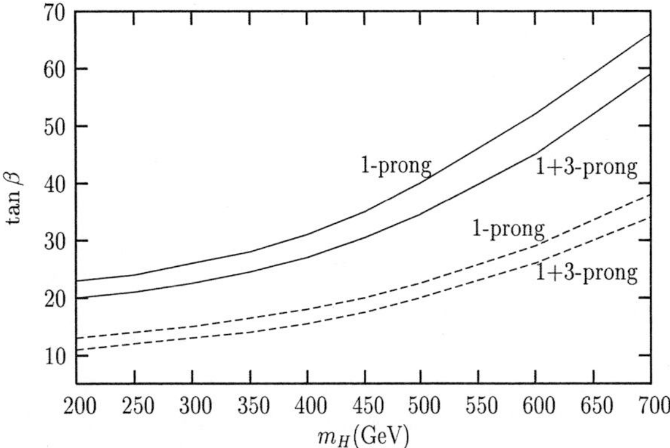

Figure 5. The 5σ $H^{\pm}$ discovery contours at the LHC shown for integrated luminosities of $30fb^{-1}$(solid) and $100fb^{-1}$(dashed) for $H^{\pm} \to \tau\nu$, with 1 and 3 prong hadronic decay of τ [6]

like the corresponding gauge couplings. Thus

$$M_1 = (\alpha_1/\alpha_G)m_{1/2} \simeq 0.4m_{1/2}\,,$$
$$M_2 = (\alpha_2/\alpha_G)m_{1/2} \simeq 0.8m_{1/2} \tag{16}$$

represent the bino $\tilde{B}$ and wino $\tilde{W}_3$ masses respectively. A very important weak scale scalar mass, appearing in the radiative EWSB condition, is

$$\mu^2 + M_Z^2/2 = \frac{M_{H_d}^2 - M_{H_u}^2 \tan^2\beta}{\tan^2\beta - 1} \simeq -M_{H_u}^2\,, \tag{17}$$

where the last equality holds at $\tan\beta > 5$, favoured by the Higgs mass limit from LEP [1]. The sign of $M_{H_u}^2$ turning negative by RGE triggers EWSB, as required by (17). The RHS is related to the GUT scale parameters by the RGE

$$\begin{aligned} -M_{H_u^2} &= C_1(\alpha_i, y_t, \tan\beta)m_0^2 + C_2(\alpha_i, y_t, \tan\beta)m_{1/2}^2 \\ &\simeq -\epsilon\, m_0^2 + 2m_{1/2}^2. \end{aligned} \tag{18}$$

The tiny co-efficient of m_0^2 results from an almost exact cancellation of the GUT scale value by a negative top Yukawa (y_t) contribution. We see from Eq. (16–18) that apart from a narrow strip of $m_0 >> m_{1/2}$, the mSUGRA parameter space satisfies the mass hierarchy

$$M_1 < M_2 < \mu. \tag{19}$$

Thus the lighter neutralinos and chargino are dominated by the gaugino components

$$\tilde{\chi}_1^0 \simeq \tilde{B}; \qquad \tilde{\chi}_2^0, \tilde{\chi}_1^{\pm} \simeq \tilde{W}_3, \tag{20}$$

while the heavier ones are dominated by the higgsino. The lightest neutralino $\tilde{\chi}_1^0(\equiv \tilde{\chi})$ is the LSP.

The ligtest sfermions are the right-handed sleptons, getting only the U(1) gauge contributions to the RGE, i.e.

$$m_{\tilde{\ell}_R}^2 \simeq m_0^2 + 0.15m_{1/2}^2. \tag{21}$$

The Yukawa coupling contribution drives the $\tilde{\tau}_R$ mass still lower. Moreover the mixing between the $\tilde{\tau}_{L,R}$ states, represented by the off-diagonal term

$$m_{LR}^2 = m_\tau(A_\tau + \mu\tan\beta) \tag{22}$$

drives the lighter mass eigenvalues further down. Thus the lighter stau mass eigenstate

$$\tilde{\tau}_1 = \tilde{\tau}_R \sin\theta_{\tilde{\tau}} + \tilde{\tau}_L \cos\theta_{\tilde{\tau}} \tag{23}$$

is predicted to be the lightest sfermion. Moreover one sees from Eqs. (16, 19 and 21) that $\tilde{\tau}_1$ is predicted to be the next to lightest superparticle (NLSP) over half of the parameter space

$$m_0 < m_{1/2}. \tag{24}$$

Thanks to the modest $\tilde{\tau}_L$ component in Eq. (23), a large part of the SUSY cascade decay signal at the LHC proceeds via

$$\tilde{\chi}_1^{\pm} \to \tilde{\tau}_1\nu \to \tau\nu\tilde{\chi}_1^0, \tag{25}$$

$$\tilde{\chi}_2^0 \to \tau'\tilde{\tau}_1 \to \tau'\tau\tilde{\chi}_1^0. \tag{26}$$

The dominance of the $\tilde{\tau}_R$ component in $\tilde{\tau}_1$ implies that the polarisation

$$P_\tau \simeq +1, \quad (27)$$

while $P_{\tau'} \simeq -1$. We shall see in the next section that the polarisation effect can be utilised to extract the SUSY signal containing a positively polarised τ from Eqs. (25, 26).

A very important part of the above mentioned parameter space is the stau co-annihilation region [11], where the $\tilde{B}$ LSP co-annihilates with a nearly degenerate $\tilde{\tau}_1$, $\tilde{\chi}_1^0\tilde{\tau}_1 \to \tau\gamma$, to give a cosmologically compatible relic density [12]. The mass degeneracy $m_{\tilde{\tau}_1} \simeq m_{\tilde{\chi}_1^0}$ is required to hold to $\sim 5\%$, since the freeze out temeparture is $\sim 5\%$ of the LSP mass. Because of this mass denegenercay the positively polarised τ lepton coming from Eqs. (25, 26) is rather soft. We shall see in the next section how the polarisation effect can be utilised to extract the soft τ signal and also to measure the tiny mass difference between the co-annihilating particles.

7. τ Polarisation in SUSY Search

The polarisation of τ coming from the $\tilde{\tau}_1$ decay of Eqs. (25) and (26) is given in the collinear approximation by [13]

$$P_\tau = \frac{\Gamma(\tau_R) - \Gamma(\tau_L)}{\Gamma(\tau_R) + \Gamma(\tau_L)} = \frac{(a_{11}^R)^2 - (a_{11}^L)^2}{(a_{11}^R)^2 + (a_{11}^L)^2} \quad (28)$$

$$a_{11}^R = -\frac{2g}{\sqrt{2}} N_{11} \tan\theta_W \sin\theta_{\tilde{\tau}} - \frac{g m_\tau}{\sqrt{2} m_W \cos\beta} N_{13} \cos\theta_{\tilde{\tau}}$$

$$a_{11}^L = \frac{g}{\sqrt{2}} [N_{12} + N_{11} \tan\theta_W] \cos\theta_{\tilde{\tau}} - \frac{g m_\tau}{\sqrt{2} m_W \cos\beta} \sin\theta_{\tilde{\tau}} N_{13}$$

where the 1st and 2nd subscript of a_{ij} refer to $\tilde{\tau}_i$ and $\tilde{\chi}_j^0$; and

$$\tilde{\chi} \equiv \tilde{\chi}_1^0 = N_{11}\tilde{B} + N_{12}\tilde{W}_3 + N_{13}\tilde{H}_d + N_{14}\tilde{H}_u \quad (29)$$

gives the composition of LSP. Thus the dominant term is $a_{11}^R \simeq -\frac{2g}{\sqrt{2}} N_{11} \tan\theta_W \sin\theta_{\tilde{\tau}}$, implying $P_\tau \equiv +1$. In fact in the mSUGRA model there is a cancellation between the subdominant terms, so that one gets $P_\tau > 0.9$ throughtout the allowed parameter space [14]. Moreover in the $\tilde{\tau}_1$ NLSP region of Eq. (24) $P_\tau > 0.95$, so that one can approximate it to $P_\tau = +1$. The polarisation of the τ' from Eq. (26) is obtained from Eq. (28) by replacing $a_{11}^{L,R}$ by $a_{1,2}^{L,R}$. The dominant contribution comes from $a_{12}^L \simeq \frac{g}{\sqrt{2}} N_{22} \cos\theta_{\tilde{\tau}}$, implying $P_{\tau'} \equiv -1$. There is a similar cancellation of the subdominant contributions, leading to $P_{\tau'} < -0.95$ in the $\tilde{\tau}_1$ NLSP region. Thus one can safely approximate $P_{\tau'} = -1$. Figure 6 shows that P_τ is > 0.9 for $\tilde{\tau}_1 \to \tau\tilde{\chi}^0$ decay throughout the mSUGRA parameter space [14]. It also shows that the branching fraction of the decay (25) is large over the $\tilde{\tau}_1$ NLSP region of Eq. (24), so that one expects a large part of the SUSY signal in the $\not{E}_T$ channel to contain a τ-jet with $P_\tau = +1$. Figure 7 shows the R distribution of this $P_\tau = +1$ τ-jet at the LHC [14]. For comparison the R-distributions are also shown for $P_\tau = 0$ and -1 for this τ-jet. Thus one can test the SUSY model or check the composition of $\tilde{\tau}_1(\tilde{\chi}^0)$ by measuring this distribution.

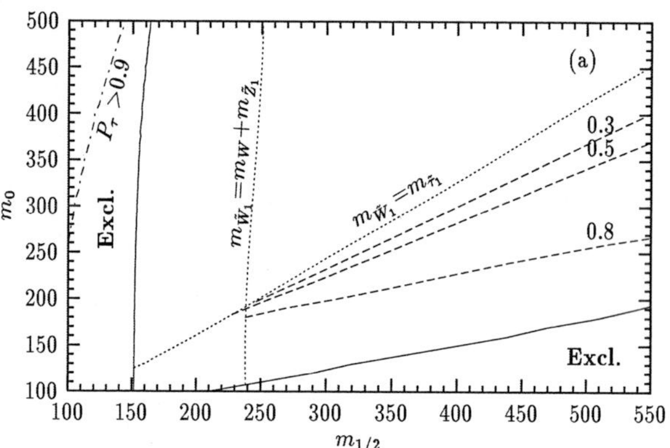

Figure 6. $BR(\tilde{W}_1 \to \tilde{\tau}_1\nu_\tau)$ is shown as contour plots (dashed lines) in m_0 and $m_{1/2}$ plane for $A_0 = 0$, $\tan\beta = 30$ and positive μ. The kinematic boundaries (dotted lines) are shown for $\tilde{W}_1 \to W\tilde{Z}_1$ and $\tilde{W}_1 \to \tilde{\tau}_1\nu_\tau$ decay. The entire region to the right of the boundary (dot-dashed line) corresponds to $P_\tau > 0.9$. The excluded region on the right is due to the $\tilde{\tau}_1$ being the LSP while that on the left is due to the LEP constraint $m_{\tilde{W}_1^\pm} > 102$ GeV [14]. Note that here $\tilde{W}_1$ and $\tilde{Z}_1$ correspond to $\tilde{\chi}_1^\pm$ and $\tilde{\chi}_1^0$ in the text

Let us conclude by briefly discussing the use of τ polarisation in probing the stau co-annihilation region at LSP, corresponding to $m_{\tilde{\tau}_1} \simeq m_{\tilde{\chi}_1^0}$ [15]. This is one of the very few regions of mSUGRA parameter space compatible with the cosmological measurement of the dark matter relic density, and the only one which is also compatible with the muon magnetic moment anomaly [16].

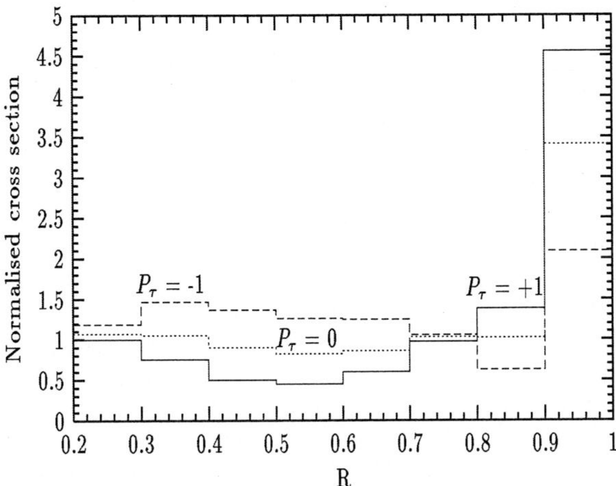

Figure 7. The normalised SUSY signal cross sections for $P_\tau = 1$ (solid line), 0 (dotted lines) and -1 (dashed lines) in the 1-prong hadronic τ-jet channel shown as functions of the τ jet momentum fraction (R) carried by the charged prong [14]

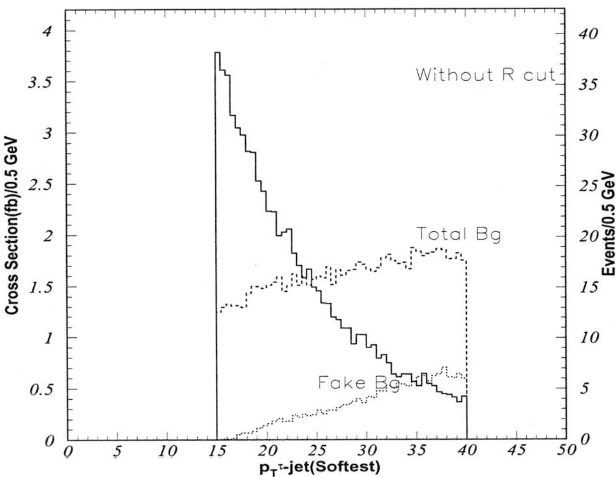

Figure 8. p_T (GeV) of softest τ-jet for signal and background processes [15]

It corresponds to a narrow strip adjacent to the lower boundary of Fig. 6 which can be totally covered at the LHC. Therefore the stau co-annihilation region is a region of special interest to the SUSY search programme at the LHC. In particular one is looking for a distinctive signature, which will identify the SUSY signal at the LHC to this region and also enable us to measure the tiny mass difference between the co-annihilating particles, $\Delta M = m_{\tilde{\tau}_1} - m_{\tilde{\chi}_1^0}$. Such a distinctive signature is provided by the presence of a soft ($P_\tau = +1$) τ-jet from the $\tilde{\tau}_1 \to \tau\tilde{\chi}_1^0$ decay of Eqs. (25, 26) in the canonical multijet+$\not{E}_T$ SUSY signal. Figure 8 [15] shows the p_T distributions of this soft ($P_\tau = +1$) τ jet signal along with the ($P_\tau = -1$) τ-jet background coming mainly from the $\tilde{\chi}_2^0$ decay of Eq. (26) and W decay. It also shows a significant fake τ background from the accompanying hadronic jets in these events. Figure 9 shows that the $R > 0.8$ cut of Eq. (5) effectively suppresses the ($P_\tau = -1$) background to a little over half the signal size and practically eliminates the fake τ background. A distinctive signal with a very steep slope is clearly sticking above the background at the low p_T end. One can use this slope to extract the signal from the background τ jets at 3σ level with a $10fb^{-1}$ luminosity run of the LHC, going upto 10σ with luminosity of $100fb^{-1}$. Moreover one can estimate ΔM to an accuracy of 50% at the $\sim 1.5\sigma$ level with $10fb^{-1}$, going upto 5σ with $100fb^{-1}$ luminosity [15].

8. Summary

The τ polarisation can be easily measured at the LHC in its 1-prong hadronic decay channel by measuring what fraction of the hadronic τ-jet momentum is carried by the charged prong. A simple cut requiring this fraction to be > 0.8 retains most of the detectable $P_\tau = +1$ τ-jet events, while effectively suppressing the $P_\tau = -1$ τ-jet events and practically eliminating the fake τ-jet events. We show with the help of Monte Carlo simulations that this cut can be effectively used for (1) Charged Higgs boson and (2) SUSY searches at the LHC. (1) The most important channel for the $H^\pm$ signal at the LHC contains a $P_\tau = +1$ τ-jet from $H^\pm \to \tau\nu$ decay. The above polarisation cut can effectively suppress the $P_\tau = -1$ τ-jet background from W decay, while retaining most of the detectable signal ($P_\tau = +1$) τ-jet events. So it can be used to extract the $H^\pm$ signal at the LHC. (2) Over half the mSUGRA parameter space the NLSP is the $\tilde{\tau}_1$, which is dominated by the right-handed component, while the LSP(χ) is dominantly bino. In this region a large part of the SUSY cascade decay is predicted to proceed via $\tilde{\tau}_1 \to \tau\chi$, giving a $P_\tau = +1$ τ-jet along with the canonical $\not{E}_T$+jets. One can use the above polarisation cut to extract this SUSY signal. A very important part of this region is the co-annihilation region, corresponding to $m_{\tilde{\tau}_1} \simeq m_\chi$. So the $P_\tau = +1$ τ-jet signal is expected to be soft in this region. However one can use this polarisation cut to extract this signal from the $P_\tau = -1$

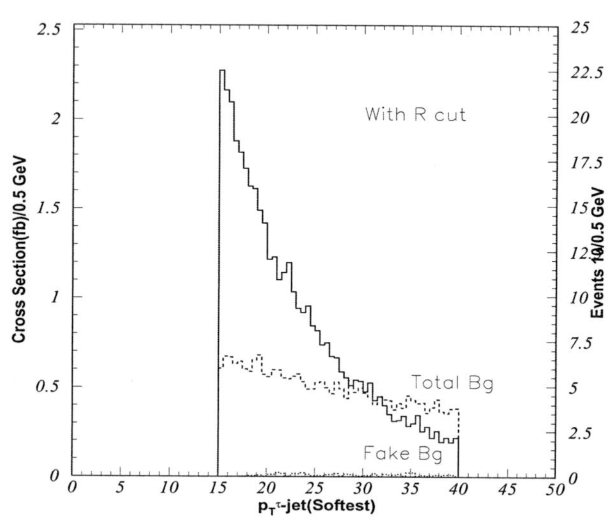

Figure 9. Same as Fig. 8 but with R cut (> 0.8) [15]

τ-jet and fake τ-jet backgrounds, and also to measure the small mass difference between the co-annihilating superparticles.

Acknowledgement: DPR was supported in part by the BRNS (DAE) through Raja Ramanna Fellowship.

REFERENCES

1. Review of Particle Properties, J. Phys. G33(2006)1
2. B.K.Bullock, K. Hagiwara and A. D. Martin, Phys. Rev. Lett. 67 (1991) 3055; Nucl. Phys. B395 (1993) 499; D.P.Roy, Phys. Lett. B277 (1992) 183
3. S.Raychaudhuri, D.P.Roy, Phys. Rev. D52 (1995) 1556; Phys. Rev. D53(1996) 4902
4. D.P.Roy Mod. Phys. Lett. A19 (2004) 1813
5. D.P.Roy Phys. Lett. B349 (1999) 607
6. M.Guchait,R.Kinnunen and D.P.Roy, Euro.Phys.J. C52 (2007) 665
7. Higgs working group report(Les Houches 2003):K.A.Assamgon et. al. (hep-ph/0406152)
8. T. Sjostrand, P. Eden, C. Friberg, L. Lonnblad, G. Miu, S. Mrenna and E. Norrbin, Computer Physics Commun. 135(2001)238
9. S. Jadach, Z. Was,R. Decker and J. H. Kuehn, Comput.Phys.Commun.76(1993) 361; P. Golonka *et al*, hep-ph/0312240 and references therein
10. For review see, e.g. *Perspectives in Supersymmetry,* ed. G.L.Kane, world scientific(1998); *Theory and Phenomenology of sparticles*, M. Drees, R.M. Godbole and P.Roy; World Scientific(2004); *Weak scale Supersymmetry: From superfields to scattering events,*, H. Baer and X. Tata, Cambridge UK, Univ. Press(2006)
11. Supersymmety parameter Analysis; SPA convention and projects, J.A. Aaguilar- Saavvedra et.al. Euro Phys.J. C46 (2006) 43; D.P.Roy, AIP Conf. Proc. 939 (2007) 63 (arXiv:0707.1949)[hep-ph]
12. WMAP collaboration, D.N.Spergel, et. al. Astrophys.J.Suppl.148(2003)175, Astrophy/0302209
13. M.M.Nojiri, Phys. Rev. D51 (1995) 6281; M.M.Nojiri, K. Fujii and T. Tsukamoto, Phys. Rev. D54 (1996) 6756
14. M.Guchait and D.P.Roy, Phys. Rev D54(1996)6756; Phys. Lett. B541(2002)356
15. R.M.Godbole, M. Guchait and D.P.Roy, arXiv:0807.2390[hep-ph]
16. Muon g-2 collaboration, G. Bennet et. al. Phys. Rev. Lett. 92(2004) 161802

Extra Spacetime Dimensions and the LHC

Sreerup Raychaudhuri

Department of Theoretical Physics, TIFR, Mumbai-400005, India

The last decade has seen an explosive revival of interest in extra spacetime dimensions. Inspired by developments in string theory, ingenious phenomenological models have been constructed in which gravity becomes strong at the scale of a few TeV, thereby solving the long-standing hierarchy problem of particle physics. Perhaps the most interesting aspect of these theories is the possibility of 'seeing' quantum gravity effects – including microscopic black holes – in experiments carried out at the TeV scale, of which the Large Hadron Collider (LHC) at CERN is the imminent one. Some of these ideas are reviewed in this article and the possibility of seeing signals for extra dimensions at the LHC are briefly discussed.

1. More Dimensions

From the earliest days one just has to look at a solid object – like the box depicted on the right – to get an idea of the three spatial dimensions normally described as length, breadth and height. Today this is usually the starting point of elementary mathematics textbooks, which then go on to say that every point in space can be described by three real numbers x, y and z, as Réne Descartés taught us three centuries ago. The overwhelming majority of humans are quite content with this simple description, but some of the more subtle minds have, over the ages, been attracted by the idea that there might be extra spatial dimensions beyond these canonical three. In fact, long before Descartés, the Hindu *Védas* enumerated ten *dishā* or directions, the Jewish *Kabbālah* talked of divine attributes being channelised to the earth through ten-dimensions called *sefirôt* and Mayan cosmogony visualised thirteen directions emanating like plumes from the cosmic serpent Kukulcán (Fig. 1).

Figure 1. Mayan sculpture of the 'plumed serpent' Kukulcán, spouting eleven feathers, purported to represent eleven directions. The remaining two directions are presumably pointed into and out of the plane of the picture

There is a dreamlike quality, however to these early speculations, and it may be contended that we are interpreting the ancient texts with the benefit of hindsight, whereas the original authors really meant something quite different. Since the precise meanings of the ancients have been lost with the extinction of the languages which they used, no one can really tell for sure. Be that as it may, even during the relatively recent times of Descartés, one had only to look across the English Channel to find the Cambridge philosopher Henry More (1614–1687) – whose language we *do* understand – speculating about the existence of a fourth space dimension. In a curious mixture of geometry and mysticism, harking back the days of Pythagoras and Plato, More claimed that the *spissitude* or fourth dimension of space represents the realm of spiritual things [1].

In a more modern context, and shorn of all mystic and spiritualistic trappings, extra space dimensions were first popularised, more or less as a mathematical curiosity, by the English polymath Charles Hinton in 1884. Hinton, who coined the word *tesseract* for a four-dimensional cube, tried to explain the three-dimensional section of a four-dimensional object by using the analogy of the time evolution of a three-dimensional object [2]. This inspired the novelist H.G. Wells to write that 'Everybody knows that time is only a kind of space' (*The Time Machine*, 1895). Wells was to prove a true prophet, for exactly ten years later, the true scientific foundation for the fourth dimension was laid by Albert Einstein's 1905 theory of Special Relativity, where space and time variables were, for the first time, allowed to *mix* non-trivially in moving frames of

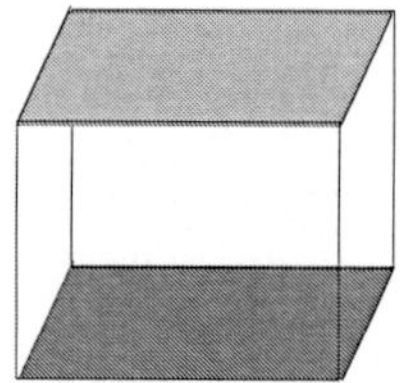

reference. Nowadays, it is usual to denote the space and time variables by $x^0 = ct$, $x^1 = x$, $x^2 = y$ and $x^3 = z$, where c is the speed of light in vacuum. In this notation, when one considers the transformed coordinates in a frame of reference moving with a relative velocity v with respect to the previous one, they are related to the older coordinates by the set of linear equations

$$\begin{aligned} x'^0 &= \Lambda_{00}x^0 + \Lambda_{01}x^1 + \Lambda_{02}x^2 + \Lambda_{03}x^3 \\ x'^1 &= \Lambda_{10}x^0 + \Lambda_{11}x^1 + \Lambda_{12}x^2 + \Lambda_{13}x^3 \\ x'^2 &= \Lambda_{20}x^0 + \Lambda_{21}x^1 + \Lambda_{22}x^2 + \Lambda_{23}x^3 \\ x'^3 &= \Lambda_{30}x^0 + \Lambda_{31}x^1 + \Lambda_{32}x^2 + \Lambda_{33}x^3 , \end{aligned} \quad (1)$$

where the coefficients $\Lambda_{\mu\nu}(\mu, \nu = 0, 1, 2, 3)$ are constants depending only on the ratio v/c. The new coordinates, therefore, are an admixture of *all* the previous ones – the new spatial coordinates x'^1, x'^2, x'^3 depend on the old time x^0, and the new time x'^0 depends on the old space coordinates x^1, x^2, x^3. In relativity, therefore time is indeed a sort of space, as Hinton and Wells had speculated.[1]

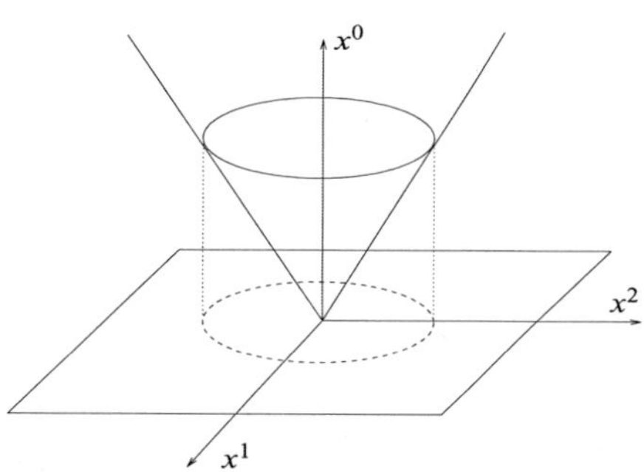

Figure 2. Three-dimensional projection of Minkowski space, with the x^3 $(= z)$ coordinate suppressed. Spherical (here circular) light wavefronts spreading out from the origin generate the so-called *light cone* (actually a hyper-cone). Events falling outside the light cone cannot be causally connected with an event at the origin

[1]Though time is still distinct because (*pace* Wells) we cannot go back in time.

It is, in fact, convenient to describe physics in terms of the geometric formulation (1911) of Hermann Minkowski, who invented the four-dimensional spacetime continuum (Fig. 2) which bears his name, and which is fundamental to all modern descriptions of relativity theory. In Minkowski space, the 'distance' δs between two neighbouring points at $(x^0, \vec{x})$ and $\left(x^0 + \delta x^0, \vec{x} + \vec{\delta x}\right)$ is given by the pseudo[2]-Pythagorean formula

$$\delta s^2 = \left(\delta x^0\right)^2 - \left(\delta \vec{x}\right)^2 \quad (2)$$

which means that light wavefronts emitted from the point $\left(x^0, \vec{x}\right)$ are just the spheres corresponding to $\delta s = 0$. One crucial demand of relativity is that the coefficients $\Lambda_{\mu\nu}$ should be such that $\delta s'^2 = \delta s^2$, which makes the transformation (1) between moving frames of reference rather like a rotation in the four-dimensional space. An obvious consequence of $\delta s = 0$ for light is that the path of a light ray satisfies

$$\left(\frac{\delta \vec{x}}{\delta x^0}\right)^2 = 1 \quad (3)$$

which is the equation of a straight line in the three euclidean dimensions. This incorporates the ancient result that light travels in straight lines.

In 1914, a young Finnish relativist, Gunnar Nordstrom, tried adding a fifth (invisible) dimension to Minkowski space in a brilliant attempt [3] to unify Newtonian gravity with electromagnetic theory as formulated by Maxwell. Nordstrom formulated the Newtonian scalar potential Φ as the fifth component of the electromagnetic potential A_μ, and started with Maxwell's equations in five dimensions, hoping that its four-dimensional projection would yield both the usual Maxwell equations as well as Newton's law of gravitation. This theory failed to work, but one can only admire the courage and prescience of the young Finn who sought to thus unite two completely different forces in so ingenious a manner.

The correct four-dimensional theory of gravity, viz. Einstein's theory of General Relativity (GR), Fig. 3, appeared a year later, in 1915 – and though the basic idea are quite simple, it turned out to be far more mathematically complicated than Newton's simple theory, replacing the simple scalar gravitational potential Φ by a bunch of ten potential functions grouped into a

[2]'Pseudo' because of the negative sign.

Figure 3. Albert Einstein's theory of Special Relativity extended the three-dimensional world in a scientific sense to four dimensions. His follow-up theory of General Relativity is still the best description we have of gravity in the classical sense

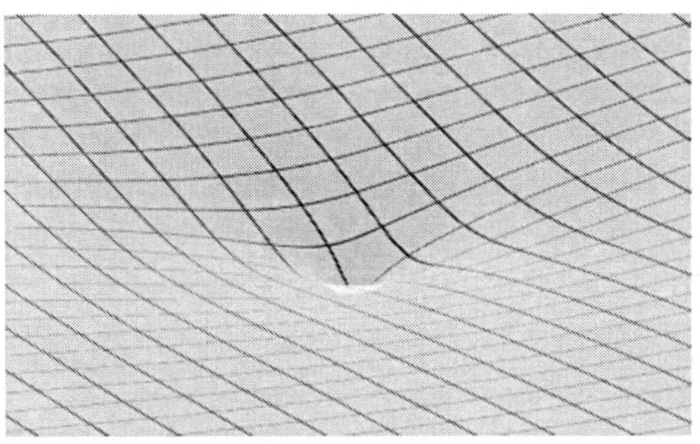

Figure 4. The 'rubber sheet analogy'. In the vicinity of the Sun, spacetime curves like a rubber sheet on which a weight has been placed, leading to the bending of light from distant stars, which follows the lines marked on the sheet. The curvature in this two-dimensional cartoon is greatly exaggerated

structure called a tensor, generally written as a matrix

$$\mathbb{G} = \begin{pmatrix} g_{00} & g_{01} & g_{02} & g_{03} \\ g_{01} & g_{11} & g_{12} & g_{13} \\ g_{02} & g_{12} & g_{22} & g_{23} \\ g_{03} & g_{13} & g_{23} & g_{33} \end{pmatrix} . \tag{4}$$

These ten functions $g_{\mu\nu} (\mu \leq \nu = 0, 1, 2, 3)$ determine the shape and curvature of the spacetime continuum, in which all material things are embedded. The path of light still follows the extremal property $\delta s = 0$, but now we have to re-define

$$\begin{aligned} \delta s^2 &= g_{00} \left(\delta x^0\right)^2 + g_{11} \left(\delta x^1\right)^2 + g_{22} \left(\delta x^2\right)^2 \\ &+ g_{33} \left(\delta x^3\right)^2 + 2g_{01}\delta x^0 \delta x^1 + 2g_{02}\delta x^0 \delta x^2 \\ &+ 2g_{03}\delta x^0 \delta x^3 + 2g_{12}\delta x^1 \delta x^2 + 2g_{13}\delta x^1 \delta x^3 \\ &+ 2g_{23}\delta x^2 \delta x^3 \end{aligned} \tag{5}$$

which reduces to the Minkowski form when $g_{00} = 1$, $g_{11} = g_{22} = g_{33} = -1$ and the rest vanish, i.e.

$$\mathbb{G} = \text{diag.} \left(1, -1, -1, -1, \right) . \tag{6}$$

It is the presence of matter which makes these functions $g_{\mu\nu}$ deviate from their Minkowski values, which correspond to the case of Special Relativity, i.e. reference frames moving with uniform velocities relative to each other. A deviation from Minkowski space may, therefore, be identified with the presence of acceleration or equivalently, gravitational fields[3]. Einstein's field equations of GR, from which one relates the $g_{\mu\nu}$ to the matter-energy density in the universe are, therefore, the field equations of the gravitational field. Unlike the field equation $\nabla^2 \varphi = -4\pi G \rho$ of Newtonian gravity, however Einstein's equations are non-linear, though they do reduce to the linear law of Newton in the limit when the gravitational field becomes very weak. This limiting behaviour explains why Newton's law has proved so successful in explaining all terrestrial and astrophysical phenomena for 300 years. The only anomaly in Einstein's day happened to be a tiny discrepancy between the calculated and observed orbits of the planet Mercury, which being nearest to the Sun, feels the strongest gravitational attraction. This tiny discrepancy – at the level of one part in a hundred *million* – disappeared when GR was used instead of Newtonian gravity. In 1916, the German mathematician Karl Schwarzschild, while soldiering on the Russian front in World War I, discovered that the solution of Einstein's equations in the neighbourhood of an isolated massive spherical body such as the Sun leads to

$$\begin{aligned} \delta s^2 &= \left(1 - \frac{2G_N M}{c^2 r}\right) \left(dx^0\right)^2 \\ &- \frac{\delta r^2}{1 - \frac{2G_N M}{c^2 r}} - r^2 \left(\delta\theta^2 + \sin^2\theta \, \delta\varphi^2\right) , \end{aligned} \tag{7}$$

where M is the solar mass and (r, θ, φ) are spherical polar coordinates chosen with the centre of the Sun as origin. Clearly $\delta s = 0$ no longer leads to a straight line in the Euclidean coordinates, and this means that light must bend away from a straight line in the neighbourhood of the Sun. This is equivalent to a curva-

[3]That gravity manifests as an acceleration rather than a force, was discovered by Galileo long ago, and demonstrated in his famous experiment: dropping cannonballs from the Leaning Tower of Pisa.

ture effect in space, illustrated (as far as is possible) in Fig. 4. Three years later, Einstein and GR was proved correct when a British team led by Sir Arthur Eddington travelled to Principé Island in the South Atlantic where a total solar eclipse made it possible to measure the bending of light from distant stars in the vicinity of the Sun and its strong gravitational field. This result not only made Einstein world-famous, but also made the scientific community sit up and take notice of relativity, which had been considered arcane and akin to metaphysics by the more hard-headed specimens in the scientific community.

Theodore Kaluza

Oskar Klein

Figure 5. Kaluza and Klein showed that electromagnetism can be unified with gravity if there is a fifth dimension

This success story of GR probably inspired the work of Theodore Kaluza, (Fig. 5) a young Polish teacher of mathematics, who, late in 1919, sent a paper [4] to Einstein in which he had shown that five-dimensional GR, with a circular fifth dimension of the kind introduced by Nordstrom, does indeed separate into four-dimensional GR plus Maxwell's equations in the limit when the circle shrinks to a point. This was, in effect, inverting the idea of Nordstrom: instead of a higher-dimensional Maxwell theory producing gravity in the four-dimensional theory, we now have a higher-dimensional theory of gravity which produces the Maxwell theory in the four-dimensional limit. Schematically we can write this as

$$\lim_{R\to 0} S^{(5)}_{\text{gravity}} = S^{(4)}_{\text{gravity}} + S^{(4)}_{\text{Maxwell}} \,, \tag{8}$$

where S denotes an action integral[4], the subscript 'Maxwell' refers to electromagnetic theory and R is the radius of the extra dimension. In order to obtain this results, we require to identify some of the components of the 5×5 metric tensor with the electromagnetic potential, i.e. $g_{05} = A^0, g_{15} = A^1, g_{25} = A^2$ and $g_{35} = A^3$, while the remaining 4×4 block is the usual metric tensor, defining the gravitational fields. Given that fact that gravity and electromagnetism are different-looking phenomena, this is a simply amazing – and unexpected – result, often known as the Kaluza 'miracle', and it is hard to believe that so beautiful a theory can be anything but the real description of Nature.

The marriage of two such disparate theories is not, however, free from its own internal dissonances. We have to account for the fact that the strength of the gravitational interaction is terribly weak, compared to the strength of the electromagnetic interaction. Mathematically, this is expressed by the oft-quoted ratio $F_{\text{grav}}/F_{\text{em}} \sim 10^{-40}$, but one can get a better physical feeling for this disparity by the realisation that when one stands on the ground, the electrostatic repulsion between two layers of atoms, viz. the upper surface of the floor, and the other the lower surface of one's shoes – is enough to balance the gravitational attraction *of the entire earth* on one's body. To obtain this enormous ratio in the Kaluza theory where both interactions have a common origin, then, we require to set the only available free parameter R to an extreme value, viz. $R \sim 10^{-33}$ cm.[5] It follows (Section 3) that the masses of elementary particles in this theory are either zero or are proportional to $1/R$, which is equivalent to the enormous Planck energy scale of 10^{19} GeV. This is obviously contrary to experience, since we have several elementary particles with masses ranging between 1 eV to 100 GeV. Even otherwise, however, Kaluza's formulations was not without its internal flaws, and the great Einstein himself blew alternately hot and cold about it until a better version of the theory was developed (1926) by Oskar Klein [5](Fig. 5). The improved Kaluza-Klein theory had Einstein's full stamp of approval. Writing to the aged Hendrik Lorentz (1927), he wrote "It appears that the union of gravitation and Maxwell's theory is achieved in a completely satisfactory way by the five-dimensional theory". Despite Einstein's enthusiasm, however the mass problem was not solved, and nobody had the slightest inkling that a solution would be seventy years in coming. By 1930, Einstein himself had moved away from Kaluza-Klein theory and had started to develop the torsion-gravity version of unification which was to occupy him till the end of

[4]Experts will readily recognise these as the Einstein-Hilbert action $S^{(4)}_{\text{gravity}} \sim \int d^4x\sqrt{-\det \mathbb{G}}\ \mathcal{R}^{(4)}$ and the Einstein-Maxwell action $S^{(4)}_{\text{Maxwell}} \sim -\frac{1}{4}\int d^4x\sqrt{-\det \mathbb{G}}\ F_{\mu\nu}F^{\mu\nu}$.

[5]This is usually known as the *Planck length* ℓ_P.

his days.

The next quarter of a century kept physicists busy with the birth of quantum mechanics, World War II and the horrific discovery of nuclear weapons, and none of the stalwarts had much time for extra dimensions. In 1953, Wolfgang Pauli, one of the earliest GR aficionados, again turned to the Kaluza-Klein model, which he extended to *six* dimensions in an abortive attempt to explain the strong nuclear force. In doing so, he discovered an early version of what we would today call non-Abelian gauge theory, about a year and a half before the pioneering work of Yang and Mills and Shaw in this regard. However this theory required all the particles to be massless, and the interactions to be long-range, i.e. $\Phi \propto 1/r$ like gravitation and electromagnetism, whereas the strong interaction is definitely short-range, i.e. $\Phi \propto e^{-r/\lambda}/r$, where $\lambda \sim 10^{-13}$ cm. Pauli clearly believed this obstacle to be insuperable, as is clear from letters he wrote to Abraham Pais and to Yang himself [3].

During the middle years of the twentieth century, quantum field theory gradually established itself as the appropriate tool to describe fundamental interactions. Most of the physics ideas were firmly rooted in the four dimensions of Minkowski space. It was not until the (eventually Nobel-winning) invention of dimensional regularisation by Gerardus 't Hooft in 1971 that the scientific world woke up to the realisation that a quantum field theory which does not work in four dimensions may make perfect sense in other dimensions — and even have the amazing phenomenological success which the Standard Model (SM) of particle physics seems to have achieved. Kaluza-Klein theories then had a revival of sorts in the 1970s, with the advent of string theories, which live in higher dimensions, and necessarily carry many features of the Kaluza-Klein type, including the huge mass gap from zero mass to the Planck scale. In a string theory, the fundamental objects are tiny (around Planck length) one-dimensional objects called *strings*, whose different oscillation modes appear at low energies as particle-like fields like the photon, electron, quarks and so on. A string whose excitations are all boson fields[6] can be consistently defined only in 26 dimensions, fermions are found in the 'spectrum' of strings living in 10 dimensions and one can concoct string models which live in all sorts of dimensions between 4 and 11. Since the main focus in string theories has always been to study physics at the Planck scale, the problem of masses could be pushed under the carpet, claiming it to be a matter of detail — to be understood when we have the final theory. For this reason, there was little interest in Kaluza-Klein theory from the point of view of the particle physicist, and there was even less interest – even a certain amount of disbelief – in strings among particle physicists. In this adverse intellectual climate, pioneering – in fact, seminal – work [6] in a low-scale version of quantum gravity by Ignatios Antoniadis, alone (1990) and in collaboration with Karim Benakli and M. Quiros (1994), though published in reputed journals, made very little impact on the high-energy physics community at the time.

Figure 6. Pioneers of brane-world models. From L to R: Nima Arkani-Hamed, Savas Dimopoulos and Gia Dvali

It was Antoniadis, however, who got Savas Dimopoulos and Gia Dvali thinking about the possibility of extra dimensions with a phenomenological twist [7]. But it was not until March 1998, when Nima Arkani-Hamed joined up with Dimopoulos and Dvali to form the collaboration now known as ADD (Fig. 6), that extra dimensions of space were invoked in an elegant solution to the notorious *hierarchy problem* plaguing the SM of particle physics. It was only after their work [8] that the idea captured the imagination of the particle physics community. Within a month, the authors ADD had teamed up again with Antoniadis to embed their own ideas in a string theoretic framework [9], thereby setting off an explosion of interest in the area. A brilliant variation of the ADD ideas, introduced [10] by Lisa Randall and Raman Sundrum in 1999, has also gone a long way in promoting the concept as a whole. The origin of all these scientists is a triumph of the internationalisation of science: Arkani-Hamed hails from Iran, Antoniadis and Dimopoulos are from Greece and

[6]Boson fields have particle-like excitations which are *bosons*, i.e. there is no restriction on the number of such particles which can have the same quantum state. By contrast, fermion fields are equivalent to swarms of *fermions*, which obey the restriction that no two particles can occupy the same quantum state.

Dvali is from Georgia, though all of them are now working in the United States. Sundrum originates from the Australian component of the Indian diaspora, while Randall is American-bred with a typical Anglo-Saxon surname. In fact, original research in this area has come from every continent on the globe, except Antarctica. The author of this article, together with K. Sridhar (TIFR) and Prakash M. Mathews (Saha Institute), wrote some of the earliest papers [11] on this subject and was among the first to introduce it to the Indian scientific community.

A decade later, with close to 3,000 papers having been written on the subject, extra dimensions have achieved complete respectability as the most popular way – after supersymmetry – to go beyond the SM and conceive of new physics at higher energies. During the inauguration of the LHC on September 10, 2008, this theory even achieved widespread notoriety, with the media trumpeting a (proven false) doomsday theory that proton-on-proton collisions at the LHC would produce deadly black holes capable of swallowing up the earth and all its inhabitants with it.

The rest of this article is mostly devoted to explaining the basic ideas of the ADD and Randall-Sundrum (RS) models, in as non-technical a manner as clarity permits.

2. Compact Dimensions

The uninitiated reader is generally puzzled as to how there could ever be extra space dimensions. This article started with the statement that one has only to look at a solid object to conclude that space has just the three dimensions denoted traditionally by x, y and z, and no more. This is certainly true of dimensions of the non-compact type, i.e. those which stretch from $-\infty$ to $+\infty$. However as discovered by Nordstrom and Kaluza, there exists the possibility of invisible *compact* dimensions. To understand this idea, imagine a sheet of paper laid out flat on a table – clearly this is a two dimensional object, and both the dimensions have the same nature. Now imagine that the sheet is rolled up into a cylinder. This is still two-dimensional, but now one of the two dimensions has the topology[7] of a circle of radius, say R, i.e. it has become *compact.* R is known as the *compactification radius.* If now, the sheet is rolled up very tightly, so that $R \to 0$, then it will appear like a one-dimensional object to an observer whose optical resolving power is much larger than R. This is illustrated in Fig. 7, where the radius of the depicted cylinder decreases successively upwards, until the topmost cylinder looks like a line. We see, then that it is possible for a compact dimension to become invisible if it becomes very small. Obviously, if we can increase the magnifying power of our observation it will eventually be revealed. Moreover, if there can be one invisible compact dimension, there can be others too – in fact any number of them. Only experimental evidence can really establish or rule out the existence of such compact dimensions. The problem is that experiments rarely make such definitive statements. Typically a null experiment simply tells us that if there are compact dimensions, they must have size R less than such-and-such a value. Thus, there is always the non-falsifiable possibility of having tinier extra dimensions, which evade the experimental constraint.

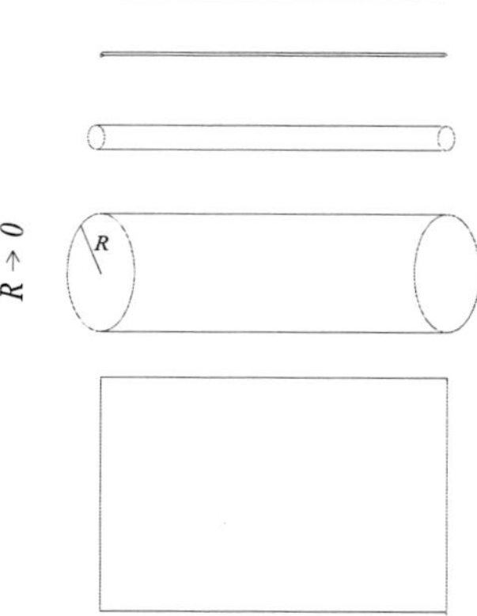

Figure 7. Compactifying two dimensions to one dimension

Stated baldly as we have done in the last paragraph, the idea of compact dimensions still appears quite bizarre. However compact dimensions may be found in any elementary textbook of quantum mechanics, and have actually been a mainstay of the quantum theory of solids for the last eight decades. The reader may recall that the wave-function of a particle in a box is described under two kinds of boundary conditions – vanishing boundary conditions, which generate standing waves, and *periodic boundary conditions*, which generate travelling waves. For an electron passing through a solid, obviously one requires a periodic boundary condition. Now, what is this periodic boundary condition?

[7]The word 'topology' is used quite deliberately, as meaning every shape which can be mapped continuously to a circle by a re-definition of coordinates using well-behaved functions. For example, a square or an ellipse has the topology of a circle, but an annulus does not.

It is the requirement that, come what may, the wave-function at a point $x = L$ will be identical with the wave-function at a point $x = 0$. There is no á priori reason to have this boundary condition[8] unless the points $x = 0$ and $x = L$ are one and the same. This makes the dimension compact, or rather the solid would be curved into a circular topology $\mathbb{S}_1$ with $2\pi R = L$, as in Fig. 8. In fact, as the periodic boundary condition is applied to all three dimensions, the typical solid studied in textbooks may be thought of as having the topology of a 3-torus $[\mathbb{S}_1]^3$. Conversely, a compact dimension may be thought of as just a periodic boundary condition in an extra degree of freedom (a.k.a. cöordinate!) of all the fields of the theory, with the compactification limit corresponding to a very small period after which the same values are repeated.

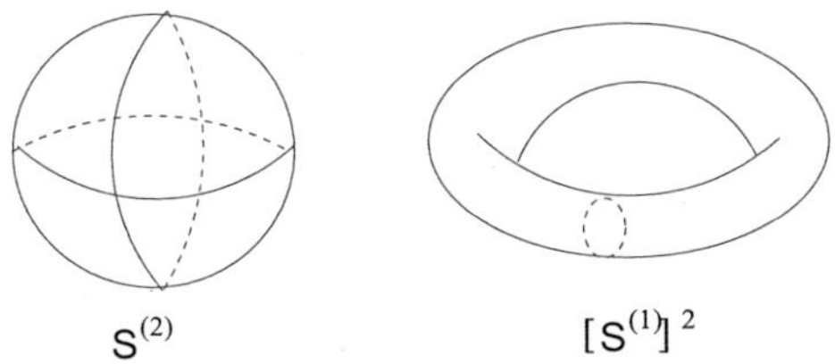

Figure 8. Compact manifolds in two dimensions: the sphere $\mathbb{S}^{(2)}$ and the two-torus $\left[\mathbb{S}^{(1)}\right]^2$

Of course, very little imagination is required to appreciate that the actual topology may not be that of a torus, but could well be that of any manifold such as a sphere or a sphere with handles, or something altogether more exotic, such as a folded or crumpled topology, or one with spikes. These would naturally correspond to mixed boundary conditions, with the period of each one being a function (or distribution) of the others. It is conceivable that one day we might learn a dynamical reason why these periodic boundary conditions develop or in other words, why some of the dimensions remain 'straight' while the others 'curl up' into tiny circles, or more complicated geometries. Till then, however we must be content with accepting compact dimensions as a phenomenological hypothesis, but not by any means an outlandish one.

[8]Most textbooks give a somewhat misleading argument that since only bulk properties matter, the wave-function over the surface does not matter, and can be taken as we wish. As early as 1912, Weyl had shown [12] that this statement is actually applicable only to the density-of-states function and specifically when the compactification takes place on a manifold.

The wave-function of a free particle propagating in five dimensions (x, y), where y is the fifth dimension in the form of a circle, will have the standard box-normalised form

$$\psi(x, y) = \frac{1}{\sqrt{\Omega_5}} e^{i(Et - \vec{p}.\vec{x} - p_5 y)}, \tag{9}$$

where Ω_5 is the volume of the five-dimensional box and p_5 is the component of momentum in the fifth (compact) direction. We choose units such that $\hbar = 1 = c$. This wave-function must satisfy the periodic boundary condition $\psi(x, y + 2\pi R) = \psi(x, y)$ in the fifth direction, which means that

$$p_5 = \frac{n}{R}, \tag{10}$$

where n is an integer. The five-dimensional relativistic energy-momentum relation is then given by

$$\begin{aligned} E^2 &= M_0^2 + \vec{p}^2 + p_5^2 \\ &= \left(M_0^2 + \frac{n^2}{R^2}\right) + \vec{p}^2. \end{aligned} \tag{11}$$

For every value of n, therefore, we can conceive of a Kaluza-Klein mass

$$M_n^2 = M_0^2 + \frac{n^2}{R^2} \tag{12}$$

and a 'four-dimensional' energy-momentum relation $E^2 = \vec{p}^2 + M_n^2$. We observe that the fifth component of momentum looks like a discrete set of four-dimensional masses, which are proportional to $1/R$. The smaller is R, the higher is the first such mass $M_1 = M_0^2 + \frac{1}{R^2}$ and the others are, naturally, even higher. Thus, if $\frac{1}{R}$ is as high as the Planck mass, we shall not be able to observe any of these modes in the laboratory, which is equivalent to saying that in all experiments performed by us, we will have $p_5 = 0$, i.e. no momentum exchange with the fifth dimension.

What do the Kaluza-Klein masses M_n correspond to? To see this, let us imagine a multi-dimensional world, in which four of the dimensions are described by the usual Minkowski variables x^μ and the remaining D compact dimensions, $y_1, y_2, \ldots, y_D$ by a D-torus, i.e. a set of periodic boundary conditions under $y_i \to y_i + 2\pi R$. In the compactification limit, as $R \to 0$, we shall have $y_i \to 0$ for all $i = 1, D$. Now, we imagine a scalar 'bulk' field, $\Phi(x^\mu, y_i)$ spread over all five dimensions, and satisfying a five-dimensional Klein-Gordon equation

$$\left(\partial_t^2 - \nabla^2 - \partial_{y_1}^2 - \ldots - \partial_{y_D}^2 + M_0^2\right) \Phi(x^\mu, y_i) = 0\ . \tag{13}$$

Because of the periodic boundary condition $\Phi(x^\mu, y_i) = \Phi(x^\mu, y_i + 2\pi R)$, we can expand the bulk scalar in a Fourier series

$$\Phi(x^\mu, y_i) = \sum_{\vec{n}=0}^{\infty} \Phi^{(\vec{n})}(x^\mu) e^{i\frac{\vec{n}.\vec{y}}{R}} , \tag{14}$$

where $\vec{n} = \{n_1, n_2, \ldots, n_D\}$ and $\vec{y} = \{y_1, y_2, \ldots, y_D\}$. Substituting this series into the bulk Klein-Gordon equation, we immediately obtain, for each four-dimensional Fourier coefficient $\Phi^{(\vec{n})}(x^\mu)$, the equation

$$\left(\partial_t^2 - \nabla^2 + \frac{\vec{n}^2}{R^2} + M_0^2\right) \Phi^{(n)}(x^\mu) = 0 , \tag{15}$$

which is simply the Klein-Gordon equation in four dimensions, with a mass

$$M_{\vec{n}} = \sqrt{M_0^2 + \frac{\vec{n}^2}{R^2}} . \tag{16}$$

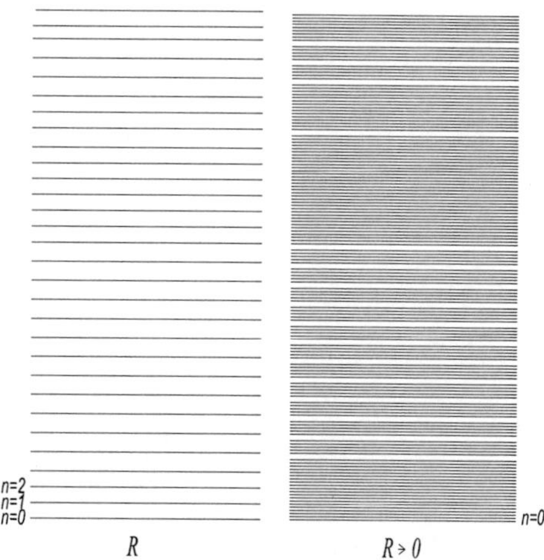

Figure 9. Illustrating a Kaluza-Klein tower of states with increasing mass. On the left, an intermediate value of R is assumed, while the right shows the case for a very small R. Note how the states become closely-spaced as $R \to 0$, forming a quasi-continuum. For this figure it is assumed that $M_0 = 0$

In the compactification limit $\vec{y} \to \vec{0}$, the bulk field $\Phi(x^\mu, y_i)$ reduces to

$$\Phi(x^\mu, y_i) = \sum_{\vec{n}=0}^{\infty} \Phi^{(\vec{n})}(x^\mu), \tag{17}$$

i.e. a sum of scalar fields with every increasing masses $M_{\vec{n}}$. This set of scalar fields, which will everywhere replace the bulk scalar field in the interaction Lagrangian when we take the compactification limit, is referred to as a *KK tower* of states (Fig. 9). The individual fields $\Phi^{(\vec{n})}(x^\mu)$ are called *KK modes*. Thus, for example, if we have a Yukawa interaction of this bulk scalar with a fermion field $\psi(x)$ which lives in four dimensions only, the interaction term in the action will look like

$$S_{\text{int}} = \int d^4x d^D y \; g_Y^{(5)} \; \bar{\psi}(x)\psi(x)\Phi(x,\vec{y}) \, (2\pi R)^d \, \delta^D(\vec{y}) \tag{18}$$

where the delta function serves to confine the interaction term to the four dimensions. Clearly, integrating over the delta functions and using Eq. (17) reduces this to

$$\begin{aligned} S_{\text{int}} &= \int d^4x \; g_Y^{(4)} \; \bar{\psi}(x)\psi(x) \sum_{\vec{n}=0}^{\infty} \Phi^{(\vec{n})}(x) \\ &= \sum_{\vec{n}=0}^{\infty} \int d^4x \; g_Y^{(4)} \; \bar{\psi}(x)\psi(x)\Phi^{(\vec{n})}(x) , \end{aligned} \tag{19}$$

which means that every KK excitation leads to a separate four-dimensional Yukawa term, with a coupling constant

$$g_Y^{(4)} = (2\pi R)^d \, g_Y^{(5)} . \tag{20}$$

While all these are technically possible, the trouble arises because of Eqs. 16 and 20. In Kaluza's original theory, where one extra dimension gives rise to the electromagnetic interaction, an analogue of Eq. 20 is applicable, i.e. the electronic charge is given by the relation $1/e = 2\pi R \frac{1}{\sqrt{16\pi G_N}}$, where $e \approx 0.3$ is the well-known electronic charge (in natural units) and G_N is the Newton constant. Recalling that $\sqrt{16\pi G_N} = 2/M_P$, we immediately obtain $1/e = \pi R M_P$, i.e. $1/R = \pi e M_P \approx M_P$. Invoking Eq. 16, this means that all the KK excitations – apart from the massless $\vec{n} = \vec{0}$ mode – will be tremendously heavy.

One can get around this argument, but at a heavy cost. If we are to have reasonable Kaluza-Klein masses, then we require $1/R \ll M_P$, i.e. $M_P R \gg 1$, which means that $e \ll 1$. In this case, the spin-1 exchange interaction obtained by Kaluza and Klein cannot be electromagnetism, but must be some fantastically weak force impossible to detect in the laboratory. This is not ruled out experimentally, nor can it ever be ruled out. However, given that Kaluza's original idea was to obtain electromagnetic theory out of a higher dimensional

GR theory, this appears, at first sight, to be a case of throwing away the baby with the bath water. This, in fact, was precisely Einstein's argument, which led to the original abandonment of Kaluza-Klein theory as a model of unification.

However, we now have to take into account the changes which have occurred in classical and quantum field theory since the early days of Kaluza. Not only have we discovered new forces (the strong and weak interactions) which cannot be derived from a higher dimensional GR theory without ascribing some absurdly artificial properties to the extended spacetime, but the weak interaction seems to be unified with electromagnetism. Today we believe that the electroweak and strong interactions are gauge theories, which are essentially a by-product of quantum mechanics. Thus, the requirement to generate electromagnetism out of a higher dimensional GR is no longer imperative, as it used to be in Kaluza's time. Ergo, we are no longer constrained to maintain small values of R and hence there is no longer a problem in having Kaluza-Klein states of intermediate mass.

Though the above fact was well-known from the 1970s, there was, quite naturally, resistance to give up as beautiful a construction as that of Kaluza and Klein, especially as that would mean removing the original motivation for higher-dimensional theories. The 1970s, however brought a new motive for more dimensions, because a relativistic string theory can be consistently developed only in 26 or 10 dimensions, depending on the boundary conditions. While string theory did borrow the idea of compactification from Kaluza-Klein theory, there was never any need to generate spin-1 gauge interactions from GR á lá Kaluza, since a vibrating string has enough spin-1 modes to be directly identified with the gauge interactions. Thus, though the ground for a Kaluza-Klein theory with accessible masses was already prepared by the time of the ADD revolution, there was no real reason to expect such masses. Thus, it was quite in accordance with one of the cardinal principles of science, viz. Occam's razor: *entities are not to be multiplied without necessity*, that no one thought it worthwhile to consider Kaluza-Klein masses of intermediate range.

3. Experimental Straightjackets

What had been largely abandoned by particle physicists and quantum field theorists continued to play a significant role in a then somewhat obscure branch of fundamental science – the so-called 'fifth force' experiments. These were the experiments devised to look for tiny deviations from Newtonian gravity which would signal corresponding deviations from Einstein gravity as well. Such deviations would be expected if there are extra compact dimensions, as well as in rival theories of gravitation, such as dilatonic gravity and the Brans-Dicke theory.

Why should one look for alternative theories of gravitation when the Einstein theory – and its Newtonian limit – work so well? The most striking reason to be dissatisfied with the Einstein theory lies in the so-called *cosmological constant problem* [13]. This is because Einstein's field equations of gravitation, which relate the gravitational potentials $\mathbb{G}$ to the energy-momentum tensor $\mathbb{T}$ can always be modified by replacing $\mathbb{T}$ by $\mathbb{T} - \Lambda\mathbb{G}$ (the negative sign is a convention). Here Λ is an unknown constant, called (by Einstein) the 'cosmological constant'. Noting that a uniform matter density ρ pervading all space would lead to $\mathbb{T} = -\rho\mathbb{G}$, it is easy to identify the cosmological constant with the vacuum energy density demanded in a *quantum* field theory, or to use a more fancy contemporary expression 'dark energy'. The vacuum energy is a purely quantum mechanical phenomenon, which can be traced ultimately to the *uncertainty principle*, one of the pivots on which the entire framework of quantum mechanics has been constructed. Since we have a superabundant wealth of evidence that the world is indeed quantum mechanical, it is hard to wish away the vacuum energy. In fact, adding this term is perfectly consistent with all of Einstein's initial assumptions, and hence, a good scientific procedure would be to keep this term and try to determine it from the observational data. If we do this, we obtain the result that, in units where $\hbar = 1 = c$

$$\Lambda \lesssim 1.3 \times 10^{-85}\ \mathrm{GeV}^2\ , \tag{21}$$

which means that the energy scale corresponding to the vacuum energy is around 1.1×10^{-33} GeV. This is amazingly small compared to the lightest known mass scale, that of the lightest neutrino, which is around 1 eV. Moreover, given that the vacuum energy will receive contributions from physics at all scales, including the Planck scale, there is no theoretical reason not to expect the vacuum energy to be of the order of the Planck scale, i.e. 10^{19} GeV. If we accept this argument, we must conclude that the observed vacuum energy is roughly 10^{-52} times smaller than the theoretical prediction. The cosmological constant problem lies in asking why the value of Λ is so tiny. The mind boggles at the smallness of the number 10^{-52} – better realised if we consider the fact that 10^{+52} is roughly the number of atoms contained in 10 million galaxies. One simple way out would be to postulate that there is some symmetry which forbids the writing of the $\Lambda\mathbb{G}$ term, but this cannot be invoked because that would mean that we cannot write $\mathbb{T} = \rho\mathbb{G}$ either, and that means that we would forbid the universe to have a uniform density. It would have to be a very strange symmetry indeed, which forbids this most symmetric of matter distributions! It is more practical to set $\Lambda \simeq 0$ by hand – for the moment, at least.

Once we accept that there are good reasons to look beyond the minimal theory of Einstein, the road to understanding gravity by doing experiments is clearly indicated. 'Fifth force' experiments perform precisely this task – they look for deviations from the inverse square law in highly sensitive (mostly terrestrial) experiments. Now it is easy to see that such deviations are directly predicted [14] if there are extra dimensions of the Kaluza-Klein type. To see this, let us consider, for simplicity, one extra dimension with the topology of a circle of radius R. We have seen that this leads to a scalar field having a four-dimensional projection $\Phi(x,0) = \sum_{n=0}^{\infty} \Phi^{(n)}(x)$, where each mode $\Phi^{(n)}(x)$ has mass $M_n = n/R$. Considering the Newtonian approximation, the gravitational potential will be just such a scalar field, which in the static limit with a point source m at the origin will satisfy the equations

$$\left(\nabla^2 + M_n^2\right) \Phi^{(n)}(\vec{x}) = -4\pi G_N\, m\, \delta^3(\vec{x}) \tag{22}$$

for each mode. This is the well-known Helmholtz equation with the solution $\Phi^{(n)}(x) = G_N m\, e^{-M_n r}/r$. Thus, the effective gravitational potential will be

$$\begin{aligned} \Phi(x,0) &= \sum_{n=0}^{\infty} G_N m \frac{e^{-nr/R}}{r} \\ &= \frac{G_N m}{r}\left[1 + \frac{e^{-r/R}}{1-e^{-r/R}}\right], \end{aligned} \tag{23}$$

summing the geometric series. In the limit $r \gg R$, the exponential term in the denominator can be neglected, so that we get the simple form

$$\Phi(x,0) = G_N m \left(\frac{1}{r} + \frac{e^{-r/R}}{r}\right), \tag{24}$$

where clearly the deviation from the Newtonian $1/r$ form becomes significant only when r begins to be comparable to R.

The most sensitive verifications of the inverse square law are achieved through *torsion balance experiments* (Fig. 10). Starting from the pioneering work of Henry Cavendish in 1797 to the sensitive experiments of Baron Loránd von Eötvös during 1906–1909, to the state-of-the-art measurements of the Eöt-Wash group, currently in progress at the University of Washington, the basic idea is always the same. A pair of heavy objects are fixed to the two ends of a horizontal rod suspended from a fine wire[9]. The objects are then brought into the proximity of two other heavy objects, as shown in Fig. 10, so that the feeble gravitational force between them rotates the rod by a tiny angle. This is measured and the force of gravity is calculated from the angle of deviation. The Eöt-Wash experiment, led by Eric Adelberger, where the apparatus is mounted on a small hill which acts as the attracting mass, the simple rod is replaced by an ingenious arrangement of a cut-away cylinder and a disc with holes, so that the effect of pure $1/r^2$ forces cancels to zero. Thus, in the Eöt-Wash experiment, any twist in the wire becomes a fifth-force effect. Similar experiments have been performed or suggested – in an Australian mine shaft, in a bore hole in the Arctic icecap in Greenland and in a torsion balance to be mounted on a satellite and sent into outer space. Some of these experiments were also performed in the 1990s, in India, at the author's home institute [15].

What are the conclusions of all these experiments? In a nutshell, all the results are completely consistent with Newton's inverse square law of gravity and with the exact principle of equivalence, i.e. that gravity manifests as an acceleration rather than a force.[10] This means that, once the dust of several incorrect results reported in the 1990s has settled, all searches for the fifth force have yielded negative results. One can therefore use these null results to constrain fifth force theories, and in particular, to put bounds on the size of Kaluza-Klein dimensions. The best state-of-art results come from the Eöt-Wash experiment [16], where the fifth force is parameterised as

$$\Phi^{(V)}(r) = \alpha G_N m \frac{e^{-r/\lambda}}{r}, \tag{25}$$

where α and λ are strength and range parameters respectively. Obviously, a null result will rule out large values of both, and hence the Eöt-Wash results are presented as a forbidden region in the α-λ plane, as shown in Fig. 11. Comparing with Eq. 24, one immediately sees that Kaluza-Klein theories may be identified with $\alpha \approx 1$ and $\lambda = R$. The current data, then, require $R \lesssim 160\ \mu$m, while the experiment will eventually probe

[9]Quartz fibres are the best for this purpose.

[10]Elevated to a definition of gravity by Einstein.

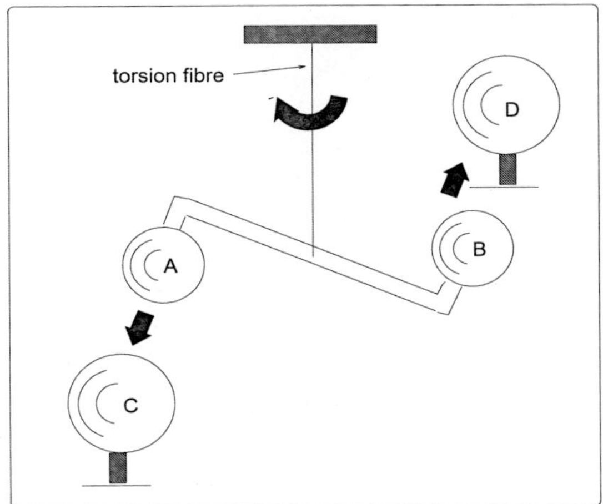

Figure 10. The basic torsion balance experiment. The masses A and B, affixed to the balance bar are attracted by the fixed masses C and D, as indicated by the arrows. The torque due to these forces is balanced by the elastic reaction of the fibre as shown. The bounding box indicates that the apparatus must be isolated from air currents and all kinds of noise, as the torsion effect is very weak

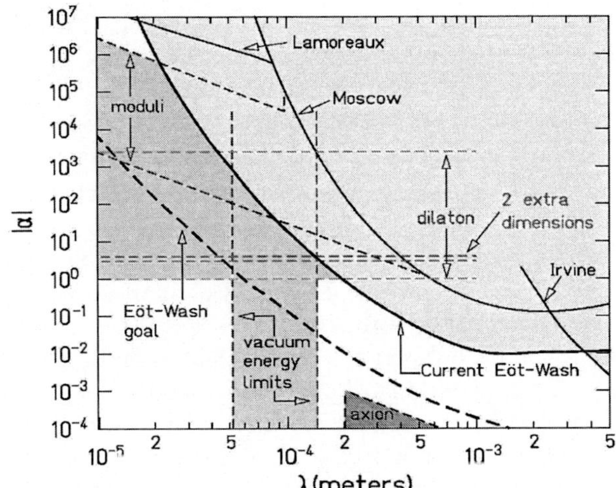

Figure 11. Results of various fifth force experiments ruling out parts of the α-λ plane. The solid boundary of the yellow-shaded region is the current result of the Eöt-Wash group, and the parallel dashed line represents the limit of sensitivity expected to be reached eventually at this experiment

as far as $R \approx 45\ \mu$m.

Once can say, therefore, that studies of gravity are consistent with the presence of extra compact dimensions so long as they are not larger than a hundred microns or so. However, we do have a wealth of microscopic probes which are smaller than this – notably atoms and molecules, which are at least a million times smaller. To such a minute particle, a 100-micron compact dimension would hardly be different from a noncompact one, and hence, we should observe their behaviour as if they live in $4 + d$ dimensions. For example, an electron moving in the electric field of a nucleus would feel an electrostatic force proportional to $r^{-(2+d)}$, instead of just r^{-2}. Now this certainly does not happen, since it is well known that anything other than an inverse square law of force would lead to splitting of the Bohr energy levels of a hydrogen atom, giving different energies[11] to different values of the azimuthal quantum number ℓ. This would change the spectrum emitted considerably from what is observed in a classic Geissler tube experiment. Even high-school learning is enough, therefore, to conclude that the atomic electrons, at least, do not 'see' compact dimensions, which pushes R down to about a nanometre.

High energy experiments probe much smaller lengths than atomic spectra. This is because quantum mechanics tells us that each particle may be associated with a wave whose wavelength λ is inversely proportional to the particle momentum, i.e.

$$\lambda = \frac{h}{p} \, , \tag{26}$$

where p is the momentum and $h = 6.678 \times 10^{-27}$ erg-s is Planck's constant. It is fairly straightforward to show that the wavelength corresponding to an electron of energy 100 GeV — which is what was achieved at the now-defunct LEP machine (1991–2001) at CERN — is around 10^{-18} m, i.e. a billion times smaller than a nanometre. Even so, there were no indications that these electrons interacted in anything but the canonical four dimensions of Minkowski and Einstein. Precision results to the level of one in 100,000 were obtained from this machine, and these would certainly have changed if there were compact dimensions of any size comparable to the wavelength $\lambda \sim 10^{-18}$ m. It is more or less experimentally certain, therefore, that there are no compact dimensions of size $> 10^{-18}$ m, and hence, no

[11]In reality, all the ℓ values have the same energy, except for some small effects associated with the electron spin, a fact referred to as *accidental degeneracy* – this is intimately connected with an $SO(4)$ symmetry of the Hamiltonian in the specific case of an inverse square law.

Kaluza-Klein masses less than a few hundred GeV/c^2. This is a matter for little concern for string theories, since they are compatible with compact dimensions as small as 10^{-35} m, but it is disappointing for extra dimension enthusiasts who would like to see their ideas verified in the present generation of experiments.

The only way to rescue compact dimensions from the tight corner into which LEP and similar experiments have driven them is to somehow ensure that electrons and such probes remain confined to four dimensions only. The extra dimensions would then be seen only by gravity, which cannot be confined to any lower number of dimensions, since it is a manifestation of the spacetime itself. But why should such a confinement occur? The motivation for this was provided by ADD, in their classic 1998 paper, and it is to this that we now turn.

4. The Hierarchy Problem

In order to appreciate the motivation for the new class of extra dimensional models, it is necessary to introduce the reader to the hierarchy problem in the SM of elementary particle physics. In the SM, it is believed that all the elementary particles were initially massless as they separated out from pure radiation in the first moments after the Big Bang. As the nascent universe cooled below a certain critical temperature, these particles now acquired masses through their interactions with a coeval elementary scalar field $H(x)$, whose excitations are called Higgs bosons. This nice mechanism was discovered in the 1960s, but till date has not passed the test of experimental verification, which demands that we find the Higgs boson. Other articles in this volume discuss this issue, but, we may note, in passing, that the introduction of this elementary scalar is the simplest and neatest solution to the mass problem, and possibly the only one which is not seriously threatened by experimental data available at the moment. The search for the Higgs boson is, therefore the most urgent purpose for which the LHC has been constructed.

This Higgs field, which plays so crucial a role in this game, happens to be the only elementary scalar field in the SM. However, as early as 1976, it was known [17] that there is a technical problem with quantum theories containing elementary scalar fields. This arises from the self-interaction term λH^4 of the scalar field, which plays a crucial role in mass generation and, therefore, cannot be wished away. The problem arises as soon as we go beyond the classical level (in a perturbation theoretic approach) and try to compute quantum corrections to the mass of the Higgs boson. In the very first order, we immediately encounter 'infinities' proportional to the inverse size of the smallest length scale to which the theory is valid. For example, if this length scale is ℓ, then $\Delta M_H^2 \propto \lambda^2/\ell^2$. Now this is a well-known phenomenon[12] in quantum field theory, and is generally handled by arranging for a cancellation of this 'infinity' with an equally large negative term $\propto -\lambda^2/\ell^2$ in the 'bare' mass parameter in the Lagrangian. What is left, after the cancellation, is the physical mass of the Higgs boson, which is a measurable quantity, unlike the 'bare' mass parameter. Once the cancellation has taken place, we allow the 'smallest' length scale to go to zero, recovering, in the limit $\ell \to 0$, the original spacetime continuum. For elementary fermions and gauge bosons, this neat trick is enough to ensure that no further 'infinities' are encountered. For elementary scalars, on the other hand, it turns out that fresh 'infinities' arise when we compute the quantum corrections at the next order, i.e. at two loops. We can, of course, arrange another cancellation at the two-loop level, but this does not really help. For one thing, new 'infinities' would arise again at three-loops (and so on), and moreover, once the bare mass is tuned to give a cancellation at two loops, it cannot give a cancellation at one-loop. The one-loop mass would then become 'infinite', making nonsense of the perturbative approximation.

One can argue, of course, that the SM is not really valid upto zero length scales, since it ignores gravitational effects altogether. Especially, at the Planck length $\ell_P = \hbar c G_N^{-1/2} \sim 10^{-35}$ m, it is known that gravity becomes strong and will begin to dominate interparticle interactions. Taking this scale as the minimum cutoff length for the SM, then, we avoid the actual infinities, but this argument does not save the situation, since we end-up with a Higgs boson as heavy as $\hbar c \ell_P^{-1} \sim 10^{19}$ GeV. As the self-coupling λ is proportional to the mass of the Higgs boson, such a heavy Higgs boson would mean an effective value of $\lambda \sim 10^{38}$, which is quite absurd in a perturbative framework. All of this goes to show that there is an internal inconsistency in the SM, considered as a quantum field theory, with a hierarchy of scales between the electroweak scale at 10^{-18} m and a smaller scale such as the Planck scale at 10^{-35} m. This inconsistency is called the *hierarchy problem*, or equivalently, the *fine-tuning problem* (Fig. 12) . Such irreducible inconsistencies are known to be typical of an incomplete or effective theory — indicating that there is a more fundamental underlying theory,

[12]The technical name for these is *ultraviolet divergences* and the cancellation trick is called *mass renormalisation*.

of which the SM is either a part, or a low-energy limit. This conjecture is, in fact, the strongest argument for searching for new physics beyond the SM.

There are, in fact, two possible solutions to the hierarchy problem, once we accept that the SM requires to be augmented by the addition of new fields and/or interactions. One way to bypass the hierarchy problem is to introduce extra fields and couplings which generate new 'infinities' *cancelling* the existing ones (at every order) – so that there is no need to tune the 'bare' mass parameter. To ensure exact cancellation, we require a higher symmetry in the theory. Among the popular symmetries which do this job are *supersymmetry* i.e. symmetry between bosons and fermions, and the rather complicated gauge symmetries seen in the so-called *little Higgs* models. Once we have ensured such cancellations, one can make the cutoff length as small as we please, even as small as the Planck length, without endangering the perturbative framework. In a sense, this postpones the inevitable breakdown of the SM to an energy scale inaccessible except in the early universe just after the Big Bang.

The other alternative is to take the bull by the horns and place the cutoff length of the theory just below the currently-accessible scale of 10^{-18} m. In this latter case, a low mass ($\sim$ 100–200 GeV/c^2) Higgs boson arises out of weak cancellations, which are neither unnatural nor unstable against quantum fluctuations. However, this approach immediately calls for a new theory at the TeV level (i.e. at 10^{-19} m), since that scale will certainly be probed by the LHC. Extra dimensional theories, as introduced by ADD, belong to this class of solutions, the new theory being one in which gravity becomes strong at the TeV scale, thereby invalidating the SM.

At the present jucture, on the eve of the LHC run, there is no *scientific* reason to prefer any one solution of the hierarchy problem over another solution. Of course, physicists have their preferences – prejudices, if you like – but here one person's philosophy is as good as another's. Almost all these models are elegant in conception but ugly in execution. This means that each starts with a simple and attractive idea, but in trying to explain the wealth of experimental data, one has to introduce extra assumptions, which are neither so simple, nor attractive. It is, of course, possible, that all this is happening because we are reasoning with insufficient information – like a newspaper speculating about a crime. The truth may turn out to be vastly more complex, with all or many of these ideas forming a part of it, just as the six blind men in the story had each grasped a piece of the elephant without realising what the whole beast looked like. Even this statement is a speculation, however and only the LHC data will show us, at least dimly, where we really stand.

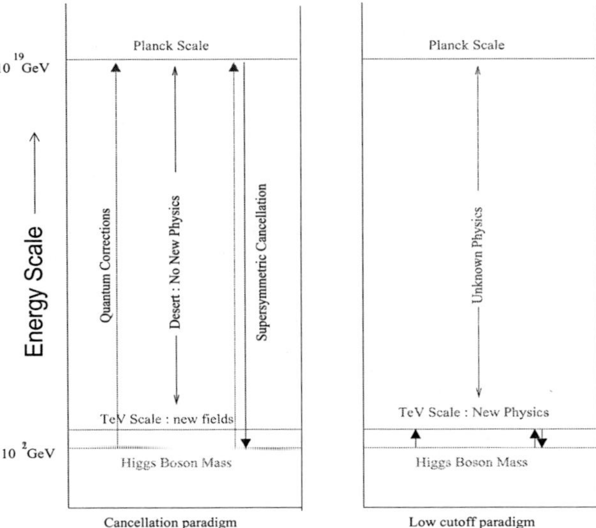

Figure 12. Illustrating the hierarchy problem and its solutions. If there is no new physics between the SM at a few hundred GeV and the Planck scale at 10^{19} GeV, quantum corrections drive the mass of the Higgs boson to the Planck scale. In cancellation-type solutions, new fields entering around a TeV cancel these quantum corrections. In low-cutoff solutions, new physics enters at the TeV scale and any cancellations which may still occur are small

5. The ADD Construction

In March 1998, the first paper [8] of ADD came out. The basic idea – like all great ideas – was startlingly simple. In a Kaluza-Klein theory, one starts by describing higher-dimensional gravity through an action integral of the form

$$S^{(4+d)} = \frac{1}{16\pi G_N^{(4+d)}} \int d^4x \; d^dy \; \mathcal{L}^{(4+d)} \; , \tag{27}$$

where $G_N^{(4+d)}$ is the actual gravitational coupling constant and $\mathcal{L}^{(4+d)}$ represents the Lagrangian density of the gravitational field. The coordinates $y_i (i = 1, d)$ represent the compact dimensions. As these shrink to a point, the action reduces to the form

$$S^{(4)} = \frac{\Omega_d}{16\pi G_N^{(4+d)}} \int d^4x \, \left[\mathcal{L}^{(4)} + \ldots \right] , \qquad (28)$$

where Ω_d is the volume of the compact space formed by these d cooordinates and the dots represent very weak interactions which will not concern us any further in this article. The appearance of the ordinary four dimensional Lagrangian density $\mathcal{L}^{(4)}$ is what we have earlier referred to as the Kaluza miracle. Comparing Eq. 28 with the standard form

$$S^{(4)} = \frac{1}{16\pi G_N} \int d^4x \, \mathcal{L}^{(4)} , \qquad (29)$$

where G_N is the usual Newton constant of gravitation, leads to the identification

$$G_N = \frac{G_N^{(4+d)}}{\Omega_d} . \qquad (30)$$

In terms of the Planck length $\ell_P = \hbar c G_N^{-1/2}$, defined earlier, and its higher dimensional equivalent $\widetilde{\ell}_P \equiv \ell_P^{(4+d)}$, this leads to the relation

$$\widetilde{\ell}_P^{2+d} = A \, \ell_P^2 \, R^d , \qquad (31)$$

where A is a constant not dramatically different from unity. If we choose $R \sim \ell_P$, as Kaluza did, then we have $\widetilde{\ell}_P \sim \ell_P$, as Kaluza found. However, if we choose R larger, we will immediately increase the length $\widetilde{\ell}_P$. For example, if we choose $d = 6$ and $R \sim 10^{-18}$ m, as allowed by LEP experiments, we get $\widetilde{\ell}_P \sim 10^{-22}$ m, which is much much larger than ℓ_P. At the length scale of 10^{-22} m, therefore, the SM will certainly break down, since we would be well inside the compact dimensions and gravity is as strong as the electroweak and other interactions at the effective Planck scale $\widetilde{\ell}_P$. Taking this length scale as the level of granularity for the SM does help in reducing the acuteness of the hierarchy problem, but it still implies Higgs boson masses of around 10^6 GeV, and a self-coupling $\lambda \sim 10^7$, showing that the problem is still there.

The observation that the hierarchy problem is ameliorated, but not solved, by a Kaluza-Klein formulation of the SM was the key observation of ADD. However, in order to actually *solve* the hierarchy problem, one has to make R larger still. To bring $\widetilde{\ell}_P \sim 10^{-19}$ m – at which stage the Higgs boson mass remains close to the expected range and λ remains perturbative – once has to choose $R \sim 10^{-14}$ m, which is about the size of a large nucleus. As we have seen there are good arguments to rule out extra dimensions of this size, unless there is a mechanism by which *they are inaccessible to matter*, such as electrons, nuclei, etc. There is, in fact, nothing crucial in this argument that was not known to Kaluza, or for that matter, the physicists of the 1970s, but they did not have the motivation to solve the hierarchy problem by expanding the Planck length to the limits of observation. Having such a strong motivation, however, ADD proceeded to create just such a mechanism, constructing a (somewhat contrived) quantum field theory in $4+d$ dimensions with a 4-dimensional domain wall[13] on which all the SM fields can be trapped. In this trapped condition, none of the SM particles can access the extra d dimensions, thus validating all the spectroscopic observations from Ångstrom to Bohr to LEP. However, gravity which is a measure of the topography of spacetime itself, cannot be thus confined, and hence, one can easily apply the arguments following Eq. 31 to this model.

Unbeknownst to themselves, ADD at that point had, in essence, rediscovered a suggestion made way back in 1983 by Misha Shaposhnikov and Valery Rubakov. In their paper [18], entitled "Do we live inside a domain wall?" the two Russian scientists had speculated that the SM fields live inside just such a narrow and deep potential well as constructed by ADD. However, they were looking for a solution to the cosmological constant problem and not to the hierarchy problem, and for reasons best known to themselves, they did not pursue the idea any further. Neither did their idea gain much currency. In fact, the idea that the universe might be a four-dimensional kink in a higher dimensional world had been suggested even earlier – in 1982 – by Keiichi Akama [19], but this did not attract any attention at all until much later, when ADD had made the concept famous. Even the original paper of ADD, which is wordy and imprecise, may have shared the fate of its precursors, had it not been for a new addition to the team. This was Ignatios Antoniadis, (Fig. 13) then at Paris, whom we have mentioned before as having pursued ideas about low-scale effects in string theory almost a decade before the ADD collaboration. In April 1998, a month after the first ADD paper, Antoniadis and ADD got together to write a paper [9] which has shaped the field of brane world physics since. This replaces the domain wall constructed so artificially by ADD with a *D-brane* – a kind of spacetime kink occurring naturally in most string theories, which had been discovered by Joseph Polchinski (Fig.13) only a

[13] A domain wall is the boundary between two different phases, such as the surface of a bubble inside a liquid or a liquid droplet suspended in air.

few years earlier. Thus, we have a very good reason for expecting the SM fields – including electrons, photons, nuclei and all the known things – to be trapped in four dimensions. If so, then the only constraints on the size of the extra dimensions come from pure gravity measurments, and these, as we have seen are pretty loose, permitting sizes as large as $R \lesssim 0.16$ mm. Even with $d = 2$, this means that ℓ_P can be large enough to remove the hierarchy problem.

What is a D-brane? To understand this, we should first note that in a string theory, the fundamental objects are tiny one-dimensional objects, which move in a space of 10 (or 26) dimensions. These 'strings' may be open or closed, depending on whether they have free end-points, or whether they form closed loops. These strings interact among themselves, joining together and breaking up, or even forming large conglomerates. At large length scales, the strings appear like point particles, and the interactions of the strings look like the scattering of point particles among themselves. The advantage of such a theory is, of course, that one never has to take the pointlike limit of zero size, so that $1/\ell$-type singularities never appear. It can be shown that the different oscillation modes of a single string appear like different particles, so that a single string can, depending on which mode is being excited, appear as a scalar, a vector boson, or a fermion, at low resolution. Even more exciting is the fact that one of the oscillation modes of a closed string appears like a spin-2 particle, which can be identified with a graviton, the quantum of the gravitational field. A string theory is, therefore a theory of gravity as well as a theory which lives in higher dimensions, and it is natural to embed the ADD construction in such a theory.

I. Antoniadis

J. Polchinski

Figure 13. Brane world pioneers: Ignatios Antoniadis of CERN, Geneva and Joseph Polchinski of the Kavli Institute, University of California at Santa Barbara

One of the interesting possibilities in an interacting string theory is that under certain circumstances, massive condensates of strings may form in a lower dimension, just as a bunch of atoms (which normally move in three dimensions) can bond together to form a flat two-dimensional plate. Such a lower dimensional object in string theory is called a D-brane, and it can be treated as a dynamical object in itself, just as a plate can be treated as an object in itself [20]. However the most important property of a D-brane is that it acts like a sticky membrane for open strings, whose open ends get stuck to the D-brane[14]. Thus, all the interactions of open strings will be confined to the neighbourhood of the D-brane, within a thickness comparable to the (tiny) length of the strings, and will appear, at low resolution, to be confined to the lower-dimensional space marked by the D-brane. As we have seen that the SM fields can be identified with different oscilation modes of open strings, this offers a natural and elegant mechanism for confining SM interactions to a lower dimension. This mechanism will not work for closed strings, since there are no ends to stick to the D-brane, and hence closed strings will be free to propagate in all the 10 (or 26) dimensions. As gravitons appear among the excitations of closed strings, this means that gravity propagates in the entire spacetime.

The exact construction of the ADD-Antoniadis model is, therefore, as follows. We assume that the fundamental underlying theory is a fermionic string theory, valid in 10 (or it can be 11) dimensions, of which 6 (or 7) are compact ones, henceforth referred to as the *bulk*. For d (we shall see presently that $d \geq 2$) of these compact dimensions, the radius of compactification R is large, maybe as large as 100 μm. Hence it is possible to have a large Planck length ℓ_P in the bulk. In addition to this, we assume that there is a D_3 brane, extending to infinity along all its three spatial directions, which we identify with the observed universe. All the SM fields correspond to oscillation modes of open strings which have one or both ends confined on the D_3 brane. This means that they are confined within a thickness δ which is indicative of the length of the strings. Closed strings are free to propagate in the brane or the bulk at will. This model is sketched in Fig. 14. The black region indicates the D_3 brane which is our universe, or rather a cutaway portion of it, since the actual brane extends to infinity in all directions. The perpendicular line penetrating the brane represents the d compact directions with large radius R, the black and white dots

[14]The name D-brane originates from this property: 'brane' is short for membrane and D refers to the fact that the ends of the string will be forced to satisfy a Dirichlet boundary condition.

being identified. This is a schematic way of indicating a d-torus, which is sketched in Fig. 8 for $d = 2$. The red squiggly lines represent strings, with the open ones stuck to the brane, while the closed ones are depicted in the bulk, away from the brane. The SM fields are, therefore confined to the box of thickness δ drawn around the brane.

One of the most elegant things about the ADD construction is that it gives us an explanation of why the gravitational interaction is so weak compared to the electroweak and other interactions. This is because electroweak interactions correspond to the interactions of open strings which are bound to the D_3 brane, and hence are closely packed together – or in the language of quantum mechanics, have overlapping wave functions. On the contrary gravity corresponds to interactions of closed strings which are free to roam around in the bulk and only occasionally cross the brane. It is only when these rare crossings occur that gravitational interactions of matter are seen. Again in the language of quantum mechanics, this means that the wavefunction of the graviton is spread over the entire bulk and has a very tiny overlap with the wavefunctions of SM particles on the brane. The gravitational interaction is, therefore suppressed by a factor governed by the brane-to-bulk size ratio, i.e. δ/R. If we take $\delta \sim 10^{-19}$ m and $R \sim 10^{-4}$ m, we will get a suppression factor of 10^{-15} which is about the ratio of the gravitational force to electroweak forces. Loosely, therefore, we may say that most of the gravitational influence of a given source is spread out through the bulk, and we on the brane, measure only a minuscule fraction of it.

d	$\widetilde{\ell}_P$(m)	$\widetilde{M}_P$	R(m)	M_1
2	10^{-20}	10 TeV	10^{-4}	10^{-3} eV
3	10^{-19}	1 TeV	10^{-8}	10 eV
4	10^{-19}	1 TeV	10^{-11}	10 keV
5	10^{-19}	1 TeV	10^{-13}	MeV
6	10^{-19}	1 TeV	10^{-14}	10 MeV
7	10^{-19}	1 TeV	10^{-15}	100 MeV

As we have seen, setting $\delta \sim 10^{-18}$ m ensures that this model is consistent with all precision tests showing that the SM fields interact in three spatial dimensions only. Setting $R \lesssim 0.1$ mm makes everything consistent with gravity experiments of the Eöt-Wash type. This means that the effective Planck length can be made as large as $\widetilde{\ell}_P \sim \delta \sim 10^{-19}$ m, by choosing R suitably, according to the formula given in Eq. 31. Taking $A \sim 1$, we obtain $R \sim 10^{-19+32/d}$ m, which is exhibited in the table below for $d = 2$–7. The choice $d = 1$ is omitted, as it leads to $R \sim 10^{14}$ m, which is patently absurd since it is as large as the solar system.

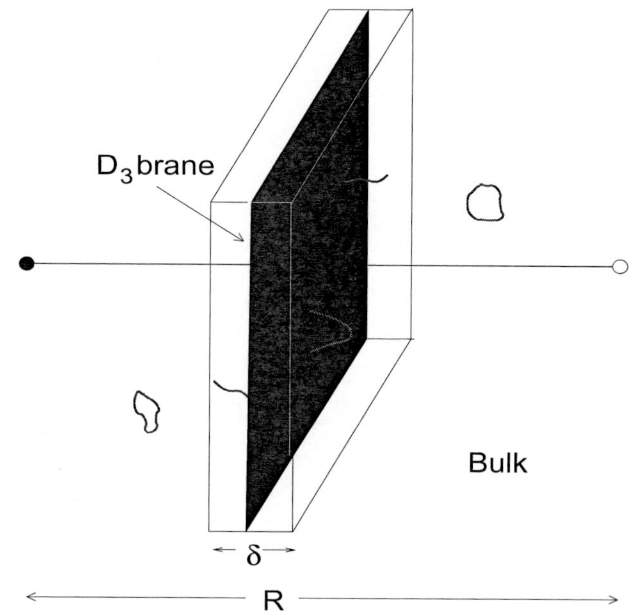

Figure 14. Sketch explaining the ADD model with a D_3-brane, shown as a dark surface

The second and third column in the above table refer to the effective Planck length and Planck mass in the bulk, i.e. the level at which the SM begins to fail. This has been kept at a TeV (= 1000 GeV), for $d \geq 3$, so that there is no chance of a hierarchy problem appearing. For $d = 2$, the same value of $\widetilde{\ell}_P$ would lead to R around a millimetre, which is ruled out by the Eöt-Wash data. However, taking $\widetilde{\ell}_P$ an order of magnitude smaller, which means cutting of the SM at 10 TeV instead of 1 TeV, leads to an acceptable value of R, as shown. The fourth column represents the required radius of compactification, and it is easy to see that such small dimensions are not likely to be probed soon in gravity experiments of the Eöt-Wash nature. In fact, as d increases, the required R rapidly shrinks to the nuclear size, i.e. a femtometre. The fifth and final column respresents the spacing between the masses of Kaluza-Klein excitations, which is proportional to $1/R$, and it can be seen that it is always very small compared to the energies (around a 100–1000 GeV) at modern colliders such as the LHC. This is an important result and it leads to the most exciting feature of the ADD model, viz. the possibility that it would lead to observable signatures at existing and upcoming collider experiments.

How do these signatures come about? We must note that all experiments take place on the brane, being part of the observable universe. Considered from the point of view of an observer located on the brane, the ADD model is mostly the SM (with a cutoff at $\widetilde{\ell}_P \sim 10^{-19}$ m) in addition to very weak gravitational interactions which occur whenever the bulk graviton crosses the brane. If we could do the experiment in all the $4+d$ dimensions, we would observe the bulk graviton having an interaction strength $\widetilde{G}_N = 2\widetilde{\ell}_P^2$, which is almost of electroweak strength, since $\widetilde{\ell}_P \gg \ell_P$. However, we ourselves are bound to the brane and hence must look at the interaction from a four-dimensional perspective. On the brane, Eq. 17 shows that the massless bulk graviton field will reduce to a tower of four-dimensional graviton fields or *KK modes*, each having mass $M_{\vec{n}}$ given by Eq. 16 with $M_0 = 0$. This means that we have a dense quasi-continuum of masses all the way up to the cutoff scale $\widetilde{M}_P$. We may recall that each massive mode corresponds to having a certain amount of momentum in the bulk, as shown by Eqs. 10–12. Now each mode will interact like a single graviton in four-dimensional Einstein gravity, analogous to the interaction shown in Eq. 19, and with the usual coupling $G_N = 2/\ell_P$ as indicated by Eq. 20. This is so weak an interaction that the probability for a single KK mode to be produced is too small to see even a single such event in the entire decade-long run of the LHC. Should such an event occur, however the single graviton mode is likely to fly off undetected since it will hardly interact at all with the matter in the detectors. Thus, one would detect the other partners in the reaction, and conclude that some energy and momentum has gone missing.[15] The real point of departure for the ADD model is that this should happen in the *same* way for *every* KK mode. This means that the total probability for such an event to happen will get multiplied by the total number of KK modes available, which must be given by the machine energy E divided by the KK mass spacing, which is around $1/R$. The table of mass spacings shown above immediately tells us that for $E \sim 10^3$ GeV – typical at the LHC – this factor varies between 10^{15} and 10^4 as d goes from 2–7. This huge number of KK modes is enough to offset the low probability of producing a single KK mode, and we end up with a decent probability of seeing events with a substantial missing energy and momentum at the LHC, especially for missing energy goes to create the mass of the KK modes, i.e. a momentum component in the bulk. This corresponds to gravitons interacting on the brane and flying off into the bulk, which is the behaviour expected of the closed strings whose excitations the gravitons correspond to.

[15]Such missing energy and momentum signals have been traditionally used to infer the presence of the weakly-interacting neutrinos, and are suggested as signals of other theories with weakly-interacting particles, such as supersymmetry.

If the above discussion seems too abstract, let us fix our ideas by considering a particular process at the LHC. This was first studied (among many other processes) by Gian Giudice and his collaborators at CERN, and almost immediately followed by similar studies by Michael Peskin and collaborators at Stanford, and by Tao Han, Joe Lykken and Ren-zie Zhang at Fermilab. All these papers [21] were made public about six months after the ADD-Antoniadis paper, within a single eventful week in November 1998, and they provided the immediate trigger for the explosion of interest that followed. The process we concentrate on here is that by which the protons in the LHC beams collide to create a photon and a KK mode of the graviton. A typical Feynman graph for this process is shown on the left of Fig. 15.

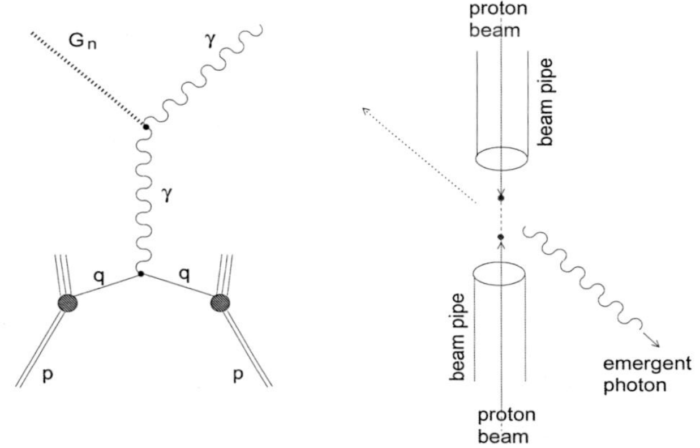

Figure 15. Single photon production in association with an invisible KK mode of the graviton in proton-proton (pp) collisions at the LHC. A typical Feynman diagram is shown on the left, with q denoting a quark or antiquark in the proton, and G_n indicating a KK mode. The three little lines indicate the rest of the proton, which continues down the beam pipe and is lost. The actual event topology is shown on the right. With a dotted line showing the hypothetical path taken by the unobserved graviton

In this graph, at the moment of collision of the two

protons, a quark from one proton and an anti-quark from the other get annihilated, producing a virtual photon, which immediately decays into a real photon and a KK mode of the graviton. Any of the modes can be produced in this way, so long as there is enough energy in the initial states to be converted into the mass of the KK mode, viz. $M_{\vec{n}}$. Now the KK mode will, as explained above, go undetected, so that the event topology will appear as sketched on the right of Fig. 15 – a single photon in the final state, with nothing visibly balancing its momentum in the transverse direction. Thus, if the cross-section (a measure of the probability of the reaction to occur) for creation of the $\vec{n}$-th mode is given by $\sigma_{\vec{n}}$, the total cross-section for seeing a single photon of this kind will be a sum over all accessible modes, i.e. $\sigma_\gamma = \sum_{\vec{n}} \sigma_{\vec{n}}$, where the sum commences from $\vec{n} = \vec{0}$ and gets cut off when $M_{\vec{n}}$ becomes greater than the machine energy.

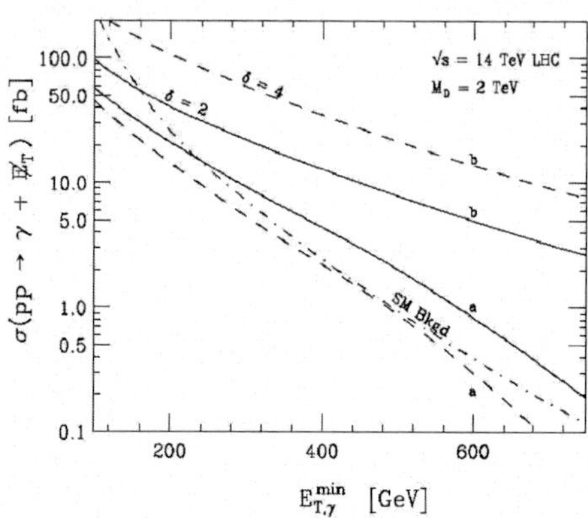

Figure 16. Single photon production at the LHC as predicted by Giudice et al [21]. To connect with the text, read d for δ and $\widetilde{M}_P$ for M_D. The symbol $\not{E}_T$ denotes undetected energy/momentum

In Fig. 16 we reproduce a graph from the work of Giudice *et al* [21], in which the cross-section for single photon production at the LHC is predicted. For a general idea, one needs to focus only on the curves marked **a**, where the solid line denotes $d = 2$ and the dashed line denotes $d = 4$. The dot-dashed horizontal line is an estimate of the SM background. This goes to show that this signal will be clearly observable only if $d = 2$ and just barely if $d = 3$. However, that should not be taken very seriously, since this is just one out of many processes, and there are others which are more viable from a phenomenological point of view.

The period immediately after November 1998 was a busy period for the high energy community. Once it had been established that one could not only solve the hierarchy problem, but that there was a possibility of observing quantum gravity effects in the laboratory, the ADD model really caught the fancy of scientists around the world. The first paper by Mathews, Sridhar and the author [11] appeared towards the end of November 1998 – within two weeks of the Giudice *et al* paper. Over the next few years, each person or group tried their hand at applying the ideas of ADD in their own area of expertise, from cosmology to supernovae to practically every conceivable process at real, upcoming and even proposed scattering experiments. As one wit remarked, large extra dimensions were being used to explain everything except the extinction of the dinosaurs.

After the initial euphoria, however, saner reflection prevailed. It turned out that a direct string-theoretic realisation of a spacetime with a D_3 brane, some large compact dimensions, and some small compact dimensions, was not so easy to construct. String theorists, confident about the physics of strings of length around 10^{-35} m, were wary of the much larger strings of length around 10^{-19} m proposed in the ADD model. There were technical problems associated with the cutting off of KK modes in virtual processes and in the calculation of quantum corrections involving graviton loops, but one could say that such technical problems plague most new theories. However the biggest blow to the ADD idea came from the realisation that it does not really solve the hierarchy problem – it simple re-formulates it! Let us see how this happens.

The easiest way to see this is to recall the origin of the hierarchy problem in the SM – it comes from the self-interactions of the Higgs scalar field. But gravitons have self-interactions too! Thus the mass of every KK mode gets a quantum correction which would be as large as the cutoff of the theory, i.e. $\widetilde{M}_P$. Since these masses are inversely related to the size R of the large compact dimensions, this means that the increase in mass corresponds to a shrinkage in the bulk. However, a shrunken bulk implies a corresponding shrinkage in $\widetilde{\ell}_P$, which leads to an even higher value of the cutoff $\widetilde{M}_P$. Re-evaluating the quantum corrections with this higher cutoff drives the masses of the light KK modes still higher and causes further shrinkage of the bulk. The process continues to bootstrap in this fashion, until stability is reached when $R \sim \widetilde{\ell}_P \sim \ell_P$ and all the graviton masses are of order M_P. This is going back to Kaluza and the original reason for abandonment of the idea of extra dimensions. Moreover the cutoff for the SM is now M_P and hence the hierarchy problem is restored in full glory. The only way to prevent this dismal scenario is to find a mechanism which allows some of the compact dimensions to remain large, while the others remain small. This can be achieved in a supersymmetric theory — but in supersymmetry, one can solve the hierarchy problem for the Higgs boson directly.

All that the ADD model has achieved, therefore, is to replace the hierarchy problem in the scalar mass by a hierarchy problem in the graviton masses, i.e. a hierarchy problem in the size of the compact dimensions. The initial u.s.p. for the model is, therefore lost. Nevertheless, it is not difficult to invoke supersymmetry, or some such idea, to save the situation, since the underlying theory is a string theory and hence necessarily supersymmetric. Nowadays, most scientists are contented with a pragmatic approach – to take the ADD model as a phenomenological possibility, assuming that the stability of the compact dimensions is achieved in some unknown way. This is similar to the way in which the SM has found universal acceptance, in spite of having the hierarchy problem in scalar masses.

6. Black Holes and Doomsday Predictions

Even though it is flawed as a solution to the hierarchy problem, the ADD model still gives rise to one of the most exciting – and bizarre – predictions ever seen in high energy physics. This is the suggestion that if this is a true picture of the world, then proton-proton collisions at the LHC would give rise to tiny *black holes*. Thus, just as the realm of string theory and quantum gravity is brought into laboratory experiments, so does the black hole – that bizarre solution of Einstein's gravitational equations, hitherto thought to be a by-product of dying stars – enter into the laboratory.

What is a black hole? If we glance at Eq. 7 it will become apparent that something bizarre must happen if the size of a massive spherical object r is decreased below the Schwarzschild radius $r_S = 2G_N M/c^2$, for at this value the radial and temporal coordinates exchange sign. This leadsto exactly the expected behaviour – motion along the radial coordinate can be unidirectional only, i.e. towards the singularity at the centre $r = 0$. Hence, once inside the distance $r = r_S$, known as the *horizon*, it is impossible for anything material – even light – to escape from this object, which is accordingly called a black hole.

If we set the mass M to the mass of a proton, then the corresponding Schwarzschild radius becomes the Planck length $\ell_P \sim 10^{-35}$ m. This, in fact, is one way to define the Planck scale. This means that so long as the proton does not come into contact with any matter at distances of this order, no black holes will form. To bring protons so close to matter, they must be accelerated to Planck energy, i.e. 10^{19} GeV, which was achievable only in the early universe, moments after the Big Bang. This could have resulted in the formation of primordial black holes, some of which may have 'evaporated' and some of which may have accreted matter to become giant black holes sitting at the centre of galaxies and quasars. Since laboratory energies, even at the vaunted LHC, are no larger than 10^4 GeV, there is no question, in four-dimensional Einstein gravity, of producing such black holes.

Large extra dimensions change the scenario completely. For now the effective Planck scale is brought down to $\widetilde{\ell}_P$, which means that the Schwarzschild radius of the proton is as large as 10^{-19} m. Protons will approach this close if their energies are of order 10^3 GeV, which will certainly be the case at the LHC. Accordingly, we expect two such protons to coalesce [22], forming a micro black hole of mass $2M_p$. If such a micro black hole is stable, it could then draw in nearby protons and grow in mass as well as horizon size, enabling it to eventually swallow up the LHC machine, the LHC tunnel, France, Switzerland, Europe and eventually the whole earth with all its inhabitants. Even the Moon would be shivered into bits and gradually sucked into the maw of this rapacious monster.

This apocalyptic vision is not correct, however, and that is because we live in a quantum world, rather than a classical one. In quantum theory, the vacuum surrounding any black hole is not an emptiness, but a bubbling ocean of virtual particle-antiparticle pairs, such as pairs of electron and positron. Such pairs are perennially being created out of the vacuum, using some of the enormous vacuum energy, and then annihilate again, returning their energy to the vacuum. Stephen Hawking, the world-famous Cambridge physicist, showed in the 1970s, that in the neighbourhood of a black hole, however, strange things begin to [23]. It can be shown that in free space an antiparticle travels backwards in time, as it were, and near a black hole, the role of time is played by the radial coordinate r. Thus, when a virtual electron-positron pair is created near a black hole horizon, the electron is drawn in, falling towards the origin with ever-increasing speed, while the positron shoots

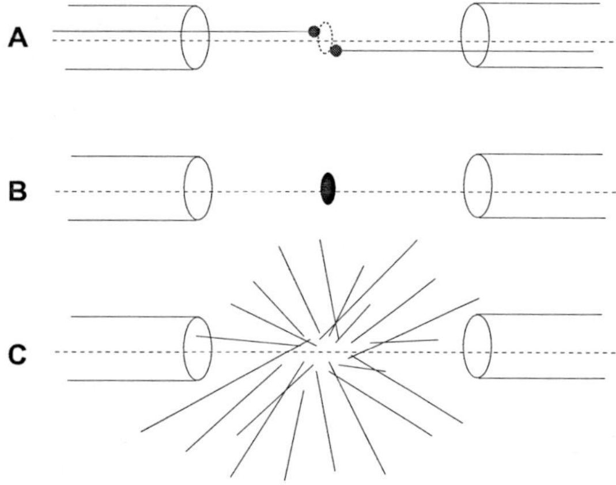

Figure 17. Micro black hole at the LHC. **A**: Two protons approach each other with impact parameter within the Schwarzschild radius. **B**: The protons coalesce into a black hole. **C**: The black hole decays by Hawking radiation, spraying particles in all directions

out, with ever-increasing speed outwards. This accelerated motion of a charged particle like the positron causes radiation, known as *Hawking radiation.* Where does the energy of this radiation come from? Eventually from the black hole itself – at the cost of its mass. Thus, if a black hole is small enough, it can eventually lose all its mass through Hawking radiation, or 'evaporate'. This is the reason why it is thought that most of the primordial black holes have disappeared.

Micro black holes at the LHC will also lose energy by Hawking radiation. The time scale is easy to calculate. It is given by the time taken by interactions to cross the black hole, i.e. $r_S/c \sim 10^{-29}$ s. Such a black hole will, for all effective purposes, be stillborn, since it will decay long before even the nearest protons in the beam (typically separated by about 10^{-6} m) can reach it. There is absolutely no danger, therefore, of such a black hole accreting any mass and growing. The world as we know it is still a safe place.

There is still a catch in the above argument, and that is the fact that it is assumed that the micro black holes at the LHC would be produced at rest. However, there is always a velocity spread in the beam so that some protons have more momentum than others. Collisions of such protons would lead to fast-moving black holes, which live much longer due to a relativistic effect called time dilation, discovered long ago by Einstein. One requires, therefore, to carry out a careful study and see if there is any chance that even a single micro black hole may be produced with a long enough lifetime to start accreting mass in the LHC experiment. After all even one accreting black hole is enough to destroy the earth. A detailed and careful analysis of this has been performed recently by Steve Giddings and Michelangelo Mangano at CERN [24], and their conclusion is that this probability is small enough to be virtually zero[16]. Moreover, any black hole with a long enough time-dilated lifetime would be a high-speed one which would pass through the earth without interacting with any matter and eventually decay harmlessly outside the earth.

No only do black holes pose no danger to us, but they also provide a unique signature of any ADD-type model [22]. For when a micro black hole decays, it will result in a spray of all sorts of particles without prejudice, which would form a near isotropic distribution of hits in the detector, as indicated in Fig. 17. At its peak, the LHC could be producing ten million black holes per year, each with this kind of spectacular signal. There is really no way in which this can be missed, so it may well be black holes which provide the first evidence that the world has more than four dimensions.

Black hole signatures are interesting and the arguments of Professors Giddings and Mangano are reassuring, but it is even more reassuring to think that the ADD model is just an idea, and not even the best one in its own genre. A much better solution of the hierarchy problem using extra dimensions was suggested by Randall and Sundrum, and it is now to their ideas that we turn.

7. The Randall-Sundrum Model

The key concept in the model of Randall and Sundrum (Fig. 18) is that of *naturalness.* This was introduced by Paul Dirac in the 1930s and is an important issue in any quantum theory. In classical mechanics, we do not bother if one parameter is very small and another is very big. Thus, the fact that a grain of dust is very much smaller than a mountain and that an elephant is much bigger than a flea does not cause any eyebrows to be raised. In the strange world of quantum mechanics, however, this happy situation is no longer true. This is

[16]We are happy to live with many such dangerous but low probabilities. For example there is a tiny probability that all the molecules of air in the room will, through random motions, collect in a corner and leave us asphyxiated; there is a tiny — well not so tiny really — probability that a piece of cosmic debris will strike the earth and cause it to break up; and so on.

because of the *completeness* property of quantum states – every state is equivalent to a linear combination of other states[17]. Thus, bizarre as it may sound, a grain of quantum dust has a component of quantum mountain in it and a quantum flea has a component of quantum elephant in it. In some experiments, therefore, which happen to probe just that component, the grain of dust will appear as big as a mountain, and similarly the flea would appear as an elephant. In more technical language, any number which is very small or very large (depending on what it is being compared with) is unstable under quantum corrections, and will tend to stabilise only when the two numbers in question are equal or nearly equal. We have already seen two examples of this happening: the small mass of the Higgs boson is driven to the Planck mass scale by quantum corrections, and the large compact dimensions of ADD are driven to the tiny Planck length, again by quantum corrections. It was Dirac who declared [25] that this kind of equality which leads to quantum stability is *natural*, and that large or small numbers in a quantum theory are *unnatural*.

We have just argued that both the SM and the ADD model share the same feature of unnaturalness, which is what the hieracrhy problem is all about. However, quantum mechanics notwithstanding, the enormous difference in strength between gravity and electroweak interactions is a fact, and cannot be wished away. In the summer of 1999, about a year after the original papers of ADD, Lisa Randall of Princeton University and Raman Sundrum, then at Boston University, proposed a model [10] with just one extra dimension which was able to create this huge difference without using any large fundamental numbers. Their ingenious construction now goes by the name RS model.

L. Randall

R. Sundrum

Figure 18. Lisa Randall and Raman Sundrum, whose proposal of warped extra dimensions provides the best solution of the hierarchy problem within extra dimensional models

The RS model assumes that there is a single extra dimension which has the topology of a circle folded along a diameter – $\mathbb{S}^{(1)}/\mathbb{Z}_2$ for the experts. For obvious reasons, such a space is called an *orbifold* – this one being the simplest of a whole class of objects which go by this name. This is pictured in Fig. 19, where we also note that at the two extreme points lie two D_3 branes, denoted as the *visible* (black) and the *invisible* (blue) branes respectively. No explanation is given or sought why this spatial dimension should be contorted in this fashion, the reason presumably lying in the underlying string theory[18]. The coordinate along the extra dimension is parametrised by an angle ϕ which is clearly limited to the range $0 \leq \phi \leq \pi$. RS then proceed to solve the five-dimensional field equations of Einstein gravity in this configuration. The solution can be done if and only if one imposes boundary conditions on the branes, which are equivalent to choosing four-dimensional cosmological constants – one positive and one negative – on the two branes, as well as a carefully-matched negative cosmological constant in the five-dimensional bulk[19], satisfying the relations

$$\Lambda_i = -\Lambda_v = -\Lambda_5 \ . \tag{32}$$

Here $\Lambda_{i,v,5}$ denote the cosmological constants on the invisible brane, the visible brane and the bulk, in that order. There is a good deal of fine-tuning in these choices of cosmological constants, since even a small deviation cannot lead to a solution of the Einstein equations. However, this may be turned to our advantage by claiming that this is these are the equilibrium values reached after the Big Bang, assuming that the universe has always satisfied some solution of Einstein's equations. Once, however, the choices are made, RS obtain a solution for the line element

$$\delta s^2 = e^{-\mathcal{K}R\phi}\left[\left(\delta x^0\right)^2 - \left(\vec{x}\right)^2\right] - R^2\,\delta\phi^2 \ , \tag{33}$$

where $\mathcal{K}$ is a constant which may be interpreted as the curvature of the fifth dimension, and is related

[17]Provided those states are linearly independent.

[18]The RS construction has proved to be very difficult to embed in any of the standard string theories, and nowadays is visualised more as a phenomenological construct. Whatever be the underlying theory, however, it must have compact dimensions and D branes, which indicates that it must be some kind of string-like theory.

[19]This makes it a five-dimensional anti-de Sitter space, generally denoted AdS_5.

to the five-dimensional Planck scale $\widetilde{M}_P$ and the five-dimensional cosmological constant Λ_5 by

$$\mathcal{K} \simeq \widetilde{M}_P^3/M_P^2 = -\Lambda_5/(24M_P^2) \ . \tag{34}$$

On the 'invisible' brane, where $\phi = 0$, this reduces to a purely Minkwoski form, noting that the length around the fifth direction is just $y = R\phi$. However, as we proceed towards the 'visible' brane at $\pi = 0$, the usual spacetime part undergoes an exponential contraction, though the fifth dimension is unchanged. Such an asymmetric contraction is normally understood as a warping and hence the exponential $e^{-\mathcal{K}R\phi}$ is referred to as a *warp factor*. A sketch of this warping effect is shown in Fig. 20. Clearly the warping will be maximum on the 'visible' brane, at $\phi = \pi$, which is identified with the observed universe.

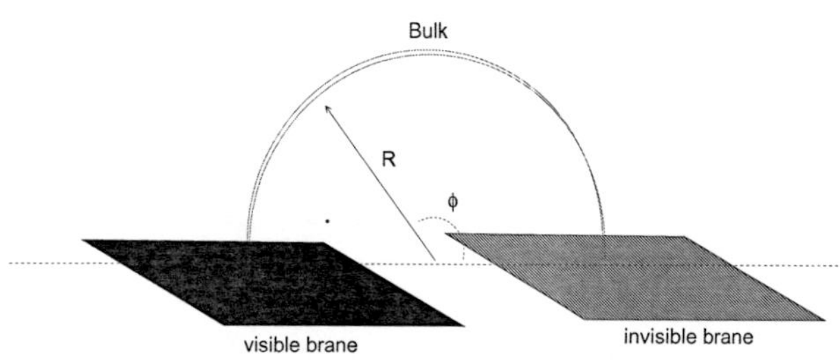

Figure 19. Sketch of the RS construction. The double red line indicates the $\mathbb{S}^{(1)}/\mathbb{Z}_2$ extra dimension. The black region marked 'visible brane' corresponds to the known universe, while the blue region marked 'invisible brane' corresponds to a 'shadow world' of strong gravity

What does all this have to do with the hierarchy problem? The answer is that we start by assuming all interactions to have the same strength, gravity as well as the electroweak interaction, and this is characterised by $1/\widetilde{\ell}^2$, where $\widetilde{\ell} \sim 10^{-19}$ m. However, gravity at this strength is an effect native to the invisible brane, where it has this strength. On the visible brane, we only see that amount of gravitational interaction which reaches us across the bulk, and this arrives after a drastic reduction of the length scale

$$\widetilde{\ell} \to e^{-\pi\mathcal{K}R}\widetilde{\ell} \ . \tag{35}$$

The very reasonable choice $\mathcal{K}R \simeq 11.73$ takes the right side of the above equation to $\ell_P \sim 10^{-35}$ m. On the other hand, electroweak interactions are native to the visible brane – where they are confined, as in the ADD case – and are of the typical strength $1/\widetilde{\ell}_P^2$. This means is that we have been able to generate the enormous difference in strength between the gravitational and electroweak strengths without having recourse to any unnaturally large or small numbers. In more picturesque language, gravity is weak not because of any inherent weakness, but because we see it shining on us very dimly across a highly opaque higher dimension, which allows only a tiny fraction of the force to get through.

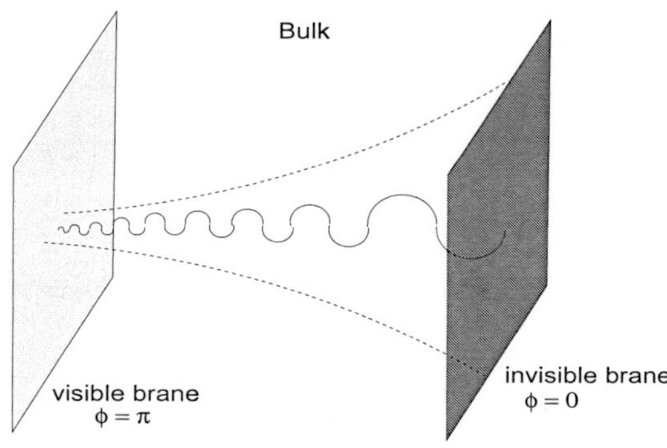

Figure 20. Warping effect in the RS model. All length scales get damped as we proceed from the invisible to the visible brane, and the graviton wavefunction damps out proportionately. Of course, the sketch is only illustrative, since the actual damping is exponential and the separation between the branes is miniscule

What is the size R of the extra dimension? A glance at Eq. 34 shows that it would be natural to choose $\widetilde{M}_P$ as well as $\mathcal{K}$ in the ballpark of M_P, which would automatically make $R \sim \ell_P$, since $\mathcal{K}R \sim 10$. All these sizes are, of course, approximate within an order of magnitude, but the important points to note are that (*a*) there is no unnatural hierarchy of sizes and (*b*) the extra dimension is really small – practically as small as originally envisaged by Kaluza and Klein. Thus, like the original KK model and the ADD Model, the RS Model also starts with two basic length scales, one much larger than the other, viz. the effective electroweak length $\widetilde{\ell}_P \sim 10^{-19}$ m, and the Planck length $\ell_P \sim 10^{-35}$ m. What this model really achieves, therefore, is to keep these two length scales apart in such a way that they cannot influence each other through quantum corrections, while generating their large ratio by exponentiating a relatively small number. This is no mean achievement, given the difficulty of the original problem.

There are many it ad hoc things in the RS model. There is the question of what happened to the other 5 compact dimensions, if the whole is embedded in a string theory of 10 dimensions. Why do they not develop similar warping? Secondly, why does the compact dimension get orbifolded into $\mathbb{S}^{(1)}/\mathbb{Z}_2$ — this must have a dynamical origin. Perhaps it was a simple circle or a non-compact dimension just after the Big Bang, and then it developed this peculiar topology as time evolved. In the absence of any model for this, we can only speculate. The other major question has to do with how the cosmological constant on the branes and on the bulk got aligned in the way they have to be for a RS solution to exist. Moreover, if the visible brane has a negative cosmological constant, why do we not see its effects in cosmology? Since, as we have seen, we have little or no understanding of the whole business of the cosmological constant, one can only hope that when we do begin to understand this, some of the mysteries of the RS model will be unravelled as well.[20]

One of the consequences of the warping effect is that the KK tower of gravitons acquires masses around a few hundred GeV – unlike the ADD case, where the masses range from very small values to the cutoff scale. This is because, the extra dimension being small (comparable to ℓ_P) the KK tower would normally have a mass gap of the order of the Planck mass, as in the original Kaluza-Klein theory. However, this large mass, when seen on the visible brane, appears with the warp factor $e^{-\pi\mathcal{K}R}$, and is reduced to the level of the electroweak scale, i.e. a few hundred GeV. When the details are worked out, the actual mass parameter turns out to be [26]

$$m_0 = \mathcal{K}e^{-\pi\mathcal{K}R} \tag{36}$$

and the masses of the heavy KK modes of the graviton are given by

$$M_n = m_0\,\xi_n \quad (n = 1, 2, \ldots). \tag{37}$$

The ξ_n are the successive zeroes of the Bessel function $J_1(x)$ of order unity, which appears in diffraction theory, and start from $\xi_1 = 1.22\pi$, $\xi_2 = 2.33\pi$, and so on.

[20]In this context one may make the somewhat cynical comment that the RS model succeeds in pushing the hierarchy problem — which we have just begun to start comprehending — on to the cosmological constant problem, which is still as big a mystery as it ever was.

Not only does the warp factor render the KK graviton modes massive, but it also makes the interactions of each KK mode much stronger. This is because the usual interaction, proportional to ℓ_P^2 gets 'warped up' according to Eq. 35, and becomes $\widetilde{\ell}_P^2$. This can eventually be parametrised in terms of a coupling constant

$$c_0 = \frac{\mathcal{K}}{M_P}, \tag{38}$$

which is a fraction ~ 0.1 and hence comparable with the electroweak coupling $g \simeq 0.6$. The KK modes of the graviton, therefore, will behave very much like weakly-interacting massive particles, being produced at the LHC if there is enough energy and having very short lifetimes, so that they appear to decay practically immediately, i.e. at the interaction vertex itself [27].

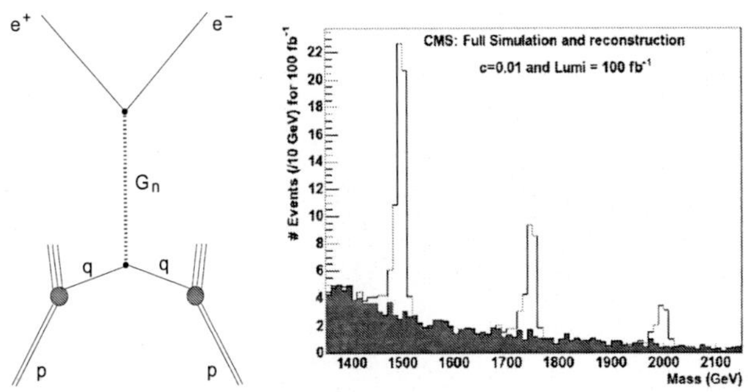

Figure 21. Di-electron signal for an RS graviton resonance. One of the Feynman diagrams is shown on the left. One the right is shown the results of a simulation using the CMS detector [28], for $c_0 = 0.01$ and with $m_0 = 400$ GeV

At the LHC, therefore, massive graviton KK modes are likely to appear as resonances in basic processes, such as the production of electron-positron pairs, or $\mu^+\mu^-$ pairs, or a pair of hadronic jets. This is illustrated in Fig. 21. On the left, a typical Feynman diagram for the production of an electron-positron (e^+e^-) pair is shown. On the right is shown the results of a simulation of this process by the CMS collaboration at the LHC [28]. On the horizontal axis, marked "Mass" is plotted the 'invariant mass' of the e^+e^- pair, i.e. the quantity

$$M_{e^+e^-} = (E_+ + E_-)^2 - (\vec{p}_+ + \vec{p}_-)^2 \ , \tag{39}$$

where $E_\mp$ and $\vec{p}_\mp$ refer to the energy and momentum of the electron and the positron respectively. The shaded histogram represents the expectations from the SM, while the peaks represent the expectations in the RS model.

As expected in a quantum resonance phenomenon, the probability of interaction shoots up when this invariant mass $M_{e^+e^-}$ matches with the mass of a real graviton mode. Three distinct peaks are predicted (for this choice of parameters), and a clinching proof that these are indeed RS graviton modes would be if the corresponding masses were found to be in the ratio $\xi_1 : \xi_2 : \xi_3$. Of course, this graph does not tell us the whole story, and there are many possible variations. For example, it is entirely possible that the coupling c_0 will be larger, in which case the resonant peaks will be shorter and fatter, and as $c_0 \to 0.1$ will simply represent small excesses over the SM histogram. It is also possible that the value of m_0 may be so large that all the graviton resonances lie beyond the kinematic reach of the LHC. In such cases we would have to look for other effects, such as those involving virtual graviton modes, to look for signals of warped gravity. Studies of this nature abound in the literature, but it would be beyond the aim of this article to take up a detailed discussion of this very interesting topic.

8. Modulus Stabilisation and the Radion

In the previous section we have discussed the RS model and shown how it provides an elegant solution of the hierarchy problem by never bringing a large and a small number together in such a way that they can be influenced by each other. In doing so, we have glossed over a major element of fine tuning in this model. Like the ADD model, this lies in the size R of the extra dimension, but here it is not the quantum stability of this size that is in question, as it is close to the Planck length anyway. What is in question is why the product $\mathcal{K}R = 11.73$ precisely, and what would happen if it varied a little. Since the warp factor $e^{-\pi\mathcal{K}R}$ is responsible for creating the factor of 10^{16} between the TeV scale and the Planck scale, clearly that ratio will be sensitive to small changes in $\mathcal{K}R$. In fact, even the choice $\mathcal{K}R = 11.0$ would make the warp factor a whole order of magnitude too small, and, conversely, choosing $\mathcal{K}R = 12.5$ would make it an order of magnitude too large. In a string theory – or any underlying theory – the size of the extra dimension R should be a dynamical variable – in string theoretic parlance, a *modulus*. Such a modulus should show time variation, and this would appear as a time variation in M_P, i.e. in the gravitational constant G_N. Given the fact that all astrophysical evidence indicates that G_N has shown no measurable variation since the Big Bang, it is clear that the modulus R must be remarkably stable, i.e. the distance between the observed universe of the visible brane, and the shadow world of the invisible brane must be remarkably constant, and have the just-so value $\mathcal{K}R = 11.73$. This is indubitably a case of fine-tuning.

The original work of Randall and Sundrum did not address the question of stabilisation of the R modulus at all. In fact, RS went on to construct a model variant where $R \to \infty$, where there is no question of fine-tuning. However, it was soon realised that the original model, fine-tuning and all, was much more relevant for high energy physics than its successor. The issue of modulus stabilisation had, therefore to be addressed seriously.

Two solutions to the modulus stabilisation problem were proposed within a few months of the original Randall and Sundrum suggestion. The first one – which has proved more popular – was by Walter Goldberger and Mark Wise (GW) from CalTech [29], and it used the high energy theorist's favourite tool, viz. a scalar field, to provide the stability effect. We shall discuss this presently. In the next spring, a supersymmetric solution was proposed by Jonathan Bagger and his collaborators, from Johns Hopkins University [30]. Though elegant, this solution never became popular, probably because the high energy community, from the beginning, has viewed extra dimensions as a *alternative* to supersymmetry as a solution to the hierarchy problem.[21]

What was the simple and attractive solution proposed by Goldberger and Wise? Like Yukawa and Higgs and Lindé before them, they had recourse to postulating the existence of a scalar field $B(x,\phi)$ – in this case, one which lives in the full five dimensional spacetime (x,ϕ). By dint of choosing B^4-type self interactions of this field in the bulk, and extra self-interactions of the form $\lambda(B^2 - v^2)^2$ on the two branes, they were able to show that the modulus $\mathcal{K}R$ is trapped in a deep potential with a minimum which can be set to 11.73 without much fine tuning at any stage. A sketch of this potential is shown in Fig. 22. The steep walls flanking the minimum show that it would take a major disturbance of the entire universe to pull the modulus out of the

[21] This is not really a logical stand, because string theories, on which brane world models are predicated, contain both supersymmetry *and* extra dimensions. However, like many collective prejudices similarly divorced from logic, it seems to have stood the test of time.

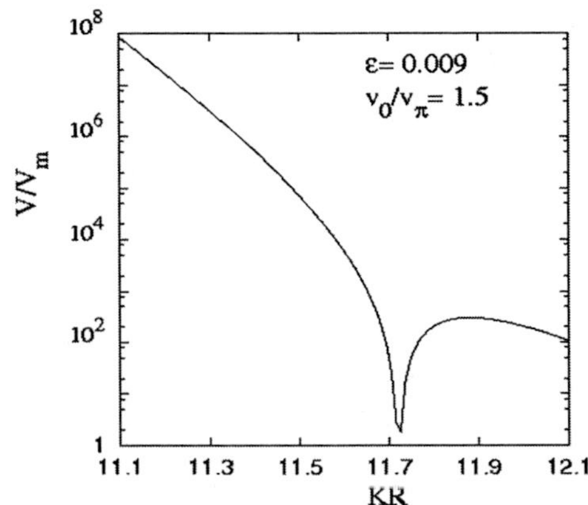

Figure 22. Illustrating the steep minimum in the Goldberger-Wise potential at the minimum $\mathcal{K}R = 11.73$. Note that the vertical axis is plotted on a logarithmic scale

minimum and set it rolling. In the absence of any such disturbance, the modulus is stable.

One important consequence of having a bulk field of this form is that fact that the warp factor $e^{-\pi\mathcal{K}R}$ on the visible brane, i.e. our universe, is no longer just a constant, but may be parametrised as $e^{-\pi[11.73+T(x)]}$, where $T(x)$ is a dynamical field. The values assumed by $T(x)$ must be very small, as indicated by the steep walls of the potential well in Fig. 22, but since they are exponentiated, they appear as a normal scalar field $\Phi(x)$ on the visible brane. For obvious reasons, this field is called a *radion.* It turns out that there are no serious theoretical constraints on the mass of this radion, and hence it can be chosen light or heavy as we wish – within reason, that is to say. Thus, it is theoretically equally possible to have a 1 GeV radion, or a 10,000 GeV radion. If the radion is light, i.e. within the kinematic access of the LHC, then it may prove to be a very distinct signal for an RS kind of model, stabilised by a GW-type mechanism.

Technically, however, there is a problem in identifying a radion signal at the LHC or any other machine. This is because a real radion couples to matter in a manner identical with the Higgs boson. The signals for both are, therefore, identical, and it will be difficult to tell whether such signals are due to a Higgs boson or a radion being produced. Of course, if both are produced, and detected, then some of the detailed behaviour in decays of these different scalars can be exploited to distinguish them. This is particularly true when the signal involves virtual quantum states, or in the language of quantum field theory, loop-level effects [31]. However all this is upset by that fact that both radion and Higgs boson are quantum mechanical states with identical quantum numbers, which permits them to form mixed states which correspond exactly to neither. Detection of such mixed states may be easy, but identification of the components of Higgs boson and radion in them is a non-trivial matter, and would require the collection of a lot of data at the LHC or any other machine before it can even be attempted [32].

9. Different Strokes

In the decade or so since extra dimensions became fashionable again, there have been many attempts to devise alternative models, using some of the basic ideas of ADD and RS, but innovating more. It would be tedious and long-winded to attempt a comrehensive listing of all the new ideas that have been suggested. However mentioning a few of the ideas may give a flavour of the kind of thinking that has been going on in this context.

There have not been all that many modifications of the ADD model, which is very simple. There have been attempts to change the compactification scheme, the simplest idea being to assume different radii of compactification for different dimensions [33]. The fact that the LHC may operate at the actual scale where gravity is strong has inspired an attempt to write down a toy string theory and work out possible signals for it. The use of supersymmetry to stabilise the large size/Flora of the bulk has been suggested [30], but has not found very many takers till date.

One of the most ingenious ideas suggested in this context is that of *dimensional deconstruction* [34]. This takes note of the fact that the observable feature of the ADD and similar models is the tower of KK modes. The proposers of this theory point out that if the world is purely four-dimensional and at a very small length scale, there is a somewhat complicated gauge theory involving replication of a particular gauge symmetry many times, then, at larger length scales this may well appear as a uniformly-spaces set of spin-2 states, which could be confused with a KK tower of gravitons. In that case, if one sees such repeated states at the HC, for example, one would jump to the conclusion that there are extra dimensions, even though the world is purely four-dimensional. This is an ingenious idea, but it does require one to postulate a very complicated gauge symmetry at small length scales. Such complicated symmetries do arise in string theories – but if we are to believe string theories, we might as well believe in extra dimensions too!

The RS model has inspired more variants, starting with the second paper of Randall and Sundrum themselves. Apart from obvious extensions like incresing the number of warped dimensins, there have been attempts to explain the observable universe as a D_3-brane which is an intersection of higher dimensional branes – this complicated construction solves some long-standing problems in the physics of flavour. This is a bit like the traditional parable of using a sledgehammer to crack a nut, but that is a criticism that may be applied, in some ways, to the RS model itself.

Perhaps the most interesting alternative idea to come out of the RS model was the idea of having an extra dimension in the form of an $\mathbb{S}^{(1)}/\mathbb{Z}_2$ orbifold and dispensing with the two D_3 branes altogether [35]. In this case, of course, *all* the SM fields live in the bulk, and, at large lenth scales, we would see each of them as a KK tower. This scenario is called a *universal extra dimension* or UED. More details may be found in the article by Dobrescu in this volume.

The other idea which has generated a lot of recent activity is that of *quantum holography.* In 1997, about six months before the first ADD paper, Juan Maldacena at Harvard had conjectured that in a model with D-branes, a theory of gravity in the bulk may appear on the brane as a theory involving gauge interactions [36]. By changing from one set of variables to another, using what is called a *duality transformation*, one theory transforms into the other[22]. Thus, a perfect fluid on the brane may appear as a particular type of black hole in the bulk theory, and similarly, there are black hole solutions of quantum gravity in the bulk which look like a theory of strong interactions on the brane. In RS-type models, the nature of quantum gravity in the bulk happens to be exactly of the type (AdS) required for the Maldacena conjecture to work, so there have been suggestions that at the LHC (i.e. on the visible brane) we may find an effective theory of composite quarks and leptons[23], which is actuallly the Maldacena dual of fundamental theory of quantum gravity in the bulk [37]. Ingenious and exciting as these ideas are, they generally lack falsifiability, since practially any theory on the brane can have an exotic gravitational counterpart in the bulk, which our brane-bound experimental equipment will not be able to test.

Most of these ideas beyond the basic ADD and RS models have to do with gaining a deeper understanding of theories with extra dimensions, or of the SM itself. Hard-headed particle physicists, especially experimental physicists, have not, therefore, shown much enthusiasm for these. Thus, apart from a few efforts, the bulk of phenomenological and experimental studies of extra dimensions have to do with just four kinds of new physics:

- Towers of invisible ADD gravitons.
- Heavy RS graviton resonances.
- A relatively light scalar radion.
- KK modes in UED models.

It remains to be seen if there are hints of more exotic new physics of the kind described above at the LHC. For this we may have to wait a few years till enough data are collected to infer backwards and pin down the nature of the new physics.

10. Valediction

At the LHC, the effort of nations and the toil of thousands is being poured into the most important quest of all – the quest for understanding the innermost working of nature. The first step would be to understand the origin of mass in the visible universe, for which it is essential to discover the Higgs boson. The next step would be to discover why the Higgs boson mass is stable, i.e. to find the correct solution to the hierarchy problem. Next after that we would seek to discover the nature of the dark matter component of the universe. It may be mentioned in passing that 95% of the universe consists of dark matter and vacuum energy – all invisible. This is a humbling thought, but, to the determined thinker, it provides an extra impetus to seek out the real nature of the universe. Extra dimensions are a part of this quest, a small, but essential piece of the jigsaw puzzle which, when solved, would unite the whole universe and all its workings into a single theory. Of this end, one cannot describe it better than in the

[22]This is technically called the AdS-CFT correspondence, where AdS stands for *Anti-de Sitter* (cosmology) and CFT stands for *Conformal Field Theory.*

[23]This goes by the name *technicolour.*

words of Jalal-ud-din Rumi, the thirteenth century Sufi poet [38]:

My place is placeless, my trace is traceless,
no body, no soul, I am from the soul of souls.
I have chased out duality, lived the two worlds as one.
One I seek, one I know, one I see, one I call –
The First and the Last, the Outer and the Inner.

On this note we conclude our story.

REFERENCES

1. R. Ward, *The Life of Henry More*, Vols. I and II, Springer (2000)
2. C.E. Hinton, *Speculations on the Fourth Dimension*, Dover (180)
3. A fascinating account of the early days of Kaluza-Klein theories is given by O. Raifeartaigh, hep-ph/9810524 (1998). See also *The Dawning of Gauge Theory* by the same author, Princeton U. Press (1997)
4. T. Kaluza, *Sitz. Preuss. Akad. Wiss. Berlin (Math.Phys.)* (1921) 966-972
5. O. Klein, *Z. Phys.* **37** (1926) 895-906
6. I. Antoniadis, *Phys. Lett.* **B246** (1990) 377-384; I. Antoniadis, K. Benakli and M. Quiros, *Phys. Lett.* **B331** (1994) 313-320
7. I. Antoniadis, S. Dimopoulos and G.R. Dvali, *Nucl. Phys.* **B516** (1998) 70-82
8. N. Arkani-Hamed, S. Dimopoulos and G.R. Dvali, *Phys. Lett.* **B429** (1998) 263-272
9. I. Antoniadis, N. Arkani-Hamed, S. Dimopoulos and G.R. Dvali, *Phys. Lett.* **B436** (1998) 257-263
10. L. Randall and R. Sundrum, *Nucl. Phys.* **B557** (1999) 79-118
11. P. Mathews, S. Raychaudhuri and K. Sridhar, *Phys. Lett.* **B450** (1999) 343-347; *ibid.* **B455** (1999) 115-119; *JHEP* **0007** (2000) 008
12. H. Weyl, *Math. Ann.* **71** (1912) 441479
13. For a very readable and up-to-date review, see S.M. Carroll, *The Cosmological Constant, Living Rev. Relativity* **4** (2001), URL:http://www.livingreviews.org/lrr-2001-1
14. E.G. Floratos and G.K. Leontaris, *Phys. Lett.* **B465** (1999) 95-100
15. For an excellent but slightly old review, see C.S. Unnikrishnan, *Pramana J. Phys. 41* (1993) Suppl. 395-411
16. S. Schlamminger et al (Eöt-Wash Experiment), *Phys. Rev. Lett.* **100** (2008) 041101
17. E. Gildener, *Phys. Rev.* **D14** (1976) 1667
18. V.A. Rubakov and M.E. Shaposhnikov, *Phys. Lett.* **B125** (1983) 136-138
19. K. Akama, *Lect. Notes Phys.* **176** (1982) 267-271
20. See, for example, J. Polchinski, *Tasi lectures on D-branes*, hep-th/9611050
21. G.F. Giudice, R. Rattazzi and J.D. Wells, *Nucl. Phys.* **B544** (1999) 3-38; E.A. Mirabelli, M. Perelstein and M. Peskin, *Phys. Rev. Lett.* **82** (1999) 2236-2239; T. Han, J.D. Lykken and R.-J. Zhang, *Phys. Rev.* **D59** (1999) 105006
22. S.B. Giddings and S.D. Thomas, *Phys. Rev.* **D65** (2002) 056010
23. S.W. Hawking, *Commun. Math. Phys.* **43** (1975) 199-220; Erratum-*ibid.* **46** (1976) 206
24. S.B. Giddings and M.L. Mangano, *Phys. Rev. D78* (2008) 035009; a very readable summary of their main arguments may be found in M.E. Peskin, *Physics* **1** (2008) 14
25. Dirac's views may be read in *A Physicist's Conception of Nature*, ed. J. Mehra, pp. 1-14
26. W.D. Goldberger and M.B. Wise, *Phys. Rev.* **D60** (1999) 107505
27. H. Davoudiasl, J.L. Hewett and T.G. Rizzo, *Phys. Rev. Lett.* **84** (2000) 2080
28. C. Collard, talk presented at *Physics at LHC*, Vienna (2004)
29. W.D. Goldberger and M.B. Wise, *Phys. Rev. Lett.* **83** (1999) 4922-4925
30. R. Altendorfer, J. Bagger and D. Nemeschansky, *Phys. Rev.* **D63** (2001) 125025
31. See, for example, P.K. Das, S.K. Rai and S. Raychaudhuri, *Phys. Lett.* **B618** (2005) 221-228
32. Some cases are discussed in G.F. Giudice, R. Rattazzi and J.D. Wells, *Nucl. Phys.* **B595** (2001) 250-276
33. J.D. Lykken and S. Nandi, *Phys. Lett.* **B485** (2000) 224-230
34. N. Arkani-Hamed, A.G. Cohen and H. Georgi, *Phys. Rev. Lett.* **86** (2001) 4757-4761
35. T. Appelquist, H.-C. Cheng and B.A. Dobrescu, *Phys. Rev.* **D64** (2001) 035002
36. J.M. Maldacena, *Int. J. Theor. Phys.* **38** (1999) 1113-1133
37. For a nice review of this and similar ideas, see I.Z. Rothstein, *Tasi Lectures on Effective Field Theories*, arXiv:hep-ph/0308266 (2004)
38. A popular translation of Rumi's poems may be found in *The Essential Rumi*, tr. Coleman Barks, HarperOne (1997)

Universal Extra Dimensions

Bogdan A. Dobrescu

Fermi National Accelerator Laboratory, Batavia, IL 60510, USA

Extra dimensions in which all particles propagate are called universal. These lecture notes present the basics of field theory in a compact dimension with flat metric, as well as the hadron collider phenomenology of universal extra dimensions.

1. Introduction

It is an obvious fact: we live in a space with three dimensions. However not everything that is obvious is true, as the development of quantum mechanics has compellingly illustrated. The possibility of extra spatial dimensions has been the subject of metaphysical speculations dating back at least to the 19th century [1]. More serious scientific inquiries into this issue have to address the question of how are the extra dimensions hidden from us. The first convincing proposal was suggested by Oskar Klein [2]: if an extra dimension is compactified on a circle, then its presence would not be observed unless the experiments have a resolution higher than the radius of the circle.

Nowadays we know that the circle compactification fails on two counts. First any gauge field, such as the photon, that propagates along the extra dimension would require a spin-0 partner of equal mass and couplings for each spin-1 particle. Second any fermion that propagates along the extra dimension would be a vectorlike fermion: its left- and right-handed components would have the same gauge charges, which is not true for any of the elementary fermions discovered so far. However if the compactification is on an interval, then the unwanted vectorlike partners of the observed fermions and the spin-0 partners of the gauge fields may be eliminated by the boundary conditions at the end of the interval.

Any particle propagating through extra dimensions, whether compactified on a circle or an interval, would appear in experiments as a tower of massive particles in 3 spatial dimensions. These massive particles are usually called Kaluza-Klein (KK) modes. Their presence can be easily understood based on the usual particle-in-a-box problems in quantum mechanics: given that space along the extra dimensions is compact, the energy states are quantised. The kinetic energy due to motion along the extra dimensions manifests itself as mass in the usual 3 spatial dimensions. For an interval of length L, the mass of the lightest KK modes is $(\pi/L)\hbar/c$ (the natural unit system, $\hbar = c = 1$ is used in what follows).

The highest partonic energies attained so far in collider experiments (at the Tevatron) are typically below 1 TeV. Therefore it is expected that extra dimensions of size 10^{-19} m are consistent with all experimental observations. In practice the situation is more complicated, especially because of the possibility that not all known particles propagate along the extra dimensions. A scenario that received extensive attention at the end of the 20th century was based on the idea that only the graviton propagates in some flat extra dimensions [3]. In that case, only measurements of gravity are sensitive to the presence of the extra dimensions and the current upper limit on their "radius" $R \equiv L/\pi$ is around 4×10^{-5} m [4]. In another scenario, all bosons propagate in extra dimensions while the fermions are localised at the end points of an interval [5,6]. As a result the KK modes of the standard model gauge bosons mediate four fermion interactions at tree-level, and the constraints from a global fit to collider data set a very stringent limit of $R \lesssim 0.3 \times 10^{-19}$ m [7].

Universal extra dimensions (UED) are arguably the simplest kind of extra dimensions: all particles propagate along some flat compact extra dimensions. The remarkable feature of UED is that a remnant of translational invariance along the extra dimensions is preserved such that a single KK mode cannot couple at tree-level to zero modes [8]. As a result, the limits are relaxed by more than an order of magnitude compared to the extra dimensions accessible only to bosons: $R \lesssim 5 \times 10^{-19}$ m. Furthermore UED lead to dramatically different phenomenological implications.

The search in collider experiments for KK modes having a spectrum and interactions consistent with a certain compactification is the best way of checking the existence of extra dimensions. Within the next few years, the ATLAS and CMS experiments at the Large Hadron Collider (LHC) are expected to discover KK modes associated with UED if the size of the extra dimensions is not far below 10^{-19} cm.

These lecture notes derive the generic properties of KK modes in the case of a flat extra dimension compactified on the interval. Then the signatures of UED at the Tevatron and the LHC are briefly reviewed.

2. Field Theory on a Flat Compact Dimension

Before discussing the phenomenology of one UED, it is necessary to study the general features of quantum field theory in a flat extra dimension. The cases of spin 0, 1/2 and 1 are analysed in turn.

2.1. Scalar Field on the Interval

Let us consider a five-dimensional spacetime: four spacetime dimensions of coordinates x^μ, $\mu = 0, 1, 2, 3$ form the usual Minkowski spacetime, and one transverse spatial dimension of coordinate x^4 is flat and compact with $0 \leq x^4 \leq L$. Thus the extra dimension is an interval (Fig. 1) and the boundary conditions at its end points determine the spectrum of KK modes.

Free scalar fields $\Phi(x^\mu, x^4)$ are described by the following action:

$$S_\Phi = \int d^4x \int_0^L dx^4 \left(\partial_\alpha \Phi^\dagger \partial^\alpha \Phi - M_0^2 \Phi^\dagger \Phi\right) \; . \quad (2.1)$$

We use letters from the beginning of the Greek alphabet to label the five-dimensional coordinates $\alpha, \beta, ... = 0, 1, 2, 3, 4$ and letters from the middle of the Greek alphabet to label the Minkowski coordinates $\mu, \nu, ... = 0, 1, 2, 3$.

Under a variation of the field $\delta\Phi(x^\mu, x^4)$ the variation of the action is given by

$$\delta S_\Phi = \delta S_\Phi^v + \delta S_\Phi^s \; , \quad (2.2)$$

where the first term is a "volume" integral

$$\delta S_\Phi^v = -\int d^4x \int_0^L dx^4 \left(\partial^\alpha \partial_\alpha \Phi^\dagger + M_0^2 \Phi^\dagger\right) \delta\Phi \quad (2.3)$$

and the second term is a "surface" integral

$$\delta S_\Phi^s = \int d^4x \left(\partial_4 \Phi^\dagger \delta\Phi\big|_{x^4=L} - \partial_4 \Phi^\dagger \delta\Phi\big|_{x^4=0}\right) \; . \quad (2.4)$$

Figure 1. The extra dimension of coordinate x^4 extends from $x^4 = 0$ to $x^4 = L$ and is transverse to the usual three spatial dimensions

Here we have assumed as usual that the field vanishes at $x^\mu \to \pm\infty$. Given that the action has to be stationary with respect to *any* variation of the field, the volume and surface terms must vanish independently. Requiring $\delta S_\Phi^v = 0$ implies that Φ is a solution to the five-dimensional Klein-Gordon equation

$$\left(\partial^\mu \partial_\mu - \partial_4^2 + M_0^2\right) \Phi = 0 \quad (2.5)$$

while $\delta S_\Phi^s = 0$ forces the boundary conditions that can be imposed on Φ to obey

$$(\partial_4 \Phi^\dagger) \delta\Phi\big|_{x^4=L} = (\partial_4 \Phi^\dagger) \delta\Phi\big|_{x^4=0} \; . \quad (2.6)$$

Given that the values of $\delta\Phi(x^\mu, x^4)$ at $x^4 = 0$ and $x^4 = L$ are in general not correlated (unless the two points are identified, which would not allow chiral fermions in the 4D theory) and Eq. (2.6) must be valid for any $\delta\Phi$, both the left- and right-handed sides of Eq. (2.6) must vanish. Therefore

$$\partial_4 \Phi\big|_{x^4=0} = 0 \quad \text{or} \quad \Phi(x^\mu, 0) = 0 \quad (2.7)$$

and

$$\partial_4 \Phi\big|_{x^4=L} = 0 \quad \text{or} \quad \Phi(x^\mu, L) = 0 \; . \quad (2.8)$$

We now solve the five-dimensional Klein-Gordon equation

$$\left(\partial^\mu \partial_\mu - \partial_4^2 + M_0^2\right) \Phi = 0 \quad (2.9)$$

subject to the boundary conditions (2.7) and (2.8). Since the boundary conditions are independent of x^μ, then Φ can be decomposed in Fourier modes as follows:

$$\Phi(x^\mu, x^4) = \sum_j \Phi^{(j)}(x^\mu) f^j(x^4) \; . \quad (2.10)$$

The four-dimensional scalar fields $\Phi^{(j)}$ satisfy

$$\left(\partial^\mu \partial_\mu + M_0^2 + M_j^2\right) \Phi^{(j)}(x^\mu) = 0 \; , \quad (2.11)$$

where M_j^2 is a positive eigenvalue. The f^j functions are solutions to the one-dimensional equation

$$\left(\partial_4^2 + M_j^2\right) f^j(x^4) = 0 \; . \quad (2.12)$$

A general solution to the above equation is

$$f^j(x^4) = C_+ e^{ijx^4/R} + C_- e^{-ijx^4/R} \; , \quad (2.13)$$

where $C_\pm$ are complex coefficients and j is a real number such that

$$M_j = \frac{j}{R} \quad (2.14)$$

and we defined the "compactification radius"

$$R \equiv \frac{L}{\pi} \; . \quad (2.15)$$

The boundary conditions (2.7) and (2.8) impose a relation between the two coefficients, $C_- = \pm C_+$ and also restrict the values of j: $e^{4ij\pi} = 1$. Furthermore the normalisation condition

$$\int_0^L dx^4 \left[f^j(x^4)\right]^* f^{j'}(x^4) = \delta_{jj'} \tag{2.16}$$

determines the last coefficient up to a phase factor which we choose to be one. Explicitly the solutions to Eq. (2.12) can be written as

$$f_0^j(x^4) = \frac{1}{\sqrt{L}(1+\delta_{j,0})}\cos\left(\frac{jx^4}{R}\right) \tag{2.17}$$

$$\text{for} \quad \partial_4\Phi|_{x^4=0} = \partial_4\Phi|_{x^4=L} = 0 \ ,$$

$$f_1^j(x^4) = \frac{1}{\sqrt{L}}\sin\left(\frac{jx^4}{R}\right) \tag{2.18}$$

$$\text{for} \quad \Phi(x^\mu, 0) = \Phi(x^\mu, L) = 0 \ ,$$

$$f_2^j(x^4) = \frac{1}{\sqrt{L}}\sin\left(\frac{(j-1/2)x^4}{R}\right) \tag{2.19}$$

$$\text{for} \quad \Phi(x^\mu, 0) = \partial_4\Phi|_{x^4=L} = 0 \ ,$$

$$f_3^j(x^4) = \frac{1}{\sqrt{L}}\cos\left(\frac{(j-1/2)x^4}{R}\right) \tag{2.20}$$

$$\text{for} \quad \partial_4\Phi|_{x^4=0} = \Phi(x^\mu, L) = 0$$

with j an integer called "KK number".

The functions f_n^j form a complete orthonormal set on the interval if

$$\sum_j \left[f_n^j(x^4)\right]^* f_n^j(x'^4) = \delta(x'^4 - x^4) \ . \tag{2.21}$$

The allowed values for j must be chosen such that the above completeness condition is satisfied. It is straightforward to check that $j \geq 0$ for $n = 0$ and $j \geq 1$ for $n = 1, 2, 3$.

The four-dimensional KK modes $\Phi^{(j)}(x^\mu)$ have masses

$$M^{(j)} = \sqrt{M_0^2 + \frac{j^2}{R^2}} \ . \tag{2.22}$$

For $n = 0$ there is a state ($j = 0$) of zero momentum ("zero mode") along the compact dimension.

In what follows we will concentrate on the KK functions f_0 and f_1. These respect a Z_2 symmetry under reflections with respect to the $x^4 = L/2$ point.

2.2. Fermions on the Interval: Chiral Boundary Conditions

We now turn to free spin-1/2 fields in five-dimensions. The Clifford algebra is generated by five anti-commuting matrices: Γ^α, $\alpha = 0, 1, 2, 3, 4$. The minimal dimensionality of these matrices is 4×4. The Γ matrices can be used to construct a spinor representation of the $SO(1,4)$ Lorentz symmetry, with the generators explicitly given by

$$\frac{\Sigma^{\alpha\beta}}{2} = \frac{i}{4}[\Gamma^\alpha, \Gamma^\beta] \ . \tag{2.23}$$

The fermions in five-dimensions have four components.

Upon compactification in the x^4, x^5 plane, the $SO(1,3)$ Lorentz symmetry generated by $\Sigma^{\mu\nu}/2$, $\mu, \nu = 0, 1, 2, 3$ remains unbroken. There are two chiralities under $SO(1,3)$ labeled as usual by L and R. These are projected by

$$P_{L,R} = \frac{1}{2}\left(1 \pm i\Gamma^0\Gamma^1\Gamma^2\Gamma^3\right) \ . \tag{2.24}$$

A five-dimensional chiral fermion $\Psi_\pm \equiv P_\pm\Psi$, decomposes into two fermions of definite chirality under $SO(1,3)$:

$$\Psi(x^\mu, x^4) = \Psi_L(x^\mu, x^4) + \Psi_R(x^\mu, x^4) \ , \tag{2.25}$$

where

$$\Psi_{L,R} \equiv P_{L,R}\Psi \ . \tag{2.26}$$

As in Section 2, we consider the compactification on an interval: $0 \leq x^4 \leq L$. The action for a free five-dimensional chiral fermion is

$$S_\Psi = \int d^4x \int_0^L dx^4 \frac{i}{2}\left[\overline{\Psi}\Gamma^\alpha\partial_\alpha\Psi - \left(\partial_\alpha\overline{\Psi}\right)\Gamma^\alpha\Psi\right] \ . \tag{2.27}$$

Under an arbitrary variation of the field $\delta\Psi(x^\mu, x^4)$, the action has to be stationary both inside the square and on its boundary:

$$\delta S_\Psi^v = -\int d^4x \int_0^L dx^4 i\left(\partial_\alpha\overline{\Psi}\right)\Gamma^\alpha\delta\Psi = 0 \ ,$$

$$\delta S_\Psi^s = \frac{i}{2}\int d^4x \int_0^L dx^4 \left(\overline{\Psi}\Gamma^4\delta\Psi\big|_{x^4=L} - \overline{\Psi}\Gamma^4\delta\Psi\big|_{x^4=0}\right) = 0 \ . \tag{2.28}$$

The first equation implies that Ψ is a solution to the five-dimensional Weyl equation, which can be decomposed into two equations:

$$\Gamma^\mu\partial_\mu\Psi_L = -\Gamma^4\partial_4\Psi_R \ ,$$
$$\Gamma^\mu\partial_\mu\Psi_R = -\Gamma^4\partial_4\Psi_L \ . \tag{2.29}$$

The second equation (2.28) restricts the values of Ψ on the boundary.

In the case of a fermion whose zero-mode is left-handed, the boundary conditions are as follows:

$$\partial_4 \Psi_L(x^\mu, 0) = \partial_4 \Psi_L(x^\mu, L) = 0 \ ,$$

$$\Psi_R(x^\mu, 0) = \Psi_R(x^\mu, L) = 0 \ . \tag{2.30}$$

The ensuing KK decomposition is given by

$$\Psi = \frac{1}{\sqrt{L}} \left\{ \chi_L^{(0)}(x^\mu) + \sqrt{2} \sum_{j \geq 1} \left[\Psi_L^{(j)}(x^\mu) \cos\left(\frac{jx^4}{R}\right) + \Psi_R^{(j)}(x^\mu) \sin\left(\frac{jx^4}{R}\right) \right] \right\} . \tag{2.31}$$

In the case of a fermion whose zero-mode is right-handed, the above equations apply with left- and right-handed labels interchanged. The conclusion is that in the case of interval compactification the boundary conditions for the left- and right-handed fermions are forced by the stationary of the action to eliminate the zero mode for one of the chiralities. All fermion KK modes for $j \geq 1$ pair up to form vectorlike fermions of Dirac masses $M^{(j)}$ as given in Eq. (2.22).

2.3. Gauge Fields on the Interval

A five-dimensional gauge boson has five components: $A_\mu(x^\nu, x^4)$, $\mu, \nu = 0, 1, 2, 3$ and $A_4(x^\nu, x^4)$ which corresponds to the polarisation along the extra dimension. From the point of view of the four-dimensional theory, A_4 is a tower of spinless KK modes. The boundary conditions are given by

$$\partial_4 A_\mu(x^\nu, 0) = \partial_4 A_\mu(x^\nu, L) = 0 \ ,$$

$$A_4(x, 0) = A_4(x, L) = 0 \ . \tag{2.32}$$

Solving the field equations with these boundary conditions yields the following KK expansions:

$$A_\mu = \frac{1}{\sqrt{L}} \left[A_\mu^{(0)}(x^\nu) + \sqrt{2} \sum_{j \geq 1} A_\mu^{(j)}(x^\nu) \cos\left(\frac{jx^4}{R}\right) \right] ,$$

$$A_4 = \sqrt{\frac{2}{L}} \sum_{j \geq 1} A_4^{(j)}(x^\nu) \sin\left(\frac{jx^4}{R}\right) . \tag{2.33}$$

The zero-mode $A_\mu^{(0)}(x^\nu)$ is one of the $SU(3)_c \times SU(2)_W \times U(1)_Y$ gauge bosons. Note that A_4 does not have a zero-mode. In the unitary gauge, the $A_4^{(j)}(x^\nu)$ KK modes are the longitudinal components of the heavy spin-1 KK modes $A_\mu^{(j)}(x^\nu)$.

2.4. One Universal Extra Dimension

Let us now discuss the case where all standard model particles propagate along one flat extra dimension compactified on an interval. The spectrum consists of equally spaced KK levels (of mass j/R), and on each level the KK modes for all standard model particles are approximately degenerate. The degeneracy is lifted by loop corrections [9] and electroweak symmetry breaking. The lightest KK particle is the first KK mode of the photon, and the heaviest particles at each level are the KK modes of the gluon and quarks. The mass spectrum is shown in Fig. 2. If the unspecified UV completion gives rise to operators localised at the ends of the interval, then the KK spectrum may change; this possibility is not considered in what follows.

Momentum conservation along the extra dimension is broken by the boundary conditions, but a remnant of it is left intact. This is reflected in a selection rule for the KK-numbers of the particles participating in any interaction. A vertex with particles of KK numbers $j_1, \ldots, j_p$ exists at tree level only if $j_1 \pm \ldots \pm j_p = 0$ for a certain choice of the $\pm$ signs. This selection rule has important phenomenological implications. First it is not possible to produce only one KK 1-mode at colliders. Second tree-level exchange of KK modes does not contribute to currently measurable quantities. There-

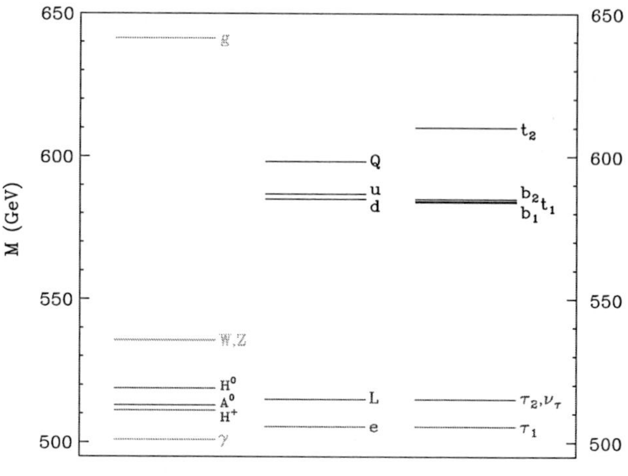

Figure 2. Mass spectrum of level-1 KK modes for a compactification scale of $1/R = 500$ GeV, from Ref. [10]

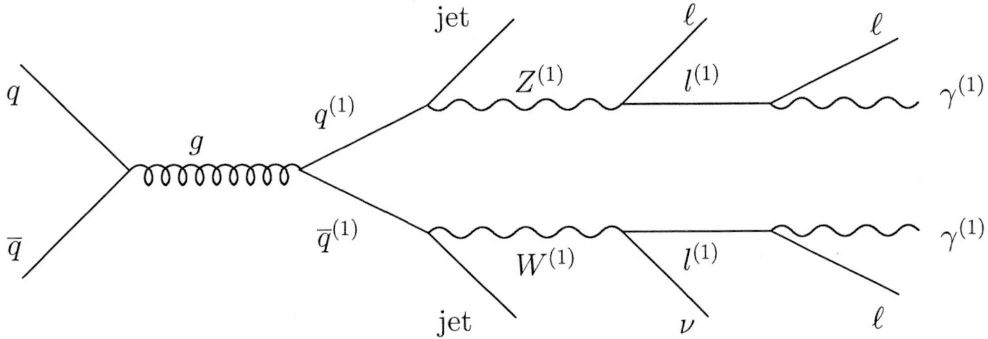

Figure 3. $3\ell + \not{E}_T$ signal from UED. $\gamma^{(1)}$ is the dark matter candidate which escapes the detector

fore the corrections to electroweak observables are loop suppressed, and the limit on $1/R$ from electroweak measurements is rather weak of the order of the electroweak scale [8].

The 1-modes may be produced in pairs at colliders. At the Tevatron and the LHC, pair production of the coloured KK modes has large cross sections [11,12] as long as $1/R$ is not too large. The coloured KK modes suffer cascade decays [10] like the one shown in Fig. 3. Note that at each vertex the KK-number is conserved and the $\gamma^{(1)}$ escapes the detector. The signal is $\ell^+\ell^-\ell^\pm + 2j + \not{E}_T$. However the approximate degeneracy of the KK modes implies that the jets are rather soft, and it is challenging to distinguish them from the background. The leptons are also soft, but usually pass some reasonably chosen cuts. Using the Run I data from the Tevatron, the CDF collaboration [13] searched for the $3\ell + \not{E}_T$ signal and has set a limit of $1/R > 280$ GeV at the 95% CL. The much larger Run II data set can be used to set a substantially improved limit, or alternatively has a fair chance of leading to a discovery.

If a signal is seen at the Tevatron or LHC, then it is important to differentiate the UED models from alternative explanations, such as superpartner cascade decays [10]. Measuring the spins at the LHC would provide an important discriminant, but such measurements are challenging [12,14]. A more promising way is to look for second level KK modes. These can be pair produced as the first level modes. However unlike the first level modes, the second level modes may decay into Standard Model particles. Such decays occur at one loop, via diagrams such as the one shown in Fig. 4.

Note that in the presence of loop corrections, the selection rule for KK numbers of the particles interacting at a vertex becomes $j_1 \pm \cdots \pm j_p = 0 \bmod 2$. This implies the existence of an exact Z_2 symmetry: the KK parity $(-1)^j$ is conserved. Its geometrical interpretation is invariance under reflections with respect to the middle of the $[0, \pi R]$ interval. Given that the lightest particle with j odd is stable, the $\gamma^{(1)}$ is a promising dark matter candidate. For $1/R$ in the 0.5–1.5 TeV range the $\gamma^{(1)}$ relic density fits nicely the dark matter density (for a review, see [15]). This whole range of compactification scales will be probed at the LHC [10].

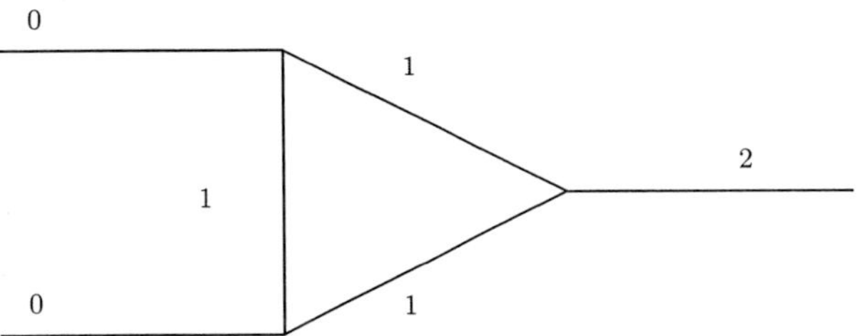

Figure 4. One-loop induced coupling of a 2-mode to two zero-modes

Another consequence of the loop-induced coupling of a 2-mode to two zero-modes is that the 2-mode can be singly produced in the s-channel [10]. The typical signal will be the cascade decay shown in Fig. 5, followed by $\gamma^{(2)}$ decay into hard leptons. The reach of the LHC in this channel has been analysed in Ref. [16].

3. Two Extra Dimensions

Gauge theories in more than four spacetime dimensions are non-renormalisable. This is not a problem as long as there is a range of scales where the higher-dimensional field theory is valid. For gauge couplings of order unity,

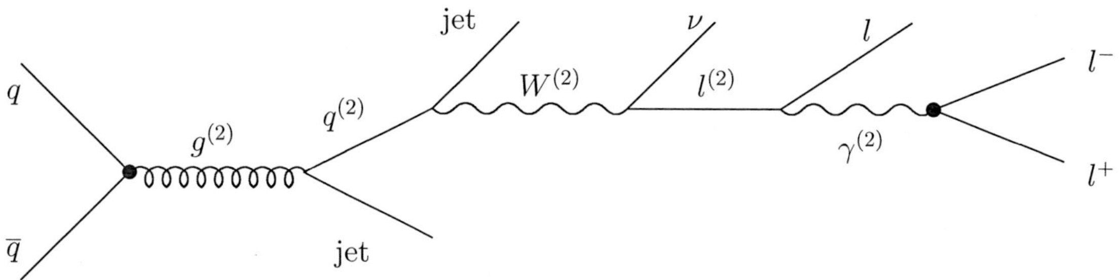

Figure 5. s-channel production of the level-2 gluon followed by cascade decay, and $\gamma^{(2)}$ decays to e^+e^- and $\mu^+\mu^-$

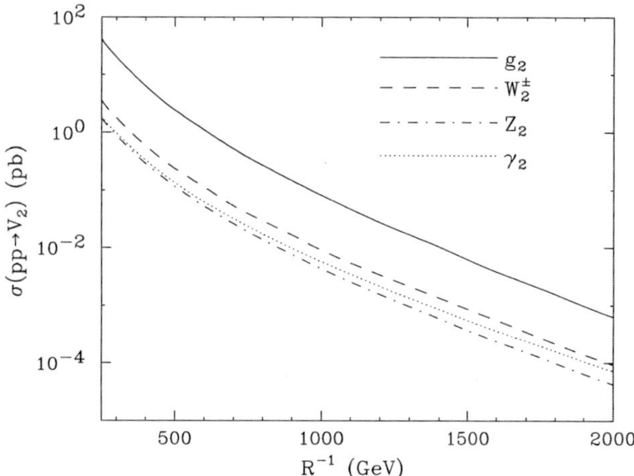

Figure 6. Production cross section for level-2 gauge bosons at the LHC, from Ref. [16]

as in the Standard Model, the range of scales is of the order of $(4\pi)^{2/n}$, so that only low values of n are interesting. Furthermore the low energy observables get corrections from loops with KK modes. The leading corrections are finite in the $n = 1$ case and logarithmically divergent for $n = 2$, while for $n \geq 3$ they depend quadratically or stronger on the cut-off. Therefore the effects of the unknown physics above the cut-off scale can be kept under control only for $n = 1$ and $n = 2$.

The case of two UED has been analysed less extensively compared to $n = 1$ UED. The general features of the Standard Model in $n = 2$ UED are presented in [17]. The hadron collider phenomenology of (1,0) modes, which are the lightest KK particles has been explored in Ref. [18]. Cascade decays of spinless adjoints proceed through tree-level 3-body decays involving leptons as well as one-loop 2-body decays involving photons. As a result, spectacular events with as many as six charged leptons, or one photon plus four charged leptons are expected to be observed at the LHC. Unusual events with relatively large branching fractions include three leptons of same charge plus one lepton of opposite charge, or one photon plus two leptons of same charge.

The cascade decays of the (1,1) modes [17], which are heavier than the (1,0) modes by a factor of $\sqrt{2}$, generate a series of closely-spaced narrow resonances in the $t\bar{t}$ invariant mass distribution.

Acknowledgements: I would like to thank the members of the Harish-Chandra Research Institute for organising a highly stimulating meeting on LHC physics. I am grateful to Biswarup Mukhopadhyaya for his constant encouragements without which this article would not have been completed.

REFERENCES

1. *E.g.*, C. H. Hinton, "What is the fourth dimension?", Scientific Romances, Vol. 1 (1884), pp 1
2. O. Klein, "Quantum theory and five-dimensional theory of relativity," Z. Phys. **37**, 895 (1926) [Surveys High Energ. Phys. **5**, 241 (1986)]
3. N. Arkani-Hamed, S. Dimopoulos and G. R. Dvali, "The hierarchy problem and new dimensions at a millimeter," Phys. Lett. B **429**, 263 (1998) [arXiv:hep-ph/9803315]
4. D. J. Kapner, T. S. Cook, E. G. Adelberger, J. H. Gundlach, B. R. Heckel, C. D. Hoyle and H. E. Swanson, "Tests of the gravitational inverse-square law below the dark-energy length scale," Phys. Rev. Lett. **98**, 021101 (2007) [arXiv:hep-ph/0611184]
5. I. Antoniadis, "A Possible new dimension at a few TeV," Phys. Lett. B **246**, 377 (1990)
6. K. R. Dienes, E. Dudas and T. Gherghetta, "Extra spacetime dimensions and unification," Phys. Lett. B **436**, 55 (1998) [arXiv:hep-ph/9803466]
7. K. M. Cheung and G. L. Landsberg, "Kaluza-Klein states of the standard model gauge bosons: Constraints from high energy experiments," Phys. Rev. D **65**, 076003 (2002) [arXiv:hep-ph/0110346]
8. T. Appelquist, H. C. Cheng and B. A. Dobrescu,

"Bounds on universal extra dimensions," Phys. Rev. D **64**, 035002 (2001) [arXiv:hep-ph/0012100]

9. H. C. Cheng, K. T. Matchev and M. Schmaltz, "Radiative corrections to Kaluza-Klein masses," Phys. Rev. D **66**, 036005 (2002) [arXiv:hep-ph/0204342]
10. H. C. Cheng, K. T. Matchev and M. Schmaltz, "Bosonic supersymmetry? Getting fooled at the LHC," Phys. Rev. D **66**, 056006 (2002) [arXiv:hep-ph/0205314]
11. C. Macesanu, C. D. McMullen and S. Nandi, "Collider implications of universal extra dimensions," Phys. Rev. D **66**, 015009 (2002) [arXiv:hep-ph/0201300]
12. J. M. Smillie and B. R. Webber, "Distinguishing spins in supersymmetric and universal extra dimension models at the Large Hadron Collider," JHEP **0510**, 069 (2005) [arXiv:hep-ph/0507170]
13. C. Lin, "A search for universal extra dimensions in the multi-lepton channel from proton antiproton collisions at $\sqrt{s} = 1.8$-TeV."
14. M. Battaglia, A. K. Datta, A. De Roeck, K. Kong and K. T. Matchev, "Contrasting supersymmetry and universal extra dimensions at colliders," *In the Proceedings of 2005 International Linear Collider Workshop (LCWS 2005), Stanford, California, 18-22 Mar 2005, pp 0302* [arXiv:hep-ph/0507284]
15. D. Hooper and S. Profumo, "Dark matter and collider phenomenology of universal extra dimensions," Phys. Rept. **453**, 29 (2007) [arXiv:hep-ph/0701197]
16. A. Datta, K. Kong and K. T. Matchev, "Discrimination of supersymmetry and universal extra dimensions at hadron colliders," Phys. Rev. D **72**, 096006 (2005) [Erratum-ibid. D **72**, 119901 (2005)] [arXiv:hep-ph/0509246]
17. G. Burdman, B. A. Dobrescu and E. Ponton, "Resonances from Two Universal Extra Dimensions," Phys. Rev. D **74**, 075008 (2006) [arXiv:hep-ph/0601186]
18. B. A. Dobrescu, K. Kong and R. Mahbubani, "Leptons and photons at the LHC: Cascades through spinless adjoints," JHEP **0707**, 006 (2007) [arXiv:hep-ph/0703231]

Selected figures presented in colour

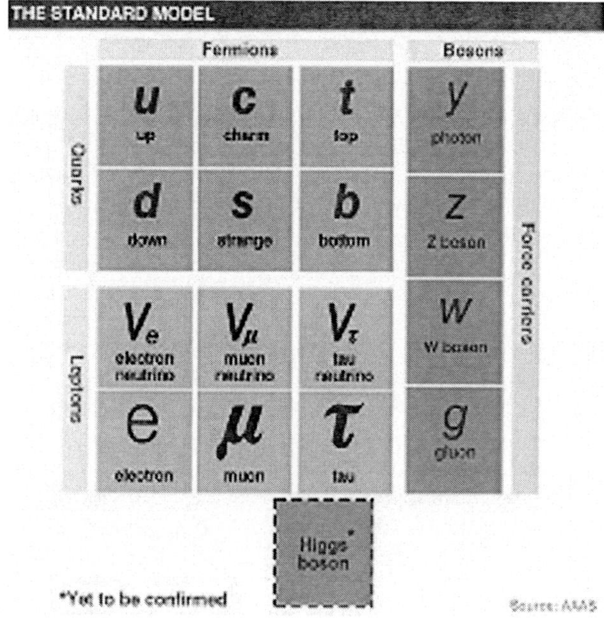

Choudhury and Mukhopadhyaya: Figure 1, Page 2

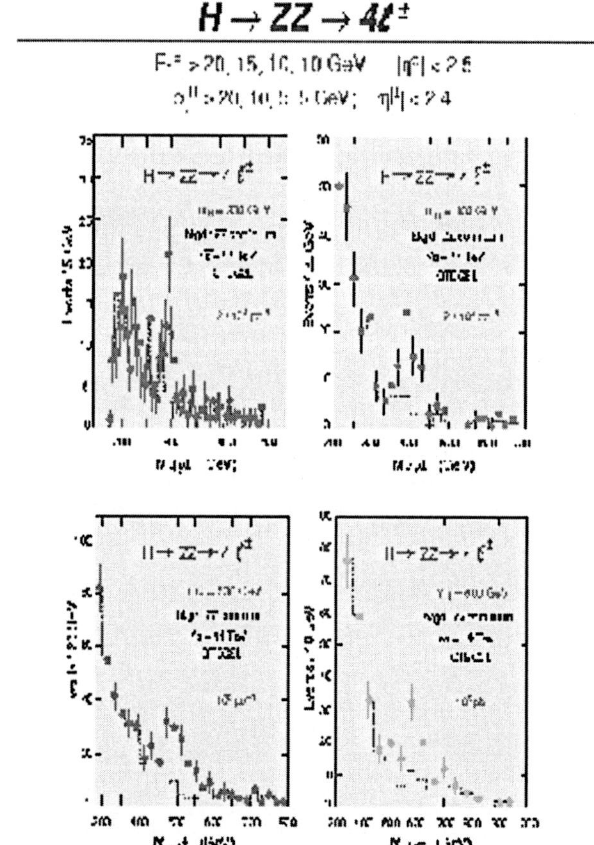

Gurtu: Figure 15, Page 21

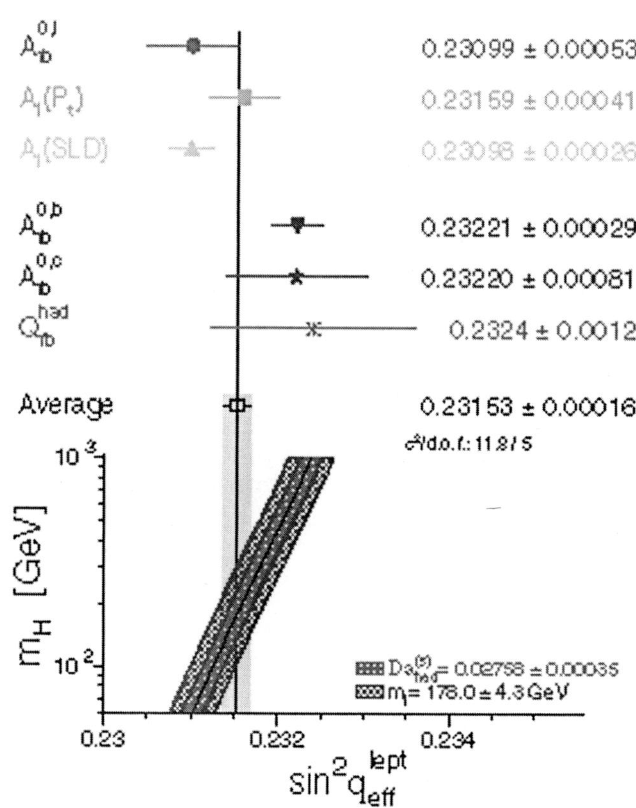

Gurtu: Figure 6, Page 16

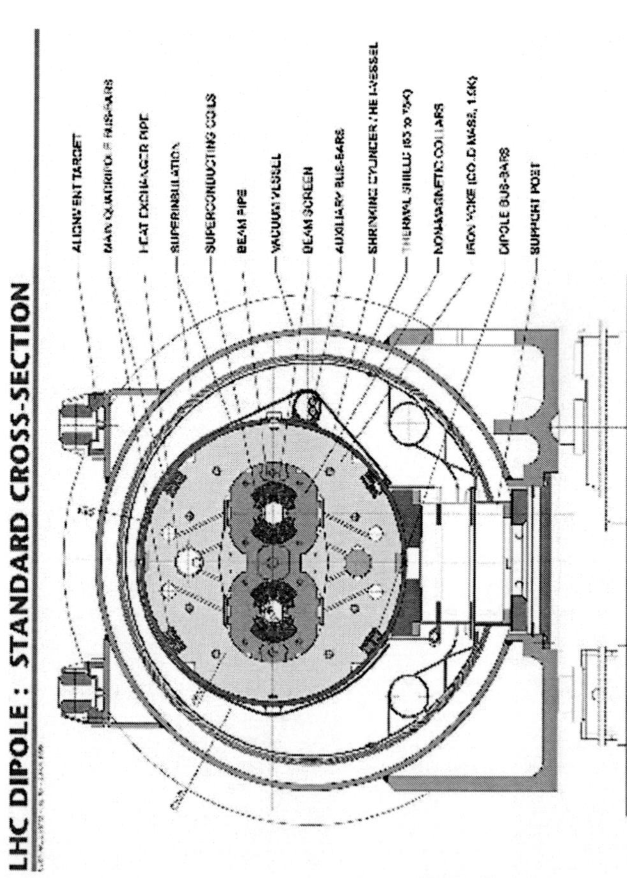

Brüning, Chohan and Chattopadhyay: Figure 3, Page 25

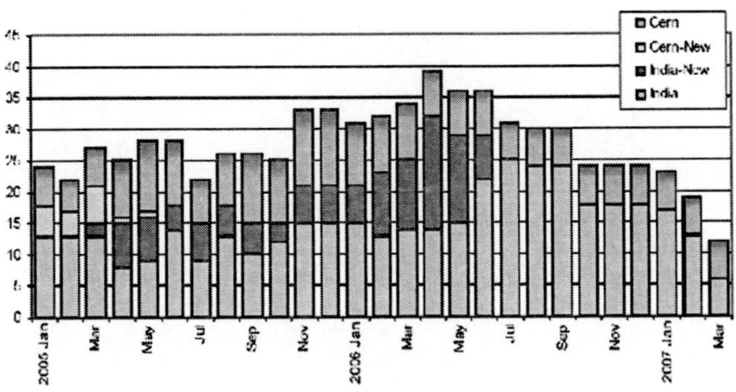

Brüning, Chohan and Chattopadhyay: Figure 10, Page 30

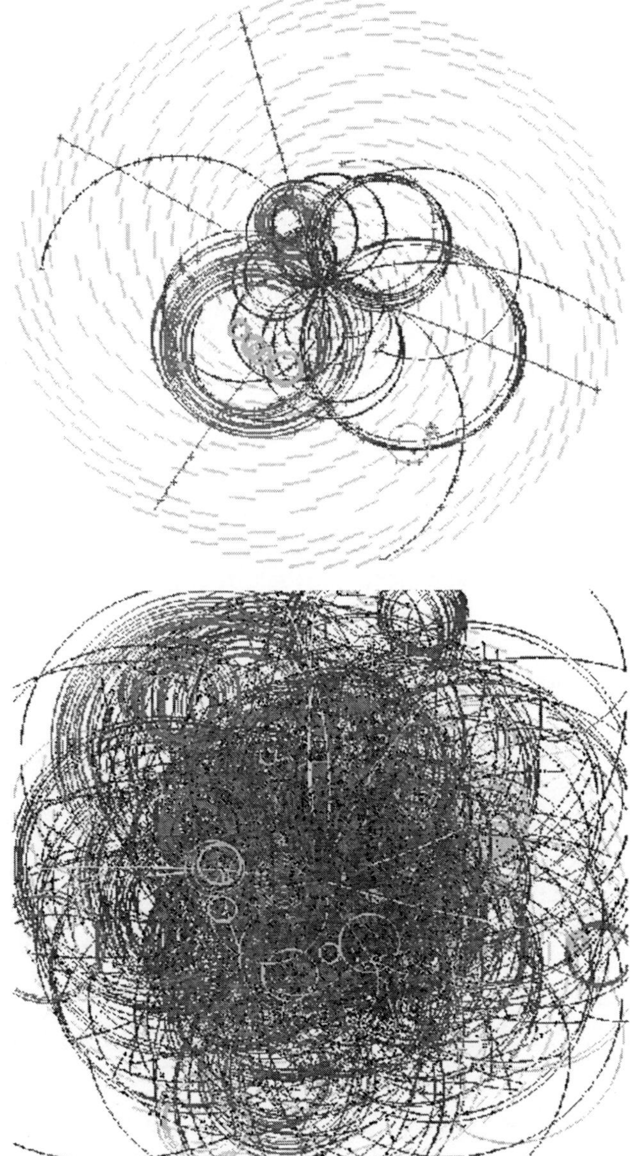

Banerjee: Figure 1, Page 37

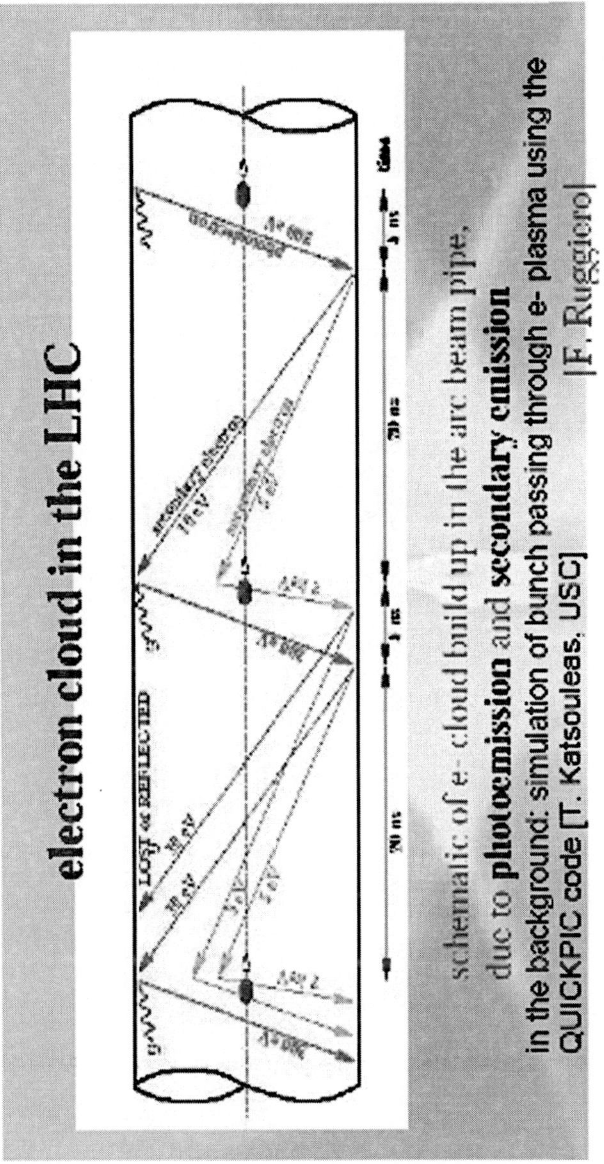

Brüning, Chohan and Chattopadhyay: Figure 17, Page 35

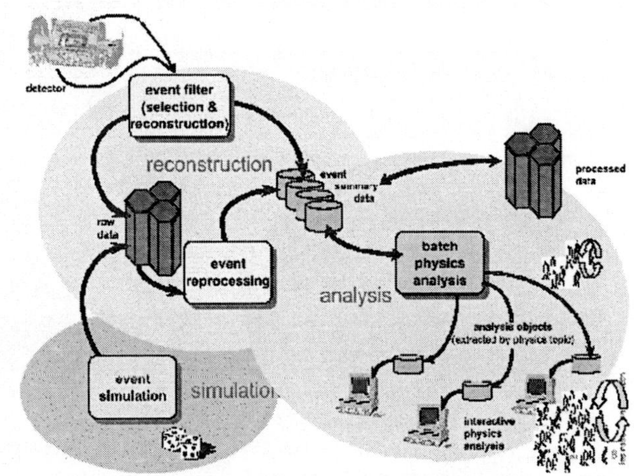

Banerjee: Figure 4, Page 39

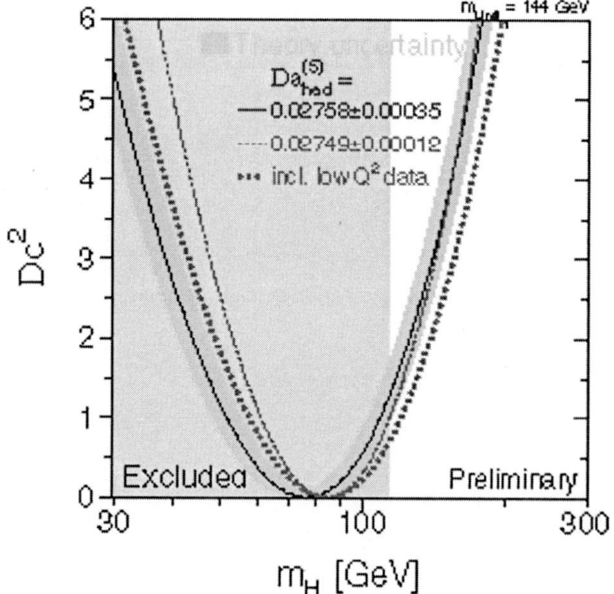

Djouadi and Godbole: Figure 1, Page 48

Djouadi and Godbole: Figure 5, Page 50

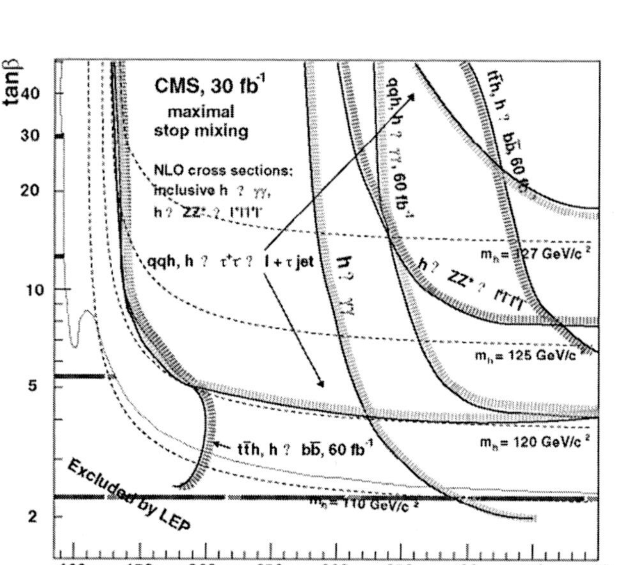

Djouadi and Godbole: Figure 22, Page 59

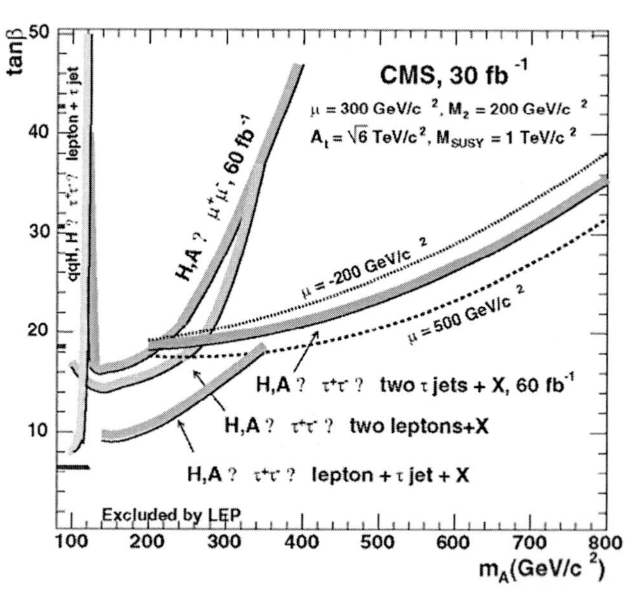

Djouadi and Godbole: Figure 23, Page 60

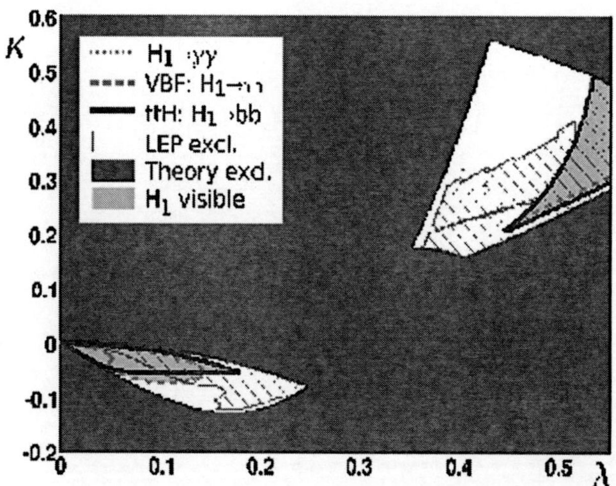

Djouadi and Godbole: Figure 32, Page 64

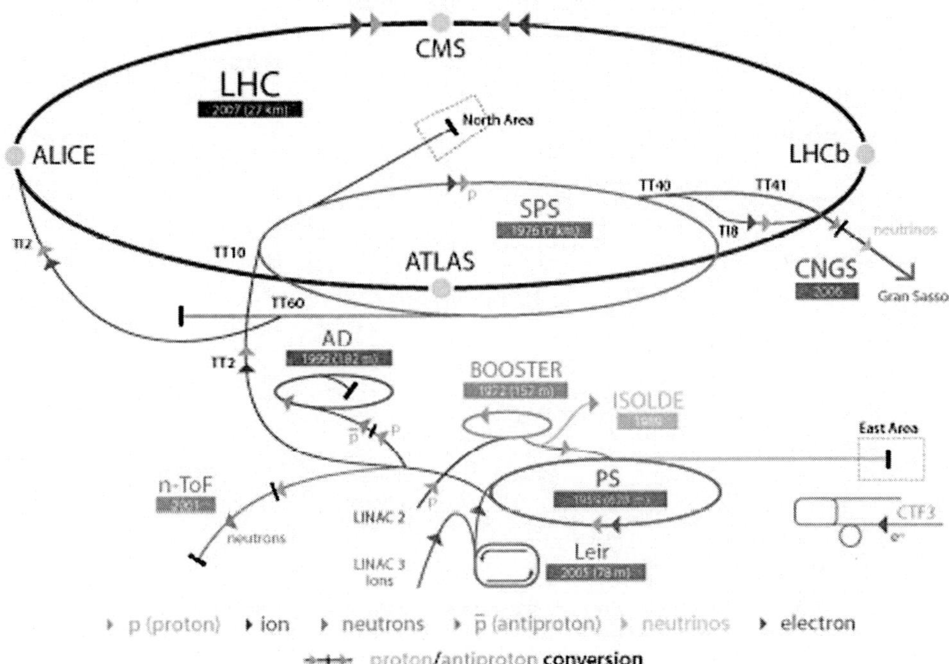

Mellado: Figure 2, Page 76

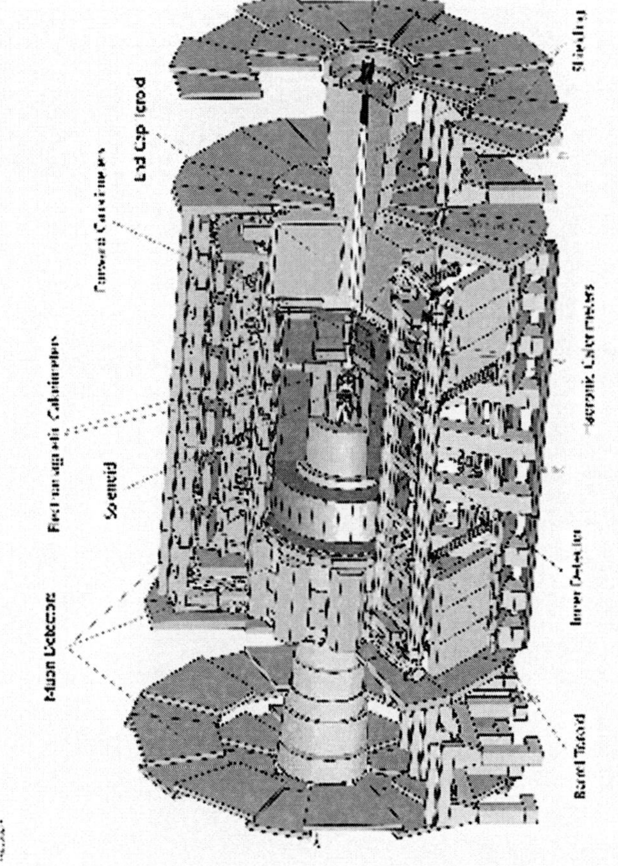

Mellado: Figure 4, Page 77

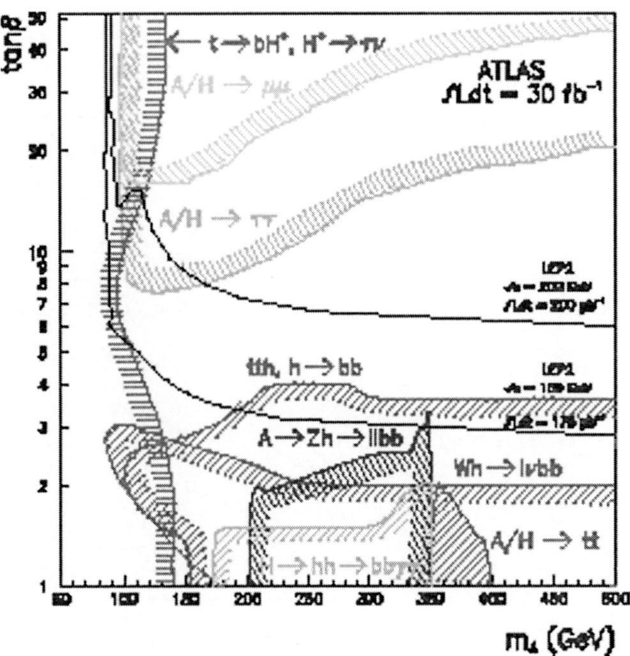

Mellado: Figure 14, Page 88

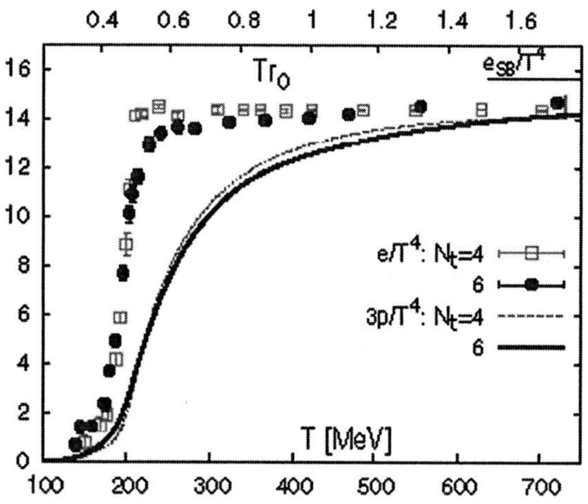

Bhalerao and Gavai: Figure 4, Page 110

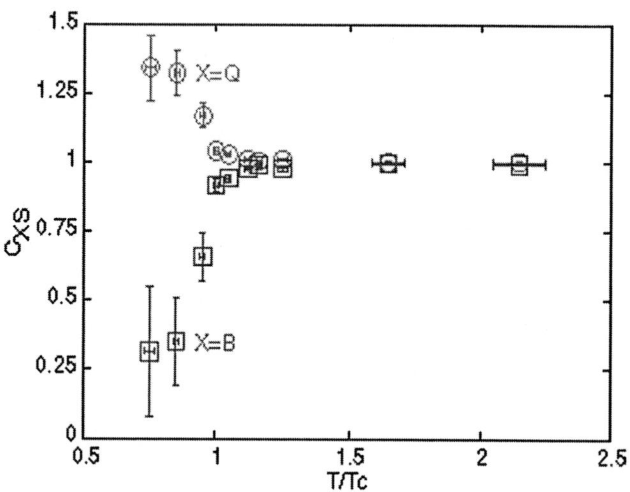

Bhalerao and Gavai: Figure 8, Page 111

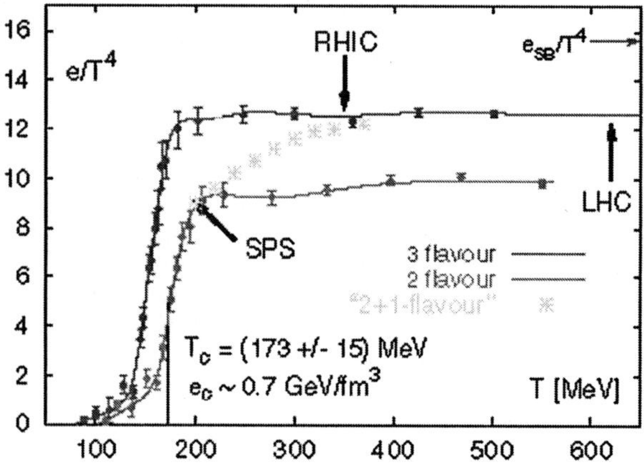

Nayak and Sinha: Figure 1, Page 132

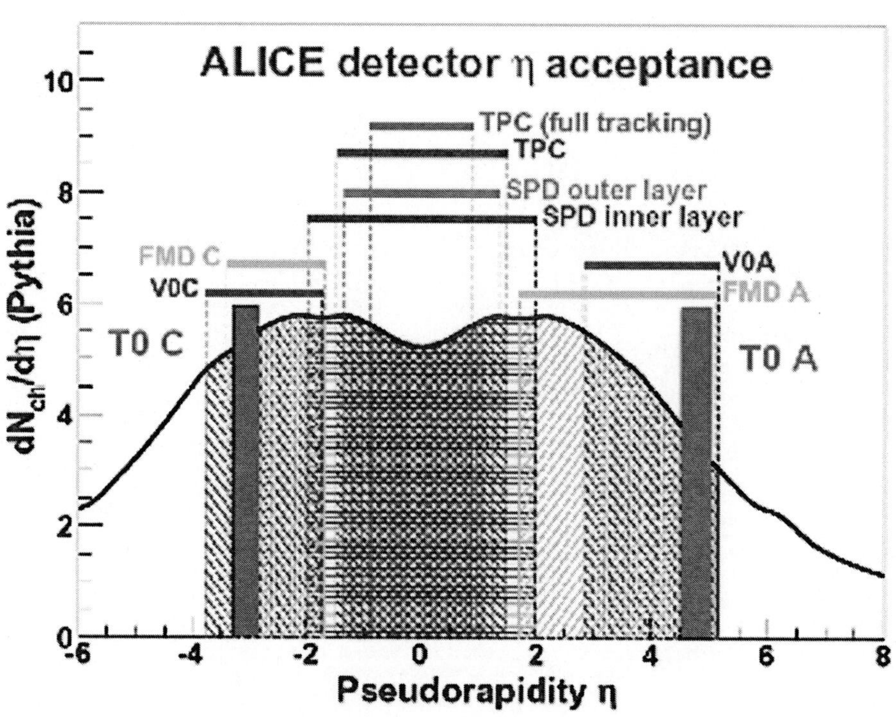

Nayak and Sinha: Figure 5, Page 134

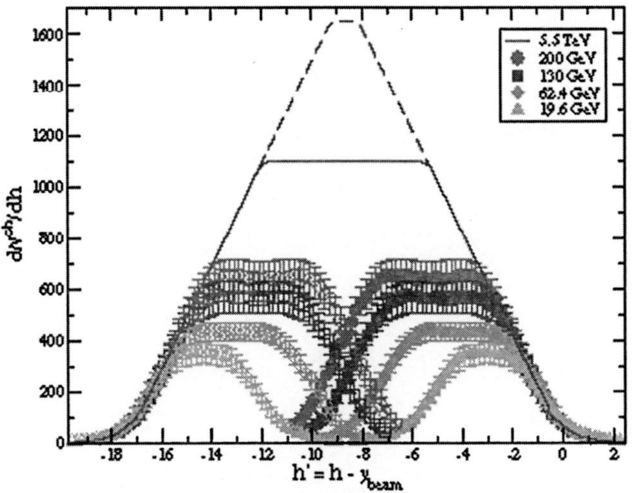

Nayak and Sinha: Figure 10, Page 137

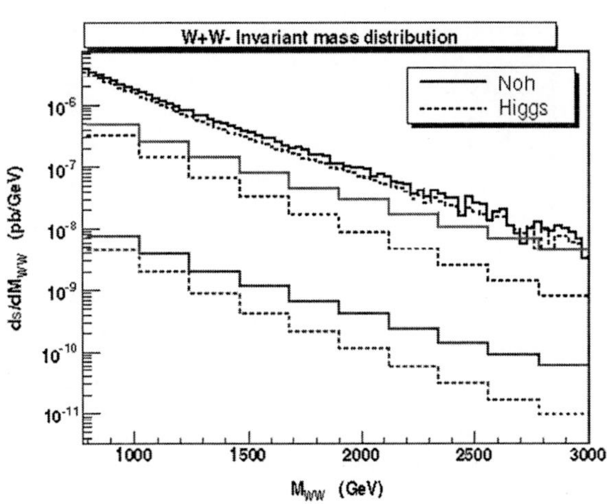

Rindani: Figure 3, Page 153

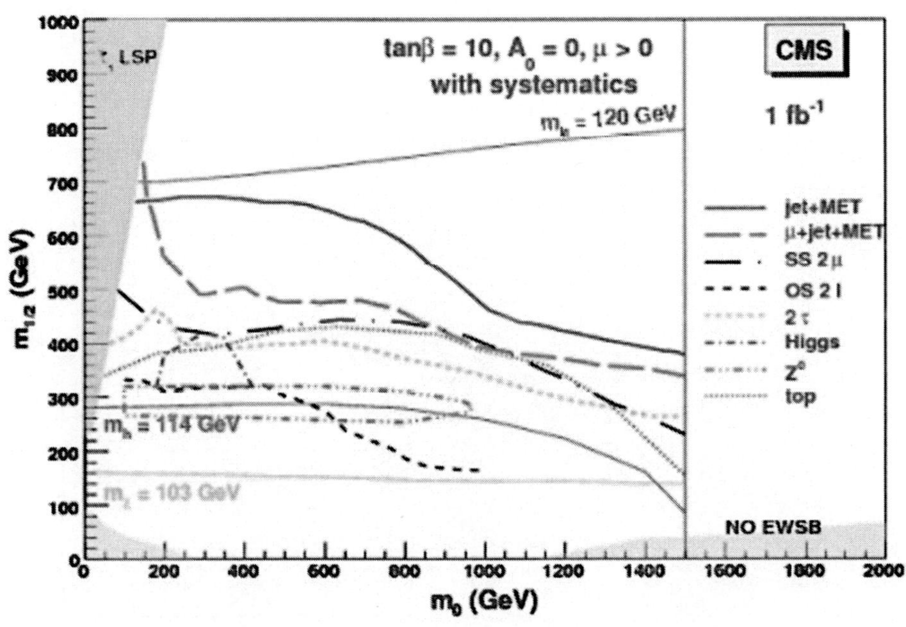

Datta, Guchait and Roy: Figure 5, Page 167

mSUGRA : tanb=10, A_0=0, m>0, m_t=171.4 GeV

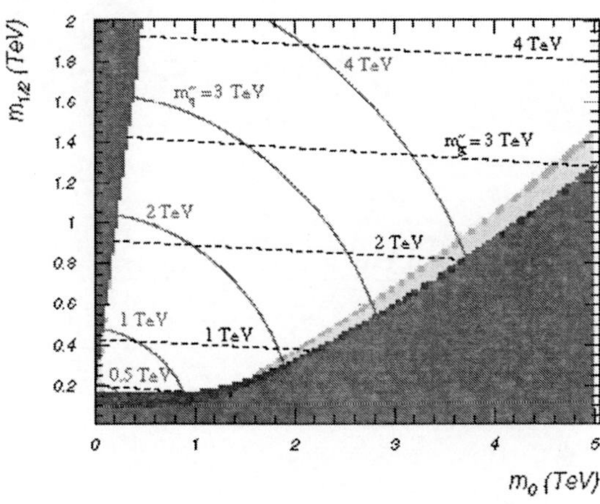

$0 < Wh^2 < 0.094$ Excluded

$0.094 < Wh^2 < 0.129$ LEP2

Baer and Tata: Figure 3, Page 188

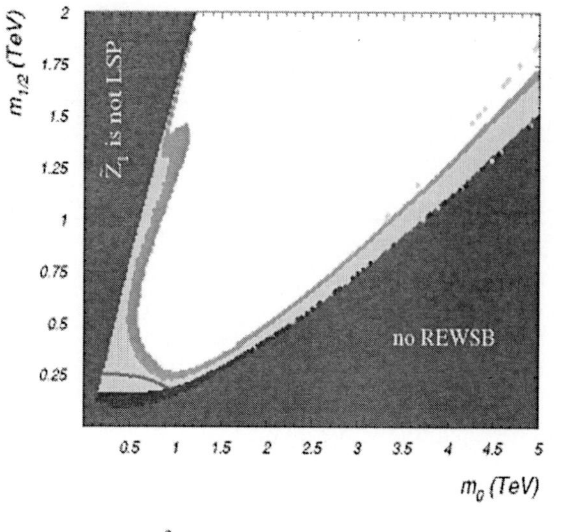

Baer and Tata: Figure 4, Page 188

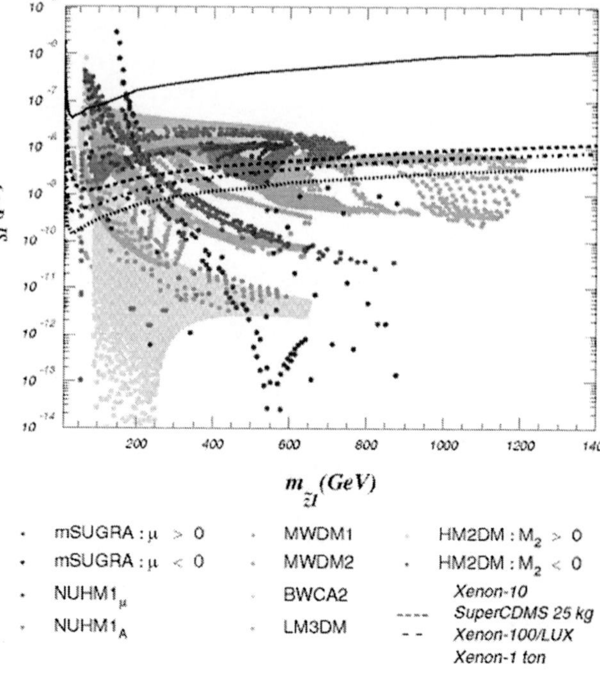

Baer and Tata: Figure 5, Page 189

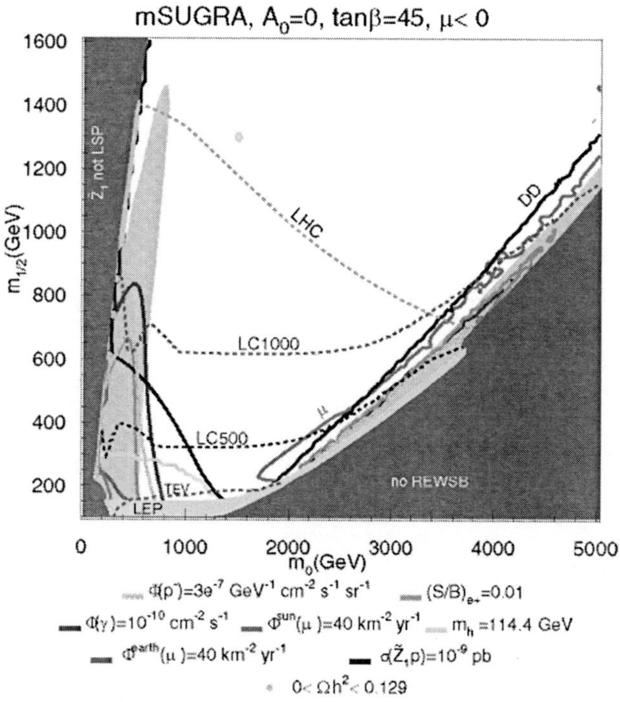

Baer and Tata: Figure 6, Page 190

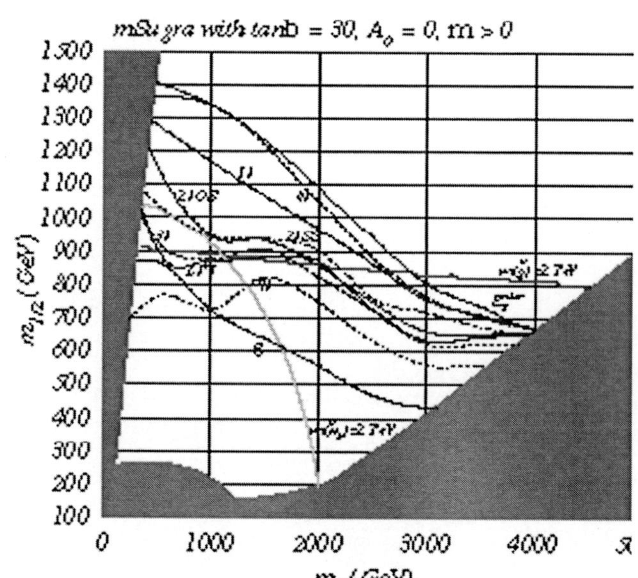

Baer and Tata: Figure 10, Page 196

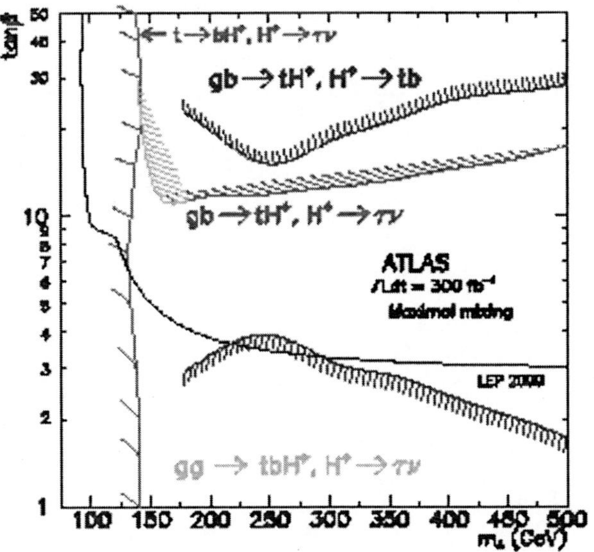

Guchait and Roy: Figure 2, Page 208

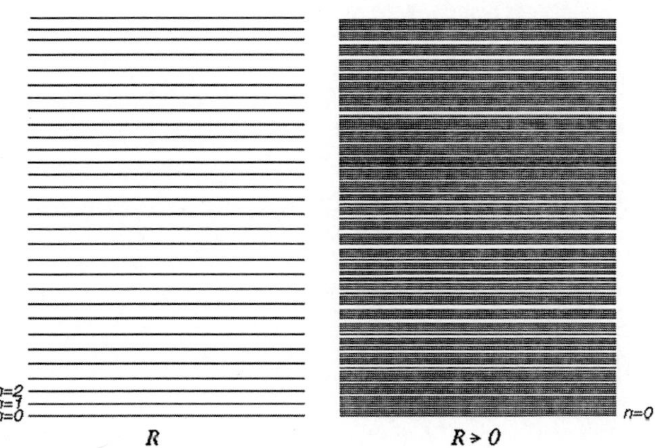

Raychaudhuri: Figure 9, Page 220

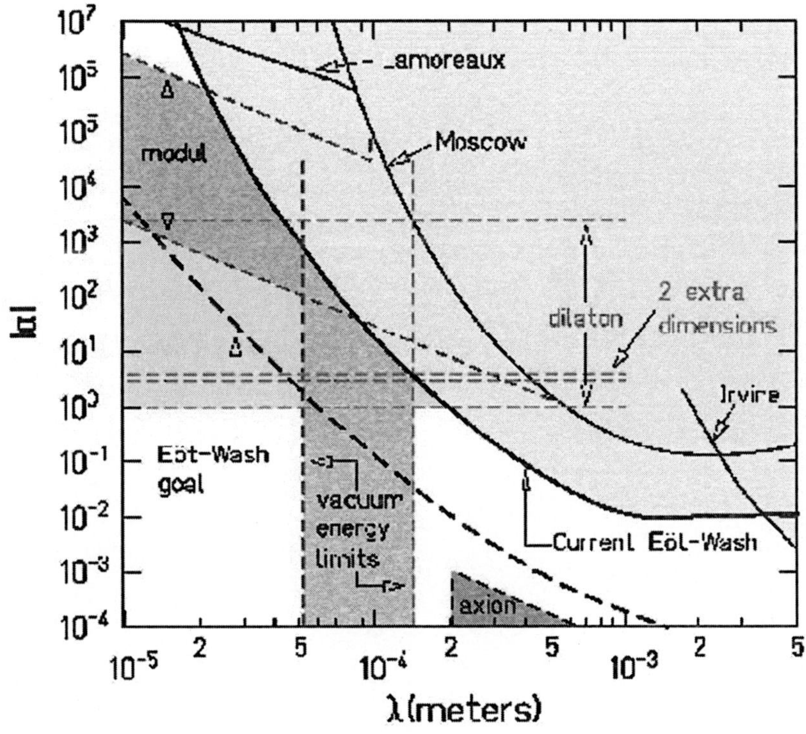

Raychaudhuri: Figure 11, Page 223

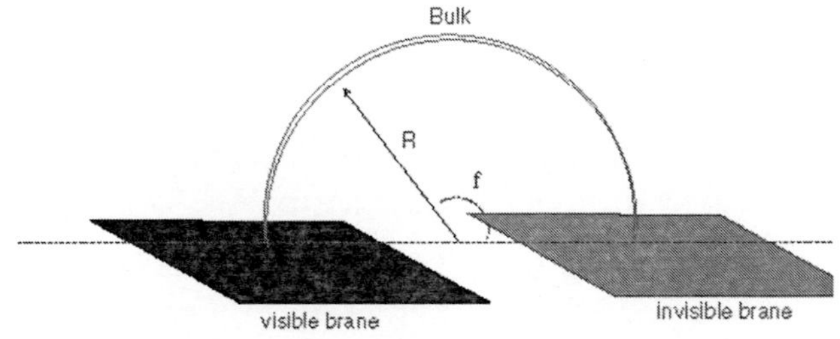

Raychaudhuri: Figure 19, Page 234